Lecture Notes in Computer Science 14818

The series Lecture Notes in Computer Science (LNCS), including its subseries Lecture Notes in Artificial Intelligence (LNAI) and Lecture Notes in Bioinformatics (LNBI), has established itself as a medium for the publication of new developments in computer science and information technology research, teaching, and education.

LNCS enjoys close cooperation with the computer science R & D community, the series counts many renowned academics among its volume editors and paper authors, and collaborates with prestigious societies. Its mission is to serve this international community by providing an invaluable service, mainly focused on the publication of conference and workshop proceedings and postproceedings. LNCS commenced publication in 1973.

Osvaldo Gervasi · Beniamino Murgante ·
Chiara Garau · David Taniar ·
Ana Maria A. C. Rocha ·
Maria Noelia Faginas Lago
Editors

Computational Science and Its Applications – ICCSA 2024 Workshops

Hanoi, Vietnam, July 1–4, 2024
Proceedings, Part IV

Springer

Editors
Osvaldo Gervasi
Department of Mathematics and Computer Science
University of Perugia
Perugia, Italy

Chiara Garau
Department of Civil and Environmental Engineering and Architecture
University of Cagliari
Cagliari, Italy

Ana Maria A. C. Rocha
Algoritmi Research Centre
University of Minho
Braga, Portugal

Beniamino Murgante
School of Engineering
University of Basilicata
Potenza, Italy

David Taniar
Faculty of Information Technology
Monash University Clayton Campus
Clayton, VIC, Australia

Maria Noelia Faginas Lago
Department of Chemistry, Biology and Biotechnology
University of Perugia
Perugia, Italy

ISSN 0302-9743 ISSN 1611-3349 (electronic)
Lecture Notes in Computer Science
ISBN 978-3-031-65272-1 ISBN 978-3-031-65273-8 (eBook)
https://doi.org/10.1007/978-3-031-65273-8

This Springer imprint is published by the registered company Springer Nature Switzerland AG
The registered company address is: Gewerbestrasse 11, 6330 Cham, Switzerland

Preface

These 11 volumes (LNCS volumes 14815–14825) consist of the peer-reviewed papers from the 55 Workshops of the 2024 International Conference on Computational Science and Its Applications (ICCSA 2024) which took place during July 1–4, 2024 in Hanoi (Vietnam). The peer-reviewed papers of the main conference tracks are published in a separate set consisting of two volumes (LNCS 14813–14814).

The conference was held in a hybrid form, with some participants present in person, hosted in Hanoi, Vietnam, by the Thuy Loi University. We enabled virtual participation for those who were unable to attend the event, due to logistical, political and economic problems, by adopting a technological infrastructure based on open source software (jitsi + riot), and a commercial Cloud infrastructure.

ICCSA 2024 was another successful event in the International Conference on Computational Science and Its Applications (ICCSA) conference series, previously held in Athens, Greece (2023), Malaga, Spain (2022), Cagliari, Italy (hybrid with few participants in presence in 2021 and completely online in 2020), whilst earlier editions took place in Saint Petersburg, Russia (2019), Melbourne, Australia (2018), Trieste, Italy (2017), Beijing. China (2016), Banff, Canada (2015), Guimaraes, Portugal (2014), Ho Chi Minh City, Vietnam (2013), Salvador, Brazil (2012), Santander, Spain (2011), Fukuoka, Japan (2010), Suwon, South Korea (2009), Perugia, Italy (2008), Kuala Lumpur, Malaysia (2007), Glasgow, UK (2006), Singapore (2005), Assisi, Italy (2004), Montreal, Canada (2003), and (as ICCS) Amsterdam, The Netherlands (2002) and San Francisco, USA (2001).

Computational Science is the main pillar of most of the present research, industrial and commercial applications, and plays a unique role in exploiting ICT innovative technologies, and the ICCSA conference series have been providing a venue to researchers and industry practitioners to discuss new ideas, to share complex problems and their solutions, and to shape new trends in Computational Science. As the conference mirrors society from a scientific point of view, this year's undoubtedly dominant theme was the machine learning and artificial intelligence and their applications in the most diverse economic and industrial fields.

The ICCSA 2024 conference is structured in 6 general tracks covering the fields of computational science and its applications: Computational Methods, Algorithms and Scientific Applications – High Performance Computing and Networks – Geometric Modeling, Graphics and Visualization – Advanced and Emerging Applications – Information Systems and Technologies – Urban and Regional Planning. In addition, the conference consisted of 55 workshops, focusing on very topical issues of importance to science, technology and society: from new mathematical approaches for solving complex computational systems, to information and knowledge in the Internet of Things, new statistical and optimization methods, several Artificial Intelligence approaches, sustainability issues, smart cities and related technologies.

In the Workshops proceedings we accepted 281 full papers, 17 short papers and 2 PhD Showcase papers. In the Main Conference Proceedings we accepted 53 full papers, 6 short papers and 3 PhD Showcase papers from 207 submissions to the General Tracks of the conference (acceptance rate 30%). We would like to express our appreciation to the workshops chairs and co-chairs for their hard work and dedication.

The success of the ICCSA conference series in general, and of ICCSA 2024 in particular, vitally depends on the support of many people: authors, presenters, participants, keynote speakers, workshop chairs, session chairs, organizing committee members, student volunteers, Program Committee members, Advisory Committee members, International Liaison chairs, reviewers and others in various roles. We take this opportunity to wholehartedly thank them all.

We also wish to thank our publisher, Springer, for their acceptance to publish the proceedings, for sponsoring part of the best papers awards and for their kind assistance and cooperation during the editing process.

We cordially invite you to visit the ICCSA website https://iccsa.org where you can find all the relevant information about this interesting and exciting event.

July 2024

Osvaldo Gervasi
Beniamino Murgante
Chiara Garau

Welcome Message from Organizers

After the very hard times of COVID, ICCSA continues its successful scientific endeavors in 2024, hosted in Hanoi, Vietnam. This time, ICCSA moved from the Mediterranean Region to Southeast Asia and was held in the metropolitan city of Hanoi, the capital of Vietnam. Hanoi is a vibrant urban environment known for the hospitality of its citizens, its rich history, vibrant culture, and dynamic urban life. Located in the northern part of the country, Hanoi is a bustling metropolis that combines the old with the new, offering a unique blend of ancient traditions and modern development.

ICCSA 2024 took place in a secure environment, allowing for safe and vibrant in-person participation. Combined with the active engagement of the ICCSA 2024 scientific community, this set the stage for highly motivating discussions and interactions regarding the latest developments in computer science and its applications in the real world for improving communities' quality of life.

Thuyloi University, also known as the Water Resources University, is a prominent institution in Hanoi, Vietnam, with a strong reputation in engineering and technical education, particularly in water resources and environmental engineering. In recent years, the University has expanded its academic offerings to include computer science, reflecting the growing importance of technology and digital skills in all sectors. This year, Thuyloi University had the honor of hosting ICCSA 2024. The Local Organizing Committee felt the burden and responsibility of such a demanding task and put all necessary energy into meeting participants' expectations and establishing a friendly, creative, and inspiring scientific and social/cultural environment that allowed for new ideas and perspectives to flourish.

Since all ICCSA participants, whether informatics-oriented or application-driven, realize the tremendous advancements in computer science over the last few decades and the huge potential these advancements offer in coping with the enormous challenges of humanity in a globalized, 'wired,' and highly competitive world, the expectations for ICCSA 2024 were high. The goal was to successfully match computer science progress with communities' aspirations, achieving progress that serves real, place- and people-based needs and paves the way towards a visionary, smart, sustainable, resilient, and inclusive future for both current and future generations.

On behalf of the Local Organizing Committee, I would like to sincerely thank all of you who contributed to ICCSA 2024.

Nguyen Canh Thai

Organization

ICCSA 2024 was organized by Thuyloi University (Vietnam), the University of Perugia (Italy), the University of Basilicata (Italy), Monash University (Australia), Kyushu Sangyo University (Japan), the University of Minho (Portugal), and the University of Cagliari (Italy).

Honorary General Chairs

Norio Shiratori	Chuo University, Japan
Kenneth C. J. Tan	Sardina Systems, UK

General Chairs

Nguyen Canh Thai	Thuyloi University, Vietnam
Osvaldo Gervasi	University of Perugia, Italy
David Taniar	Monash University, Australia

Program Committee Chairs

Beniamino Murgante	University of Basilicata, Italy
Chiara Garau	University of Cagliari, Italy
Ana Maria A. C. Rocha	University of Minho, Portugal
Bernady O. Apduhan	Kyushu Sangyo University, Japan

International Advisory Committee

Jemal Abawajy	Deakin University, Australia
Dharma P. Agarwal	University of Cincinnati, USA
Rajkumar Buyya	Melbourne University, Australia
Claudia Bauzer Medeiros	University of Campinas, Brazil
Manfred M. Fisher	Vienna University of Economics and Business, Austria
Pierre Frankhauser	University of Franche-Comté/CNRS, France
Marina L. Gavrilova	University of Calgary, Canada

Sumi Helal	University of Florida, USA & University of Lancaster, UK
Bin Jiang	University of Gävle, Sweden
Yee Leung	Chinese University of Hong Kong, China

International Liaison Chairs

Ivan Blečić	University of Cagliari, Italy
Giuseppe Borruso	University of Trieste, Italy
Elise De Donker	Western Michigan University, USA
Maria Noelia Faginas Lago	University of Perugia, Italy
Maria Irene Falcão	University of Minho, Portugal
Robert C. H. Hsu	Chung Hua University, Taiwan
Yeliz Karaca	University of Massachusetts Medical School, Worcester, USA
Tae-Hoon Kim	Zhejiang University of Science and Technology, China
Vladimir Korkhov	Saint Petersburg University, Russia
Takashi Naka	Kyushu Sangyo University, Japan
Rafael D. C. Santos	National Institute for Space Research, Brazil
Maribel Yasmina Santos	University of Minho, Portugal
Anastasia Stratigea	National Technical University of Athens, Greece

Workshop and Session Organizing Chairs

Beniamino Murgante	University of Basilicata, Italy
Chiara Garau	University of Cagliari, Italy

Award Chair

Wenny Rahayu	La Trobe University, Australia

Publicity Committee Chairs

Elmer Dadios	De La Salle University, Philippines
Nataliia Kulabukhova	Saint Petersburg University, Russia

Daisuke Takahashi	Tsukuba University, Japan
Shangwang Wang	Beijing University of Posts and Telecommunications, China

Local Organizing Committee Chairs

Doan Quang Tu	Academy of Military Science and Technology, Hanoi, Vietnam
Ho Sy Tam	Thuyloi University, Vietnam
Le Quang Tuan	Thuyloi University, Vietnam
Nguyen Huu Quynh	Thuyloi University, Vietnam
Ta Quang Chieu	Thuyloi University, Vietnam

Technology Chair

Damiano Perri	University of Perugia, Italy

Program Committee

Vera Afreixo	University of Aveiro, Portugal
Vladimir Alarcon	Northern Gulf Institute, USA
Filipe Alvelos	University of Minho, Portugal
Debora Anelli	Polytechnic University of Bari, Italy
Hartmut Asche	Hasso-Plattner-Institut für Digital Engineering, Germany
Ginevra Balletto	University of Cagliari, Italy
Socrates Basbas	Aristotle University of Thessaloniki, Greece
David Berti	ART SpA, Italy
Michela Bertolotto	University College Dublin, Ireland
Debnath Bhattacharyya	Koneru Lakshmaiah University, India
Sandro Bimonte	CEMAGREF, TSCF, France
Ana Cristina Braga	University of Minho, Portugal
Tiziana Campisi	Kore University of Enna, Italy
Yves Caniou	Lyon University, France
José A. Cardoso e Cunha	Universidade Nova de Lisboa, Portugal
Rui Cardoso	University of Beira Interior, Portugal
Leocadio G. Casado	University of Almeria, Spain
Mete Celik	Erciyes University, Turkey
Maria Cerreta	University of Naples Federico II, Italy
Mauro Coni	University of Cagliari, Italy

Florbela Maria da Cruz Domingues Correia	Polytechnic Institute of Viana do Castelo, Portugal
Roberto De Lotto	University of Pavia, Italy
Marcelo De Paiva Guimaraes	Federal University of Sao Paulo, Brazil
Frank Devai	London South Bank University, UK
Joana Matos Dias	University of Coimbra, Portugal
Laila El Ghandour	Heriot Watt University, UK
Rafida M. Elobaid	Canadian University Dubai, United Arab Emirates
Maria Irene Falcao	University of Minho, Portugal
Florbela P. Fernandes	Polytechnic Institute of Bragança, Portugal
Paula Odete Fernandes	Polytechnic Institute of Bragança, Portugal
Adelaide de Fátima Baptista Valente Freitas	University of Aveiro, Portugal
Valentina Franzoni	University of Perugia, Italy
Andreas Fricke	University of Potsdam, Germany
Raffaele Garrisi	Centro Operativo per la Sicurezza Cibernetica, Italy
Maria Giaoutzi	National Technical University, Athens, Greece
Salvatore Giuffrida	University of Catania, Italy
Teresa Guarda	Universidad Estatal Peninsula de Santa Elena, Ecuador
Sevin Gümgüm	Izmir University of Economics, Turkey
Malgorzata Hanzl	Technical University of Lodz, Poland
Maulana Adhinugraha Kiki	Telkom University, Indonesia
Clement Ho Cheung Leung	Chinese University of Hong Kong, China
Andrea Lombardi	University of Perugia, Italy
Marcos Mandado Alonso	University of Vigo, Spain
Ernesto Marcheggiani	Katholieke Universiteit Leuven, Belgium
Antonino Marvuglia	Luxembourg Institute of Science and Technology, Luxembourg
Michele Mastroianni	University of Salerno, Italy
Hideo Matsufuru	High Energy Accelerator Research Organization, Japan
Alfredo Milani	University of Perugia, Italy
Fernando Miranda	Universidade do Minho, Portugal
Giuseppe Modica	University of Reggio Calabria, Italy
Louiza de Macedo Mourelle	State University of Rio de Janeiro, Brazil
Nadia Nedjah	State University of Rio de Janeiro, Brazil
Paolo Nesi	University of Florence, Italy
Rajdeep Niyogi	Indian Institute of Technology Roorkee, India
Suzan Obaiys	University Malaya, Malaysia
Marcin Paprzycki	Polish Academy of Sciences, Poland

Eric Pardede	La Trobe University, Australia
Ana Isabel Pereira	Polytechnic Institute of Bragança, Portugal
Damiano Perri	University of Perugia, Italy
Massimiliano Petri	University of Pisa, Italy
Telmo Pinto	University of Coimbra, Portugal
Maurizio Pollino	ENEA, Italy
Alenka Poplin	Iowa State University, USA
Marcos Quiles	Federal University of São Paulo, Brazil
Humberto Rocha	University of Coimbra, Portugal
Marzio Rosi	University of Perugia, Italy
Manna Sheela Rani Chetty	Koneru Lakshmaiah University, India
Lucia Saganeiti	University of L'Aquila, Italy
Tamie Salter	Citizen Alerts Inc., Canada
Francesco Scorza	University of Basilicata, Italy
Marco Paulo Seabra dos Reis	University of Coimbra, Portugal
Jie Shen	University of Michigan, USA
Chien Sing Lee	Sunway University, Malaysia
Francesco Tajani	Sapienza University of Rome, Italy
Rodrigo Tapia-McClung	Centro de Investigación en Ciencias de Información Geoespacial, Mexico
Eufemia Tarantino	Polytechnic of Bari, Italy
Sergio Tasso	University of Perugia, Italy
Ana Paula Teixeira	Universidade do Minho, Portugal
Maria Filomena Teodoro	IST ID, Instituto Superior Técnico, Portugal
Yiota Theodora	National Technical University of Athens, Greece
Carmelo Torre	Polytechnic of Bari, Italy
Giuseppe A. Trunfio	University of Sassari, Italy
Toshihiro Uchibayashi	Kyushu University, Japan
Marco Vizzari	University of Perugia, Italy
Frank Westad	Norwegian University of Science and Technology, Norway
Fukuko Yuasa	High Energy Accelerator Research Organization, Japan
Ljiljana Zivkovic	Republic Geodetic Authority, Serbia

Workshops

Advances in Artificial Intelligence Learning Technologies: Blended Learning, STEM, Computational Thinking and Coding (AAILT 2024)

Workshop Organizers

Alfredo Milani	University of Perugia, Italy
Valentina Franzoni	University of Perugia, Italy
Sergio Tasso	University of Perugia, Italy

Workshop Program Committee Members

Li Yuanxi	Hong Kong Baptist University, Hong Kong, China
Valentino Santucci	University for Foreigners, Perugia, Italy
Krassimir Markov	Institute of Information Theories and Applications, Sofia, Bulgaria
Clement Leung	Chinese University of Hong Kong, China
Rajdeep Niyogi	Indian Institute of Technology, Rooorkee, India

Advanced and Innovative Web Apps 2024 (AIWA 2024)

Workshop Organizers

Osvaldo Gervasi	University of Perugia, Italy
Damiano Perri	University of Perugia, Italy

Workshop Program Committee Members

Tamie Salter	Citizen Alert Inc., Canada
Paolo Nesi	University of Florence, Italy
Tae Hoon Kim	Zhejiang University of Science and Technology, China
Debnath Bhattacharyya	Koneru Lakshmaiah Education Foundation, India
David Berti	ART SpA, Italy

Advanced Processes of Mathematics and Computing Models in Complex Computational Systems (ACMC 2024)

Workshop Organizers

Yeliz Karaca	University of Massachusetts Chan Medical School and Massachusetts Institute of Technology, USA
Dumitru Baleanu	Cankaya University, Turkey
Osvaldo Gervasi	University of Perugia, Italy
Yudong Zhang	University of Leicester, UK
Majaz Moonis	University of Massachusetts Medical School, USA

Workshop Program Committee Members

Tamie Salter	Citizen Alert Inc., Canada
TaiHoon Kim	Konkuk University, South Korea
Debnath Bhattacharyya	Koneru Lakshmaiah Education Foundation, India
Massimiliano Nicolini	OliTec Foundation, Italy
Martin Bohner	Missouri University of Science and Technology, Missouri, USA
René Lozi	Université Côte d'Azur, France
Ravi P. Agarwal	Texas A&M University-Kingsville, USA
Aziz Dursun	Virginia Tech, USA
Muhammad Attique Khan	HITEC University, Pakistan and Lebanese American University, Lebanon
Rasulova Mukhayo	National University Uzbekistan, Uzbekistan

Advances in Information Systems and Technologies for Emergency Management, Risk Assessment and Mitigation Based on the Resilience Concepts (ASTER 2024)

Workshop Organizers

Maurizio Pollino	Italian National Agency for New Technologies, Energy and Sustainable Economic Development, Italy
Marco Vona	University of Basilicata, Italy

Alessandro Giocoli	Italian National Agency for New Technologies, Energyand Sustainable Economic Development, Italy
Chiara Ormando	Italian National Agency for New Technologies, Energy and Sustainable Economic Development, Italy
Angelo Anelli	Italian National Research Council, Italy

Workshop Program Committee Members

Gregorio D'Agostino	Italian National Agency for New Technologies, Energy and Sustainable Economic Development, Italy
Christian Bignami	INGV, Italy
Paolo Clemente	Anti-Seismic Systems International Society, ASSISI, USA
Sonia Giovinazzi	Italian National Agency for New Technologies, Energy and Sustainable Economic Development, Italy

Advances in Web-Based Learning 2024 (AWBL 2024)

Workshop Organizers

Mustafa Inceoglu	Ege University, Turkey
Birol Ciloglugil	Ege University, Turkey

Workshop Program Committee Members

Birim Balci	Celal Bayar University, Turkey
Burak Galip Aslan	Izmir Yuksek Teknoloji University, Turkey
Mahmut Unan	University of Alabama, USA
Nisha S. Raj	Global Indian International School, Japan
Oylum Alatlı	Ege University, Turkey

Blockchain and Distributed Ledgers: Technologies and Applications (BDLTA 2024)

Workshop Organizers

Vladimir Korkhov	Saint Petersburg State University, Russia
Elena Stankova	Saint Petersburg State University, Russia
Nataliia Kulabukhova	Saint Petersburg State University, Russia

Workshop Program Committee Members

Vasily Solodkov	Higher School of Economics Banking Institute
Alexander Bogdanov	St. Petersburg State University, Russia
Alexander Degtyarev	St. Petersburg State University, Russia
Dmitry Vasiunin	Deutsche Telekom Cloud Services E.P.E., Greece
Serob Balyan	National Academy of Sciences of Armenia, Armenia
Suren Abrahamyan	Osensus Arm LLC, Armenia
Ashot Gevorkyan	National Academy of Sciences of Armenia, Armenia
Martin Vala	Univerzita Pavla Jozefa Šafárika Košiciach, Košice, Slovakia
Nodir Zaynalov	Tashkent University of Information Technologies, Uzbekistan
Michail Panteleyev	St. Petersburg Electrotecnical University, Russia
Nikolay Peryazev	Irkutsk State University, Irkutsk, Russia

Computational and Applied Mathematics (CAM 2024)

Workshop Organizers

Maria Irene Falcao	University of Minho, Portugal
Fernando Miranda	University of Minho, Portugal

Workshop Program Committee Members

Ricardo Severino	University of Minho, Portugal
Helmuth Malonek	University of Aveiro, Portugal
Isabel Cacao	University of Aveiro, Portugal
Graça Tomaz	Polytechnic of Guarda, Portugal

Rosário Fernandes	University of Minho, Portugal
Lidia Aceto	University of Piedmont Oriental, Italy
Patrícia Beites	University of Beira Interior, Portugal
Regina de Almeida	University of Trás-os-Montes e Alto Douro, Portugal
Luís Ferrás	University of Porto, Portugal
Helmuth Malonek	University of Aveiro, Portugal
João Morais	Autonomous Technological Institute of Mexico, Mexico
Yilmaz Simsek	Akdeniz University, Turkey

Computational and Applied Statistics (CAS 2024)

Workshop Organizer

Ana Cristina Braga	University of Minho, Portugal

Workshop Program Committee Member

Adelaide Freitas	University of Aveiro, Portugal
Ângela Silva	University of Minho, Portugal
Arminda Manuela Gonçalves	University of Minho, Portugal
Arunasalam Rahunanthan	Central State University, USA
Elisete Correia	University of Trás-os-Montes e Alto Douro, Portugal
Frank Westad	Norwegian University of Science and Technology, Norway
Humberto Rocha	University of Coimbra, Portugal
Ivan Rodriguez Conde	University of Arkansas, USA
Lino Costa	University of Minho, Portugal
Marco Reis	University of Coimbra, Portugal
Maria Filipa Mourão	Polytechnic Institute of Viana do Castelo, Portugal
Martín Pérez Pérez	University of Vigo, Spain
Michal Abrahamowicz	McGill University, Canada
Vanda Lourenço	New University of Lisbon, Portugal
Vera Afreixo	University of Aveiro, Portugal

Cyber Intelligence and Applications (CIA 2024)

Workshop Organizer

Gianni D'Angelo	University of Salerno, Italy

Workshop Program Committee Members

Massimo Ficco	University of Salerno, Italy
Marcello Trovati	Edge Hill University, UK
Michal Choras	Bydgoszcz University of Science and Technology, Poland

Computational Methods, Statistics and Industrial Mathematics (CMSIM 2024)

Workshop Organizers

Maria Filomena Teodoro	University of Lisbon and Portuguese Naval Academy, Portugal
Marina A. P. Andrade	University Institute of Lisbon, Portugal
Maria de Fátima Alves de Pina	University Institute of Lisbon, Portugal

Workshop Program Committee Members

Ricardo Moura	University of Lisboa and FCT, Portugal
Aldina Correia	Escola Superior de Tecnologia e Gestão, Portugal
Eliana Costa	Escola Superior de Tecnologia e Gestão, Portugal
Ana Borges	Escola Superior de Tecnologia e Gestão, Portugal
Cristina Lopes	ISCAP, Portugal
Fernando Carapau	University of Evora, Portugal
Fernanda Costa	University of Minho, Portugal
Maria Luísa Morgado	University of Trás os Montes e Alto Douro and University of Lisboa, Portugal
Ana Isabel Pereira	Instituto Politécnico de Bragança, Portugal
Bruna Ramos	University of Minho, Portugal
Rosário Ramos	Universidade Aberta, Portugal
Elisabete Carolino	Escola Superior da Saúde de Lisboa, Portugal
Sofia Rézio	Instituto Politécnico de Lisboa, Portugal
Matteo Sacchet	University of Turin, Italy

Marina Marchisio Conte	University of Turin, Italy
António Seijas-Macias	University of Coruña, Spain
Luís F. A. Teodoro	University of Glasgow, UK and University of Oslo, Norway
Christos Kitsos	University of West Attica, Greece

Computational Optimization and Applications (COA 2024)

Workshop Organizers

Ana Maria A. C. Rocha	University of Minho, Portugal
Humberto Rocha	University of Coimbra, Portugal

Workshop Program Committee Members

Eligius M. T. Hendrix	University of Málaga, Spain
Joana Matos Dias	University of Coimbra, Portugal
Ana Isabel Pereira	Polytechnic Institute of Bragança, Portugal
Filipe Alvelos	Univeristy of Minho, Portugal
Emerson José de Paiva	Federal University of Itajubá, Brazil

Computational Astrochemistry (CompAstro 2024)

Workshop Organizers

Marzio Rosi	University of Perugia, Italy
Nadia Balucani	University of Perugia, Italy
Daniela Ascenzi	University of Trento, Italy
Stefano Falcinelli	University of Perugia, Italy

Workshop Program Committee Members

Albert Rimola	Universitat Autònoma de Barcelona, Spain
Dimitrios Skouteris	Master-Tec, Perugia, Italy
Piero Ugliengo	University of Turin, Italy
Andrea Lombardi	University of Perugia, Italy
Piergiorgio Casavecchia	University of Perugia, Italy
Fernando Pirani	University of Perugia, Italy
Costanza Borghesi	University of Perugia, Italy

Gabriella Digenova	University of Perugia, Italy
Lisa Giani	Grenoble Alpes University, France
Marco Parriani	University of Perugia, Italy
Cecilia Coletti	University of Chieti, Italy
Massimiliano Aschi	University of L'Aquila, Italy
Emilia Dearagao	University of Perugia, Italy
Giacomo Pannacci	University of Perugia, Italy
Gianmarco Vanuzzo	University of Perugia, Italy

Computational Methods for Porous Geomaterials (CompPor 2024)

Workshop Organizers

Vadim Lisitsa	Russian Academy of Science, Russia
Evgeniy Romenski	Russian Academy of Science, Russia

Workshop Program Committee Members

Kirill Gadylshin	Russian Academy of Science, Russia
Tayana Khachkova	Russian Academy of Science, Russia
Dmitri Prokhorov	Russian Academy of Science, Russia
Mikahil Novikov	Russian Academy of Science, Russia

Workshop on Computational Science and HPC (CSHPC 2024)

Workshop Organizers

Elise De Doncker	Western Michigan University, USA
Fukuko Yuasa	High Energy Accelerator Research Organization, Japan
Hideo Matsufuru	High Energy Accelerator Research Organization, Japan

Workshop Program Committee Members

Tadashi Ishikawa	High Energy Accelerator Research Organization, Japan
Hiroshi Daisaka	Hitotsubashi University, Japan
Naohito Nakasato	University Aizu, Japan

Stefano Carboni	University of Sassari, Italy
Francesca Sinatra	University of Cagliari, Italy
Salvatore Dore	University of Sassari, Italy
Andrea Gallo	University of Trieste, Italy

Workshop Program Committee Members

Giuseppe Borruso	University of Trieste, Italy
Silvia Battino	University of Sassari, Italy
Beniamino Murgante	University of Basilicata, Italy
Mara Ladu	University of Cagliari, Italy
Luigi Mundula	University of Perugia, Italy
Maria Attard	University of Malta, Malta
Enrico Dagostini	University of Malta, Malta
Emilio Ghiani	University of Cagliari, Italy
Marco Naseddu	University of Cagliari, Italy
Balázs Kulcsaár	University of Debrecen, Hungary
Tu Anh Trinh	College of Technology and Design for UEH University, Vietnam
Giovanni Mauro	University of Campania Luigi Vanvitelli, Italy
Maria Ronza	University of Naples, Federico II, Italy
Massimiliano Bencardino	University of Salerno, Italy

Evaluating Inner Areas Potentials (EIAP 2024)

Workshop Organizers

Diana Rolando	Polytechnic of Turin, Italy
Alice Barreca	Polytechnic of Turin, Italy
Manuela Rebaudengo	Polytechnic of Turin, Italy
Giorgia Malavasi	Polytechnic of Turin, Italy

Workshop Program Committee Members

John Accordino	Virginia Commonwealth University, USA
Sara Favargiotti	University of Trento, Italy
Maddalena Ferretti	Polytechnic University of Marche, Italy
Daniel Laven	Mid Sweden University, Sweden
Barbara Lino	University of Palermo, Italy
Umberto Mecca	Polytechnic University of Turin, Italy

Lorenzo Savio	Polytechnic University of Turin, Italy
Asja Aulisio	Polytechnic University of Turin, Italy

Econometrics and Multidimensional Evaluation of Urban Environment (EMEUE 2024)

Workshop Organizers

Carmelo Maria Torre	Polytechnic of Bari, Italy
Francesco Tajani	Sapienza University of Rome, Italy
Pierluigi Morano	Polytechnic of Bari, Italy
Simona Panaro	University of Sussex, UK
Maria Cerreta	University of Naples Federico II, Italy
Debora Anelli	Polytechnic of Bari, Italy

Workshop Program Committee Members

Pierfrancesco de Paola	University of Naples Federico II, Italy
Daniela Tavano	University of Calabria, Italy
Giuseppe Cerullo	Sapienza University of Rome, Italy
Francesco Paolo del Giudice	Sapienza University of Rome, Italy
Marco Locurcio	Polytechnic University of Bari, Italy
Maria Rosa Trovato	University of Catania, Italy
Felicia di Liddo	Polytechnic University of Bari, Italy
Maria Paz Saez Perez	University of Granada, Spain
Yasmine Selim	Ain Shams University, Cairo
Hasan Mara	Indian Institute of Technology Roorkee, India
Philipp Wiesner	Technical University of Berlin, Germany
Maria Gamboa Perez	Complutense University of Madrid, Spain
Manuel Yanez	Universidad Autónoma de Madrid, Spain
Lucia Ika Fitriastuti	Solusi Bisnis Indonesia, Indonesia
Frank Devai	London South Bank University, UK
Frank Westad	Norwegian University of Science and Technology, Norway
Eugenio Muccio	University of Naples Federico II, Italy
Chiara Mazzarella	TU Delft, The Netherlands
Daniele Cannatella	TU Delft, The Netherlands
Sabrina Sacco	University of Naples Federico II, Italy
Piero Zizzania	University of Naples Federico II, Italy
Stefano Cuntò	University of Naples Federico II, Italy

Sveva Ventre	University of Naples Federico II, Italy
Caterina Loffredo	University of Naples Federico II, Italy
Giuseppe Ciciriello	University of Naples Federico II, Italy
Maria Somma	University of Naples Federico II, Italy
Ludovica La Rocca	University of Naples Federico II, Italy
Gaia Daldanise	National Research Council, Italy
Giuliano Poli	University of Naples Federico II, Italy

Environmental, Social, Governance of Energy Planning (ESGEP 2024)

Workshop Organizers

Ginevra Balletto	University of Cagliari, Italy
Emilio Ghiani	University of Cagliari, Italy
Roberto De Lotto	University of Pavia, Italy
Alessandra Marra	University of Salerno, Italy
Riccardo Trevisan	University of Cagliari, Italy
Balázs Kulcsár	University of Debrecen, Hungary

Workshop Program Committee Members

Jacopo Torriti	University of Reading, UK
Roberto Gerundo	University of Salerno, Italy
Luigi Mundula	University of Perugia, Italy
Mara Ladu	University of Cagliari, Italy
Giuseppe Borruso	University of Trieste, Italy
Tu Anh Trinh	College of Technology and Design for UEH University, Vietnam
Giovanni Mauro	University of Campania Luigi Vanvitelli, Italy
Maria Ronza	University of Naples, Federico II, Italy
Massimiliano Bencardino	University of Salerno, Italy

Ecosystem Services in Spatial Planning for Resilient Urban and Rural Areas (ESSP 2024)

Workshop Organizers

Sabrina Lai	University of Cagliari, Italy
Corrado Zoppi	University of Cagliari, Italy
Francesco Scorza	University of Basilicata, Italy

Beniamino Murgante	University of Basilicata, Italy
Floriana Zucaro	University of Naples Federico II, Italy
Carmela Gargiulo	University of Naples Federico II, Italy

Workshop Program Committee Members

Carmen Guida	University of Naples Federico II, Italy
Irina Di Ruocco	Insubria University, Italy
Silvia Rossetti	University of Parma, Italy
Barbara Caselli	University of Parma, Italy
Gloria Pellicelli	University of Parma, Italy
Federica Leone	University of Cagliari, Italy
Federica Isola	University of Cagliari, Italy
Francesca Leccis	University of Cagliari, Italy
Francesca Perrone	Sapienza University of Rome, Italy
Angela Pilogallo	Italian National Research Council, Italy
Alfonso Annunziata	University of Basilicata, Italy
Alessandro Marucci	University of L'Aquila, Italy
Francesco Zullo	University of L'Aquila, Italy
Gizem Dinç	Süleyman Demirel University, Turkey
Atila Gul	Süleyman Demirel University, Turkey
Sarah Scheiber	University of Malta, Malta
Matteo Caglioni	Université Côte d'Azur, France

Ethical AI Applications for a Human-Centered Cyber Society (EthicAI 2024)

Workshop Organizers

Valentina Franzoni	University of Perugia, Italy
Alfredo Milani	University of Perugia, Italy
Jordi Vallverdu	University Autonoma Barcelona, Spain

Workshop Program Committee Members

Sergio Tasso	University of Perugia, Italy
Yuanxi Li	Hong Kong Baptist University, Hong Kong, China
Daniele Mezzetti	Santa Maria della Misericordia Hospital of Perugia, Italy
Abeer Dyoub	L'Aquila University, Italy

14th International Workshop on Future Computing System Technologies and Applications (FiSTA 2024)

Workshop Organizers

Bernady Apduhan	Kyushu Sangyo University, Japan
Rafael Santos	National Institute for Space Research, Brazil

Workshop Program Committee Members

Agustinus Borgy Waluyo	Monash University, Australia
Andre Ricardo Abed Grégio	Federal University of Parana, Brazil
Earl Ryan Aleluya	MSU-Iligan Institute of Technology, Philippines
Kai Cheng	Kyushu Sangyo University, Japan
Alvaro Fazenda	Federal University of São Paulo, Brazil
Yusuke Gotoh	Okayama University, Japan
Eric Pardede	La Trobe University, Australia
Yasuaki Sumida	Kyushu Sangyo University, Japan
Kazuaki Tanaka	Kyushu Institute of Technology, Japan
Toshihiro Uchibayashi	Kyushu University, Japan
Toshihiro Yamauchi	Okayama University, Japan
Fenghui Yao	Tennessee State University, USA

Geographical Analysis, Urban Modeling, Spatial Statistics (Geog-An-Mod 2024)

Workshop Organizers

Beniamino Murgante	University of Basilicata, Italy
Giuseppe Borruso	University of Trieste, Italy
Harmut Asche	Hasso-Plattner-Institut für Digital Engineering, Germany
Andreas Fricke	Hasso-Plattner-Institut für Digital Engineering, Germany
Rodrigo Tapia McClung	Centro de Investigación en Ciencias de Información Geoespacial, Mexico

Workshop Program Committee Members

Ginevra Balletto	University of Cagliari, Italy
Silvia Battino	University of Sassari, Italy
Mara Ladu	University of Cagliari, Italy
Marco Mazzarino	IUAV Univeristy Venice, Italy
Maria del Mar Munoz Leonisio	Univeristy of Cadiz, Spain
Ahinoa Amaro Garcia	University of Las Palmas of Gran Canaria, Spain
Veronica Camerada	University of Sassari, Italy
Maria Attard	University of Malta, Malta
Enrico Dagostini	University of Malta, Malta
Francesca Krasna	University of Trieste, Italy
Malgorzata Hanzl	Lodz University of Technology, Poland
Anastasia Stratigea	National Technical University of Athens, Greece
Tu Anh Trinh	College of Technology and Design for UEH University, Vietnam
Giovanni Mauro	University of Campania Luigi Vanvitelli, Italy
Maria Ronza	University of Naples, Federico II, Italy
Massimiliano Bencardino	University of Salerno, Italy

Geomatics for Resource Monitoring and Management (GRMM 2024)

Workshop Organizers

Alessandra Capolupo	Polytechnic of Bari, Italy
Eufemia Tarantino	Polytechnic of Bari, Italy
Alberico Sonnessa	Polytechnic of Bari, Italy

Workshop Program Committee Members

Umberto Fratino	Polytechnic of Bari, Italy
Valeria Monno	Polytechnic of Bari, Italy
Antonino Maltese	University of Palermo, Italy
Athos Agapiou	Cyprus University of Technology, Cyprus
Michele Mangiameli	University of Catania, Italy
Angela Gorgoglione	University of la República de Uruguay, Uruguay
Roberta Ravanelli	Sapienza University of Rome, Italy
Alessandra Mascitelli	D'Annunzio University of Chieti–Pescara, Italy
Ester Scotto di Perta	University of Naples, Federico II, Italy

Giacomo Caporusso	National Research Council, Italy
Andrea Montanino	University of Naples, Federico II, Italy
Antonino Iannuzzo	University of Sannio, Benevento, Italy
Silvano Dal Sasso	University of Basilicata, Potenza, Italy
Laura Mirra	National Research Council, Water Research Institute, Italy
Alessandro Pagano	National Research Council, Water Research Institute, Italy
Francesco Chiaravalloti	National Research Council, Water Research Institute, Italy
Francesco Di Capua	University of Basilicata, Italy
Stefania Santoro	National Research Council, Water Research Institute, Italy
Cinzia Albertini	National Research Council, IREA, Italy
Alessandra Saponieri	University of Salento, Italy

International Workshop on Information and Knowledge in the Internet of Things (IKIT 2024)

Workshop Organizers

Teresa Guarda	Peninsula State University of Santa Elena, Ecuador
José María Díaz Nafría	Madrid Open University, Spain
Filipe Portela	University of Minho, Portugal

Workshop Program Committee Members

Arnulfo Alanis	Instituto Tecnologico de Tijuana, Mexico
Bruno Sousa	University of Coimbra, Portugal
Carlos Balsa	Instituto Politécnico de Bragança, Portugal
Filipe Mota Pinto	Instituto Politécnico de Leiria, Portugal
Gustavo Gatica	Universidad Andrés Bello, Chile
Isabel Lopes	Instituto Politécnico de Bragança, Portugal
José María Díaz Nafría	Universidad a Distancia, Spain
Maria Fernanda Augusto	BiTrum Research Group, Spain
Maria Isabel Ribeiro	Instituto Politécnico Bragança, Portugal
Modestos Stavrakis	University of the Aegean, Greece
Simone Belli	Universidad Complutense de MadridSpain
Walter Lopes Neto	Instituto Federal de Educação, Brazil

Regenerating Brownfields Enhancing Urban Resilience Appeal (INFERENCE 2024)

Workshop Organizers

Francesca Moraci	Mediterranea University of Reggio Calabria, Italy
Maurizio Oddo	University of Enna Kore, Italy
Antonella Versaci	University of Enna Kore, Italy
Celestina Fazia	University of Enna Kore, Italy
Tiziana Campisi	University of Enna Kore, Italy
Kh Md Nahiduzzaman	University of British Columbia, Canada

Workshop Program Committee Members

Alessandro Baracco	University of Enna Kore, Italy
Kaoutare Amini Alaoui	Mohammed VI Polytechnic University, UM6P, Morocco
Maurizio Errigo	Sapienza University of Rome, Italy
Marsia Marino	Sapienza University of Rome, Italy
Nessrine Moumen	Mohammed VI Polytechnic University, UM6P, Morocco
Francesca Perrone	Sapienza University of Rome, Italy
Pasquale Pizzimenti	Mediterranea University of Reggio Calabria, Italy
Barbara Scala	University of Brescia, Italy
Clarastella Vicari Aversa	Mediterranea University of Reggio Calabria, Italy

International Workshop on Territorial Planning to Integrate Risk and Urban Ontologies (IWPRO 2024)

Workshop Organizers

Elisabetta Maria Venco	University of Pavia, Italy
Beniamino Murgante	University of Basilicata, Italy
Roberto De Lotto	University of Pavia, Italy
Caterina Pietra	University of Pavia, Italy

Workshop Program Committee Members

Stefano Borgo	National Research Council, Italy
Valentina Costa	University of Genoa, Italy
Pajouh Danesh	Middle East Technical University, Turkey

Ilaria Delponte	University of Genoa, Italy
Lorena Fiorini	University of L'Aquila, Italy
Veronica Gazzola	Polytechnic of Milan, Italy
Ghazaleh Goodarzi	Islamic Azad University, Iran
Michele Grimaldi	University of Salerno, Italy
Alessandra Marra	University of Salerno, Italy
Naghmeh Mohammadpourlima	Akademi University, Finland
Francesca Pirlone	University of Genoa, Italy
Silvia Rossetti	University of Parma, Italy
Lucia Saganeiti	University of L'Aquila, Italy
Bahareh Shahsavari	University of Minnesota, USA
Ilenia Spadaro	University of Genoa, Italy
Maria Rosaria Stufano Melone	Polytechnic of Bari, Italy

MaaS Solutions for Airports, Cities and Regional Connectivity (MaaS 2024)

Workshop Organizers

Gianfranco Fancello	University of Cagliari, Italy
Francesco Piras	University of Cagliari, Italy
Tanja Congiu	University of Sassari, Italy
Mara Ladu	University of Cagliari, Italy
Martina Sinatra	University of Cagliari, Italy
Ginevra Balletto	University of Cagliari, Italy

Workshop Program Committee Members

Italo Meloni	University of Cagliari, Italy
Tu Anh Trinh	College of Technology and Design for UEH University, Vietnam
Giuseppe Borruso	University of Trieste, Italy
Luigi Mundula	University of Perugia, Italy
Francesca Sinatra	University of Trieste, Italy
Salvatore Dore	University of Trieste, Italy
Andrea Gallo	University of Trieste, Italy
Marcello Tadini	University of Eastern Piedmont, Italy
Marco Mazzarino	IUAV Univeristy Venice, Italy
Maria del Mar Munoz Leonisio	University of Cadiz, Spain
Veronica Camerada	Univeristy of Sassari, Italy

Brunella Brundu	Univerisity of Sassari, Italy
Maria Attard	University of Malta, Malta
Enrico Dagostini	University of Malta, Malta
Giovanni Mauro	University of Campania Luigi Vanvitelli, Italy
Maria Ronza	University of Naples, Federico II, Italy
Massimiliano Bencardino	University of Salerno, Italy

Development of Urban Mobility Management and Risk Assessment (MAINTAIN 2024)

Workshop Organizers

Tiziana Campisi	University of Enna Kore, Italy
Massimo Di Gangi	University of Messina, Italy
Antonio Comi	University of Rome Tor Vergata, Italy
Grigorios Fountas	Aristotle University of Thessaloniki, Greece
Jesús González-Feliu	Excelia Business School, La Rochelle, France

Workshop Program Committee Members

Paola Panuccio	Mediterranea University of Reggio Calabria, Italy
Domenica Savia Pellicano	Mediterranean University of Reggio Calabria, Italy
Alexandros Nikitas	University of Huddersfield, UK
Antonio Polimeni	University of Messina, Italy
Orlando Belcore	University of Messina, Italy
Marinella Giunta	Mediterranea University of Reggio Calabria, Italy
Borja Alonso	University of Cantabria, Spain
Luigi Dall'Olio	University of Cantabria, Santander, Spain
Kh Md Nahiduzzaman	UBC, Canada

Multidimensional Evolutionary Evaluations for Transformative Approaches (MEETA 2024)

Workshop Organizers

Maria Cerreta	University of Naples Federico II, Italy
Giuliano Poli	University of Naples Federico II, Italy
Daniele Cannatella	TU Delft, The Netherlands

Ludovica Larocca	University of Naples Federico II, Italy
Maria Somma	University of Naples Federico II, Italy
Gaia Daldanise	National Research Council, Italy

Workshop Program Committee Members

Maria Lucia Raiola	University of Naples Federico II, Italy
Eugenio Muccio	University of Naples Federico II, Italy
Chiara Mazzarella	TU Delft, The Netherlands
Sabrina Sacco	University of Naples Federico II, Italy
Piero Zizzania	University of Naples Federico II, Italy
Stefano Cuntò	University of Naples Federico II, Italy
Sveva Ventre	University of Naples Federico II, Italy
Caterina Loffredo	University of Naples Federico II, Italy
Giuseppe Ciciriello	University of Naples Federico II, Italy
Laura Di Tommaso	University of Naples Federico II, Italy
Benedetta Grieco	University of Naples Federico II, Italy
Simona Panaro	University of Sussex, UK

Building Multi-dimensional Models for Assessing Complex Environmental Systems (MES 2024)

Workshop Organizers

Vanessa Assumma	University of Bologna, Italy
Caterina Caprioli	Politechnic of Turin, Italy
Giulia Datola	Politechnic of Turin, Italy
Federico Dell'Anna	Politechnic of Turin, Italy
Marta Dell'Ovo	Politechnic of Milan, Italy
Marco Rossitti	Politechnic of Milan, Italy

Workshop Program Committee Members

Maksims Feofilovs	Riga Technical University, Latvia
Ossama Abdelwahab	University of Bari, Italy
Mariarosaria Angrisano	Pegaso Telematic University, Italy
Francesca Torrieri	Polytechnic of Milan, Italy
Maurizio Pioletti	University of Padua, Italy
Daniela Tavano	University of Calabria, Italy
Simone Persico	Polytechnic of Turin, Italy

Chiara D'Alpaos	University of Padua, Italy
Ezgi Şahin	Mersin University, Turkey
Giorgia Sugoni	LINKS Foundation, Italy
Rubina Canesi	University of Padua, Italy
Giulia Marzani	University of Bologna, Italy
Danny Casprini	Polytechnic of Milan, Italy
Simona Barbaro	University of Palermo, Italy
Giulio Cavana	Polytechnic of Turin, Italy
Diana Rolando	Polytechnic of Turin, Italy
Giuliano Poli	University of Naples Federico II, Italy
Francesco Sica	University of Rome La Sapienza, Italy
Sabrina Lai	University of Cagliari, Italy

Models and Indicators for Assessing and Measuring the Urban Settlement Development in the View of Zero Net Land Take by 2050 (MOVEto0 2024)

Workshop Organizers

Lucia Saganeiti	University of L'Aquila, Italy
Lorena Fiorini	University of L'Aquila, Italy
Angela Pilogallo	University of L'Aquila, Italy
Francesco Zullo	University of L'Aquila, Italy
Alessandro Marucci	University of L'Aquila, Italy

Workshop Program Committee Members

Ginevra Balletto	University of Cagliari, Italy
Giuseppe Borruso	University of Trieste, Italy
Chiara Garau	University of Cagliari, Italy
Beniamino Murgante	University of Basilicata, Italy
Ljiljana Zivkovic	MBA, Republic Geodetic Authority, Serbia
Ilaria Del Ponte	University of Genoa, Italy
Carmen Guida	University of Naples Federico II, Italy
Chiara Di Dato	University of L'Aquila, Italy

4th Workshop on Privacy in the Cloud/Edge/IoT World (PCEIoT 2024)

Workshop Organizers

Michele Mastroianni	University of Salerno, Italy
Mauro Iacono	University of Campania Luigi Vanvitelli, Italy
Lelio Campanile	University of Campania Luigi Vanvitelli, Italy

Workshop Program Committee Members

Maria Ganzha	Warsaw University of Technology, Poland
Armando Tacchella	University of Genoa, Italy
Alessio Merlo	School for Advanced Defense Studies, Italy
Antonio Iannuzzi	Roma Tre University, Italy
Arcangelo Castiglione	University of Salerno, Italy
Daniel Grzonka	Cracow University of Technology, Poland
Davide Cerotti	University of Piedmont Oriental, Italy

Scientific Computing Infrastructure (SCI 2024)

Workshop Organizers

Vladimir Korkhov	St. Petersburg University, Russia
Elena Stankova	St. Petersburg State University, Russia

Workshop Program Committee Members

Adam Belloum	University of Amsterdam, the Netherlands
Dmitry Vasiunin	Deutsche Telekom Cloud Services E.P.E., Greece
Serob Balyan	National Academy of Sciences of Armenia, Armenia
Suren Abrahamyan	Osensus Arm LLC, Armenia
Ashot Gevorkyan	National Academy of Sciences of Armenia, Armenia
Michal Hnatic	Univerzita Pavla Jozefa Šafárika v Košiciach, Slovakia
Martin Vala	Univerzita Pavla Jozefa Šafárika v Košiciach, Slovakia
Nodir Zaynalov	Tashkent University of Information Technologies, Uzbekistan
Michail Panteleyev	St. Petersburg Electrotecnical University, Russia

Nikolay Peryazev	Irkutsk State University, Irkutsk, Russia
Alexander Degtyarev	St. Petersburg State University, Russia
Alexander Bogdanov	St. Petersburg State University, Russia
Nataliia Kulabukhova	SberAutoTech, Russia

Downscale Agenda 2030 (SDGscale 2024)

Workshop Organizers

Anna Richiedei	University of Brescia, Italy
Michele Pezzagno	University of Brescia, Italy
Ginevra Balletto	University of Cagliari, Italy
Francesca Sinatra	University of Trieste, Italy
Federico Martellozzo	University of Florence, Italy
Tú Anh Trinh	University of Economics Ho Chi Minh City, Vietnam

Workshop Program Committee Members

Giuseppe Borruso	University of Trieste, Italy
Elisabetta Maria Venco	University of Pavia, Italy
Riccardo Privitera	University of Catania, Italy
Elisa Conticelli	University of Bologna, Italy
Giovanni Marinelli	Polytechnic University of Marche, Italy
Francesca Sinatra	University of Trieste, Italy
Salvatore Dore	University of Trieste, Italy
Maria Attard	University of Malta, Malta
Giovanni Mauro	University of Campania Luigi Vanvitelli, Italy
Maria Ronza	University of Naples, Federico II, Italy
Massimiliano Bencardino	University of Salerno, Italy

Socio-Economic and Environmental Models for Land Use Management (SEMLUM 2024)

Workshop Organizers

Debora Anelli	Polytechnic of Bari, Italy
Pierluigi Morano	Polytechnic of Bari, Italy
Benedetto Manganelli	University of Basilicata, Italy
Francesco Paolo Del Giudice	Sapienza University of Rome, Italy

Workshop Program Committee Members

Laura Gabrielli	University of Ferrara, Italy
Sergio Copiello	University of Venice, Italy
Antonio Nesticò	University of Salerno, Italy
Pierfrancesco De Paola	University of Napoli, Italy
Elena Fregonara	Polytechnic of Turin, Italy
Paola Amoruso	LUM, Italy

Ports of the Future - Smartness and Sustainability (SmartPorts 2024)

Workshop Organizers

Giuseppe Borruso	University of Trieste, Italy
Gianfranco Fancello	University of Cagliari, Italy
Patrizia Serra	University of Cagliari, Italy
Silvia Battino	University of Sassari, Italy
Marco Petrelli	Roma Tre University, Italy

Workshop Program Committee Members

Ginevra Balletto	University of Cagliari, Italy
Beniamino Murgante	University of Basilicata, Italy
Marco Mazzarino	IUAV University Venice, Italy
Maria del Mar Munoz Leonisio	University of Cadiz, Spain
Ahinoa Amaro Garcia	University of Las Palmas of Gran Canaria, Spain
Veronica Camerada	University of Sassari, Italy
Brunella Brundu	University of Sassari, Italy
Maria Attard	University of Malta, Malta
Enrico Dagostini	University of Malta, Malta
Tu Anh Trinh	College of Technology and Design for UEH University, Vietnam
Giovanni Mauro	University of Campania Luigi Vanvitelli, Italy
Maria Ronza	University of Naples, Federico II, Italy
Massimiliano Bencardino	University of Salerno, Italy

Smart Transport and Logistics - Smart Supply Chains (SmarTransLog 2024)

Workshop Organizers

Giuseppe Borruso	University of Trieste, Italy
Marcello Tadini	University of Eastern Piedmont, Italy
Maria del Mar Munoz Leonisio	University of Cádiz, Spain
Maria Attard	University of Malta, Malta
Veronica Camerada	University of Sassari, Italy
Brunella Brundu	University of Sassari, Italy

Workshop Program Committee Members

Ginevra Balletto	University of Cagliari, Italy
Silvia Battino	University of Sassari, Italy
Gianfranco Fancello	University of Cagliari, Italy
Mara Ladu	University of Cagliari, Italy
Martina Sinatra	University of Cagliari, Italy
Francesca Sinatra	University of Trieste, Italy
Salvatore Dore	University of Trieste, Italy
Andrea Gallo	University of Trieste, Italy
Marco Mazzarino	IUAV University Venice, Italy
Enrico Dagostini	University of Malta, Malta
Marco Naseddu	University of Cagliari, Italy
José Ángel Hernández Luis	University of Las Palmas de Gran Canaria, Spain
Maurizio Cociancich	Adriafer, Italy
Giovanni Longo	University of Trieste, Italy
Luca Toneatti	University of Trieste, Italy
Giovanni Mauro	University of Campania Luigi Vanvitelli, Italy
Maria Ronza	University of Naples, Federico II, Italy
Massimiliano Bencardino	University of Salerno, Italy

Smart Tourism (SmartTourism 2024)

Workshop Organizers

Silvia Battino	University of Sassari, Italy
Francesca Krasna	University of Trieste, Italy
Maria del Mar Munoz Leonisio	University of Cadiz, Spain

Ginevra Balletto	University of Cagliari, Italy
Brisol García García	Polytechnic University of Quintana Roo, Mexico
Ainhoa Amaro Garcia	University of Las Palmas of Gran Canarias, Spain

Workshop Program Committee Members

Giuseppe Borruso	University of Trieste, Italy
Beniamino Murgante	University of Basilicata, Italy
Gianfranco Fancello	University of Cagliari, Italy
Mara Ladu	University of Cagliari, Italy
Martina Sinatra	University of Cagliari, Italy
Salvatore Dore	University of Trieste, Italy
Veronica Camerada	University of Sassari, Italy
Brunella Brundu	University of Sassari, Italy
Maria Attard	University of Malta, Malta
Tu Anh Trinh	College of Technology and Design for UEH University, Vietnam
Giovanni Mauro	University of Campania Luigi Vanvitelli, Italy
Maria Ronza	University of Naples, Federico II, Italy
Massimiliano Bencardino	University of Salerno, Italy

Sustainable Evolution of Long-Distance Freight Passenger Transport (SOLIDEST 2024)

Workshop Organizers

Tiziana Campisi	University of Enna Kore, Italy
Socrates Basbas	Aristotle University of Thessaloniki, Greece
Massimo Di Gangi	University of Messina, Italy
Antonio Comi	University of Rome Tor Vergata, Italy
Francesco Russo	Mediterranea University of Reggio Calabria, Italy
Giovanni Tesoriere	University of Enna Kore, Italy

Workshop Program Committee Members

Antonio Polimeni	University of Messina, Italy
Orlando Belcore	University of Messina, Italy
Filippo G. Praticò	Mediterranea University of Reggio Calabria, Italy
Borja Alonso	University of Cantabria, Spain
Luigi Dall'Olio	University of Cantabria, Spain
Antonio Russo	University of Enna Kore, Italy

Sustainability Performance Assessment: Models, Approaches, and Applications Toward Interdisciplinary and Integrated Solutions (SPA 2024)

Workshop Organizers

Francesco Scorza	University of Basilicata, Italy
Sabrina Lai	University of Cagliari, Italy
Jolanta Dvarioniene	Kaunas University of Technology, Lithuania

Workshop Program Committee Members

Carmela Gargiulo	University of Naples Federico II, Italy
Floriana Zucaro	University of Naples Federico II, Italy
Corrado Zoppi	University of Cagliari, Italy
Federica Leone	University of Cagliari, Italy
Federica Isola	University of Cagliari, Italy
Francesca Leccis	University of Cagliari, Italy
Francesca Perrone	Sapienza University of Rome, Italy
Angela Pilogallo	Italian National Research Council, Italy
Alfonso Annunziata	University of Basilicata, Italy
Luigi Santopietro	Italian National Research Council, Italy
Beniamino Murgante	University of Basilicata, Italy
Alessandro Marucci	University of L'Aquila, Italy
Francesco Zullo	University of L'Aquila, Italy
Monica Salvia	Italian National Research Council, Italy
Fulvia Pietrapertosa	Italian National Research Council, Italy
Valentin Grecu	Lucian Blaga University, Romania
Georgia Pozoukidou	Aristotle University of Thessaloniki, Greece
Giampiero Lombardini	University of Genoa, Italy

Specifics of Smart Cities Development in Europe (SPEED 2024)

Workshop Organizers

Chiara Garau	University of Cagliari, Italy
Katarína Vitálišová	Matej Bel University, Slovakia
Paolo Nesi	University of Florence, Italy
Anna Vaňová	Matej Bel University, Slovakia
Kamila Borsekova	Matej Bel University, Slovakia
Paola Zamperlin	University of Florence, Italy

Workshop Program Committee Members

Tiziana Campisi	University of Enna Kore, Italy
Giovanna Concu	University of Cagliari, Italy
Claudia Loggia	University of KwaZulu-Natal, South Africa
Alessandro Plaisant	University of Sassari, Italy
Francesca Maltinti	University of Cagliari, Italy
Silvia Rossetti	University of Parma, Italy
Gerardo Carpentieri	University of Naples Federico II, Italy
Carmen Guida	University of Naples Federico II, Italy
Anastasia Stratigea	National Technical University of Athens, Greece
Yiota Theodora	National Technical University of Athens, Greece
Marco Fanfani	University of Florence, Italy
Emanuele Bellini	University of Rome 3, Italy

Smart, Safe and Health Cities (SSHC 2024)

Workshop Organizers

Chiara Garau	University of Cagliari, Italy
Gerardo Carpentieri	University of Naples Federico II, Italy
Carmen Guida	University of Naples Federico II, Italy
Floriana Zucaro	University of Naples Federico II, Italy
Tanja Congiu	University of Sassari, Italy
Martina Carra	University of Brescia, Italy
Benedetto Barabino	University of Brescia, Italy

Workshop Program Committee Members

Rosaria Battarra	Italian National Research Council, Italy
Barbara Caselli	University of Parma, Italy
Francesca Maltinti	University of Cagliari, Italy
Romano Fistola	University of Naples Federico II, Italy
Sabrina Lai	University of Cagliari, Italy
Claudia Loggia	University of KwaZulu-Natal, South Africa
Alessandro Plaisant	University of Sassari, Italy
Silvia Rossetti	University of Parma, Italy
Michela Tiboni	University of Brescia, Italy
John Zacharias	Peking University, China
Thomas Buhler	University of Franche-Comté, France

Aynaz Lotfata	University of California, USA
Lucia Saganeiti	University of L'Aquila, Italy
Francesco Zullo	University of L'Aquila, Italy
Paola Zamperlin	University of Florence, Italy
Katarína Vitálišová	Matej Bel University, Faculty of Economics, Slovakia
Tiziana Campisi	University of Enna Kore, Italy
Caterina Pietra	University of Pavia, Italy
Elisabetta Maria Venco	University of Pavia, Italy
Vincenza Torrisi	University of Catania, Italy

Smart and Sustainable Island Communities (SSIC 2024)

Workshop Organizers

Chiara Garau	University of Cagliari, Italy
Anastasia Stratigea	National Technical University of Athens, Greece
Yiota Theodora	National Technical University of Athens, Greece
Giovanna Concu	University of Cagliari, Italy

Workshop Program Committee Members

Milena Metalkova-Markova	University of Portsmouth, UK
Tarek Teba	University of Portsmouth, UK
Despina Dimelli	Technical University of Crete, Greece
Gerardo Carpentieri	University of Naples Federico II, Italy
Carmen Guida	University of Naples Federico II, Italy
Floriana Zucaro	University of Naples Federico II, Italy
Anestis Gourgiotis	University of Thessaly, Greece
George Tsilimigkas	University of the Aegean, Greece
Maria Panagiotopoulou	National Technical University of Athens, Greece
George Somarakis	National Technical University of Athens, Greece
Harry Kyriakidis	National Technical University of Athens, Greece
Apostolos Lagarias	Univ. of Thessaly, Greece
Paola Zamperlin	University of Florence, Italy
Giuseppina Vacca	University of Cagliari, Italy
Roberto Minunno	Curtin University, Perth, Australia
Claudia Loggia	University of KwaZulu-Natal, South Africa
Francesca Maltinti	University of Cagliari, Italy
Elisabetta Maria Venco	University of Pavia, Italy

From Street Experiments to Planned Solutions (STEPS 2024)

Workshop Organizers

Silvia Rossetti	University of Parma, Italy
Angela Ricciardello	University of Enna Kore, Italy
Francesco Pinna	University of Cagliari, Italy
Chiara Garau	University of Cagliari, Italy
Tiziana Campisi	University of Enna Kore, Italy
Vincenza Torrisi	University of Catania, Italy

Workshop Program Committee Members

Martina Carra	University of Brescia, Italy
Barbara Caselli	University of Parma, Italy
Tanja Congiu	University of Sassari, Italy
Gabriele D'Orso	University of Palermo, Italy
Matteo Ignaccolo	University of Catania, Italy
Md Kh Nahiduzzaman	University of British Columbia, Canada
Muhammad Ahmad Al-Rashid	University of Malaya, Malesia
Alessandro Plaisant	University of Sassari, Italy
Marianna Ruggieri	University of Enna Kore, Italy
Michele Zazzi	University of Parma, Italy

Sustainable Development of Ports (SUSTAINABLEPORTS 2024)

Workshop Organizers

Giuseppe Musolino	Mediterranea University of Reggio Calabria, Italy
Tiziana Campisi	University of Enna Kore, Italy
Efstathios Bouhouras	Aristotle University of Thessaloniki, Greece
Elena Cocuzza	University of Catania, Italy
Gianfranco Fancello	University of Cagliari, Italy
Massimo Di Gangi	University of Messina, Italy

Workshop Program Committee Members

Georgios Vaggelas	National and Kapodistrian University of Athens, Greece
Antonio Polimeni	University of Messina, Italy

Orlando Belcore	University of Messina, Italy
Lucia Dalla Spina	Mediterranea University of Reggio Calabria, Italy
Elen Twerdy	University of Lubljana, Slovenia
Claudia Caballini	Polytechnic University of Turin, Italy

Theoretical and Computational Chemistry and Its Applications (TCCMA 2024)

Workshop Organizers

Noelia Faginas-Lago	University of Perugia, Italy
Andrea Lombardi	University of Perugia, Italy
Marcos Mandado	University of Vigo, Spain

Workshop Program Committee Members

Ángel Martín Pendás	University of Oviedo, Spain
Ernesto Garcia Para	University of Basque Country, Spain
Giovanni Bistoni	University of Perugia, Italy
Juan Jose Nogueira Perez	University Autonoma of Madrid, Spain
Jose Hermida	Universidade di Vigo, Spain
Luca Mancini	University of Perugia, Italy

Transport Infrastructures for Smart Cities (TISC 2024)

Workshop Organizers

Francesca Maltinti	University of Cagliari, Italy
Mauro Coni	University of Cagliari, Italy
Francesco Pinna	University of Cagliari, Italy
Chiara Garau	University of Cagliari, Italy
Nicoletta Rassu	University of Cagliari, Italy
James Rombi	University of Cagliari, Italy

Workshop Program Committee Members

Benedetto Barabino	University of Brescia, Italy
Martina Carra	University of Brescia, Italy

Mauro D'Apuzzo	University of Cassino, Italy
Roberto Minunno	Curtin University, Australia
Claudia Loggia	University of KwaZulu-Natal, South Africa
Amir Kavussi	Tarbiat Modares University, Iran
Tiziana Campisi	University of Enna Kore, Italy
Alessandro Plaisant	University of Sassari, Italy
Elena Cocuzza	University of Catania, Italy
Vincenza Torrisi	University of Catania, Italy

From Structural to Transformative-Change of City Environment: Challenges and Solutions and Perspectives (TRACE 2024)

Workshop Organizers

Maria Rosaria Guarini	Sapienza University of Rome, Italy
Francesco Sica	Sapienza University of Rome, Italy
Carmelo Maria Torre	Polytechnic University of Bari, Italy
Francesco Tajani	Polytechnic University of Bari, Italy

Workshop Program Committee Members

Angela Pilogallo	Italian National Research Council, Italy
Maria Cerreta	University of Naples Federico II, Italy
Giuseppe Cerullo	Sapienza University of Rome, Italy

Transport and Digital Multiscale Sustainable Network for Circular Economy (TransNet 2024)

Workshop Organizers

Andrea Gallo	University of Trieste, Italy
Mara Ladu	University of Cagliari, Italy
Enrico Dagostini	University of Malta, Malta
Salvatore Dore	University of Trieste, Italy
Martina Sinatra	University of Cagliari, Italy
Francesca Sinatra	University of Trieste, Italy

Workshop Program Committee Members

Martina Carra	University of Brescia, Italy
Ginevra Baletto	University of Cagliari, Italy
Giuseppe Borruso	University of Trieste, Italy
Beniamino Murgante	University of Basilicata, Italy
Maria del Mar Munoz Leonisio	Univeristy of Cadiz, Spain
Ahinoa Amaro Garcia	Univeristy of Las Palmas of Gran Canaria, Spain
Malgorzata Hanzl	Lodz University of Technology, Poland
Balázs Kulcsár	University of Debrecen, Hungary

Temporary Real Estate Management: Approaches and Methods for Time-Integrated Impact Assessments and Evaluations (TREAT 2024)

Workshop Organizers

Chiara Mazzarella	TU Delft, The Netherlands
Hilde Remoy	TU Delft, The Netherlands
Maria Cerreta	University of Naples Federico II, Italy

Workshop Program Committee Members

Maria Lucia Raiola	University of Naples Federico II, Italy
Eugenio Muccio	University of Naples Federico II, Italy
Daniele Cannatella	TU Delft, The Netherlands
Sabrina Sacco	University of Naples Federico II, Italy
Piero Zizzania	University of Naples Federico II, Italy
Stefano Cuntò	University of Naples Federico II, Italy
Sveva Ventre	University of Naples Federico II, Italy
Caterina Loffredo	University of Naples Federico II, Italy
Giuseppe Ciciriello	University of Naples Federico II, Italy
Maria Somma	University of Naples Federico II, Italy
Ludovica La Rocca	University of Naples Federico II, Italy
Laura Di Tommaso	University of Naples Federico II, Italy
Benedetta Grieco	University of Naples Federico II, Italy
Gaia Daldanise	National Research Council, Italy
Simona Panaro	University of Sussex, UK
Giuliano Poli	University of Naples Federico II, Italy

Urban Regeneration: Innovative Tools and Evaluation Model (URITEM 2024)

Workshop Organizers

Fabrizio Battisti	University of Florence, Italy
Giovanna Acampa	University of Florence and University of Enna Kore, Italy
Orazio Campo	La Sapienza University of Rome, Italy

Workshop Program Committee Members

Alice Barreca	Polytechnic of Turin, Italy
Marco Rossitti	Polytechnic of Milan, Italy
Celestina Fazia	University of Enna Kore, Italy

Urban Space Accessibility and Mobilities (USAM 2024)

Workshop Organizers

Chiara Garau	University of Cagliari, Italy
Alessandro Plaisant	University of Sassari, Italy
Barbara Caselli	University of Parma, Italy
Gabriele D'Orso	University of Palermo, Italy
Mauro D'Apuzzo	University of Cassino, Italy
Matteo Ignaccolo	University of Catania, Italy

Workshop Program Committee Members

Ivan Blečić	University of Cagliari, Italy
Mauro Coni	University of Cagliari, Italy
Martina Carra	University of Brescia, Italy
Tiziana Campisi	University of Enna Kore, Italy
Tanja Congiu	University of Sassari, Italy
Francesca Maltinti	University of Cagliari, Italy
Silvia Rossetti	University of Parma, Italy
Angela Pilogallo	Italian National Research Council, Italy
Lorena Fiorini	University of L'Aquila, Italy
Francesco Pinna	University of Cagliari, Italy

Aime Tsinda	University of Rwanda, Rwanda
Nicoletta Rassu	University of Cagliari, Italy
Youssef El Ganadi	International University of Rabat, Morocco
Marco Migliore	University of Palermo, Italy
Elena Cocuzza	University of Catania, Italy
Alessio Salvatore	Italian National Research Council, Italy
Giuseppe Stecca	Italian National Research Council, Italy
Gerardo Carpentieri	University of Naples Federico II, Italy
Carmen Guida	University of Naples Federico II, Italy
Floriana Zucaro	University of Naples Federico II, Italy

Virtual Reality and Augmented Reality and Applications (VRA 2024)

Workshop Organizers

Osvaldo Gervasi	University of Perugia, Italy
Damiano Perri	University of Florence, Italy
Sergio Tasso	University of Perugia, Italy

Workshop Program Committee Members

Massimiliano Nicolini	OliTec Foundation, Italy
JungYoon Kim	Gachon University, South Korea
Paolo Nesi	University of Florence, Italy
David Berti	ART SpA, Italy
Tamie Salter	Citizen Alert Inc., Canada
Marcelo Guimarães	Federal University of São Paulo, Brazil

Workshop on Advanced and Computational Methods for Earth Science Applications (WACM4ES 2024)

Workshop Organizers

Luca Piroddi	University of Malta, Malta
Patrizia Capizzi	University of Palermo, Italy
Marilena Cozzolino	Università Del Molise, Italy
Sebastiano D'Amico	University of Malta, Malta
Chiara Garau	University of Cagliari, Italy
Giuseppina Vacca	University of Cagliari, Italy

Workshop Program Committee Members

Andrea Angelini	ISPC, CNR, Italy
Raffaele Martorana	University of Palermo, Italy
Paolo Mauriello	University of Molise, Italy
Raffaele Persico	University of Calabria, Italy
Eufemia Tarantino	Polytechnic of Bari, Italy
Roberto Ricciu	University of Cagliari, Italy
Stefano Sfarra	University of L'Aquila, Italy
Peter Iregbeyen	University of Malta, Malta
Elisa Pilia	University of Cagliari, Italy
Emanuele Colica	Ministry for Transport, Infrastructure and Public Works, Malta
Luciano Galone	University of Malta, Malta
Sergio Vincenzo Calcina	University of Cagliari, Italy
Maria Brovelli	Polytechnic of Milan, Italy
Francesca Trevisiol	University of Bologna, Italy

Additional Reviewers

Burak Aslan	Izmir Institute of Technology, Turkey
Birim Balcı	Manisa Celal Bayar University, Turkey
Atrin Barzegar	Università degli Studi della Campania L. Vanvitelli, Italy
Manuel Carlos Figueiredo	University of Minho, Portugal
Chiara Chioni	University of Trento, Italy
Annalisa Contato	University of Palermo, Italy
Aurora Costales	Universidad de Oviedo, Spain
Hamid Daneshpajouh	Middle East Technical University, Turkey
Vito Dario Camiola	Università di Catania, Italy
Roberta De Fazio	Università degli Studi della Campania Luigi Vanvitelli, Italy
Michele Di Giovanni	Università degli Studi della Campania, Italy
David Ferro Costas	University of Santiago de Compostela, Spain
Michal Hnatič	P.J. Šafarik University in Košice, Slovak Republic
Francesco Loddo	Henge S.r.l., Italy
Caterina Loffredo	University of Naples Federico II, Italy
Vanda Lourenco	Universidade Nova de Lisboa, Portugal
Ángel Martín Pendás	Universidad de Oviedo, Spain
Maria Mourao	Polytechnic Institute of Viana do Castelo, Portugal

Martin Perez Perez	University of Vigo, Spain
Marco Pezzella	University of Perugia, Italy
Maria Polidoro	ESTG IPP, Portugal
Josef Rebenda	Brno University of Technology, Czech Republic
Donatella Rita Fiorino	University of Cagliari, Italy
Evgeniy Romenskiy	Sobolev Institute of Mathematics, Russia
Ezgi Sahin Toptal	Mersin University, Turkey
Domenica Savia Pellicanò	Mediterranean University of Reggio Calabria, Italy
Carina Silva	ESTeSL-IPL, Portugal
Paula Simões	CMA FCT UNL CINAMIL, Portugal
Francesca Sinatra	University of Trieste, Italy
Ana Sofia Rézio	Iscal, Portugal
Alexandra Stankulova	Politecnico di Torino, Italy
Maria Stella De Biase	Università della Campania Luigi Vanvitelli, Italy
Guan Yue Hong	Western Michigan University, USA

Sponsoring Organizations

ICCSA 2024 would not have been possible without the tremendous support of many organizations and institutions, for which all organizers and participants of ICCSA 2024 express their sincere gratitude:

Springer Nature Switzerland AG, Switzerland
(https://www.springer.com)

Computers Open Access Journal
(https://www.mdpi.com/journal/computers)

Thuyloi University, Hanoi, Vietnam
(https://en.tlu.edu.vn/)

University of Perugia, Italy
(https://www.unipg.it)

University of Basilicata, Italy
(http://www.unibas.it)

Monash University, Australia
(https://www.monash.edu/)

Kyushu Sangyo University, Japan
(https://www.kyusan-u.ac.jp/)

Universidade do Minho
Escola de Engenharia

University of Minho, Portugal
(https://www.uminho.pt/)

University of Cagliari, Italy
(https://en.unica.it/en)

Venue

ICCSA 2024 took place on the main campus of Thuyloi University in Hanoi, Vietnam.

Plenary Lectures

Harnessing Artificial Intelligence for Enhanced Spatial Analysis of Natural Hazard Assessments

Prof. Dr. Biswajeet Pradhan

Director - Centre for Advanced Modelling and Geospatial Information Systems (CAMGIS), School of Civil and Environmental Engineering, Faculty of Engineering and IT, University of Technology Sydney, Australia

Abstract. In the realm of natural hazard assessments within spatial domains, the advent of Artificial Intelligence (AI) represents a paradigm shift, revolutionizing the way we conceptualize, model, and interpret environmental risks. This keynote address illuminates the profound impact of AI technologies, particularly machine learning algorithms and data-driven approaches, in reshaping our understanding and prediction capabilities concerning natural disasters.

By assimilating and scrutinizing vast spatial datasets, AI-driven models offer unparalleled accuracy and efficiency, facilitating timely and precise hazard assessments. Real-time processing of geospatial information not only enables rapid predictions but also forms the cornerstone of proactive disaster management strategies. Furthermore, AI's capacity lies in its adeptness at deciphering intricate spatial patterns inherent to natural hazards, unraveling subtle cues and previously unnoticed correlations within the data fabric.

This keynote delves into how AI's nuanced interpretation, coupled with advanced algorithms, elevates hazard modeling, providing deeper insights into the spatial dynamics of environmental risks. By augmenting traditional methodologies and revealing hidden patterns, AI fosters comprehensive risk assessments, fostering informed decision-making processes. The fusion of AI and natural hazard assessments in spatial domains heralds a more resilient approach to disaster preparedness and response.

Join us in embracing this transformative era, where AI's sophisticated modeling techniques and precise spatial interpretations converge, heralding proactive and effective mitigation strategies amidst the ever-evolving landscape of environmental challenges.

Short Bio. Distinguished Professor Dr. Biswajeet Pradhan is an internationally established scientist in the field of Geospatial Information Systems (GIS), remote sensing and image processing, complex modelling/geo-computing, machine learning and soft-computing applications, natural hazards and environmental modelling. He is the Director of the Centre for Advanced Modelling and Geospatial Information Systems (CAMGIS) at the Faculty of Engineering and IT at the University of Technology, Sydney (Australia). He was listed as the World's Most Highly Cited Researcher by the Clarivate Analytics Report for five consecutive years, 2016–2020, as one of the world's most influential minds.

He ranked number one (1) in the field of "Geological & Geomatics Engineering" during the calendar year 2021–2023, according to the list published by Stanford University Researchers, USA. This list ranks the world's top 2% most highly cited researchers based on Scopus data. In 2018–2020, he was awarded as World Class Professor by the Ministry of Research, Technology and Higher Education, Indonesia. He is a recipient of the Alexander von Humboldt Research Fellowship from Germany. Between 2015–2021, he served as "Ambassador Scientist" for the Alexander Humboldt Foundation, Germany.

Professor Pradhan has received 58 awards since 2006 in recognition of his excellence in teaching, service and research. Out of his more than 850 articles (Google Scholar citation: 70,000, H-index: 129), more than 750 have been published in science citation index (SCI/SCIE) technical journals. He has authored/co-authored ten books and thirteen book chapters.

Software Engineering Research in a New Situation

Prof. Carl K. Chang

Professor Emeritus, Iowa State University, USA

Abstract. With the rise of Generative Artificial Intelligence (GAI), epitomized by Large Language Models (LLMs), a profound shift has unfolded in software engineering research. In this presentation, I will traverse my four-decade journey in software engineering research, focusing on situational awareness in the era of the Internet of Things (IoT). I have witnessed the turbulence brought forth by the AI community that demands changes in our approaches. Meanwhile, owing to the pervasiveness of services computing, services became the first-class citizen in modern-day software engineering methodologies.

I argue that situational awareness must permeate the entire lifecycle to consistently deliver software services that align with the dynamic needs of users and the ever-evolving environments. I will elucidate this argument by reviewing the Situ framework, offering a comprehensive illustration of my perspective. Furthermore, I will outline my vision regarding the formidable research challenges considering the rapidly shifting landscape dominated by an irresistible and profoundly disruptive generative AI tsunami.

Short Bio. Carl K. Chang is a former department chair and Professor Emeritus of Computer Science at Iowa State University. His research interests include requirements engineering, net-centric computing, situational software engineering and digital health. Chang was the 2004 President of the IEEE Computer Society. Previously he served as the Editor-in-Chief for IEEE Software (1991–1994), and as the Editor-in-Chief of IEEE Computer (2007–2010). He was the 2012 recipient of the Richard E. Merwin Medal from the IEEE Computer Society. Chang is a Life Fellow of IEEE, a Fellow of AAAS, and a Life Member of the European Academy of Sciences (EurASc).

Interpretability and Privacy Preservation in Large Language Models (LLMs)

Prof. My Thai

University of Florida (UF) Research Foundation Professor
Associate Director of UF Nelms Institute for the Connected World

Abstract. Large Language Models (LLMs) have transformed the AI landscape, captivating researchers and practitioners with their remarkable ability to generate human-like text and perform complex tasks. However, this transformative power comes with a set of critical challenges, particularly in the realms of interpretability and privacy preservation. In this keynote, we embark on an exploration of these pressing issues, shedding light on how LLMs operate, their limitations, and the strategies we can employ to mitigate risks. We begin by examining the interpretability in LLMs, which often function as enigmatic "black boxes." Their complex neural architectures make it challenging to understand how they arrive at specific outputs. This lack of transparency raises questions of trust and accountability. When deploying LLMs in real-world applications—whether for chatbots, content generation, or decision-making—it becomes crucial to demystify their decision paths.

We will use explainable AI (XAI) to offer faithful explanations, from the black-box to white-box models, and from feature-based [1, 2] to neuron circuits-based [3, 4] explanations. By visualizing attention mechanisms, feature importance, and saliency maps, we empower users to comprehend LLM predictions. XAI not only fosters trust but also encourages responsible utilization of LLMs.

We next turn our attention to one of the utmost concerns and challenges: data privacy. LLMs process vast amounts of data, raising risks of data leakage, model inversion, the right to be forgotten, and inadvertent exposure of sensitive information. Furthermore, the integration of LLMs into diverse applications also significantly brings these challenges to the next level [5]. This talk explores strategies to protect privacy, including differential privacy, federated learning, and data encryption.

Short Bio. My T. Thai is a University of Florida (UF) Research Foundation Professor, Associate Director of UF Nelms Institute for the Connected World, and a Fellow of IEEE and AAIA. Dr. Thai is a leading authority who has done transformative research in Trustworthy AI and Optimization, especially for complex systems with applications to healthcare, social media, critical networking infrastructure, and cybersecurity. The results of her work have led to 7 books and 350+ publications in highly ranked international journals and conferences, including several best paper awards from the IEEE, ACM, and AAAI.

In responding to a world-wide call for responsible and safe AI, Dr. Thai is a pioneer in designing deep explanations for black-box ML models, while defending against explanation-guided attacks, evident by her Distinguished Papers Award at the Association for the Advancement of Artificial Intelligence (AAAI) conference in 2023. At the same year, she was also awarded an ACM Web Science Trust Test-of-Time award, for her landmark work on combating misinformation in social media. In 2022, she received an IEEE Big Data Security Women of Achievement Award. In 2009, she was awarded the Young Investigator (YIP) from the Defense Threat Reduction Agency (DTRA), and in 2010 she won the NSF CAREER Award. She is presently the Editor-in-Chief of the Springer Journal of Combinatorial Optimization and the IET Blockchain Journal, and editor of the Springer book series Optimization and Its Applications.

References

1. Vu, M., Thai, M.T.: PGM-explainer: probabilistic graphical model explanations for graph neural networks. In: Advances in Neural Information Processing Systems (NeurIPS) (2020)
2. Nguyen, T., Lai, P., Phan, H., Thai, M.T.: XRand: differentially private defense against explanation-guided attacks. In: AAAI Conference on Artificial Intelligence (AAAI) (2023)
3. Vu, N., Nguyen, T., Thai, M.T.: NeuCEPT: learn neural networks' mechanism via critical neurons with precision guarantee. In: IEEE International Conference on Data Mining (ICDM) (2022)

4. Conmy, A., Mavor-Parker, A., Heimersheim, S., Garriga-Alonso, A.: Towards automated circuit discovery for mechanistic interpretability. In: Advances in Neural Information Processing Systems (NeurIPS) (2023)
5. Vu, M., Nguyen, T., Jeter, T., Thai, M.T.: Analysis of privacy leakages in federated large language models. In: International Conference on Artificial Intelligence and Statistics (AISTATS) (2023)

Contents – Part IV

Workshop on Computational Science and HPC (CSHPC 2024)

Performance Evaluation of Accelerated Complex Multiple-Precision LU Decomposition

Tomonori Kouya(✉)

Shizuoka Institute of Science and Technology, 2022-2 Toyosawa, Fukuroi, Shizuoka 437-8555, Japan
kouya.tomonori@sist.ac.jp
https://www.sist.ac.jp

Abstract. The direct method is one of the most important algorithms for solving linear systems of equations, with LU decomposition depleting a significant portion of the computation time. This study explores strategies for accelerating complex LU decomposition, using multiple-precision floating-point arithmetic of the multiple-component type. Specifically, the potential efficiency gains are explored using a combination of single instruction and multiple data stream (SIMD) and the 3M method for complex matrix multiplication. This approach was compared with the implementation of the direct method in MPLAPACK, using benchmark tests focusing on computation time and numerical errors.

Keywords: LU decomposition · multiple-precision floating-point arithmetic · complex arithmetic · parallelization

1 Introduction

Linear systems of equations are frequently encountered in scientific computations; therefore, various solvers have been developed to address the computational environment and mathematical properties of these equations. The direct method, including LU decomposition, is considered standard because it can solve equations with a limited number of algebraic operations, maintaining numerical stability using a pivoting strategy. The high performance linpack (HPL) benchmark test [9] employs the direct method to select the top 500 supercomputers.

In recent years, considerable research has been conducted on mixed-precision computing, thereby tailoring the number of precision digits (the length of the mantissa) of the floating-point (FP) numbers to the requirements of the problem. This approach improves computational performance by combining various types of hardware-oriented IEEE754 FP arithmetic, such as binary16 (11-bit precision), binary32 (24-bit), and binary64 (53-bit). However, the inadequate accuracy of the calculation poses a significant disadvantage. Thus, there is an

O. Gervasi et al. (Eds.): ICCSA 2024 Workshops, LNCS 14818, pp. 3–19, 2024.
https://doi.org/10.1007/978-3-031-65273-8_1

increasing demand for "reproducible computing" to minimize the environmental dependence of the calculation results. For well-conditioned problems, speed-up should be pursued using fast binary 16, 32, and 64 FP operations with relatively short precision and hardware-oriented computation. Conversely, for ill conditioned problems, multiple-precision FP operations exceeding binary64 should be employed to ensure accuracy and reproducibility. In particular, stable solutions to nonlinear problems based on linear computation, require minimizing the rounding errors in the intermediate results. Utilizing a standard multiple-precision FP arithmetic library that is known for user-friendly interface and high performance, such as QD [1] and multiple-precision floating-point reliable (MPFR) [10], is essential. These libraries are incorporated into MPLAPACK/MPBLAS [7], serving as a de facto standard multiple-precision linear computation environment. We developed three high-performance multiple-precision real basic linear subprograms (BLAS) of double-double (DD, 106 bits), triple-double (TD, 159 bits), and quadruple-double (QD, 212 bits) arithmetic using advanced vector extensions two (AVX2) [5]. The results have been compiled into BNCmatmul [4] and released as open-source software.

Based on these results, this paper discusses the performance evaluation of complex multiple-precision LU decomposition, particularly focusing on multi-component DD, TD, and QD precision, further employing 3M, a type of complex multiplication method. Benchmark tests were conducted to compare the acceleration provided by complex arithmetic using AVX2, LU decomposition using the Strassen algorithm [11] and Ozaki scheme [8], and parallelization using OpenMP.

2 Acceleration of Complex Basic Linear Computation

As mentioned above, we implemented real BLAS and LU decomposition in purely real values using single instruction and multiple data stream (SIMD) operations such as AVX2 and a library of multiple-precision linear computations that supported parallelization using open multi-processing (OpenMP). Additionally, we developed arbitrary precision floating-point operations based on MPFR and multi-component fixed-precision operations supporting DD, TD, and QD precision, which are faster than MPLAPACK/MPBLAS. Building on this foundation, we implemented a complex BLAS [6]. Notably, despite publishing a collection of results on arbitrary-precision complex LU decomposition [6], we have yet to parallelize it. In the following section, strategies for improving the performance of complex-valued linear computation, which serves as the basis for complex LU decomposition, are explained.

2.1 3M Method for Complex Linear Computation

For any $a, b \in \mathbb{C}$, the standard complex multiplication $ab \in \mathbb{C}$ is calculated by using the 4M (multiplication) method, which involves four real multiplications:

$$\begin{aligned} ab := (\mathrm{Re}(a)\mathrm{Re}(b) - \mathrm{Im}(a)\mathrm{Im}(b)) + \{\mathrm{Im}(a)\mathrm{Re}(b) \\ +\mathrm{Re}(a)\mathrm{Im}(b)\} \cdot \mathrm{i}. \end{aligned} \tag{1}$$

In contrast, the 3M method is executed as follows:

$$\begin{aligned} t_1 &:= \mathrm{Re}(a)\mathrm{Re}(b),\ t_2 := \mathrm{Im}(a)\mathrm{Im}(b) \in \mathbb{R} \\ ab &:= (t_1 - t_2) + \{(\mathrm{Re}(a) + \mathrm{Im}(a)) \\ &\quad (\mathrm{Re}(b) + \mathrm{Im}(b)) - t_1 - t_2\} \cdot \mathrm{i} \in \mathbb{C}. \end{aligned} \tag{2}$$

The 3M method retains the same computation in the real part as the 4M method but reduces the number of real multiplications in the imaginary part and hence its name. The efficiency of the 3M method cannot be ascertained without benchmark testing of its implementation because it includes three additional additions compared with the 4M method. Nevertheless, The 3M method has proven to be valuable for tasks such as matrix multiplication [12] and arbitrary-precision complex arithmetic [2,3], particularly in scenarios with extensive computation times dedicated to multiplication.

We implemented a multiple-precision matrix multiplication based on the 3M method such that for any $A \in \mathbb{C}^{m\times n}$ and $B \in \mathbb{C}^{n\times l}$,

$$\begin{aligned} T_1 &:= \mathrm{Re}(A)\mathrm{Re}(B),\ T_2 := \mathrm{Im}(A)\mathrm{Im}(B) \in \mathbb{R}^{m\times l} \\ AB &:= (T_1 - T_2) + \{(\mathrm{Re}(A) + \mathrm{Im}(A)) \\ &\quad (\mathrm{Re}(B) + \mathrm{Im}(B)) - T_1 - T_2\} \cdot \mathrm{i} \in \mathbb{C}^{m\times l}. \end{aligned} \tag{3}$$

This approach allows for complex matrix multiplication based on existing high-performance real matrix multiplication techniques. Although the efficiency of this implementation remains uncertain because of the three additional matrix additions and subtractions, benchmark tests are necessary to evaluate the performance of the multiple-precision complex matrix multiplication.

2.2 Acceleration Using AVX2 SIMDization

The data structures for the implemented multiple-precision complex vectors and matrices are shown in Fig. 1.

As described above, the complex basic linear computation, which included the 3M method, was constructed using a real multiple-precision BLAS developed through SIMDization, such as AVX2 in our library. Complex vectors and matrices were allocated as separate arrays for the real and imaginary parts of multi-component multiple-precision FP data types, as depicted on the right of Fig. 1. However, as depicted on the left of Fig. 1 for arbitrary-precision FP arithmetic, the same data type (a set of real and imaginary parts) in multple-precision complex arithmetic (MPC) [3] was adopted for a single element. The MPC based on MPFR employed the 3M method for element-wise complex multiplication to ensure high speed.

This approach simplifies the implementation of fast complex matrix multiplication based on the formula in (3) for multi-component-type multiple-precision calculations. Having already implemented four matrix multiplication algorithms: simple matrix multiplication using triple loops, block matrix multiplication,

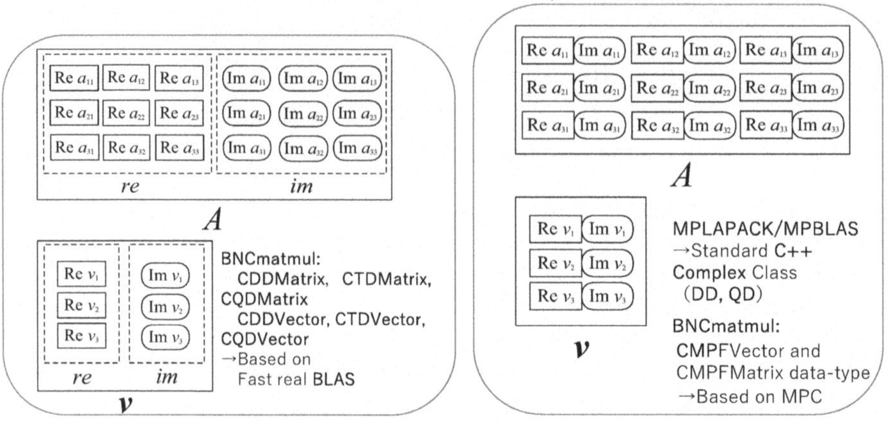

Fig. 1. Two types of data structure for multiple-precision complex vectors $\mathbf{v} \in \mathbb{C}^3$ and matrices $A \in \mathbb{C}^{3\times3}$: multi-componet-type multiple-precision (left) and arbitrary precision standard C++ complex class used in MPBLAS (right)

Strassen matrix multiplication (and the Winograd algorithm), and the Ozaki scheme, complex matrix multiplication was achieved by combining them.

Therefore, SIMDization of complex arithmetic units is not necessary for basic matrix multiplication. However, to speed up the complex LU decomposition, we decided to SIMDize the arithmetic, similar to real LU decomposition, to speed up the element-by-element forward elimination calculation. For multi-component complex vectors and matrices, SIMD Load/Store instructions can be used to retrieve and store elements because the real and imaginary parts are separated. Using this approach, complex DD operations using AVX2 (`_bncavx2_rcdd_add`, `sub`, `mul`, `div`, `inv`(reciprocal)) can be implemented as shown in Fig. 2. SIMDized complex TD and QD operations are implemented in the same way.

The SIMDized complex arithmetic method was based on a combination of SIMDized real arithmetic methods.

2.3 Parallelization with OpenMP

For parallelization, the multi-component multiple-precision complex linear computations were based on real linear computation units. Real matrix multiplication based on the Strassen algorithm was parallelized in sections, as illustrated in Fig. 3. The nested nature of parallelization precluded performance improvement beyond eight threads. Arbitrary-precision complex matrix multiplication was implemented using MPC operations, including parallelization.

The Ozaki scheme divides the computation into shorter length elements to avoid introducing rounding errors into the short-precision matrix multiplication base of general matrix multiplications (GEMM), specifically in this case, xGEMM using binary64 DGEMM. The product obtained by xGEMM was then combined with multiple-precision matrix addition to obtain a high-precision

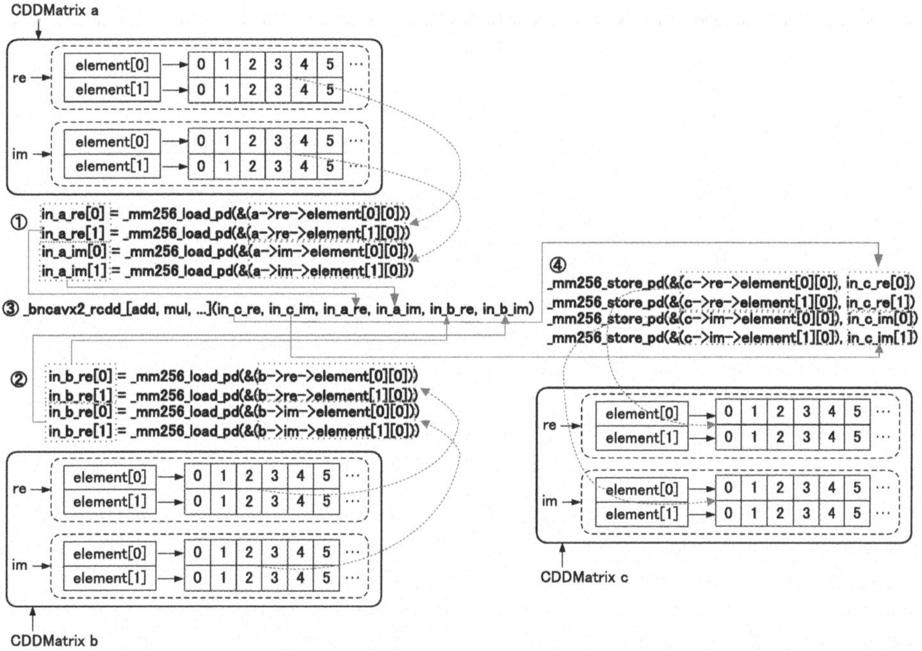

Fig. 2. Complex double-double arithmetic with AVX2 utilized in complex BLAS and LU decomposition

matrix product. To parallelize this process, parallelized xGEMMs are used or OpenMP loop parallelization can be performed for each xGEMM, as illustrated in Fig. 4. In this study, the latter approach was used for parallelization.

2.4 Benchmark Test for Complex Matrix Multiplication

The AMD EPYC environment was used to present the results obtained from the benchmark tests.

CPU and OS AMD EPYC 9354P 3.7 GHz 32 cores, Ubuntu 20.04.6 LTS
C/C++ Intel Compiler version 2021.10.0
Compiler Option `-O3 -std=c++11 -fp-model precise`
with AVX2 `-axCORE-AVX2 -march=skylake -mtune=skylake -mcpu=skylake`
MPLAPACK 2.0.1, GNU MP 6.2.1, MPFR 4.1.0, MPC 1.2.1

In addressing the problem of computing the matrix product $C := AB$ for the given complex square matrices A and $B \in \mathbb{C}^{n\times n}$, the elements of A and B used for benchmarking were uniformly distributed random numbers in the interval $[0, 1]$. For the Ozaki scheme, to ensure sufficient accuracy, a maximum of six divisions were used for DD accuracy, eight divisions for TD accuracy, and 12 divisions for QD accuracy.

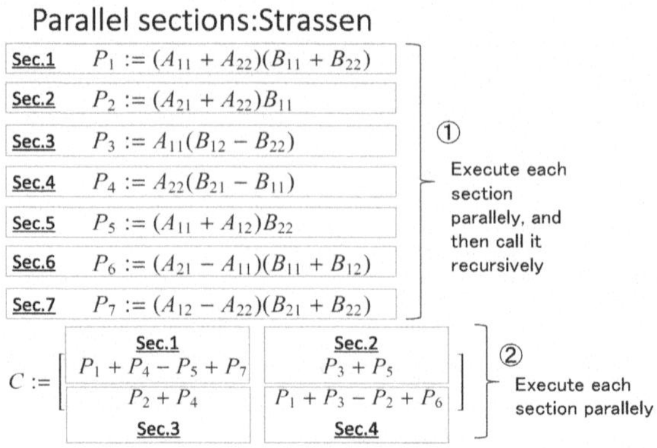

Fig. 3. Parallelized Strassen algorithm

For all the benchmark tests used in this study, the block size of the matrix multiplication was fixed at 32×32, and 32 was chosen as the minimum number of matrix rows and columns when stopping the recursive loop in the Strassen algorithm. The Ozaki scheme used the DGEMM of Intel math kernel library banded in the Intel compiler kit.

The results of the DD-precision matrix multiplication are presented in Fig. 5. The left figure shows the computation time in seconds, and the right figure shows the corresponding speed-up ratio achieved by parallelization using OpenMP. In the left figure, our implementation of the Strassen and Ozaki schemes achieved up to 100 times performance improvement compared to the CGEMM routine of MPBLAS in serial calculations. Additionally, the 3M method exhibited a speed improvement of approximately 1.3 times over the 4M method, whereas the Strassen method achieved a speed-up of 3x-4x. The computation time of the Ozaki scheme was nearly equivalent to that of the AVX2 Strassen scheme because of the small number of partitions, whereas the block matrix multiplication based on the 3M method was slightly slower than that based on the Strassen matrix multiplication.

In OpenMP parallelization, the Strassen matrix multiplication did not exhibit any acceleration, even when 32 threads were used. Similarly, the Ozaki scheme demonstrated only a modest 2x speed-up. In contrast, block matrix multiplication achieved a significant improvement, with speeds of up to 12 times faster.

Next, the results obtained by TD-precision matrix multiplication are presented in Fig. 6.

The AVX2 SIMDized TD-precision Strassen block matrix multiplication was slower than that of the non-SIMDized version although the reason for this discrepancy remains unclear. Therefore, only non-SIMDized multiplications were employed in TD precision. The 3M method demonstrated a speed improvement of approximately 1.3 times compared to the 4M method, which is common with DD precision. The speed-up achieved by the Ozaki scheme was approximately eight times faster than that of the Strassen 3M method.

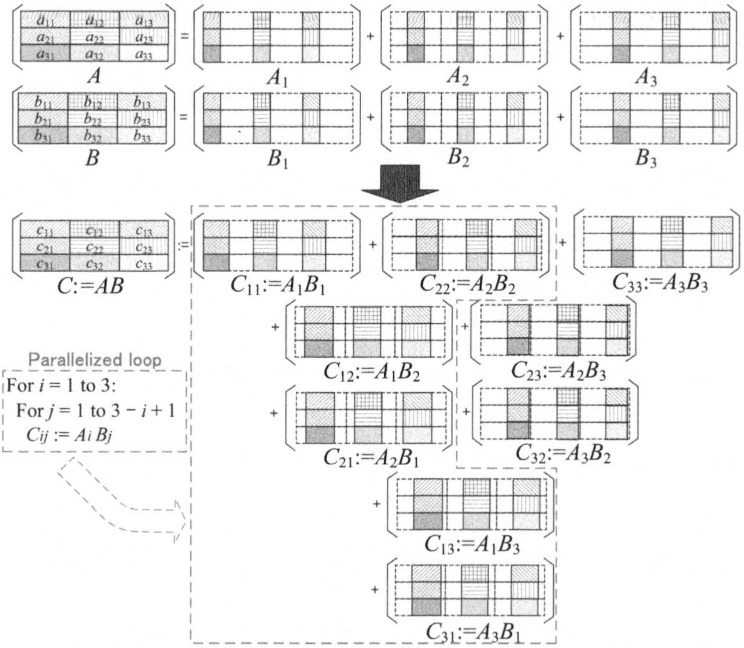

Fig. 4. Parallelized Ozaki scheme

In OpenMP parallelization, the Strassen matrix multiplication was observed to be slower, and the Ozaki scheme was only approximately twice as fast, even when utilizing 32 threads, as in DD accuracy. Block matrix multiplication was up to 20 times faster.

Finally, the results of QD-precision matrix multiplication are shown in Fig. 7.

Similar to DD and TD precision, the Strassen matrix multiplication was approximately twice faster in performance using AVX2 optimization and 1.3 times faster in performance using the 3M method.

In OpenMP parallelization, the Strassen matrix multiplication demonstrated faster performance, whereas the Ozaki scheme achieved only a speed-up of 2x compared to the DD- and TD-precision schemes despite utilizing 32 threads. Block matrix multiplication was 20 times faster than TD precision.

These results demonstrate that while the 3M method can speed up serial complex matrix multiplication, it does not speed up parallelization in relation to the number of threads, except in the case of block matrix multiplication.

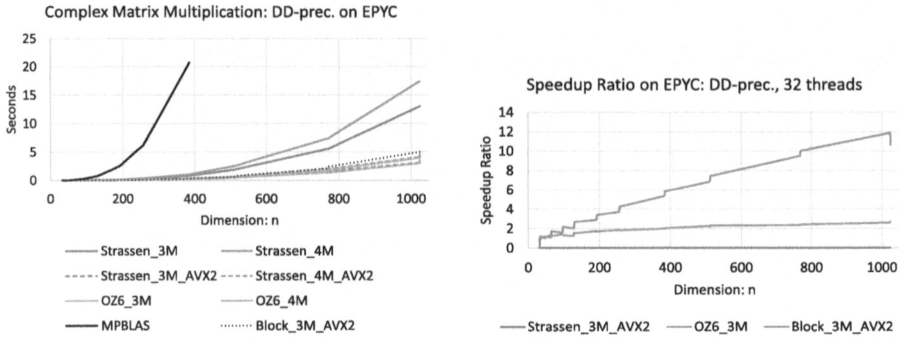

Fig. 5. DD-precision: computation time in seconds (left) and speed-up ratio when using 32 threads (right)

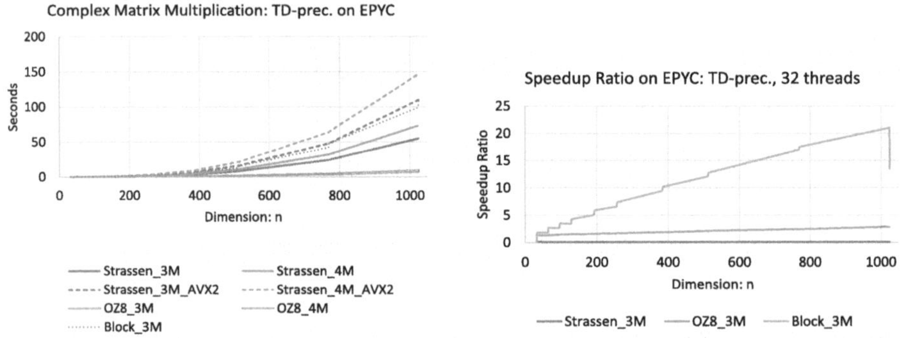

Fig. 6. TD-prec. : Computation time in second (left) and speed-up ratio when using 32 threads (right)

3 Acceleration of Multiple-Precision Complex LU Decomposition

Complex LU decomposition was implemented based on a complex basic linear computation that integrates all the implemented speed-up techniques (3M method, SIMD, and OpenMP parallelization) to achieve high performance. The following is a concise description of the implementation:

The target n-dimensional complex linear system is expressed as

$$A\mathbf{x} = \mathbf{b}, \tag{4}$$

where $A \in \mathbb{C}^{n \times n}$, $\mathbf{x} \in \mathbb{C}^n$, and $\mathbf{b} \in \mathbb{C}^n$. The (4) was solved to determine the unknown vector $\mathbf{x}$ given A and b.

The current standard implementation of the LU decomposition uses fast matrix multiplication. As shown in Fig. 8, a certain block size $K(=\text{NDIM})$ is set in advance, and $A - L_{21}U_{12}$ is performed to compute A_{22}. Consequently, the

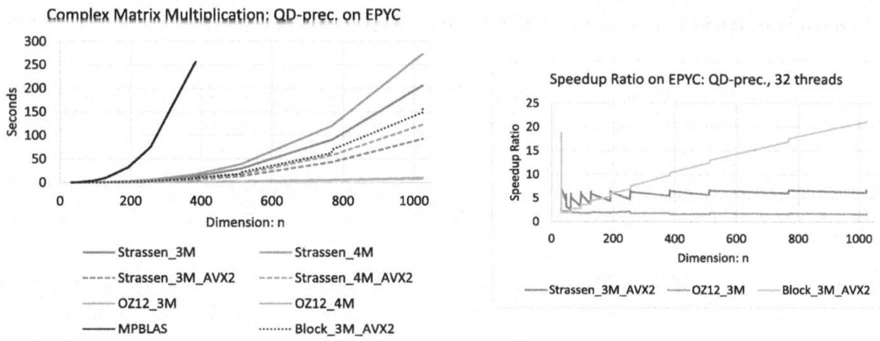

Fig. 7. QD-precision: Computation time in seconds (left) and speed-up ratio when using 32 threads (right)

computational performance varies depending on the value of K. The LU decomposition (Cgetrf function) of MPLAPACK in the comparisons, also used complex matrix multiplication (CGEMM function). In all cases, LU decomposition was computed using a partial pivoting strategy.

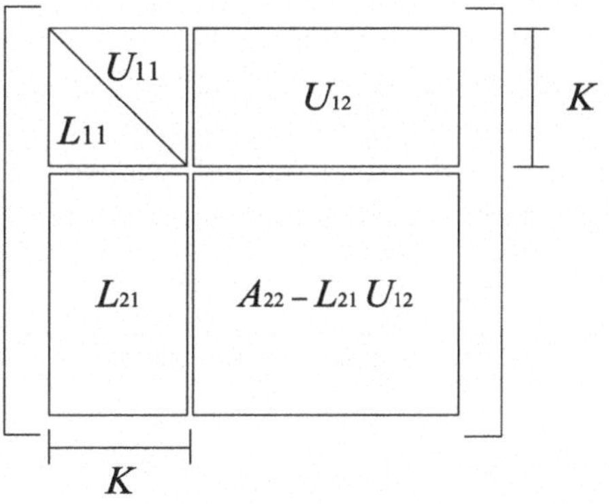

Fig. 8. LU decomposition using matrix multiplication

In addition, a standard LU decomposition approach (referred to as "Normal LU" or "Normal"), which eliminates rows sequentially without updating or performing matrix multiplication, was implemented using complex AVX2 arithmetic, as illustrated in Fig. 2, to benefit from the ease of speeding up the process using SIMD instructions. AVX2 was effective in real LU decomposition, with the Ozaki scheme emerging as the fastest. However, the question remains: will the same effect be obtained for complex LU decomposition?

4 Benchmark Test of Complex LU Decomposition

During the benchmark test, the computation time and maximum error of the numerical solution of the LU decomposition were measured by varying K while calculating DD, TD, and QD precision. The graphs in Fig. 9 to Fig. 17 show the computation time for normal LU and LU decomposition using matrix multiplication, as well as the relative error of the numerical solutions obtained through forward and backward substitutions.

We focused on solving the n-dimensional complex linear system depicted in Equation (4). For benchmarking purposes, the elements of A were generated as uniform random numbers within the range $[0, 1]$.

The true solution $\mathbf{x}$ is $(\mathbf{x})_k = k + k \cdot \mathrm{i}$ and the constant vector were derived by computing $\mathbf{b} := A\mathbf{x}$ using 2048-bit precision calculation.

4.1 DD-Precision Computation

First, we present the results of the DD accuracy calculation, which involved measuring the relative error and computation time while increasing $1 \le K \le n$ by 32.

In examining the relative error (Fig. 9), the MPLAPACK and AVX2 SIMDized normal LU decompositions demonstrated superior performance. The normal LU and LU decompositions using the Ozaki scheme and Strassen matrix multiplication, respectively, yielded similar results. Overall, the results show that the accuracy tends to decrease when Strassen is used. In this problem, the Ozaki scheme with DGEMM and five divisions did not achieve optimal accuracy, necessitating more than six divisions. Therefore, the results are presented for five, six, and seven divisions.

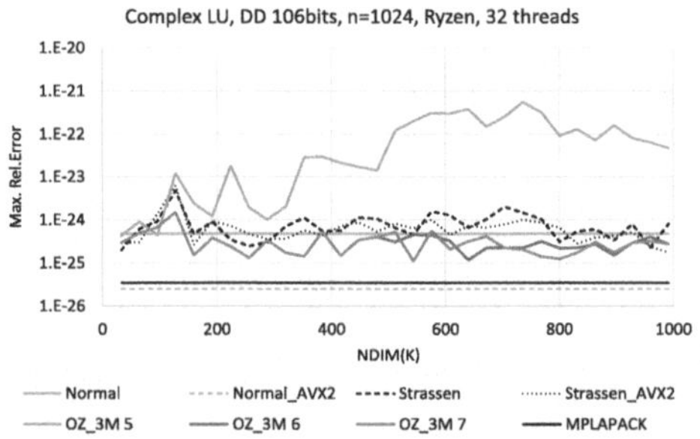

Fig. 9. Complex LU decomposition with DD-precision: Max. relative error

Figure 10 demonstrates the measured serial computation time (left) and 32-thread parallel computation time (right). The results with normal LU decomposition and AVX2 acceleration are represented by the solid orange and dashed lines, respectively. The relative error of MPLAPACK is depicted as a solid black

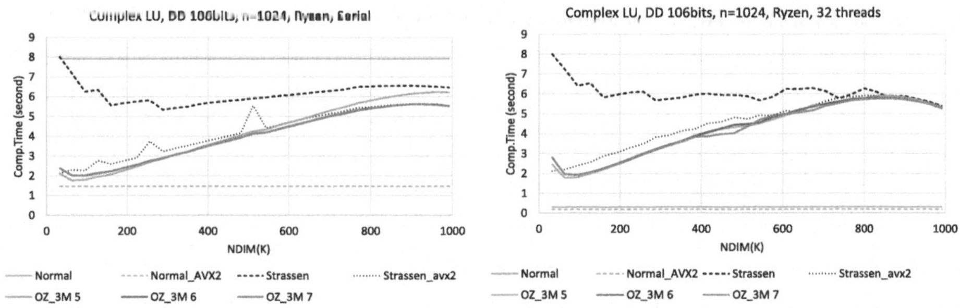

Fig. 10. Complex LU decomposition with DD-precision: serial computation in seconds (left), 32-thread computation in seconds (right)

line although it was not included in the computation time graph because of the significant difference in scale (130.7 s).

The AVX2 normal LU decomposition exhibited the fastest serial computation, whereas the AVX2 Strassen LU decomposition and Ozaki scheme showed similar computation times, with the smaller K values requiring less time. However, they were not as fast as the AVX2 normal LU decomposition.

In the case of DD-precision calculations, OpenMP parallelization of matrix multiplication initially increased the computation time with two threads, after which it became faster in proportion to the number of threads. Therefore, with 32 threads, the computation time was approximately the same as that for serial computation, and as a result, the parallelization effect was not enhanced. However, as depicted in Fig. 11, the normal LU decomposition achieved a speed-up in proportion to the number of threads, and the AVX2 version remained the fastest, even with 32 threads.

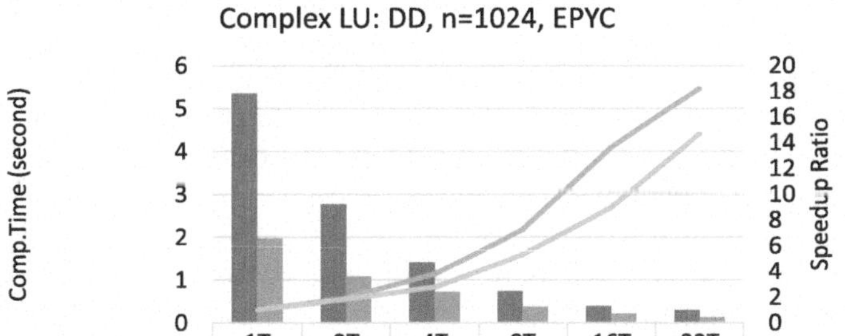

	1T	2T	4T	8T	16T	32T
Normal	5.34	2.77	1.41	0.74	0.39	0.29
Normal AVX2	1.97	1.08	0.71	0.37	0.22	0.13
Normal	1.00	1.93	3.78	7.25	13.64	18.24
Normal AVX2	1.00	1.83	2.76	5.30	8.98	14.71

Fig. 11. Parallelization of normal complex LU decomposition: DD-precision

The reason for this phenomenon is believed to be the higher computational complexity inherent in complex LU decomposition compared with real LU decomposition despite the enhanced performance of complex operations achieved through SIMD. As previously mentioned, although SIMD algorithms are not essential for complex basic linear computation implementations, they may facilitate acceleration of the Fast Fourier Transform, which is required locally for LU decomposition.

4.2 QD-Precision Computing

Next, the results of QD accuracy calculations are presented.

Figure 12 displays the relative errors obtained from QD-precision benchmark tests. As shown in this figure, MPLAPACK slightly outperforms the AVX2 SIMDized normal LU decomposition. LU decomposition using the Ozaki scheme and Strassen matrix multiplication exhibit larger relative errors than normal LU decomposition.

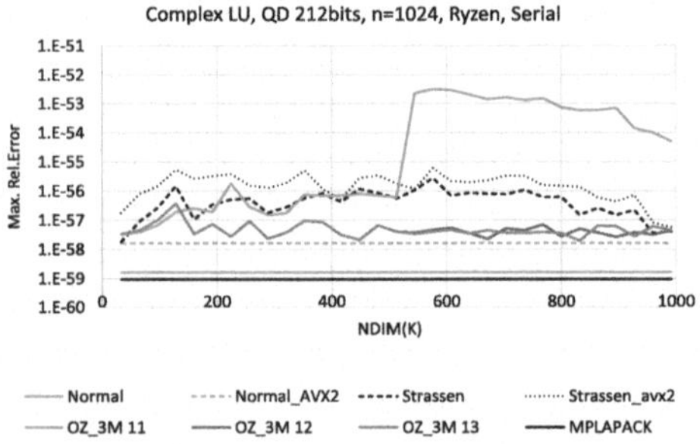

Fig. 12. Complex LU decomposition with QD precision: Max. relative error

Figure 13 illustrates the computation time obtained from QD-precision benchmark tests. The graph shows the measured serial computation time (left) and 32-thread parallel computation time (right) using the standard LU decomposition and AVX2-accelerated approach, represented by solid and dashed orange lines, respectively. For this particular problem, the Ozaki scheme using DGEMM does not achieve the highest QD accuracy with 11 divisions and requires more than 12 divisions; hence, the results for 11, 12, and 13 divisions are displayed. The relative error for MPLAPACK is represented by the solid black line; however, the serial computation time is 595.4 s; therefore, it is not included in the graph of the computation time because it is on a different scale.

For serial calculations, AVX2 SIMDized normal LU decomposition generally exhibited the fastest performance. However, the Ozaki LU decomposition scheme was slightly faster when K was small. The Strassen LU decomposition, even with AVX2 SIMDized, was not as fast as the AVX2 SIMDized normal LU decomposition.

In QD-precision calculations, OpenMP parallelization of matrix multiplication accelerated the computation linearly with the number of threads. However,

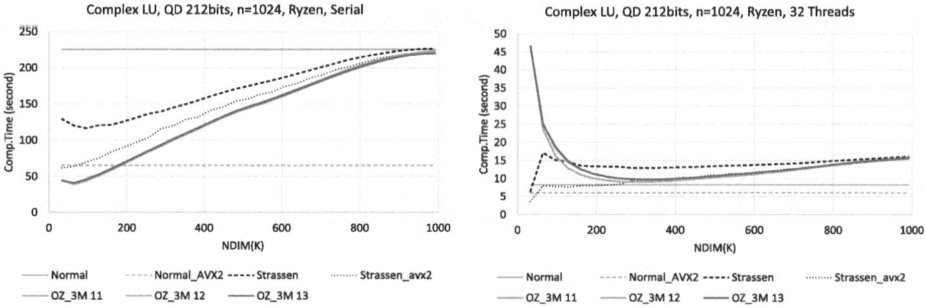

Fig. 13. Complex LU decomposition in QD precision: serial computation in seconds (left), 32-thread computation in seconds (right)

the parallelization effect of the normal LU decomposition surpassed this, with the AVX2 SIMDized normal LU decomposition remaining the fastest overall, except in certain cases where it was outperformed by AVX2 SIMDized Strassen LU decomposition. As illustrated in Fig. 14, although parallelization efficiency is high, AVX2 slows it down.

4.3 TD-Precision Computing

Finally, we present the results of the TD-precision calculation, which represents an intermediate accuracy level between DD and QD.

In this case, as depicted in Fig. 15, for TD accuracy, the Ozaki scheme with DGEMM did not achieve the highest accuracy with seven divisions, requiring eight or more divisions.

The relative error analysis revealed that similar to the case of QD accuracy, the AVX2 SIMDized normal LU decomposition performed the best. Further-

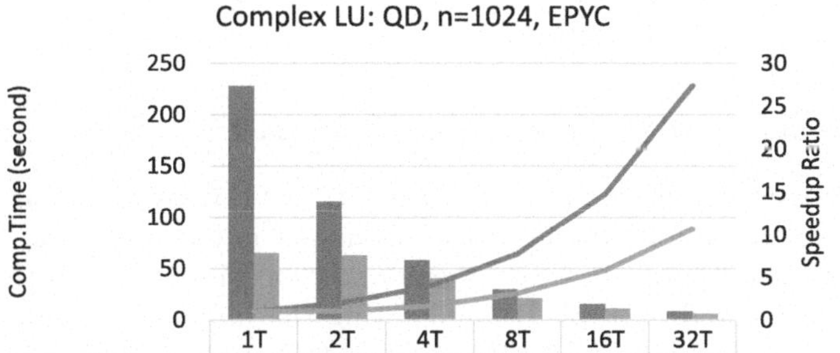

	1T	2T	4T	8T	16T	32T
Normal	227.63	115.16	57.90	29.53	15.46	8.31
Normal AVX2	64.73	62.41	40.39	21.06	11.19	6.10
Normal	1.00	1.98	3.93	7.71	14.73	27.39
Normal AVX2	1.00	1.04	1.60	3.07	5.78	10.62

Fig. 14. Parallelization of normal complex LU decomposition: QD-precision

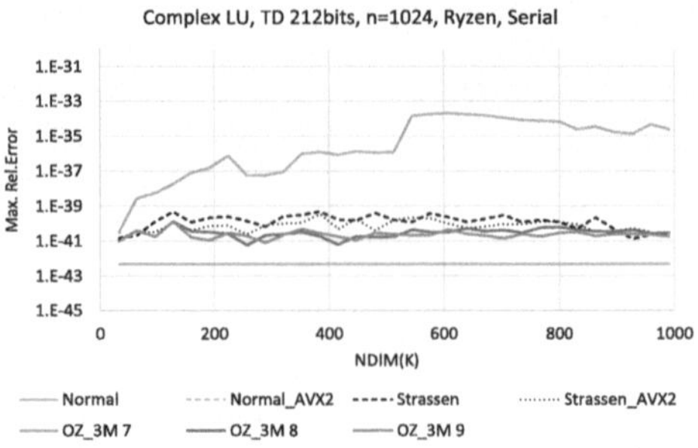

Fig. 15. TD-precision complex LU decomposition: Max. relative error

more, the relative errors observed in the LU decomposition using the Ozaki scheme and Strassen matrix multiplication were greater than that of normal LU decomposition.

A graph depicting the computation time is presented in Fig. 16, where the 32-thread parallelization using TD accuracy was omitted because of the significant increase in computation time for small K.

In serial calculations, the AVX2 SIMDized normal LU decomposition in general did not outperform the others, with the LU decomposition using the Ozaki scheme being significantly faster for small K values. In TD-precision calculations, OpenMP parallelization of matrix multiplication displayed a straightforward speed-up proportional to the number of threads. However, the parallelization effect of the normal LU decomposition outweighed this effect. Normal LU decomposition with AVX2 remained the fastest, except in the case of LU decomposition with the Ozaki scheme performing better. As shown in Fig. 17, the parallelization efficiency is considerable; however, the ratio is reduced compared to DD and QD normal LU decomposition. The reason the parallel efficiency of LU decomposition with TD precision is lower than that of DD and QD is unclear; however, the fact that the precision length is not a power of 2 for binary64 may have hindered compiler optimization.

4.4 Summary of High-Performance Complex LU Decomposition

These benchmark results indicate that the AVX2 SIMDized normal LU decomposition generally exhibits the fastest performance in terms of both DD and QD accuracy with good parallelization. A summary focusing solely on the computation time is presented in Table 1, highlighting the minimum computation time achieved for Strassen with AVX2 for DD and QD, and without AVX2 in TD.

Finally, the 32-thread AVX2 normal LU decomposition was demonstrated to outperform MPLAPACK with serial computations by approximately 726 times in DD accuracy and approximately 91 times in QD accuracy. These findings suggest that the default setting for complex LU decomposition should be the AVX2 SIMDized normal LU decomposition.

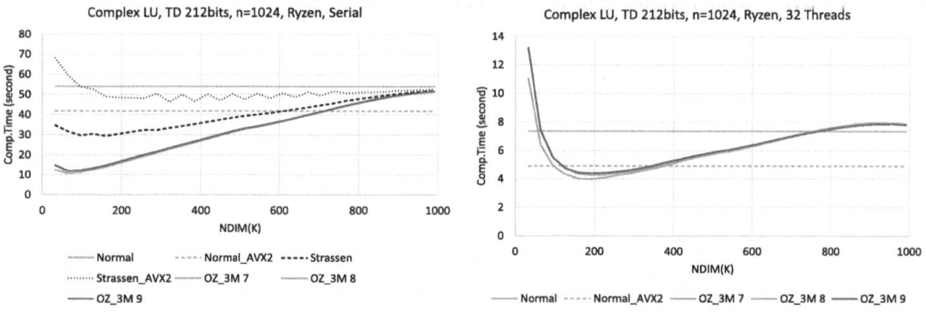

Fig. 16. Complex LU decomposition with TD-precision: serial computation in seconds (left) and 32-thread computation in seconds (right)

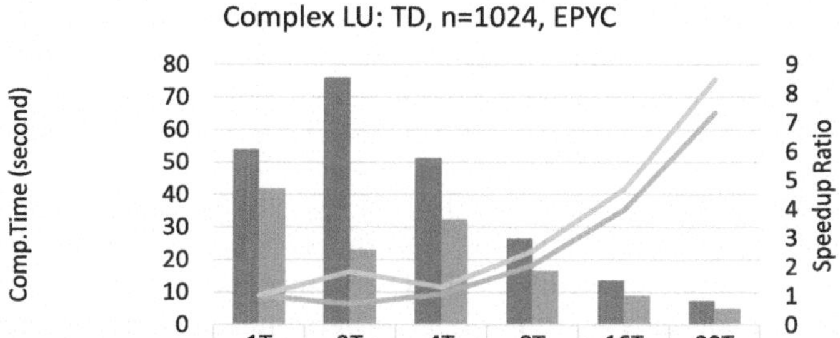

	1T	2T	4T	8T	16T	32T
Normal	53.95	75.97	51.18	26.41	13.59	7.35
Normal AVX2	41.76	22.91	32.26	16.47	8.92	4.91
Normal	1.00	0.71	1.05	2.04	3.97	7.34
Normal AVX2	1.00	1.82	1.29	2.53	4.68	8.50

Fig. 17. Parallelization of normal complex LU decomposition with TD-precision

Table 1. Minimum computation time of complex LU decomposition in second

DD precision, $n = 1024$					
	MPLAPACK	Normal	Normal AVX2	Strassen AVX2	OZ
Serial	130.7	5.3	2.0	2.1 ($K = 32$)	2.0 ($K = 64$)
Parallel(32T)	–	0.29	0.13	2.1 ($K = 32$)	1.9 ($K = 96$)
TD precision, $n = 1024$					
Serial	–	54.0	41.8	46.7 ($K = 384$)	11.9($K = 64$)
Parallel(32T)	–	7.4	4.9	8.5 ($K = 992$)	4.3($K = 192$)
QD precision, $n = 1024$					
Serial	595.4	227.6	64.7	60.1($K = 32$)	40.7($K = 64$)
Parallel(32T)	–	8.3	6.1	3.6($K = 32$)	9.8($K = 352$)

5 Conclusion and Future Works

The implementation of the multiple-precision complex basic linear computation based on the enhanced multiple-precision real basic linear computation demonstrated improved speed using the 3M method for matrix multiplication. Additionally, the complex LU decomposition was found to be sufficiently rapid through the utilization of SIMD complex operations and parallelization using OpenMP. However, the acceleration of LU decomposition was observed to be significant enough to surpass conventional LU decomposition algorithms in DD- and QD-precision, using the current complex matrix multiplication both in serial and parallel computations.

Future work will focus on enhancing the performance of numerical algorithms for ill-conditioned nonlinear problems by improving the following functions.

1. Implementation of a multiple-precision sparse matrix and extension of the current implementation to real and complex sparse matrices.
2. Acceleration of reproducible computations for binary64 and other types of precision arithmetic.
3. Extending fast multi-precision computation to the Python environment.

This includes seeking further speed-up of real and complex basic linear computations of dense matrices and LU decomposition. Of particular interest is confirming the potential speed-up of SIMD-enhanced multi-component multiple-precision real and complex arithmetic for other numerical algorithms.

Acknowledgment. This research was supported by JSPS Grant-in-Aid for Scientific Research 23K11127. We appreciate the sincere and important comments of the unknown reviewers who helped improve the overall presentation of this paper.

References

1. Bailey, D.: QD. https://www.davidhbailey.com/dhbsoftware/
2. Bini, D.A., Fiorentino, G.: Design, analysis, and implementation of a multiprecision polynomial rootfinder. Numer. Algorithms **23**, 127–173 (2000)
3. Enge, A., Théveny, P., Zimmermann, P.: MPC. http://www.multiprecision.org/mpc/
4. Kouya, T.: BNCmatmul. https://github.com/tkouya/bncmatmul
5. Kouya, T.: Acceleration of multiple-precision matrix multiplication based on multi-component floating-point arithmetic using AVX2. In: Gervasi, O., et al. (eds.) Computational Science and Its Applications - ICCSA 2021, pp. 202–217. Springer International Publishing, Cham (2021). https://doi.org/10.1007/978-3-030-86976-2_14
6. Kouya, T.: Acceleration of Complex Matrix Multiplication Using Arbitrary Precision Floating Point Arithmetic (2023)
7. MPLAPACK/MPBLAS: Multiple precision arithmetic LAPACK and BLAS. https://github.com/nakatamaho/mplapack
8. Ozaki, K., Ogita, T., Rump, S.M., Oishi, S.: Fast algorithms for floating-point interval matrix multiplication. J. Comput. Appl. Math. **236**, 1795–1814 (2012)

9. Petitet, A., Whaley, R.C., Dongarra, J., Cleary, A.: HPL - a portable implementation of the high-performance linpack benchmark for distributed-memory computers
10. Project, M.: The MPFR library. https://www.mpfr.org/
11. Strassen, V.: Gaussian elimination is not optimal. Numer. Math. **13**(4), 354–356 (1969). https://doi.org/10.1007/BF02165411
12. Van Zee, F.G., Smith, T.M.: Implementing high-performance complex matrix multiplication via the 3M and 4M methods. ACM Trans. Math. Softw. **44**(1), 1–36 (2017). https://doi.org/10.1145/3086466

Innovative Integration of Machine Learning Techniques for Early Prediction of Metabolic Syndrome Risk Factors

Shendry Balmore Vásquez Rosero[1,2](✉)

[1] Universidad Estatal Península de Santa Elena, Av Principal Santa Elena, La Libertad, Ecuador
srosero@upse.edu.ec

[2] Universidad Politécnica Salesiana, Campus El Vecino, Sede Cuenca, Ecuador

Abstract. Over the past two decades, chronic degenerative diseases have risen to prominence in global and national morbidity and mortality statistics. Notably, type 2 diabetes mellitus, arterial hypertension, and metabolic syndrome have been highlighted for their prevalence and have been identified by the World Health Organization (WHO) as potential causes of 50% of worldwide fatalities. Despite increased awareness driven by internet dissemination about risks associated with sedentary lifestyles and poor diets, and the subsequent shift in public perception towards healthier living, it remains a reality that individual concern typically arises following the initial symptomatology of these conditions. In response to this situation, the current study proposes the development of an early warning system, underpinned by advanced machine learning algorithms such as LightGBM, XGBoost, and ensemble methods based on Random Forests that employ gradient boosting techniques to enhance predictive accuracy. This model processes data efficiently, requiring minimal computational resources, to provide personalized risk predictions based on categorical characteristics, as well as biometric and clinical variables. In light of the findings, the deployment of a web platform is anticipated, which will enable individuals to assess their health risk using readily available indicators, thereby promoting greater proactivity in personal health management

Keywords: Síndrome metabólico · Machine Learning · LightGBM · XGBoost y Random Forest

1 Introduction

Sedentary behavior, understood as the lack of regular physical activity, has become an increasing concern in modern society. Specifically, in Ecuador, as in many parts of the world, sedentariness is a public health issue affecting diverse population groups, including students, workers, and university faculty. This sedentary trend carries with it a series of negative health consequences,

O. Gervasi et al. (Eds.): ICCSA 2024 Workshops, LNCS 14818, pp. 20–36, 2024.
https://doi.org/10.1007/978-3-031-65273-8_2

one of the most prominent being the emergence and rise of metabolic syndrome. It is necessary to understand that metabolic syndrome is a complex medical condition characterized by the presence of various cardiovascular risk factors, including abdominal obesity, high blood pressure, elevated blood glucose levels, low levels of high-density lipoprotein (HDL), and elevated blood triglyceride levels. This combination of factors significantly increases the risk of cardiovascular diseases and type 2 diabetes.

Within the Ecuadorian university context, this issue worsens due to the long hours of sedentary study and work faced by students, workers, and faculty. The lack of regular physical activity, combined with unhealthy eating habits, creates a conducive environment for the development of metabolic syndrome. That is why many universities in Ecuador, especially the State University of Santa Elena Peninsula since 2010, have been carrying out a program and various occupational health proposals aimed at understanding the relationship between cardiovascular risk indicators and the sedentariness of the main actors in the university community. Therefore, the present study aims to gather part of that research conducted between 2018 and 2022 to implement a risk prediction model based on robust machine learning models that later serve as a diagnostic and prediction guide with the aim of achieving an early diagnosis of metabolic syndrome based on risk indicators. For this purpose, the following sections have been proposed, which have allowed the implementation of a novel risk prediction method for metabolic syndrome based on risk indicators:

1. Theoretical Framework, which includes the foundations of Machine Learning and the justification for applying these methods in predicting the risk of metabolic syndrome
2. Explanation of the Proposed Method
3. Experimental Design
4. Experimentation
5. Results and Discussion
6. Conclusions and, finally,
7. Bibliographic References

2 Theoretical Framework

During the last twenty years, it has been observed that chronic and degenerative diseases lead the statistics of morbidity and mortality both globally and nationally. Among these diseases are type 2 diabetes, hypertension, metabolic syndrome, and lipid disorders, which are widely distributed and have a high global prevalence. The World Health Organization (WHO) has indicated that these diseases could be responsible for 50% of all deaths globally, anticipating a 17% increase in the next decade [5,11].

In America, it is estimated that non-communicable chronic diseases are responsible for about 5.5 million deaths annually, which represents 81% of all deaths. Of these, 2.2 million occur in people under 70 years of age. Breaking down by regions, North America records approximately 2.7 million deaths; in Mexico, Latin America, and the Caribbean, deaths total 868,600; and in South America, about 1.94 million. Cardiovascular diseases and diabetes, along with other ailments, account for 28.1% and 20.8% of these deaths, respectively [12]. These conditions have also contributed to an increase in premature mortality, with the risk of dying from one of these diseases between the ages of 30 and 70 estimated at 15% to 18% in men and 13% in women throughout the American continent as evidenced in [4].

On the other hand, within the field of computer science, the art of programming has played an important role over the last 20 years [2,10], a role that has been characterized by the introduction of variables in a structured way into the computer, which then delivers results to its programmer. It can be said that the computer program has enabled data processing to obtain information, but this process would not be executed without the programmer's intervention. Therefore, in recent years at a practical level, new paradigms have emerged in which the programmer plays a passive role since it is intended that the computer "learn" based on the analysis of the set of variables that the programmer has entered. It is then said that this is the basis of "*Machine Learning*" (ML) [3] or automated learning, where the computer learns by itself. For decades, traditional programming both in computers and online has been preeminent, with the development of algorithms and techniques to optimize program efficiency. But recently, machine learning has revolutionized this field. This approach has modified the previously explained paradigm to a new one. Now, instead of simply inputting data into the program, these data are used along with the obtained results to create the program, or 'model'. Hence, through techniques such as prediction, regression, clustering, and classification, names like supervised learning and unsupervised learning appear as the main algorithms of this ML.

Within these supervised and unsupervised techniques or algorithms [10], some models like *LightGBM*, *XGBoost*, and *Random Forest* [1] stand out for their exceptional performance in a variety of classification and regression tasks [8]. They have proven their effectiveness in numerous studies and competitions, making them solid options for tackling complex problems such as the prediction of risk factors [9], together with their ability to handle complex data and extract relevant patterns from multiple variables. It's also worth mentioning that while *LightGBM* and *XGBoost* are complex ensemble models, model interpretation techniques can be used to understand their decision-making. This is crucial for healthcare professionals and researchers to understand the predictions and take appropriate actions.

3 Proposed Method

To develop a novel approach that assesses the risk of individuals based on their occupation (workers, teachers, students) and gender, while using machine learning, a predictive model was considered that takes into account the interaction between these categorical characteristics and biometric and health variables. Below, the set of proposed steps for achieving the objectives is listed:

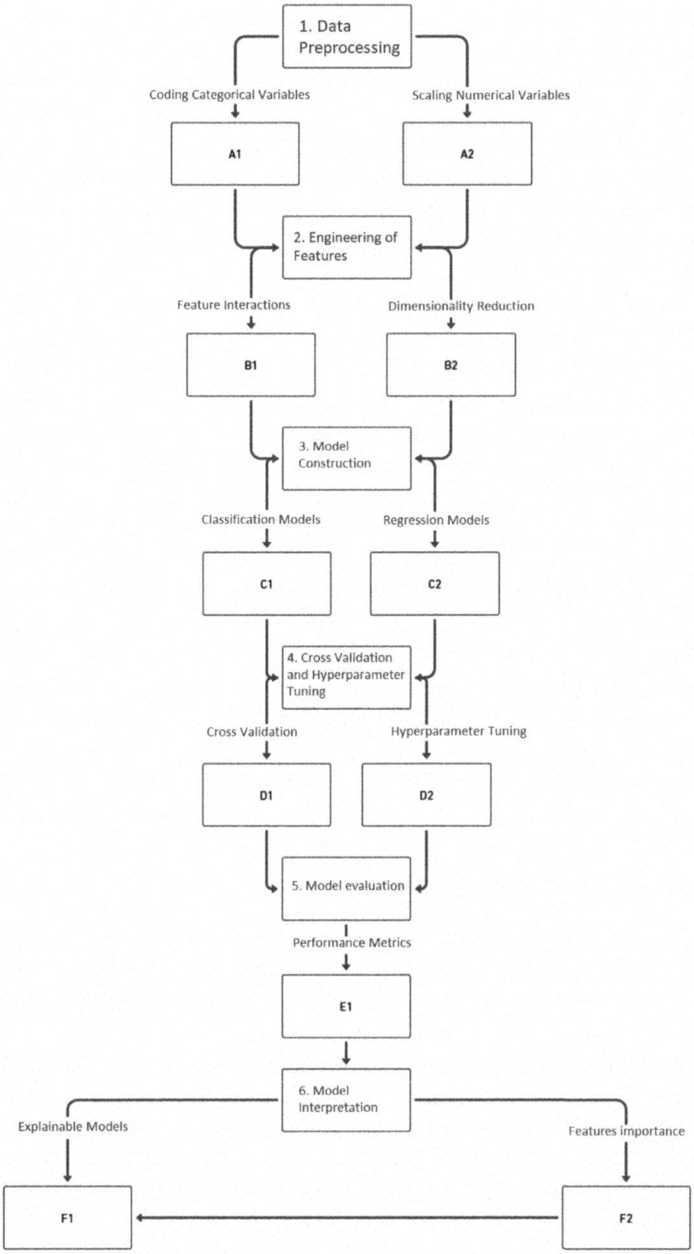

Fig. 1. Summary of the proposed model.

- Data Preprocessing: This step involved the encoding of categorical variables, i.e., converting categorical variables (gender, type of person) into numeric formats using one-hot encoding or label encoding. Additionally, scaling numeric variables, which entailed ensuring that all numeric variables are appropriately scaled so that the model is not biased towards variables with greater magnitude.
- Feature Engineering: This involves the interactions of features and the possibility of generating new features that represent the interaction between gender and occupation, since the risk might vary between a female student compared to a male worker. Likewise, dimensionality reduction with techniques such as Principal Component Analysis (PCA).
- Model Construction: Work was done on classification models, given the risk classification, classification models such as Random Forest and Gradient Boosting were used to predict the risk category. Regression models were also used because the risk is measured on a continuous scale, making regression models potentially the most appropriate.
- Cross-Validation and Hyperparameter Tuning: Cross-Validation to robustly evaluate the model's performance and Hyperparameter Tuning such as grid search or random search to find the best hyperparameters for the selected models.
- Model Evaluation: Performance Metrics such as recall, F1-score, ROC-AUC for classification, or R-squared, RMSE for regression, depending on the type of problem.
- Model Interpretation: Finally, the importance of features was analyzed to determine which are the most influential factors in predicting risk.

A summary of the proposed model can be seen in Fig. 1 on the following page.

3.1 Simplified Model Algorithm

- PHASE 1.- Data Set Loading, starts by loading the data set to inspect it and understand its structure; an Exploratory Data Analysis (EDA) is generated to gain a deep understanding of the data, including variable distributions, descriptive statistics, and possible correlations.
- PHASE 2.- Data Preprocessing, this involves cleaning the data before determining missing values or anomalies; proceeding to convert categorical variables to numeric formats using appropriate encoding. Normalize or standardize numerical variables to pave the way for Feature Engineering, which implies creating feature interactions if relevant. Finally, consider dimensionality reduction for cases with many variables.
- PHASE 3.- Construction and Selection of Models; Divide the data into training and test sets. Select candidate models and train them on the training set. Adjust hyperparameters using cross-validation. Finally, Model Evaluation: Evaluate each model on the test set. Compare performance metrics and select the most promising model.

- PHASE 4.- Model Interpretation: Analyze the importance of features. Use explainability techniques to understand the model's predictions.

Algorithm 1. General Algorithm

Input: Dataset loading
1: Phase 1: EDA Analysis
2: Phase 2: Preprocessing
3: Phase 3: Construction, selection, and optimization of models:
   Repeat this phase until conclusive results are obtained
Output: Presentation (Accuracy, Precision, Recall, F1-Score)

In [6], similar works can be found where a similarity measure within the conventional process of the Forest and others algorithm is proposed. In fact they say:*"When the risk for new-onset diabetes was examined for the Framingham cohort, in both men and women, the presence of metabolic syndrome was highly predictive of new-onset diabetes. Almost half of the population-attributable risk for diabetes could be explained by the presence of ATP III metabolic syndrome."*. In the other hand the main idea is that a rating is unique for a group or for a user when that rating is different from the majority of ratings issued by other users, similar to [13] where the study collects data from a chain of health examination institution in Urumqi from 2017–2019, and performs 39,134 remaining data after preprocessing such as deletion and filling, concluding that *"The interpretability methods can be used as a novel means of identifying variables that are more likely to be good predictors. These predictors can be evaluated as features in other models developed with more appropriate datasets. In addition, it can also provide a methodological reference for the feasibility of applying the model interpretability method in health examination data mining"* .

4 Experimental Design

4.1 Dataset Characteristics

The dataset to be used consists of data collected between August 23, 2021, and May 31, 2022, as part of the Occupational Health program of the State University of Santa Elena (UPSE) led by Dr. Bertha Maggi, in charge of the Department of Well-being and Occupational Health at the University; to maintain the confidentiality of the data, all types of identifiers of the individuals to whom the control was directed have been removed in advance, so unless otherwise indicated, the data can be considered for general purposes without there being variables or indicators that could generate a bias that allows precisely determining a characteristic of the institution, characteristics that the author and collaborators in this preliminary study reserve for future formal and detailed analyses of occupational health at the University of Santa Elena. Under this context and based on a preliminary analysis, the data found in the dataset include:

- The"gender" column indicates the gender of the person (M/F).
- The column "type_person(W/T/S)" indicates whether the person is a Worker, Teacher, or Student.
- QS appears to be a qualitative marker, possibly an indicator of health status or a specific test; until the conclusion of this study, it was not possible to interview the person in charge of the data to provide a detailed definition of the QS field.
- Glycosylated is a value that does not seem to be correctly defined since all records show 0.0, which could be a data error; similarly to the previous field, until the conclusion of this study, it was not possible to interview the person in charge of the data to provide a detailed definition of the glycosylated field, from the literature reviewed and for subsequent analysis this field will be ignored, focusing on the "glycosylated hemoglobin" field which potentially represents greater importance when generating correlations with other variables.
- Glucose mg/dL shows the blood glucose level.
- Total Cholesterol, HDL (High-Density Lipoproteins), LDL (Low-Density Lipoproteins), and Triglycerides are measures related to the lipid profile.
- Glycosylated Hemoglobin is a measure of long-term blood glucose control.
- Weight Kg, Height meters, BMI (Body Mass Index), Abdominal Perimeter are anthropometric measures.
- SysBP and DiaBP are the systolic and diastolic blood pressure, respectively.
- Pulse and SO2 (probably oxygen saturation) are vital parameters.
- Diagnosis indicates whether the person is at risk or not at risk, probably in relation to a specific health condition.

The detail of the descriptive statistics is shown in Table 1.

Table 1. Descriptive statistical analysis of the dataset in general

	Blood Glucose	Total Cholesterol	HDL	LDL	Triglycerides	Glycated Hemoglobin	Weight Kg	Height mts	BMI	Abdominal Perimeter	SBP (Systolic Blood Pressure)	DBP (Diastolic Blood Pressure)	Pulse	Oxygen Saturation
count	1079	1079	1079	1079	1079	1079	1079	1079	1079	1079	1079	1079	1079	1079
mean	96,48	180,31	40,67	124,63	127,97	4,90	58,28	1,70	22,21	79,40	112,40	68,70	74,05	97,78
std	15,28	35,78	8,61	37,24	72,59	0,63	11,30	4,88	7,09	8,05	8,22	5,04	7,85	1,20
min	0,00	0,00	0,00	0,00	0,00	0,00	37,10	0,00	0,00	61,00	82,00	50,00	47,00	76,00
0,25	90,00	155,40	35,00	101,60	80,90	4,80	54,00	1,54	22,21	77,00	112,00	68,70	72,00	97,78
0,50	95,20	176,50	39,20	117,10	113,00	4,80	54,00	1,54	22,21	77,00	112,00	68,70	72,00	97,78
0,75	100,80	201,00	45,05	145,20	152,40	4,80	58,00	1,54	24,05	77,95	112,00	68,70	73,00	98,00
max	352,80	478,00	100,00	423,40	791,00	13,80	132,70	161,70	62,10	132,00	193,00	99,00	116,00	99,00

Summary of Numerical Variables: Most of the numerical variables appear to have no missing values. Some variables have a minimum value of 0, which could be a missing or atypical value for certain biometric measures such as Glucose mg/dL, Total Cholesterol, HDL, LDL, and Triglycerides. This may require further investigation. The HEIGHT meters variable has a maximum value of 161.7, which is evidently incorrect and may be a data entry error. The variables WEIGHT Kg, BMI, and ABDOMINAL PERIMETER seem to have values that could be realistic, but we would need to confirm if the range of values is expected for the study population. The blood pressure variables (systolic, diastolic) and PULSE have very consistent values for the 25%, 50%, and 75% percentiles, suggesting that they could have been imputed or that the data is not as varied as expected. The SO2 (oxygen saturation) variable has a very low standard deviation and an average very close to the maximum of 99%, which could indicate a lack of variability or possible outliers.

Summary of Categorical Variables:
There are two genders (M, F) and three types of person (Worker, Teacher, Student). The diagnosis category has two values (At Risk, Not At Risk), and most individuals are labeled as Not At Risk.

Missing Values:
There do not appear to be any explicit missing values in the dataset.

Next Steps:

- Verify outliers and errors: It's important to check for outliers or possible entry errors, especially for values that are 0 or exceptionally high where they shouldn't be.
- Distribution analysis: Proceed to visualize the distribution of both numerical and categorical variables to better understand the characteristics of the study population.
- Investigate Consistent Variables: Decide to thoroughly analyze variables with little variability to determine if these are imputed values or if they reflect the reality of the dataset.
- Verify the Scale of Variables: Given that some variables might need to be rescaled or transformed to be effectively used in a machine learning model, proceed with the analysis of outliers and potential entry errors, and then visualize the distribution of some of the key variables to gain a better understanding of the data Figs. 3, 4, 5.

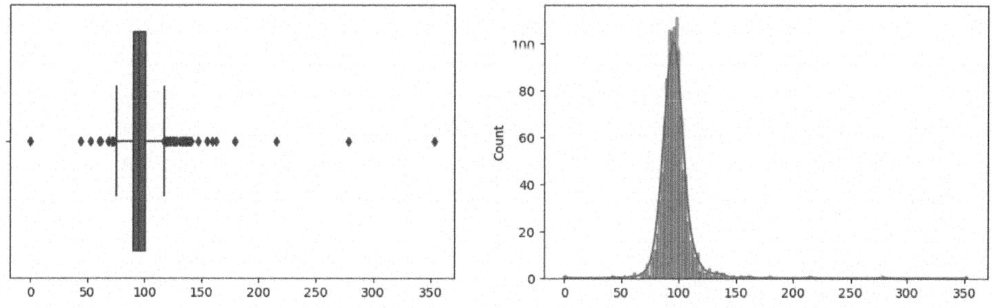

Fig. 2. Glucose analysis

The visualization of variables can be seen in the Figs. 10, 7, 8, 9 among the most important:

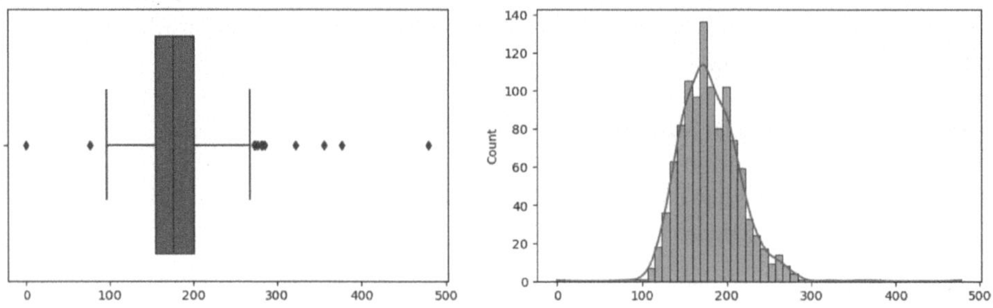

Fig. 3. Cholesterol analysis

The visualizations of the distributions of the selected variables show:

- Glucose mg/dL: Most values are within a normal range, but there are some very high outliers that could represent real cases of elevated glucose levels or possible entry errors.
- Total Cholesterol: Similar to glucose, there is a relatively normal distribution with some high outliers.
- HEIGHT meters: Most heights are within the expected range, though there are extremely high values surpassing realistic values for human height, which clearly indicates a data entry error.
- WEIGHT Kg: Weight appears to have a normal distribution with some outliers at the upper end.

5 Experimentation

As part of the experimentation, possible errors in the data were corrected. This involved correcting the extremely high values in HEIGHT (meters) which could obviously be considered as errors in data entry during digitization. On the other hand, for the outliers, it was necessary to decide whether to trim them, transform them, or keep them depending on whether they are representative of the population or not. Additionally, the consistency of variables was verified, that is, reviewing variables with little variability, such as SO2, to make the decision whether to keep them in their current form. These corrections can be seen in Figures 10, 7, 8, 9:

After this process, the corrections and adjustments were applied successfully to the data, which involved:

- Height (meters): After correcting values that seemed to be in centimeters, the distribution now appears more realistic and aligns with what is expected for human height.

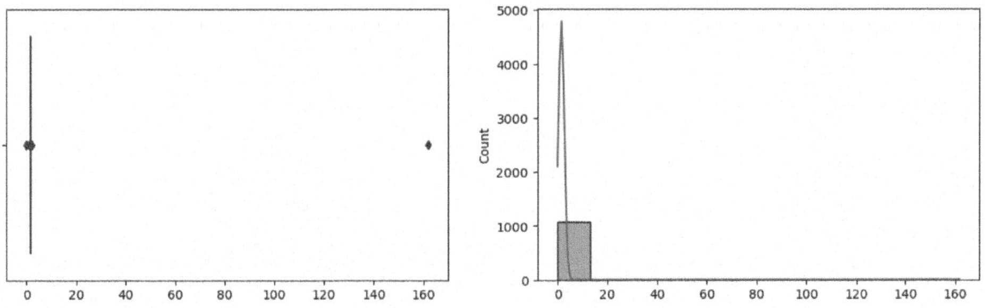

Fig. 4. Height analysis

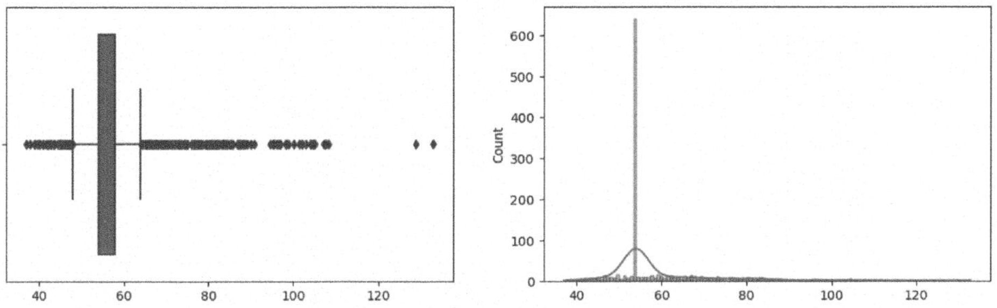

Fig. 5. Weight analysis

- Glucose mg/dL, Total Cholesterol, Weight Kg: By trimming values at the 1st and 99th percentiles, we managed outliers without eliminating corresponding data. The distributions of these variables now show less extreme variation and are more consistent with what we would expect in a typical population.

The next step was to investigate the SO2 variable due to its low variability and any other variables that might require further inspection. Additionally, preparing categorical variables for modeling through appropriate encoding and normalizing numerical variables as necessary.

Below are the data for the SO2 variable in the Fig. 10:

The visualization of the SO2 variable reveals very low variability, and it has been confirmed that it is not a variable with variance close to zero, which means that there could be some useful variability in it. However, the low variability suggests that it may not be very informative for machine learning models.

Regarding the normalization of numerical variables, errors were encountered when attempting to exclude certain variables that do not exist under the'float64' or'int64' data types. Steps were taken to correct this and ensure that only relevant numerical variables were normalized. Additionally, the presence of the'QS' and'glycosylated' variables was verified before proceeding with the normalization of the remaining numerical variables.

Once the numerical variables were successfully normalized, and the categorical variables were encoded, the SO2 variable was reanalyzed, where it was found

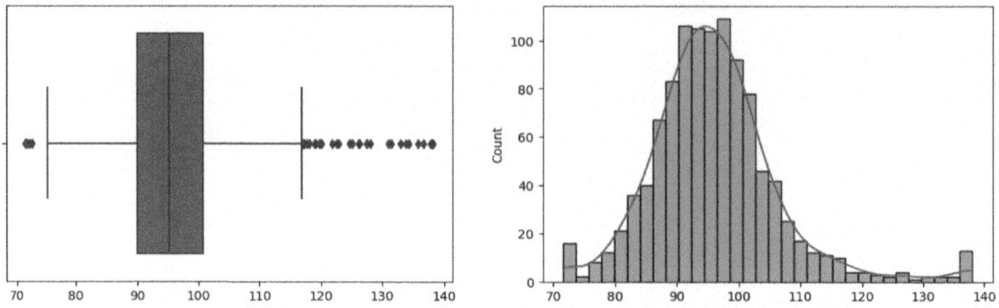

Fig. 6. Glucose analysis

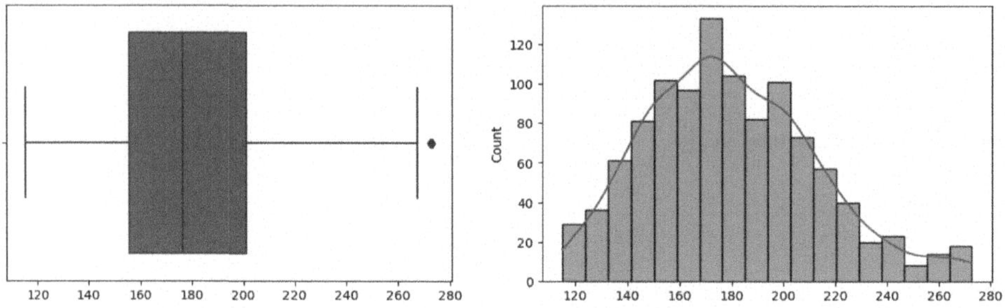

Fig. 7. Cholesterol analysis

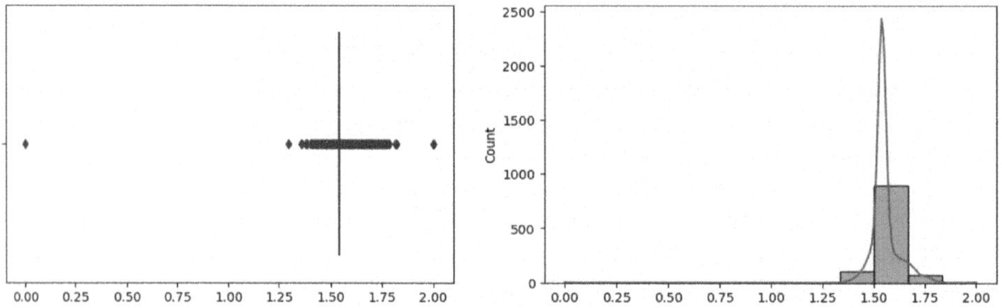

Fig. 8. Height analysis

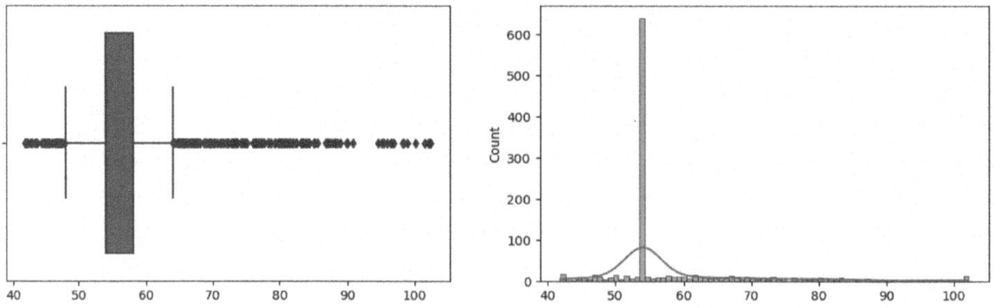

Fig. 9. Weight analysis

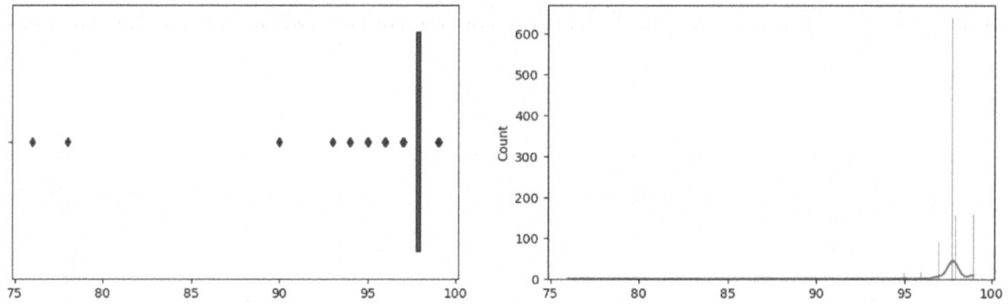

Fig. 10. Oxygen Saturation Analysis

to have low variability and not have variance close to zero, so it was estimated that it could still provide information to the models. With this background, the following steps were taken:

- Handling the target variable; that is, encoding the target diagnostic variable appropriately for modeling. It should be noted that if it is a binary classification problem, it must be ensured that it is encoded as 0 and 1.
- Splitting the data into training and test sets to validate the performance of the machine learning models.
- Selecting and training models after choosing the appropriate machine learning models for the binary classification task and training them using the training data.

Once the data has been successfully divided into training and test sets, we obtained 863 samples in the training set and 216 samples in the test set. The features (X) have 19 columns, which are the independent variables after encoding and normalization. The target variable (y) is a binary vector representing the presence or absence of risk.

The next step was to select and train suitable machine learning models for the binary classification task. Ensemble models of decision trees were chosen, specifically the classifier *Random Forest*, *LightGBM* and *XGBoost*.

The choice of *Random Forest, LightGBM,* and *XGBoost* as training models for classification was made because *Random Forest* maintains robustness and accuracy as this method is known for its high accuracy and robustness. By using multiple decision trees and averaging their predictions, this model reduces the risk of overfitting, which is especially useful in datasets with a lot of variability. Similarly, concerning the handling of nonlinear data, it is effective for classifying nonlinear data, something common in many real-world problems. Additionally, due to the feature importance, it provides a good indication of feature importance, which was useful for understanding the model and the problem, not to mention that it is less sensitive to outliers, as by its ensemble nature, it is less sensitive to outliers compared to other algorithms. On the other hand, *LightGBM* maintains efficiency in large datasets as it uses gradient-based algorithms. Likewise, due to its low memory usage and training speed, much faster than

Table 2. Comparison of Classifier Results

Method	Accuracy	Precision	Recall	F1-Score
Random Forest	0.9352	0.94	0.89	0.94
XGBoost	0.9537	0.96	0.95	0.96
LightGBM	0.9630	0.97	0.95	0.96

other models, which is crucial when working with large volumes of data. It also efficiently handles categorical features without extensive preprocessing, simplifying the data preparation process, and finally, due to its high performance with shallow trees, as it performs well even with shallow trees, which is beneficial for avoiding overfitting. Lastly, *XGBoost* was chosen for its performance and speed as *XGBoost* is another gradient boosting algorithm, known for its performance and speed in machine learning model competitions. Alongside this, it was chosen because it includes regularization (L1 and L2), which helps prevent overfitting and improves the overall model accuracy, as well as its flexibility because it allows users to define custom optimization objectives and evaluation criteria, increasing its flexibility to address different classification problems.

6 Results and Discussion

In this section, the most relevant results following the phases of the proposed method are presented. The calculation of singular values and measures was carried out using the Train set. The best parameter values for the methods have been highlighted in the Table.2.

Random Forest results

Overall Accuracy (accuracy): 94.44%, indicating that the model has correctly predicted the risk classification in approximately 94.44% of the cases in the test set.

For the"No Risk" (sin riesgo) class (encoded as 0):

- Accuracy: 97%, meaning that when the model predicts "No Risk," it is correct 97% of the time.
- Recall: 96%, indicating that the model captures 96% of the actual "No Risk" cases.
- F1-Score: 97%, a weighted average of precision and recall.

For the "Risk" class (encoded as 1):

- Accuracy: 81%, meaning that when the model predicts "Risk," it is correct 81% of the time.
- Recall: 85%, indicating that the model captures 85% of the actual "Risk" cases.
- F1-Score: 83%, a weighted average of precision and recall.

Conclusion. These results suggest that the model performs well overall, with excellent ability to identify "No Risk" cases. However, there is a slight room for improvement in identifying "Risk" cases, as indicated by the lower precision and recall for that class

LightGBM Results

The LightGBM model has shown outstanding performance in binary classification, as detailed below: **Overall Accuracy**

The model's accuracy is 96.30%, indicating a high efficacy in correctly predicting classification in the test set.

For the "No Risk" class (encoded as 0):

- Precision: 99% (The model correctly predicts 99% of instances).
- Recall: 97% (The model identifies 97% of all actual instances of the class).
- F1-Score: 98% (High weighted average of precision and recall).
- Support: 182 (Number of actual instances in the test set).

For the "Risk" class (encoded as 1):

- Precision: 84% (The model correctly predicts 84% of instances).
- Recall: 94% (The model identifies 94% of all actual instances of the class).
- F1-Score: 89% (Good performance but not as high as for class 0).
- Support: 34 (Number of actual instances in the test set).

Averages

- Macro Avg:
  - Precision: 0.92
  - Recall: 0.95
  - F1-Score: 0.93
- Weighted Avg:
  - Precision: 0.97
  - Recall: 0.96
  - F1-Score: 0.96

Conclusion These results suggest that the **LightGBM** model possesses a remarkable ability to identify classes, with particularly strong performance in the "No Risk" class. The effectiveness of gradient boosting tree-based ensemble models is demonstrated for this type of classification tasks.

XGBoost Results

The results of the XGBoost model in binary classification are detailed below:

Overall Accuracy

The accuracy of the model is 95.83%, indicating high efficacy in correctly predicting classification in the test set.

For the "No Risk" Class (coded as 0):

- Precision: 0.99 (The model correctly predicts 99% of instances).
- Recall: 0.96 (The model identifies 96% of all real instances of the class).
- F1-Score: 0.97 (High weighted average of precision and recall).
- Support: 182 (Number of real instances of class 0 in the test set).

For the "Risk" Class (coded as 1):

- Precision: 0.82 (The model correctly predicts 82% of instances).
- Recall: 0.94 (The model identifies 94% of all real instances of the class).
- F1-Score: 0.88 (Good performance but not as high as for class 0).
- Support: 34 (Number of real instances of class 1 in the test set).

Averages

- Macro Avg:
  - Precision: 0.90
  - Recall: 0.95
  - F1-Score: 0.93
- Weighted Avg:
  - Precision: 0.96
  - Recall: 0.96
  - F1-Score: 0.96

Conclusion. The XGBoost model has proven to be highly effective in classifying the "No Risk" class and quite effective in classifying the "Risk" class. Although the precision for the "Risk" class is lower than for the "No Risk" class, the high recall indicates that the model is capable of identifying most real risk cases, which is positive especially if it is critical not to overlook risk cases. The difference between precision and recall for the "Risk" class could be an area of focus for improving the model.

Discussion

These results indicate that the ***LightGBM*** model has an even greater ability than the ***XGBoost*** model to correctly identify both risk and "No Risk" cases, with slightly better precision and F1-Score for the "Risk" class.

This demonstrates the effectiveness of ensemble models based on gradient boosting trees for this type of classification task. Both ***XGBoost*** and ***LightGBM*** are known for their high precision and performance in classification tasks and are robust choices for binary classification.

Given the high performance of both models, the choice between them could be based on other factors such as training time, interpretability, or performance on an additional validation dataset if available.

7 Conclusions

In this study, the performance of three machine learning models (*LightGBM, XGBoost, and Random Forest*) for predicting the risk of metabolic syndrome in workers, students, and teachers from an institution with a rigorous occupational health program has been evaluated. The results obtained reveal a high degree of effectiveness in classifying workers into risk and no-risk categories for developing metabolic syndrome.

The *LightGBM* model stands out as the most effective in the classification task, achieving an overall accuracy of 96.30%. Furthermore, it demonstrates exceptional performance in identifying risk cases, with a precision of 99% and a recall of 97% for the "No Risk" class. This indicates that the model is highly precise and capable of identifying the majority of real "No Risk" cases. For the "Risk" class, the *LightGBM* model achieves a precision of 84% and a recall of 94%, indicating significant capability in identifying risk cases.

The *XGBoost* model also presents notable results, with an overall accuracy of 95.83%. Although its precision for the "Risk" class is slightly lower than that of the *LightGBM* model, its high recall of 94% indicates a solid ability to capture real "Risk" cases. This is essential for early detection of potential cases of metabolic syndrome.

Lastly, the *Random Forest* model shows respectable performance, with an overall accuracy of 94.44%. Its results indicate a reasonable ability to classify workers into risk and no-risk categories. However, its precision and recall for the "Risk" class are lower than those of the other evaluated models.

Collectively, these results support the effectiveness of ensemble models based on gradient boosting trees (*LightGBM and XGBoost*) in predicting the risk of metabolic syndrome in workers from institutions with occupational health programs. These models have high potential to enhance early detection and risk management of metabolic syndrome, which can be crucial for the health and well-being of workers.

Furthermore, it is worth noting that the choice between the *LightGBM* and *XGBoost* models may depend on other factors such as training time, interpretability, and performance on an additional validation dataset if available. Both models represent solid and reliable options for binary classification in this context.

This study provides a solid foundation for future research and practical applications in the field of occupational health and metabolic syndrome risk management.

References

1. Ahn, J.M., Kim, J., Kim, K.: Ensemble machine learning of gradient boosting (XGBoost, LightGBM, CatBoost) and attention-based CNN-LSTM for harmful algal blooms forecasting. Toxins **15**(10), 608 (2023). https://doi.org/10.3390/toxins15100608
2. Canelo Sotelo, C.A., Espinoza Haro, P.C. (Dir.).: Redes neuronales artificiales y máquina con soporte vectorial para clasificar a los solicitantes de microcrédito (ed.). Universidad Nacional de Ingeniería (2021)
3. Dell'Aversana, P.: A Global Approach to Data Value Maximization: Integration, Machine Learning and Multimodal Analysis (ed.). Cambridge Scholars Publishing (2019)
4. Gómez Mendoza, R.A.(Il.)., Carrillo Esper, R. (Il.)., González Chávez, A.(Il.).: (2022). Síndrome metabólico y su impacto, CMMI Vol. 08: (1 ed.). Editorial Alfil, S. A. de C. V
5. González Chávez, A., López Espinosa, R., Mesa Pérez, J.A.: Manual práctico para la detección y el tratamiento integral del síndrome metabólico: (2 ed.). Editorial Alfil, S.A. de C.V (2019)
6. Horvat, T., Havaš, L., Srpak, D., Havaš, J., Srpak, J., Crčić, D.: predicting metabolic syndrome risks using machine learning methods. In: 2023 IEEE 21st Jubilee International Symposium on Intelligent Systems and Informatics (SISY), Pula, Croatia, 2023, pp. 000017–000022. https://doi.org/10.1109/SISY60376.2023.10417918. keywords: Sociology;Medical services;Predictive models;Prediction algorithms;Statistics;Biomedical monitoring;Medical diagnostic imaging;artificial intelligence;machine learning;metabolic syndrome;WEKA
7. Lee, W.: Python Machine Learning: (ed.). Wiley (2019). https://bibliotecas.ups.edu.ec:3488/es/lc/bibliotecaups/titulos/186090
8. Liang, W., et al.: Predicting Hard Rock Pillar Stability Using GBDT, XGBoost, and LightGBM Algorithms. Mathematics (2227-7390), [s. l.], vol. 8, n. 5, pp. 765 (2020). https://doi.org/10.3390/math8050765
9. Ortega Candel, J.M.: Big data, machine learning y data science en Python: (1 ed.). RA-MA Editorial (2022)
10. Restrepo Leal, D.A., Viloria Porto, J.P., Robles Algarín, C.A.: El camino a las redes neuronales artificiales: (1 ed.). Editorial Unimagdalena (2021)
11. Samaniego Bañuelos, F.: ¿De qué diabetes estamos hablando?: (ed.). Editorial ITACA (2019)
12. Villegas Abrill, C.B., Paz Aliaga, A. (Dir.).: Efecto de una intervención educativa-nutricional en el síndrome metabólico premórbido y estado nutricional de docentes de la Universidad Nacional del Altiplano de Puno: (ed.). D - Universidad Nacional San Agustín de Arequipa (2022)
13. Zhang, Y., Zhang, X., Razbek, J., et al.: Opening the black box: interpretable machine learning for predictor finding of metabolic syndrome. BMC Endocr. Disord. **22**(1), 214 (2022). https://doi.org/10.1186/s12902-022-01121-4

Computational Astrochemistry (CompAstro 2024)

The $He^{+\cdot}$ + CH_3CN Charge-Exchange Reaction: A Combined Semi-empirical and Quantum Chemical Study

Luca Mancini[1(✉)], Emília Valença Ferreira de Aragão[1], Fernando Pirani[1,2], Marzio Rosi[2,4], Noelia Faginas-Lago[1,4], and Daniela Ascenzi[3]

[1] Dipartimento di Chimica, Biologia e Biotecnologie, Università degli Studi di Perugia, 06123 Perugia, Italy
luca.mancini@unipg.it
[2] Dipartimento di Ingegneria Civile ed Ambientale, Università degli Studi di Perugia, 06125 Perugia, Italy
[3] Dipartimento di Fisica, niversità di Trento, Trento, Italy
[4] Istituto CNR di Scienze e Tecnologie Chimiche "Giulio Natta" (CNR-SCITEC), 06123 Perugia, Italy

Abstract. The reaction with the $He^{+\cdot}$ cation is considered one of the most important processes in the destruction of interstellar complex organic molecules (iCOMs). The development of accurate astrochemical databases requires the detailed knowledge of the rate constants and branching ratios of the processes. The dynamics of the charge exchange process between $He^{+\cdot}$ and methyl cyanide (CH_3CN) can be studied through a Landau-Zener-St*ü*ckelberg approach, which needs a detailed analysis of the potential energy surface (PES). In the present contribution, the Improved Lennard-Jones (ILJ) formulation is used to unveil the main features of the PES for $He^{+\cdot}$ + CH_3CN reaction, in order to set the stage for an accurate investigation of the dynamics of the charge-exchange process. The low computational cost and the high portability of the ILJ formulation allows to extend the approach to the study of different systems, with relevance in several fields, ranging from astrochemistry to nanomaterials.

Keywords: Astrochemistry · Improved Lennard-Jones · Reaction Dynamics · Theoretical Chemistry · Force Field

1 Introduction

Interstellar complex organic molecules (iCOMs), defined as organic molecules containing at least 6 atoms [1], have been detected through different stages of stellar evolution, from pre-stellar clouds to protostars and protoplanetary disks, and they act as important tracers of the physical conditions and of the evolution of the regions where they are observed. For an updated list of interstellar

O. Gervasi et al. (Eds.): ICCSA 2024 Workshops, LNCS 14818, pp. 39–54, 2024.
https://doi.org/10.1007/978-3-031-65273-8_3

and circumstellar molecules, now comprising over 300 species [2], see the Cologne Database for Molecular Spectroscopy [3–5]. The correct implementation of astrochemical models, built to reproduce the observed iCOM abundances, strongly depends on the kinetics of the chemical reactions implemented as possible formation and destruction mechanisms of iCOMs. To this purpose astrochemical databases (*e.g.* the KIDA [6] and UMIST [7,8] databases) are widely used to obtain kinetic parameters and there is a strong need to keep such databases updated with the most recent experimental/theoretical estimates of rate coefficients and branching ratios.

A particular class of interstellar molecules is represented by nitriles, a set of N-bearing organic compounds, which accounts for ca. 15 % of the total species detected in the interstellar medium (ISM) [9] and represent an interesting category of compounds in prebiotic chemistry [10], being possible precursors of complex structures of amino acids and RNA [11–13]. Acetonitrile (CH_3CN), one of the simplest nitriles, is also one of the most abundant iCOM observed. After the first detection towards the SgrB2 and SgrA molecular clouds [14], it has been revealed in different regions associated to various stages of the formation of stars and stellar systems [15–24] and in comets [25–27]. According to the main astrochemical databases, the most important destruction mechanisms of acetonitrile are due to collisions with ions, including $He^{+\cdot}$ (that is considered one of the main destruction routes for iCOMs), formed through the interaction of He atoms with cosmic rays. A detailed chemical investigation is therefore pivotal to unveil the reaction dynamics and derive the rate constant of the process.

To this purpose, a theoretical analysis can help to elucidate the main features of the potential energy surface (PES) which triggers the dynamics of the charge-exchange process. In this respect, semi-empirical models represent a feasible and accurate method to treat non-covalent interactions and provide useful parameters for the development of the PES, given in a proper analytical form, which is a necessary condition to perform molecular dynamics simulations. Among them, the Improved Lennard-Jones (ILJ) functional [28] has been developed in order to have a refined analysis of both long-range attraction and short-range repulsion. Recently, the ILJ formulation has been used to investigate the long range component of the interaction in the case of bimolecular (neutral-neutral and ion-neutral) gas-phase reactions [29–31], helping to elucidate the reaction mechanism at the microscopic level. Additionally, the ILJ formulation represents a feasible method, characterized by a high degree of portability and low computational demand. Therefore, it can be used for several purposes, ranging from astrochemistry and reaction dynamics to material sciences, including the development of new generation nano-materials for energy storage and carbon neutrality. In the present contribution, the ILJ potential has been used, in conjunction with *ab initio* calculations, to study the PES for the $He^{+\cdot}$ + CH_3CN reaction, in order to provide the building blocks for the complete treatment of the reaction dynamics.

2 Theoretical Methods

2.1 Electronic Structure Calculations

Electronic structure calculations have been performed in order to derive the main geometrical parameters of the interacting systems. The geometry of the CH_3CN molecule has been optimized using Density Functional Theory (DFT) calculations, with the B3LYP functional [32,33], together with the correlation consistent valence polarized set aug-cc-pVTZ [34–36]. The same computational strategy has been used to calculate harmonic vibrational frequencies, in order to check the nature of the stationary point, as well as the Mulliken atomic charges. Finally, the same level of theory has been used to perform an analysis of the atomic orbitals, in order to derive the shape and the energy of the molecular orbitals (MO's), to be used in the dynamical effects treatment of the interaction with $He^{+\cdot}$.

2.2 Semi-empirical Analysis of the Non-covalent Interactions

The non-covalent interaction between the $He^{+\cdot}$ radical cation and CH_3CN has been analysed adopting a strategy already used for similar systems [29,37,38]. In details, the PES has been built for both the entrance ($He^{+\cdot}$ + CH_3CN) and the exit (He + $CH_3CN^{+\cdot}$) channels. The PES is given in an analytical form and depends on the distance (defined as R) between $He^{+\cdot}$ (or He in the case of the exit channel) and the center of mass of the acetonitrile molecule (or its radical cation, for the exit channel), and on the relative orientation of the involved partners. In this respect, it is important to notice that, due to the symmetry of the CH_3CN molecule, the relative orientation can be defined considering only θ, which is the angle defined by the $\mathbf{R}$ vector and the z axis (taken to coincide with the C_{3v} symmetry axis of CH_3CN).

In order to have a simpler representation of the system, CH_3CN has been represented considering three moieties: the C and N atoms of the cyano group and an effective atom (indicated as C_{eff}), having the mass and polarizability of a methyl group (CH_3). The global potential describing the interaction is composed of two contributions: a first term representing the electrostatic potential (V_{elect}) and a second term (V_{nelect}), accounting for the non electrostatic contribution of the interaction.

PES feature of the entrance channel:

For the **entrance channel**, the electrostatic term is defined as:

$$V_{elect}(R,\theta) = 14400 \cdot \left(\frac{q_C}{r_C} + \frac{q_N}{r_N} + \frac{q_{C_{eff}}}{r_{C_{eff}}}\right) \tag{1}$$

where q_C, q_N and $q_{C_{eff}}$ are the effective atomic charges, calculated as in Sect. 2.1. The variables r_C, r_N and $r_{C_{eff}}$ represent the distances between one of the three moieties (C, N, C_{eff}) of CH_3CN and $He^{+\cdot}$, evaluated by exploiting the Carnot's theorem. In details, considering the reference system reported in Fig. 1, the three distances are formulated as:

$$r_C^2 = R^2 + d_C^2 - 2R \cdot d_C \cdot cos\theta \tag{2}$$

$$r_N^2 = R^2 + d_N^2 - 2R \cdot d_N \cdot cos\theta \tag{3}$$

$$r_{C_{eff}}^2 = R^2 + d_{C_{eff}}^2 - 2R \cdot d_{C_{eff}} \cdot cos(\pi - \theta) \tag{4}$$

where d_C, d_N and $d_{C_{eff}}$ are the distances between the CM (at the axes origin) and the position of the C, N and C_{eff} atoms/effective atoms within the CH_3CN molecule . The factor 14400 allows to obtain $V_{elect}(R,\theta)$ in meV when the charges are given in atomic units and the distances in Å.

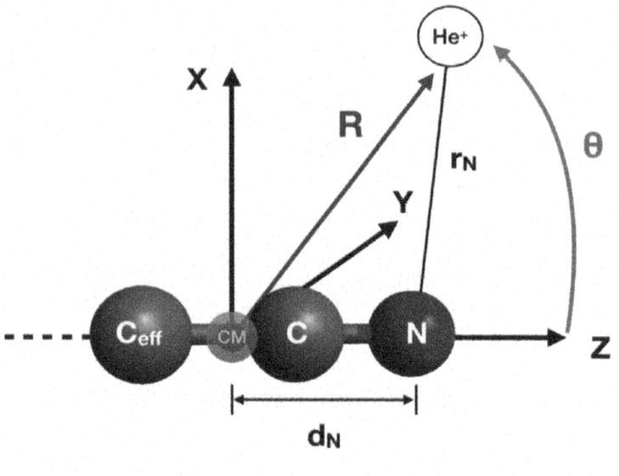

Fig. 1. Coordinate system used for the representation of the (He - $CH_3CN)^+$ system. For simplicity, only the relation between r_N and R, θ is reported, while the other parameters are defined in the same way.

The non-electrostatic contribution has been evaluated considering the Improved Lennard-Jones (ILJ) model [28,39–41] formulation. V_{nelect} is defined as balance of size repulsion contributions, dominant at short distance, with induction and dispersion attraction contributions, prevalent at large intermolecular distances. Such balance is obtained as additivity of components due to the interaction of $He^{+\cdot}$ with the three effective atoms of the molecule (C, N and C_{eff}). Each component is described by an ILJ function (V_{ILJ}), a simple-reliable potential model that simultaneously includes the combination of the long-range attraction with short-range repulsion effects. For any of the three pairs, interacting at a separation distance r (with r equal to r_C, r_N or $r_{C_{eff}}$, according to the moiety under consideration) the general formulation of V_{ILJ} is:

$$V_{ILJ}(r) = \varepsilon \left[\frac{m}{n(r) - m} \left(\frac{r_m}{r} \right)^{n(r)} - \frac{n(r)}{n(r) - m} \left(\frac{r_m}{r} \right)^{m} \right] \tag{5}$$

where, for each interaction pair, ε and r_m represents the well depth of the potential and its location, respectively, while the ratio r_m/r is the reduced distance. The ε and r_m values are estimated on the basis of the polarizability components associated to each interacting pair. For an ion-neutral pair the m parameter is equal to 4, while the $n(r)$ term, that modulates both strength and radial dependence of the repulsion and the strength of the attraction, is expressed considering the hardness of the two interacting fragments, taken into account in the β parameter:

$$n(r) = \beta + 4 \left(\frac{r}{r_m} \right)^2 \tag{6}$$

Therefore, in the entrance channel the overall non electrostatic component is the sum of three contributions:

$$V_{nelect}(R, \theta) = V_{ILJ}(r_C) + V_{ILJ}(r_N) + V_{ILJ}(r_{C_{eff}}) \tag{7}$$

For all the interacting pairs the relevant parameters are given in Sect. 3.2.

PES feature of the exit channel:

The PES in the exit channel arises from a balance between a long range attraction, due to dispersion and induction components, and a short range repulsion, associated to size effects. The combination of the dispersion attraction and size repulsion effects is defined as a van der Waals interaction, V_{vdW}, which is expressed, similarly to what already done for the entrance channel, in terms of three V'_{ILJ} functions, one for each interaction pair due to He and each effective atom of $CH_3CN^{+\cdot}$molecular ion:

$$V_{vdW}(R, \theta) = V'_{ILJ}(r_C) + V'_{ILJ}(r_N) + V'_{ILJ}(r_{C_{eff}}) \tag{8}$$

For such pairs, m has been fixed to 6, to correctly represent the asymptotic behaviour of the dispersion. The potential parameters associated with the three different V'_{ILJ} contributions are reported in Sect. 3.2. It should be emphasized that the $CH_3CN^{\cdot+}$ molecular ion is formed in an excited state, by ejection of one electron from one of the inner valence MO's, as it will be detailed in Sect. 3.1. As a consequence, its electronic polarizability, mostly determined by the behaviour of electrons populating outer MO's, is assumed to be the same of the neutral CH_3CN and it is therefore partitioned in the same way.

Additionally, a component V_{ind} has been added, to take into account the ion-induced dipole contribution, not directly enclosed in V_{vdW}, and whose strength is comparable with that of the dispersion. Such component depends on the separation distance R between the He atom and the CM of the molecular cation $CH_3CN^{+\cdot}$. Therefore V_{ind} is formulated, considering the polarizability α of He (equal to 0.2 Å^3) and expressing R in units of Å, by the semi-empirical relation:

$$V_{ind}(R) = -7200\frac{\alpha_{He}}{R^4} \tag{9}$$

where the factor 7200 provides V_{ind} in meV units.

3 Results

3.1 Electronic Structure Calculations

In Fig. 2, the optimized geometry obtained for the acetonitrile molecule is reported, including bond distances (in Å) and angles (in degrees). The distances obtained at the B3LYP/aug-cc-pVTZ level of theory can be compared with the data reported in the NIST database [42], showing a very good agreement. The main deviation is indeed represented by the C-H bond length, which is reported to be equal to 1.104 Å (differing by 0.015 Å from the calculated data). As already noticed in the analysis of previous systems [43–47], the proposed methodology can be considered a good approximation balancing between chemical accuracy and computational cost.

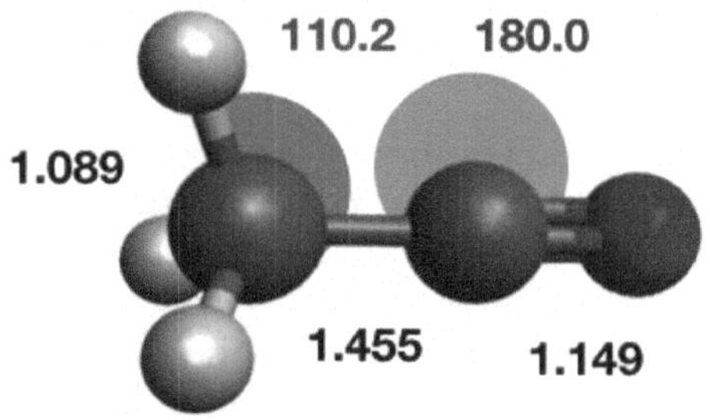

Fig. 2. Bond distances (in black, in Å) and angles (in colours, in degrees) for the CH_3CN molecule obtained at the B3LYP/aug-cc-pVTZ level of theory.

The optimized geometry has been used as a starting point for the analysis of the molecular orbitals of the system. In Table 1, the energies of some of the MO's calculated for the CH_3CN molecule are reported.

Table 1. Energies (in eV) calculated for the main MO's of the CH_3CN molecule necessary for the analysis of the charge-exchange process. The zero level is represented by the electron separated from the molecule.

Molecular Orbital	Energy (eV)
LUMO	0.643
HOMO	–11.420
HOMO-5	–16.555
HOMO-6	–23.988
HOMO-7	–27.569

The results obtained must be compared with the ionization energy of the He atom, equal to 24.587 eV [48], the highest value in the whole Periodic Table. As a consequence, an electron transfer from the HOMO or HOMO-5 to the 1s atomic orbital of $He^{+\cdot}$ would be highly exothermic (see the data reported in Table 1), and such large differences in the asymptotic values of the entrance and exit PESs would lead to an inefficient non-adiabatic coupling between the entrance and exit PESs at distances that are relevant for the electron transfer. As a consequence, the best candidate for the charge exchange via non-adiabatic coupling of the PESs appears to be the HOMO-6 orbital, having an energy of -23.998 eV. Figure 3 shows the shapes of the main MO's considered in the analysis. As can be seen, the HOMO-6 orbital involved in the charge exchange process is characterized by the presence of a nodal plane in correspondence of the C atom of the cyano group.

It should be noted that the energies of the MO's obtained at the DFT level of theory provide only an estimation of the ionization energy and have been used to establish which is the orbital involved in the ionization process. In absence of clear experimental data we decided to rely on the results coming from the DFT analysis which provides values closer to the accurate ones, without making use of multi-reference methods, which are extremely costly in terms of CPU time and disk space requirements.

In Table 2 are reported the effective atomic charges (q) calculated for the three moieties in which the CH_3CN molecule has been partitioned. The CH_3CN molecule is a symmetric top rotor that exhibits rotational components both along its C_{3v} symmetry axis, where is aligned its high permanent electric dipole (3.919 D), and perpendicular to it. As previously reported (see Sect. 2.2), the molecule can be considered as formed by three effective atoms. The polarizability of CH_3CN (equal to 4.30 Å^3), is partitioned in three effective components, one assigned to C_{eff} (equal to 2.20 Å^3), another assigned to C (1.00 Å^3), and the third associated to N (1.10 Å^3). The partial charges calculated for CH_3CN, and reported in Table 2, appear to be consistent with its permanent electric dipole and quadrupole moment. In Table 3 are reported the ILJ parameters used for to derive the PES for the entrance and exit channel.

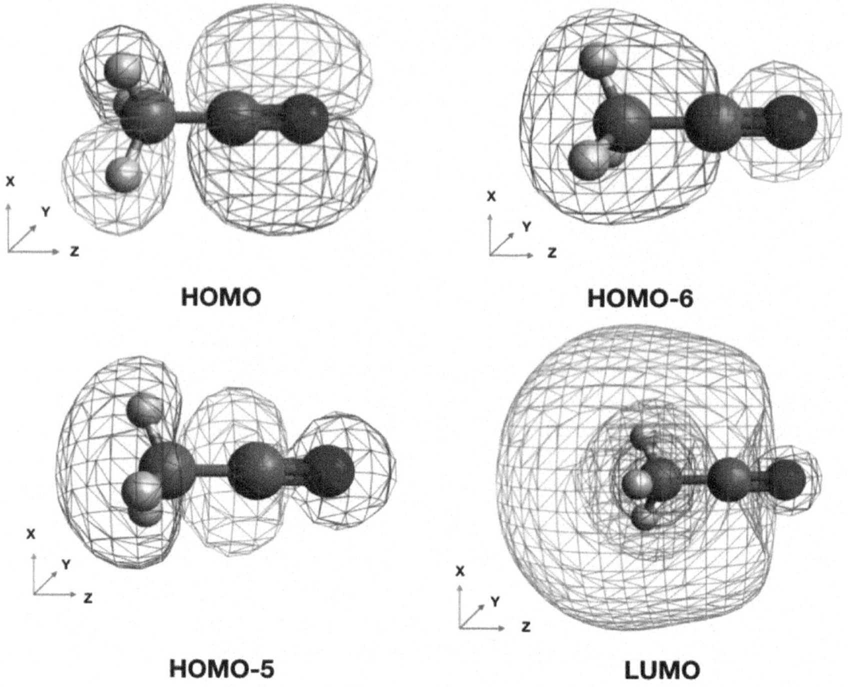

Fig. 3. Shape of the main MO's for the CH_3CN molecule, obtained at the B3LYP/aug-cc-pVTZ level of theory. Blue regions indicate the positive densities, while red regions are the negatively charged ones. (Color figure online)

Table 2. Atomic charges (q) obtained for the CH_3CN molecule, evaluated at the B3LYP/aug-cc-pVTZ level of theory. The charge of C_{eff} is representative of the methyl group and calculated considering the atomic charge of C and three atomic charges of H atoms.

Atom	**Charge q (a.u.)**
C	0.4501
N	–0.5006
C_{eff}	0.0505

Table 3. List of the ILJ parameters, used in Eq. 5 and 6, derived for the modeling of the non-covalent interaction in both the entrance and exit channel.

interacting pairs	r_m(Å)	ϵ (meV)	m	β
Entrance channel				
$He^{+\cdot}$ - C_{eff}	2.58	312	4	7
He^{+} - C	2.31	189	4	7
$He^{+\cdot}$ - N	2.33	198	4	7
Exit channel				
He - C_{eff}	3.69	2.3	6	8
He - C	3.30	2.1	6	8
He - N	3.33	2.2	6	8

3.2 PES in the Entrance and Exit Channel

The data obtained through the analysis reported in Sect. 2.2 have been used to calculate the analytical PES for the entrance and exit channels. To provide an easier representation of the interacting system, polar coordinates are employed. Specifically, the CH_3CN molecule (and consequently the $CH_3CN^{+\cdot}$ radical cation) exhibit cylindrical symmetry. As a consequence (as mentioned above), the potential representing the interaction can be described using only two coordinates: R (the distance between the $He^{+\cdot}$ cation and the CM of the CH_3CN molecule) and θ, representing the angle between the **R** vector and the z axis. This significantly reduces the complexity of the treatment. Figure 4 shows some cuts of the entrance PES for specific values of θ angle.

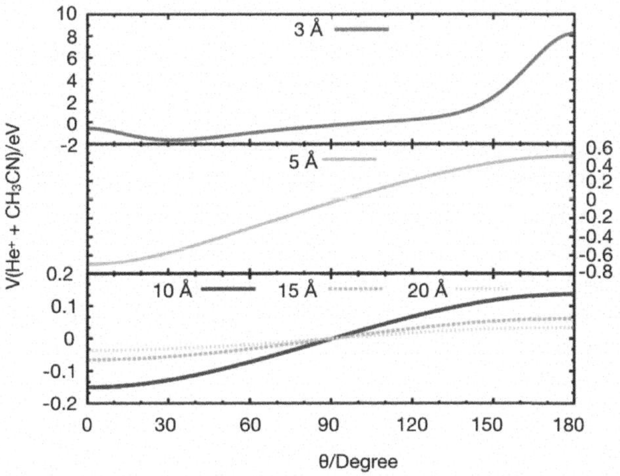

Fig. 4. Cuts of the PES for the $He^{+\cdot}$ + CH_3CN entrance channel for five different values of R, as a function of θ

As observed from the cuts reported in Fig. 4, entrance barriers are evident, with heights increasing as R decreases. These barriers, under thermal and sub-thermal conditions, are insurmountable due to the rotational motion of the CH_3CN molecule, which changes the θ angle. Consequently, the rotational component perpendicular to the C_3v axis is significantly hindered during the collision event. In Fig. 5 some cuts of the PES in the exit channel are presented.

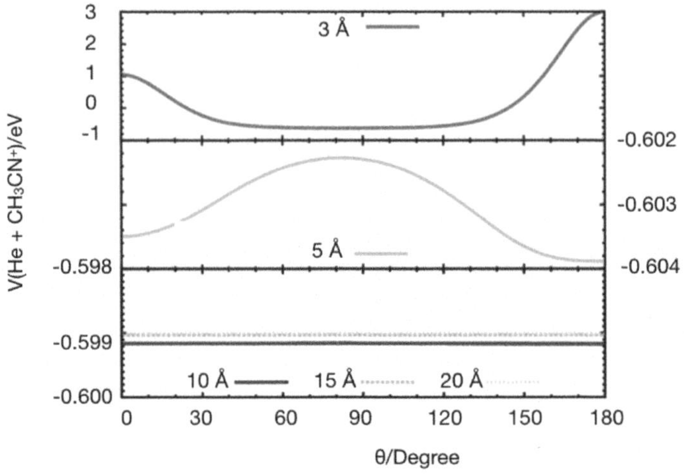

Fig. 5. Cuts of the PES for the He + $CH_3CN^{+\cdot}$ exit channel for five different values of R, as a function of θ

To properly characterize the reaction dynamics, it is pivotal to analyze the *non-adiabatic transitions* at the crossing between the entrance and exit channel of the PES, which are located in a wide range of separation distances. Additionally, the significant disparity in the ionization energy (IE) of the two species (IE_{He}= 24.587 eV and IE_{CH3CN} = 12.20 ± 0.01 eV [48]) does not permit a crossing between the entrance and exit channels in the case of an electron being ejected by the HOMO of CH_3CN. In Fig. 6, the crossing points between the entrance $He^{+\cdot}$ – CH_3CN (green curves) and exit He – $CH_3CN^{\cdot+}$ (red continuous curve) channels are illustrated for six different orientations, determined by θ values equal to 0°, 25°, 50°, 75°, 90° and 110° with the CH_3CN radical cation formed in its electronic ground state. Additionally, the red dashed curves represent the exit channels assuming the removal of an electron from the inner valence orbital HOMO-6.

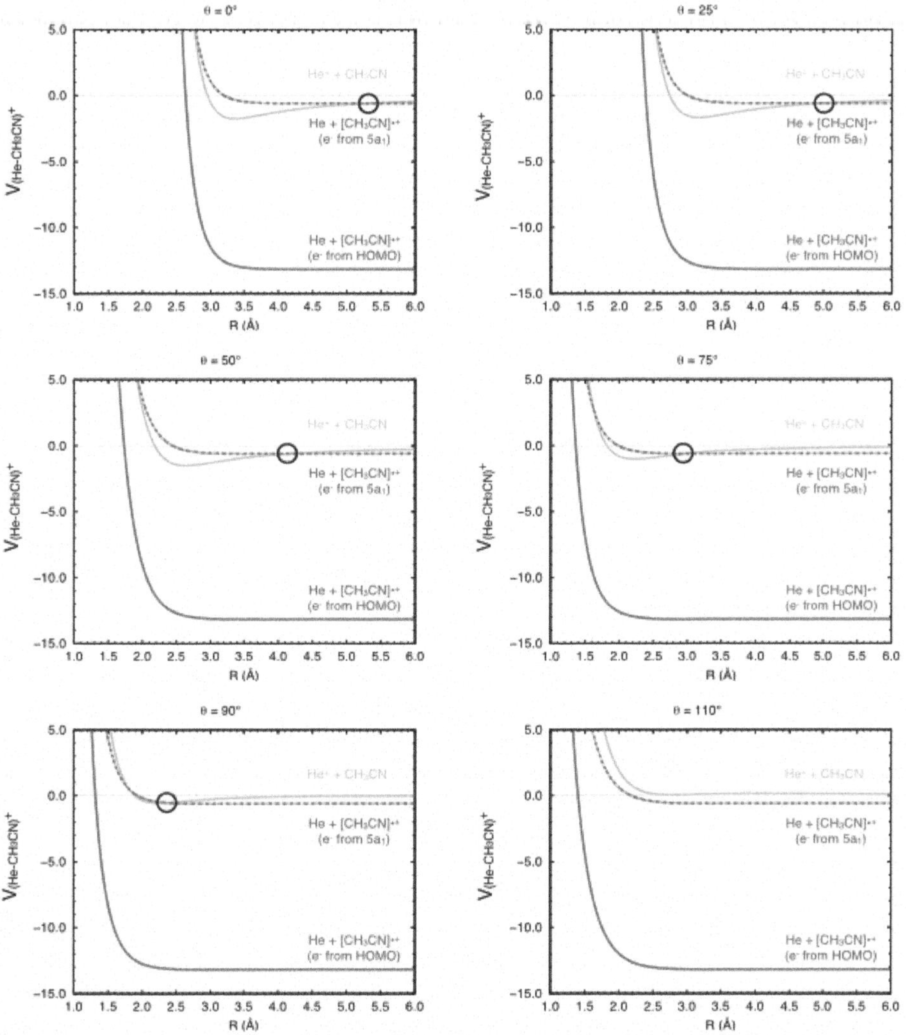

Fig. 6. PES crossings: PE curves of the entrance He$^+$ – CH_3CN (green curves) and exit He – $CH_3CN^{\cdot+}$ (red continuous curve) channels for six different orientation determined by θ values equal to 0°, 25°, 50°, 75°, 90° and 110° with the CH_3CN radical cation formed in its electronic ground state. The red dashed curves represent exit channels assuming the removal of an electron from the HOMO-6 orbital (indicated in figure as $5a_1$, according to the photoelectron spectra).

As emphasized above, in order to have effective crossings between entrance and exit interaction potentials, the electron must be removed from an inner orbital, with a higher ionization potential. Photoelectron spectra [49] and the present electronic state calculations suggest that the inner valence molecular orbital HOMO-6, is the one having an energy comparable with the IE of He, thus being the proper MO to promote the electron exchange process. The experimental findings are in agreement with the energy of the MO calculated (and reported above). Assuming the IE of HOMO-6 (indicated as $5a_1$ in the photoelectron spectra of acetonitrile [49]) as the proper asymptotic level for the exit channels, a sequence of crossings is obtained (see the red dashed lines in Figs. 6) where *non-adiabatic* charge transfer effects are essential to trigger the ion-molecule reaction. It is of relevance to note that the most effective crossing points manifest for θ varying from 0° up to about 90°, while no crossings occur for θ significantly larger than 90°.

4 Conclusions

In this study, a theoretical investigation of the interaction between the $He^{+\cdot}$ radical cation and the CH_3CN molecule has been conducted to offer valuable insights for future analysis of the dynamics involved in the charge exchange process, which is particularly significant in astrochemical models. Reactions involving cosmic rays generated $He^{+\cdot}$ ions are indeed among the primary destruction mechanisms proposed for iCOMs.

The method proposed here involves a phenomenological approach to construct the PESs, in an analytical form, for the entrance and exit channels of the reactions, namely $He^{+\cdot}$ + CH_3CN and He + $CH_3CN^{+\cdot}$. The phenomenological approach identifies the primary interaction components and describes them exploiting empirical or semi-empirical formulas defined by a limited number of parameters, *e.g.* via semi-empirical Improved-Lennard Jones potentials. These potential parameters can be adjusted within a restricted range of variation since they are linked to fundamental chemico-physical properties of the interacting partners, such as electronic polarizabilities (which can be partitioned into atomic and group components), ionization potentials, electron affinity, permanent charges, anisotropic charge distributions determining permanent electric multipoles, energy and shape of the atomic and molecular orbitals involved in charge (electron) transfer effects (CT). The next step of the treatment involves an accurate investigation of the reaction dynamics. In particular, the PESs presented in this work will enable the identification of a series of crossings between entrance and exit channels where the reactive process is triggered by non-adiabatic couplings due to the charge transfer process. Subsequently, the Landau-Zener-Stückelberg approach will be employed to evaluate the crossing probability and, ultimately, the reactive cross sections. Guided ion beam scattering experiments involving $He^{+\cdot}$ - CH_3CN collisions have recently been conducted in the laboratories of the University of Trento, and the analysis of the experimental findings, exploiting the results of this study, is currently underway.

Acknowledgements. This work has been funded by the European Union - NextGenerationEU under the Italian Ministry of University and Research (MUR) National Innovation Ecosystem grant ECS00000041 - VITALITY - CUP J97G22000170005. N. Faginas Lago and M. Rosi also acknowledge financial support under the National Recovery and Resilience Plan (NRRP), Mission 4, Component 2, Investment 1.1, Call for tender No. 104 published on 2.2.2022 by the Italian Ministry of University and Research (MUR), funded by the European Union - NextGenerationEU- Project Title 2022JC2Y93 ChemicalOrigins: linking the fossil composition of the Solar System with the chemistry of protoplanetary disks - CUP J53D23001600006 - Grant Assignment Decree No. 962 adopted on 30.06.2023 by the Italian Ministry of Ministry of University and Research (MUR). D. Ascenzi acknowledges financial support from European Union's Horizon 2020 research and innovation programme under the Marie Sklodowska Curie grant agreement N. 811312 (project 'Astro-Chemical Origins' ACO), from MUR PRIN 2020 project n. 2020AFB3FX 'Astrochemistry beyond the second period elements' and from the National Recovery and Resilience Plan (NRRP), Mission 4, Component 2, Investment 1.1, Call for tender No. 1409 published on 14.9.2022 by the Italian Ministry of University and Research (MUR), funded by the European Union - NextGenerationEU- Project Title P20223H8CK Degradation of space-technology polymers by thermospheric oxygen atoms and ions: an exploration of the reaction mechanisms at an atomistic level - CUP E53D23015560001. Views and opinions expressed are however those of the author(s) only and do not necessarily reflect those of the European Union or European Commission. Neither the European Union nor the granting authority can be held responsible for them.

References

1. Herbst, E., Van Dishoeck, E.F.: Complex organic interstellar molecules. Ann. Rev. Astron. Astrophys. **47**, 427–480 (2009)
2. McGuire, B.A.: 2021 census of interstellar, circumstellar, extragalactic, protoplanetary disk, and exoplanetary molecules. Astrophys. J. Suppl. Ser. **259**(2), 30 (2022)
3. Müller, H.S.P., Thorwirth, S., Roth, D.A., Winnewisser, G.: The cologne database for molecular spectroscopy. CDMS. Astron. Astrophys. **370**, L49–L52 (2001)
4. Müller, H.S.P., Thorwirth, S., Roth, D.A., Winnewisser, G.: The cologne database for molecular spectroscopy, CDMS: Useful tool for astronomers and spectroscopists. J. Mol. Struct. **742**, 215–227 (2005)
5. https://cdms.astro.uni-koeln.de/classic/
6. Wakelam, V., et al.: A kinetic database for astrochemistry (KIDA). Astrophys. J. Suppl. Ser. **199**(1), 21 (2012)
7. Millar, T.J., Walsh, C., de Sande, M.V., Markwick, A.J.: The UMIST database for astrochemistry 2022. Astron. Astrophys. **682**, A109 (2024). https://doi.org/10.1051/0004-6361/202346908
8. McElroy, D., Walsh, C., Markwick, A., Cordiner, M., Smith, K., Millar, T.: The UMIST database for astrochemistry 2012. Astron. Astrophys. **550**, A36 (2013)
9. Cabezas, C., et al.: The millimeter-wave spectrum and astronomical search of succinonitrile and its vibrational excited states. Astron. Astrophys. **629**, A35 (2019)
10. Brack, A.: The Molecular Origins of Life: Assembling Pieces of the Puzzle. Cambridge University Press (1998)
11. Zhao, Q., Garimella, S.S., Savoie, B.M.: Thermally accessible prebiotic pathways for forming ribonucleic acid and protein precursors from aqueous hydrogen cyanide. J. Am. Chem. Soc. **145**(11), 6135–6143 (2023)

12. Meisner, J., Zhu, X., Martínez, T.J.: Computational Discovery of the Origins of Life (2019)
13. Walker, S.I.: Origins of life: a problem for physics. arXiv preprint arXiv:1705.08073 (2017)
14. Solomon, P., Jefferts, K., Penzias, A., Wilson, R.: Detection of millimeter emission lines from interstellar methyl cyanide. Astrophys. J. vol. 168, p. L107 **168**, L107 (1971)
15. Matthews, H.E., Sears, T.J.: Detection of the j= 1-0 transition of CH3CN. Astrophys. J. Part 2-Lett. Editor, vol. 267, Apr. 1, 1983, p. L53–L57. **267**, L53–L57 (1983)
16. Minh, Y., Irvine, W.M., Ohishi, M., Ishikawa, S., Saito, S., Kaifu, N.: Measurement of the methyl cyanide E/A ratio in TMC-1. Astron. Astrophys. (ISSN 0004-6361), vol. 267, no. 1, p. 229–232. **267**, 229–232 (1993)
17. Megías, et al.: The complex organic molecular content in the L1517B starless core. Mon. Not. R. Astron. Soc. **519**(2), 1601–1617 (2023)
18. Vastel, C., Loison, J.C., Wakelam, V., Lefloch, B.: Isocyanogen formation in the cold interstellar medium. Astron. Astrophys. **625**, A91 (2019)
19. Cazaux, S., et al.: The hot core around the low-mass protostar IRAS 16293–2422: scoundrels rule! Astrophys. J. **593**(1), L51 (2003)
20. Bottinelli, S., et al.: Complex molecules in the hot core of the low-mass protostar NGC 1333 IRAS 4A. Astrophys. J. **615**(1), 354 (2004)
21. Calcutt, H., et al.: The ALMA-PILS survey: complex nitriles towards IRAS 16293–2422. Astron. Astrophys. **616**, A90 (2018)
22. Mercimek, S., et al.: Chemical survey of class i protostars with the IRAM-30 m. Astron. Astrophys. **659**, A67 (2022)
23. Bianchi, E., et al.: SOLIS-XV. CH3CN deuteration in the SVS13-A class i hot corino. Astron. Astrophys. **662**, A103 (2022)
24. Öberg, K.I., et al.: The comet-like composition of a protoplanetary disk as revealed by complex cyanides. Nature **520**(7546), 198–201 (2015)
25. Rodgers, S., Charnley, S.: Organic synthesis in the coma of comet hale-bopp? Mon. Not. R. Astron. Soc. **320**(4), L61–L64 (2001)
26. Le Roy, L., et al.: Inventory of the volatiles on comet 67P/churyumov-gerasimenko from rosetta/ROSINA. Astron. Astrophys. **583**, A1 (2015)
27. Biver, N., et al.: Observations of comet C/2020 F3 (NEOWISE) with IRAM telescopes. Astron. Astrophys. **668**, A171 (2022)
28. Pirani, F., Brizi, S., Roncaratti, L.F., Casavecchia, P., Cappelletti, D., Vecchiocattivi, F.: Beyond the lennard-jones model: a simple and accurate potential function probed by high resolution scattering data useful for molecular dynamics simulations. Phys. Chem. Chem. Phys. **10**(36), 5489–5503 (2008)
29. Richardson, V., et al.: Fragmentation of interstellar methanol by collisions with he˙+: an experimental and computational study. Phys. Chem. Chem. Phys. **24**(37), 22437–22452 (2022)
30. Marchione, D., et al.: Unsaturated dinitriles formation routes in extraterrestrial environments: a combined experimental and theoretical investigation of the reaction between cyano radicals and cyanoethene (C2H3CN). J. Phys. Chem. A **126**(22), 3569–3582 (2022)
31. Giani, L., et al.: Revised gas-phase formation network of methyl cyanide: the origin of methyl cyanide and methanol abundance correlation in hot corinos. Mon. Not. R. Astron. Soc. **526**(3), 4535–4556 (2023)

32. Becke, A.D.: A new mixing of Hartree-Fock and local density-functional theories. J. Chem. Phys. **98**(2), 1372–1377 (1993)
33. Stephens, P.J., Devlin, F.J., Chabalowski, C.F., Frisch, M.J.: Ab initio calculation of vibrational absorption and circular dichroism spectra using density functional force fields. J. Phys. Chem. **98**(45), 11623–11627 (1994)
34. Dunning Jr, T.H.: Gaussian basis sets for use in correlated molecular calculations. I. the atoms boron through neon and hydrogen. J. chem. phys. **90**(2), 1007–1023 (1989)
35. Woon, D.E., Dunning Jr, T.H.: Gaussian basis sets for use in correlated molecular calculations. III. the atoms aluminum through argon. J. chem. phys. **98**(2), 1358–1371 (1993)
36. Kendall, R.A., Dunning Jr, T.H., Harrison, R.J.: Electron affinities of the first-row atoms revisited. systematic basis sets and wave functions. J. chem. phys. **96**(9), 6796–6806 (1992)
37. Cernuto, A., Tosi, P., Martini, L.M., Pirani, F., Ascenzi, D.: Experimental investigation of the reaction of helium ions with dimethyl ether: stereodynamics of the dissociative charge exchange process. Phys. Chem. Chem. Phys. **19**(30), 19554–19565 (2017)
38. Cernuto, A., Pirani, F., Martini, L.M., Tosi, P., Ascenzi, D.: The selective role of long-range forces in the stereodynamics of ion-molecule reactions: The He++ methyl formate case from guided-ion-beam experiments. ChemPhysChem **19**(1), 51–59 (2018)
39. Liuti, G., Pirani, F.: Regularities in van der Waals forces: correlation between the potential parameters and polarizability. Chem. Phys. Lett. **122**(3), 245–250 (1985)
40. Pirani, F., Maciel, G.S., Cappelletti, D., Aquilanti, V.: Experimental benchmarks and phenomenology of interatomic forces: open-shell and electronic anisotropy effects. Int. Rev. Phys. Chem. **25**(1–2), 165–199 (2006)
41. Pirani, F., Albertı, M., Castro, A., Teixidor, M.M., Cappelletti, D.: Atom-bond pairwise additive representation for intermolecular potential energy surfaces. Chem. Phys. Lett. **394**(1–3), 37–44 (2004)
42. Johnson, R.D., et al.: NIST Computational Chemistry Comparison and Benchmark Database. http://srdatanist.gov/cccbdb (2006)
43. Vanuzzo, G., et al.: Reaction N(2D) + CH2CCH2 (Allene): an experimental and theoretical investigation and implications for the photochemical models of titan. ACS Earth Space Chem. **6**(10), 2305–2321 (2022)
44. Mancini, L., et al.: The reaction n (2d)+ CH3CCH (methylacetylene): a combined crossed molecular beams and theoretical investigation and implications for the atmosphere of titan. J. Phys. Chem. A **125**(40), 8846–8859 (2021)
45. Pannacci, G., et al.: A combined crossed molecular beam and theorerical study of the O (3 P, 1 D)+ acrylonitrile (CH 2 CHCN) reactions and implications for combustion and extraterrestrial environments. Phys. Chem. Chem. Phys. **25**(30), 20194–20211 (2023)
46. Vanuzzo, G., et al.: The N (2D)+ CH2CHCN (vinyl cyanide) reaction: a combined crossed molecular beam and theoretical study and implications for the atmosphere of titan. J. Phys. Chem. A **126**(36), 6110–6123 (2022)
47. Rosi, M., et al.: Possible scenarios for SiS formation in the interstellar medium: electronic structure calculations of the potential energy surfaces for the reactions of the SiH radical with atomic sulphur and S2. Chem. Phys. Lett. **695**, 87–93 (2018)

48. NIST Chemistry Webbook. NIST Standard Reference Database Number 69, Eds. P.J. Linstrom and W.G. Mallard, National Institute of Standards and Technology, Gaithersburg MD, 20899 https://webbook.nist.gov/chemistry/. March 26 2024
49. Asbrink, L., Von-Niessen, W., Bieri, G.: 30.4 nm He(D) Photoelectron Spectra of Organic Molecules. J. Electron Spectro. Relat. Phenom. **21**, 93–101 (1980)

Data Analysis of Molecular Dications Fragmentation Processes by Double Photoionization with Synchrotron Radiation

Marco Parriani[1(✉)], Andrea Giustini[1], Michele Alagia[2], Robert Richter[3], Stefano Stranges[2,4], Raimund Feifel[5], and Stefano Falcinelli[1]

[1] Department of Civil and Environmental Engineering, University of Perugia, Via G. Duranti 93, 06125 Perugia, Italy
marco.parriani@dottorandi.unipg.it, {andrea.giustini, stefano.falcinelli}@unipg.it

[2] IOM CNR Laboratorio TASC, 34012 Trieste, Italy
alagia@iom.cnr.it, stefano.stranges@uniroma1.it

[3] Sincrotrone Trieste, Area Science Park, 34149 Basovizza, Trieste, Italy
robert.richter@elettra.trieste.it

[4] Department of Chemistry and Drug Technology, University of Rome Sapienza, 00185 Rome, Italy

[5] Department of Physics, University of Gothenburg, Origovägen 6B, 412 58 Gothenburg, Sweden
raimund.feifel@physics.gu.se

Abstract. In this paper is presented a summarizing description of the collection and computational analysis of the relevant data recorded following a double photoionization experiment on molecular species of interest in astrochemistry. In particular, the computational procedure used to analyze experimental data collected in the double photoionization of allene molecules by tunable radiation in the 25.0–45.0 photon energy range of the GASPHASE and CiPo beamlines of the ELETTRA synchrotron facility of Basovizza (Trieste, Italy) is outlined and discussed. Data presented are recorded using the ARPES (Angle Resolved PhotoEmission Spectroscopy) end station, which is a molecular beam apparatus coupled with a PEPIPICO (Photoelectron-Photoion-Photoion Coincidence) technique and TOF (Time-of-Flight) mass spectrometry, in three different beamtimes.

Keywords: photoionization · astrochemistry · dications

1 Introduction

1.1 Double Photoionization Studies

Ionic species are of great interest in astrochemistry and in planetary atmospheres [1–5], and they can be formed in space in various way, such as by chemi-ionization and photoionization [6–10]. In particular, the absorption of UV and EUV (Extreme

O. Gervasi et al. (Eds.): ICCSA 2024 Workshops, LNCS 14818, pp. 55–70, 2024.
https://doi.org/10.1007/978-3-031-65273-8_4

Ultraviolet) photons and cosmic rays can induce multiple ionization with formation of doubly charged molecular species, which are called molecular dications. They were predicted to exist in the ionospheres of Mars, Venus and Titan, and then considerable amounts have been detected in such environments highlighting their relevant role, as for example, in the ion escape process from the atmosphere of Mars and Titan [11–15]. They were also found in interstellar clouds, comet tails and planetary ionospheres. Moreover, dications are involved in flames and plasma chemistry and physics [16]. The importance of coincidence studies of dications produced by the direct ejection of a couple of valence-shell electrons in determining the basic chemical properties of molecules has been pointed out by the pioneering work completed by J.H.D. Eland [17, 18]. In particular, the phenomenon of photoionization refers to the interaction of electromagnetic radiation with matter resulting in the dissociation of that matter into electrically charged particles [19, 20]. Our focus is studying the double photoionization of gas-phase small molecules producing dications that are important species involved in planetary ionospheres, astrochemistry and astrophysics [17–20], since they are optimal models for their simple and isolated structure, which permits to understand such a complex process as photoionization [21], in which several effects can play a major role.

Molecular dications fragmentation. Molecular dications have been subject of study due to their role in upper planetary atmosphere and interstellar medium, and in general for the role played in high energy environments [12–14]. We refer to double photoionization (DPI) as the process in which two electrons are removed simultaneously from an atom or a molecule by the absorption of a single photon, this leading to the formation of a dication (doubly charged atomic or molecular ion). Analysis of the emitted electrons and the ionic fragments produced by the dissociation of these dications makes it possible to extract fundamental information about the process, such as energy thresholds, kinetic energy releases (KERs) and angular distributions.

1.2 Electron-Ion-Ion Coincidence Measurements

A coincidence measurement is a method used to determine whether two or more particles originate from the same event to correlate them, when detecting the ion fragments emerged from the dissociation of multiply charged molecular ions. In a coincidence experiment the product ions and the photoelectrons emitted are detected and eventually correlated, providing a set of information on the energetic and mechanism of the photodissociation process. There are several different techniques developed during the years, based on the fundamental principle of measuring the arrival time of each charged particle [17]. In this paper will be presented an example of angle-resolved photoelectron spectroscopy (ARPES) setup, used by our group in collaboration with the staff of the GASPHASE and CiPo beamlines of the Elettra synchrotron facility, in Trieste, Italy.

2 Experimental Setup

2.1 The Allene Sample

Allene, C_3H_4, is the first member of the cumulene series C_nH_4. It is well known that allene and the other odd cumulenes belong to the D2d symmetry group, and exhibit stereoisomerism (axial chirality) [21]; The ground state electronic structure is of special interest since the HOMO level is doubly degenerate, having two helical shape Mos of opposite chirality [22]. It is thought to be present in the interstellar medium at least as an intermediate [23, 24], but as it lacks a dipole moment it has not yet been detected there directly. The data presented as example are the result of three different beam times performed at the Gas-phase and CiPo beamlines of the Elettra synchrotron facility (a detailed description of the beamline can be found at [25].

2.2 ARPES 3D Ion Imaging End Station

The ARPES end station used for the experimental measurements is an ion imaging spectroscopy apparatus, with a differential pumping system and the possibility to have two interchangeable molecular beam systems, a pyrolitic source and a hypodermic needle; This apparatus allows photoionization studies to be performed using a tunable and high photon flux of a third-generation synchrotron source. Such experimental studies were mainly based on the use of a PEPIPICO (PhotoElectron–PhotoIon–PhotoIon Coincidence) detector coupled with TOF (Time-of-Flight) mass spectrometer specially designed and built to measure the spatial momentum components of product fragments ions in the double photoionization experiments as we have done in the past [16, 23]. This specific endstation has been successfully used before applied to HX (where X = Cl, Br, and I) double photoionization experiments [6, 26, 27]. in the case of the allene, very recently, a detailed experimental and theoretical investigation of double and triple ionization, by single photon absorption, has been carried out [28], providing the energy pattern of dication electronic states, as well as some details on their nuclear dissociation dynamics. Our experimental investigation has been carried out at the GASPHASE beamline of the ELETTRA synchrotron facility of Basovizza (Trieste, Italy), using a synchrotron radiation in the 25.0 – 45.0 eV photon energy range to conduct a detailed study of the nuclear dissociation dynamics of allene dications by single photon valence electron double ionization by means of the 3D-ion imaging spectroscopy. This different multi-coincidence technique allows higher mass resolution and much higher sensitivity than before [28, 29], and allows to investigate metastable dissociation channels where ions differ by a single unit of mass. The investigated process can be schematized as follows below in Eq. (1):

$$C_3H_4 + h\nu \rightarrow (C_3H_4)^{2+} + 2e^- \rightarrow \text{fragment ions} \quad (1)$$

where, $(C_3H_4)^{2+}$ is the intermediate molecular dication of allene, which can lead to different species according to the relative position of the intermolecular potentials describing the fragment ion-ion and neutral doubly charged ion interactions characterizing the system [23]. In general, taking as an example the simple case of an AB^{2+} diatomic molecular dication, three different cases can be distinguished, the case of the thermodynamically stable dications, the case of unstable species, that cannot be detected in gas phase, and the case of molecular dications in a metastable state, which are the result of an avoided crossing of the potential energy curves involved. In this third case, the resulting fragments A^+ and B^+ have in general a high relative kinetic energy release, with a process named Coulomb explosion of the metastable dication, leading to monocations with a translational energy content that range between 1–10 eV. In the case reported of allene, an analysis of metastable traces is shown in the data treatment section. The allene sample used was commercial as in previous experiments [28]; a stainless-steel needle has been used to introduce the sample into the chamber; in the experiment, the synchrotron radiation, with an energy range of 25–45 eV, passes through a monochromator, specifically a 400 lines/mm spherical grating in first order diffraction, with in and out slits adjusted to obtain a photon energy resolution of around 1.5 meV, and through a magnesium filter used to avoid spurious effects due to ionization by photons from higher orders of diffraction; then the radiation crosses the beam of the target molecule at the extraction region; the electrons are then accelerated towards a multi-channel plate (MCP) detector, and used as a trigger for the ions time of flight. Ions are extracted, accelerated towards a time of flight (TOF) spectrometer and mass selected. The photon flux crossing the target

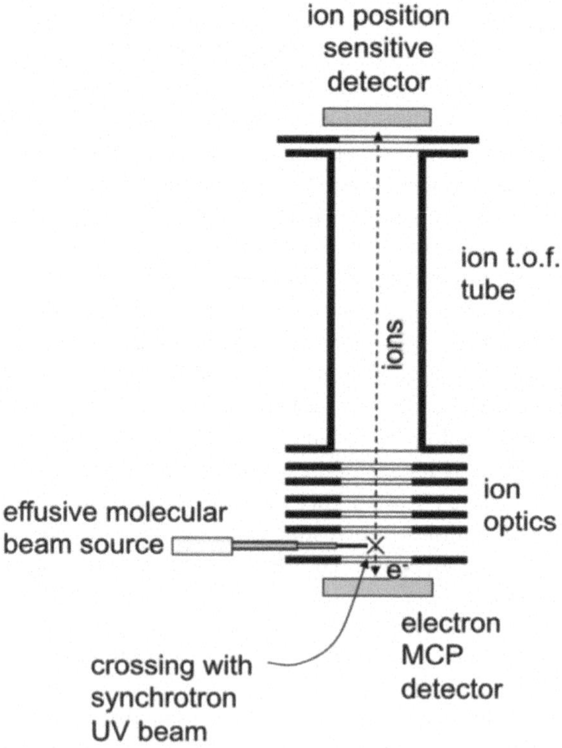

Fig. 1. Scheme of the experimental electron-ion-ion extraction and detection system installed in the ARPES Ion imaging spectrometer (MCP stands for multi-channel plate)

beam and the gas pressure are constantly monitored, thus ion yields are corrected for pressure and photon flux variations after every change in photon energy, moreover the gas inlet source is supplied with a mixture of the target molecule plus He, to allow the normalization of all ion signals at each photon energy to the total ion yield of helium. The ion impacts produce electron clouds collected by a position sensitive detector, which consists of 64 strip anodes (32 vertical and 32 horizontal), individually connected with an independent time-to-digital converter (TDC), allowing to measure several ion fragments and electrons from the same photoionization event. Moreover, a discriminator is connected to the 64 detectors. After an event, the signal is read from the TCD by a Versa Module Eurocard (VME) and sent to the acquisition computer which registers the data; specifically, every x and y of each particle hit is recorded, meaning that for each ion two space and one temporal coordinate are registered. The measured ions are also sorted in groups of single, double, triple,... Coincidences for every start electron [26]. The correlation can then be visualized by plotting a two-dimensional histogram, in which is usually reported the TOF of the first ion versus the TOF of the second ion (Fig. 3).Fig. 2.The coincidence plot recorded in the double photoionization of allene molecule at a photon energy of 30.0 eV (upper panel) and 40.8 eV (lower panel), as an example to appreciate the appearance of coincidence islands over the energy range used.

3 Data Analysis

By the double photoionization of allene using synchrotron radiation in the 25–45 eV photon energy range various fragmentation channel have been detected with the related energy threshold:

$$C_3H_4 + hv \rightarrow H^+ + C_3H^+ + H_2 + 2e^- \quad hv \geq 30.0 \pm 0.1 \text{ eV} \tag{2}$$

$$H_2^+ + C_3H^+ + H + 2e^- \quad hv \geq 30.0 \pm 0.1 \text{ eV} \tag{3}$$

$$H_3^+ + C_3H^+ + 2e^- \quad hv \geq 30.0 \pm 0.1 \text{ eV} \tag{4}$$

$$H^+ + C_3H_2^+ + H + 2e^- \quad hv \geq 35.0 \pm 0.1 \text{ eV} \tag{5}$$

$$H_2^+ + C_3H_2^+ + 2e^- \quad hv \geq 34.0 \pm 0.1 \text{ eV} \tag{6}$$

$$H^+ + C_3H_3^+ + 2e^- \quad hv \geq 27.8 \pm 0.1 \text{ eV} \tag{7}$$

$$CH^+ + C_2H_2^+ + 2e^- \quad hv \geq 37.5 \pm 0.1 \text{ eV} \tag{8}$$

$$CH^+ + C_2H_3^+ + 2e^- \quad hv \geq 33.0 \pm 0.1 \text{ eV} \tag{9}$$

$$CH_2^+ + C_2H^+ + 2e^- \quad hv \geq 38.0 \pm 0.1 \text{ eV} \tag{10}$$

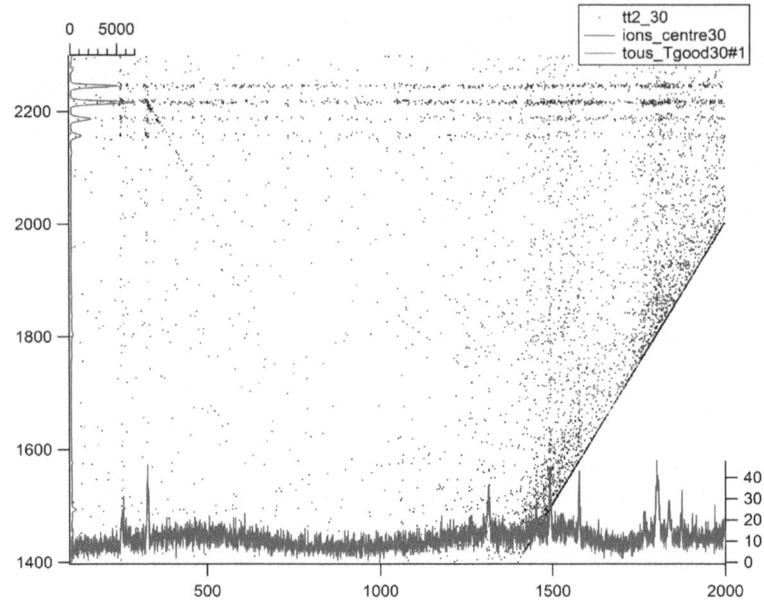

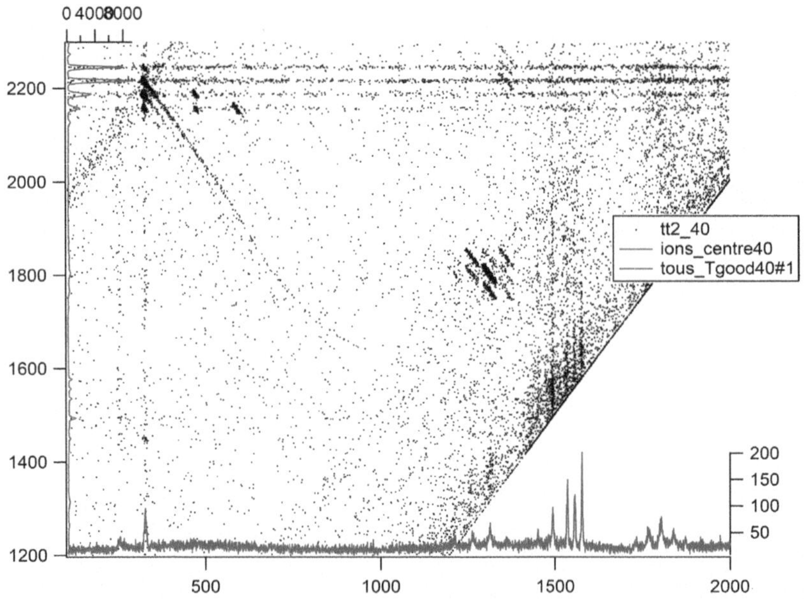

Fig. 2. The coincidence plot recorded in the double photoionization of allene molecule at a photon energy of 30.0 eV (upper panel) and 40.8 eV (lower panel), as an example to appreciate the appearance of coincidence islands over the energy range used.

$$CH_2^+ + C_2H_2^+ + 2e^- \quad h\nu \geq 32.5 \ \pm 0.1 \text{ eV} \tag{11}$$

$$CH_3^+ + C_2H^+ + 2e^- \quad h\nu \geq 33.5 \ \pm 0.1 \text{ eV} \tag{12}$$

The measured threshold energies reported above were obtained by accumulating for a minimum of 6–8 h at each investigated photon energy with a 200 meV photon energy step. Figure 2 reports the two time-of-flight of the corresponding ions and is usually referred as the coincidence map. It is possible to observe the so-called islands in correspondence of the time-of-flight crossing of two correlated ions; each specific dot in these islands represents a coincidence event and contains important information about the dynamics and kinetics of the dissociation process leading to the correlated ions [28]. Specifically, the slope and shape of the islands are related to the dissociation mechanism and energetics of the molecular ion photofragmentation, instead the intensity is directly related to the ionization probability of the molecular species and thus is fundamental to determine the photoionization thresholds and the appearing potentials of the fragments.

3.1 Ionization Thresholds

The threshold energies for each open ionization channel are usually enlightened by measuring the appearance of the so-called islands in the coincidences plot; the two time of flight values t_1 and t_2 define a point, since they are produced by the same photoionization event. By analyzing the number of events in a specific island over the energy range in which the measurements are performed (the case for allene is from 25 eV to 45 eV) one can investigate the appearance of various islands, related to the opening of the corresponding ionization channel, which can be deducted by the two peaks of the mass spectrum that are generating the island. Specifically, the procedure consists in defining the boundaries around the islands and obtaining the number of coincidences inside them; raw data are normalized over the total number of events detected over the time of the measurement, moreover these events need to be cleaned of the background, indeed in every coincidence experiment a random background is present, arising from false coincidences, random events that occur in the case of two not correlated stop signals, or for uncorrelated ion stop and electron start signals, this background usually depends to both electrons and ions counting rates. There are several methods to subtract this background [30], one of the simplest is to use a low counting rate with a long time measurement, but for the case of the detection of the ionization thresholds, in which a high counting rate is often necessary to avoid extremely long data acquisitions, one empirical method is to calculate the background in a relatively empty area of the plot, in which no event is taking place. Using the same island boundaries for the ionization channel detection and for the background calculation allows to empirically obtain the subtraction of false coincidences. Coincidence events are in every case normalized over photon flux to be comparable. It is important to say that in the case of allene, the normalization of raw data has not been done over the total number of events on the coincidence plot, this because a small portion of the spectrum, not overlapped in any case with important peaks, presented a recurring form in every spectrum at every energy; this was due to the fact that the MCP's used were nearly exhausted, so this particular shape was not considered during the data treatment. To have a more fine detection, as in the case of the incipient formation of the molecular dication for allene, the variation in energy for each measurement has been varied on the order of the 200 meV, and the focus switched from the islands to the mass spectrum; this because the difference between the background and the events detected in the island becomes smaller, and it is more accurate to trying to identify the dication appearance by looking at the spectrum. Data collected has been then verified and confirmed by using a Wannier-based model to evaluate the experimental values for the ionization thresholds obtained. The evaluation of the threshold energy for the formation of $(C_3H_4)^{2+}$ allene dication confirmed the value reported in literature, which was 28 ± 0.3 eV. Specifically, we estimated an appearance energy value of 27.8 ± 0.1 eV for the allene dication, whose appearance can be more easily noticed through the observation of various coincidence plots over the energy range of 25.0–45.0 eV (graphic examples below and over the dication energy threshold are reported in Sect. 3.3).

3.2 Kinetic Energy Releases

The measurement of the impact coordinates and the arrival time for each particle investigated allows to calculate its linear momentum in the setup [24]:

$$p_{i,x} = p \sin\theta \cos\varphi = k_1\big(m\big(x_i - x_g\big)/t_i\big) \tag{13}$$

$$p_{i,y} = p \sin\theta \cos\varphi = k_1\big(m\big(y_i - y_g\big)/t_i\big) \tag{14}$$

$$p_{i,z} = p \cos\varphi = -k_2(t_i - t_0) \tag{15}$$

where x_g and y_g are the coordinates of the center of mass of the n-th PEPIPICO (photoelectron-photoion-photoion coincidence) dissociation event (z_g coordinate does not enter the equations if the WileyMacLaren time focusing is employed), m is the ion mass and t_0 is the arrival time for an ion with no initial momentum along the z axis, p is the magnitude of the momentum, θ and ϕ are respectively the polar and azimuth angles with respect to the radiation polarization vector; k_1 and k_2 are geometrical factors. If a complete two (or more) body dissociation is detected, we have enough measurements to solve the system of kinematical equations. In particular, one can calculate the x_g and y_g coordinates by applying the linear momentum conservation law, that is $\sum p_{i,x} = 0$ and $\sum p_{i,y} = 0$. The knowledge of the center of mass coordinates allows to transform the measured x_i, y_i, and t_i coordinates into the linear momenta components following the previous mentioned equations, and then the laboratory angles and energies can be computed. Thus, from the momenta distribution in the center of mass coordinate system it is possible to obtain the Total or Partial kinetic energy released (KER) in the dissociation process.

$$KER = p^2/2m \tag{16}$$

Angular distributions. Moreover, the photofragments angular distribution, with respect to the radiation polarization plane, can be obtained from the knowledge of the ion momenta distribution and can provide valuable information about the dissociation dynamics according to Eq. 17:

$$I(\theta)\sin(\theta) = (\sigma_{tot}/4\pi)[1 + \beta/2(3\cos^2\theta - 1)] \quad (17)$$

in which I(θ) represent the differential cross section, θ is the angle between the velocity vector of the fragment ion and the light polarization vector, σ is the total cross section and β represent the anisotropy parameter; beta can assume different values, β = -1 if there is an emission of product ions along a direction perpendicular to the polarization vector, β = 0 if there is an isotropic distribution of fragment ions, β = 2 if there is an emission of product ions along a direction parallel to the polarization vector. In the case of the data taken as example, two data sets from two different experiments have been collected, the data reported in this paper are from the second data set: the first set refers to data taken with old MCPs in which their depletion caused the initial impossibility to determinate angular distributions, due to the appearance of a hole in the angular distribution of the fragments that could resemble an anisotropic distribution, misleading the analysis due to an incorrect estimation of the anisotropic parameter β. After the substitution of the old MCPs and the installation and conditioning of the new ones, several KERs have been measured over the energy range from 25.0 to 45.0 eV (with examples reported in Fig. 3 and 4), although there are no indications of an anisotropic distribution in the energy range we investigated. This can be highlighted by the estimation of the anisotropy parameter β, which is calculated through an iterative fit using Eq. 17. In our case of study, the best estimation of the parameter at different energies seems to confirm the absence of any anisotropic behavior in the energy range investigated. To examine a case in which an anisotropic distribution has been highlighted, the reader could refer to previous works as the one regarding propylene oxide [31], in which the best fit of experimental data was obtained using the sum of two different Gaussian functions highlighting a bimodality generated by the presence of two different reaction mechanisms. It is important to note that an important amount of the beamtime in which KERs have been measured has been devoted to low gain acquisitions to study the metastable tails originating from several fragmentation channels, for this reason the angular distribution for allene fragmentation could be a future, finer and deeply conducted investigation, to try adding more statistic to the study to further fortify the assumptions.

3.3 Metastable Lifetime

Doubly charged molecular dications are usually unstable species, because of their high formation energy. In the gas phase, the two charged parts of the dication tend to repel each other, making this specie in the majority of cases unstable, determining the fragmentation in two positive ions often with high KER; there are other cases in which the species can be metastable, leading to the formation of a so-called tail on the plot, indicating that the fragmentation reaction occurs in a time longer than the characteristic window time of the apparatus (in our case 50 ns). By analyzing the tail, it is possible to derive the lifetime for the dications leading to the dissociation reaction. In our case of study regarding allene, using a low gain acquisition for a minimum of 48 h, we collected data at an energy of 36.0 eV, energy in which we highlighted the presence of four distinct metastable tails,

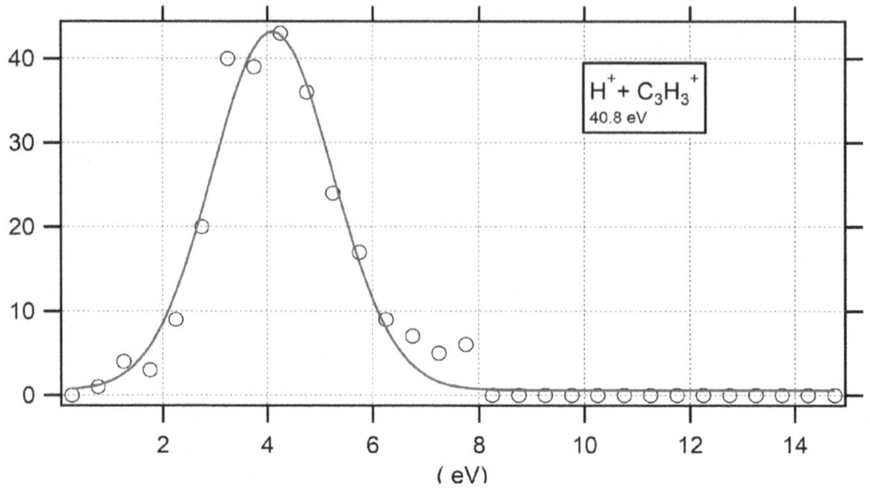

Fig. 3. A portion of the KER distribution for the production of $H^+ + C_3H_3{}^+$ ion pair products formed in the double dissociative photoionization of propylene oxide at a photon energy of 40.8 eV.

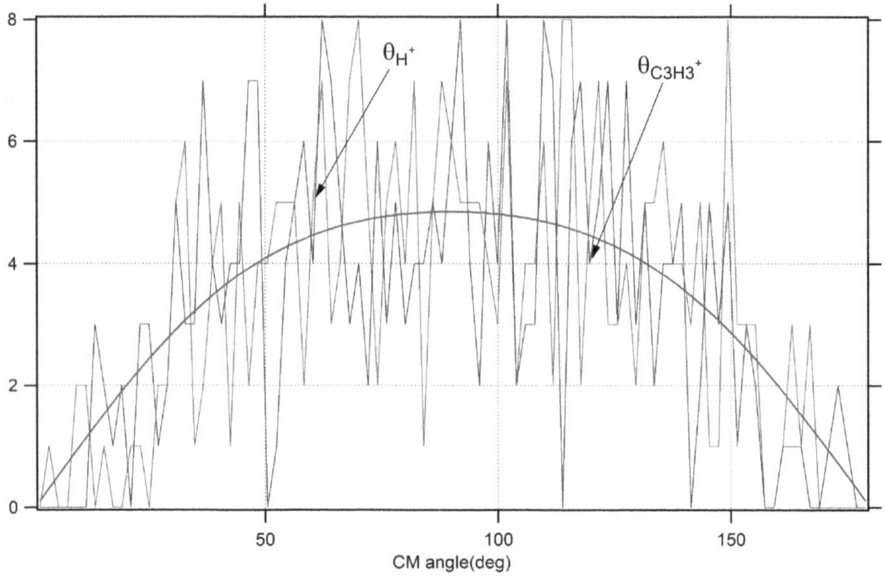

Fig. 4. The angular distribution of $H^+ + C_3H_3{}^+$ product ions in the center of the mass frame recorded at a photon energy of 40.8 eV, with an estimated $\beta = 0.15 \pm 0.13$.

reported in Figs. 5 and 6, originating from the fragmentation channels of Eqs. (2), (4), (5) and (7).

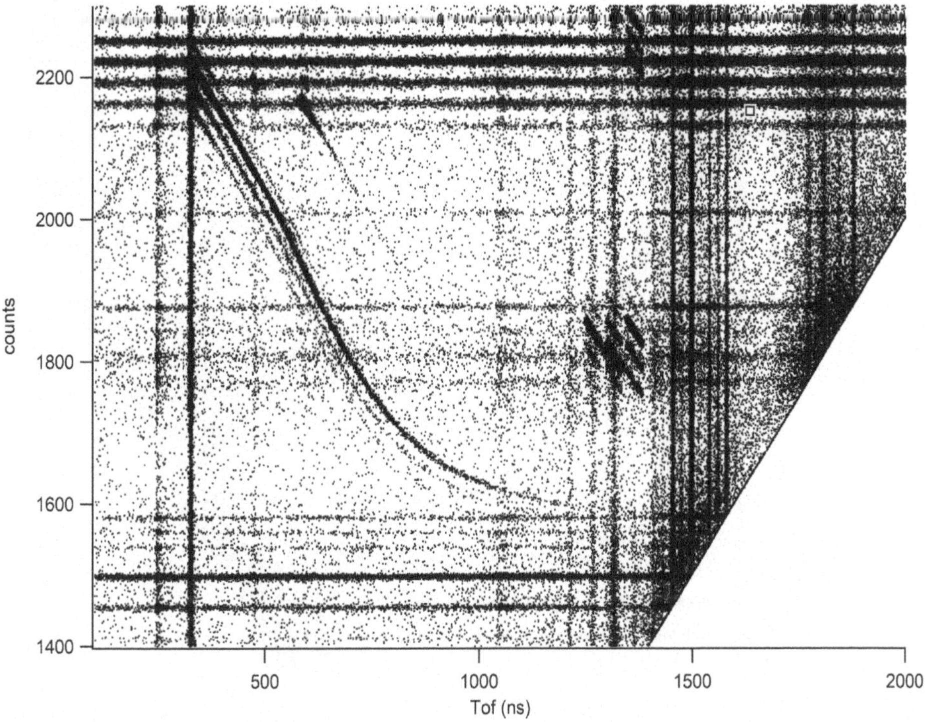

Fig. 5. The Coincidence plot collected with a low gain with a photon energy of 36 eV, to highlight the appearance of the metastable tails.

The plot has been analyzed to obtain the lifetime of the metastable dications during the dissociations. In the first place, a mask containing the trace has been created, to calculate all the events that laid inside the tail, and a background has been subtracted by translating the mask in an empty region of the coincidences plot to subtract the values inside it. To obtain the lifetime of the dications we have plotted the number of coincidence points along the track of the tails, as a function of the $(t_2 - t_1)$ difference and have fitted the data by a Monte Carlo simulation of the ion trajectories in the mass spectrometer. To give an example, the behavior for the $H^+ + C_3H_3^+ + 2e^-$ channel, with a photon energy of 36.0 eV, is reported in Fig. 7, where the dots are the experimental results, and the line is obtained with a simulation with 10^6 trajectories.

This way, the experimental data are compared with the best Monte Carlo fit simulation, and the reduced main square values χ^2 of the difference between experimental data and simulation are plotted as a function of the dication lifetime, the minimum obtained deviation gives the value we are looking for. In our case, a lifetime of 160 ± 20 ns resulted the best value to fit the experimental data; it is important to notice in any case that the Monte Carlo simulation we used is not able to perfectly fit the end of the metastable tail; while this does not seem to have a huge impact for the estimation of the lifetime of the main metastable tail, it could be of fundamental importance for the estimation of the weaker tails, especially for the one regarding the $H_3^+ + C_3H^+$ ion products, which

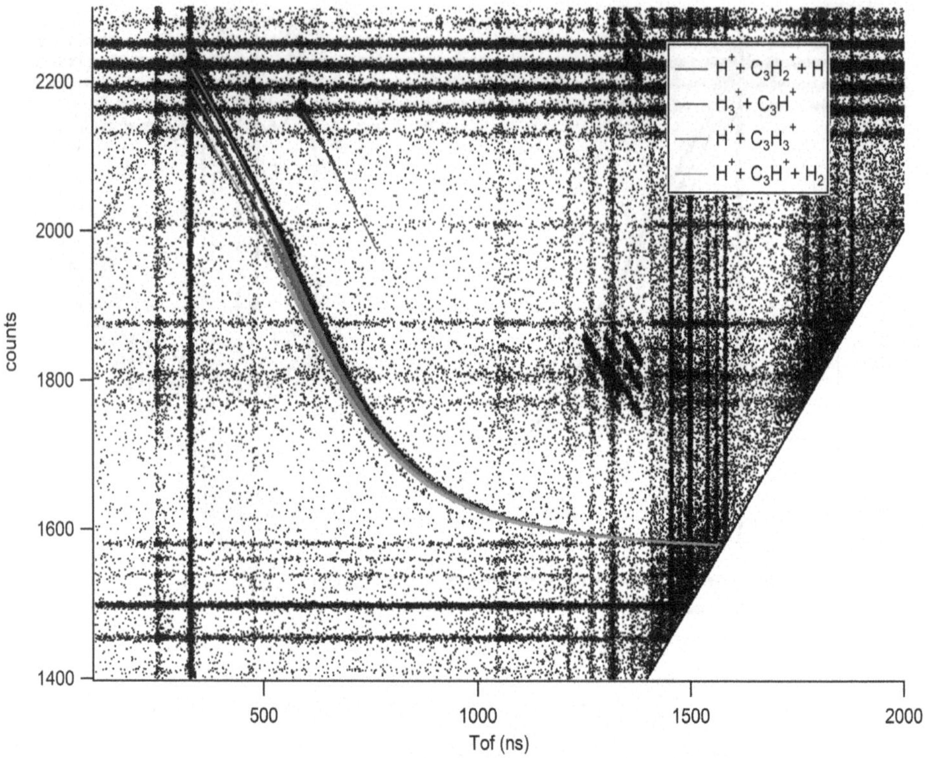

Fig. 6. The Coincidence plots collected with a low gain with a photon energy of 36 eV; the four different metastable tails simulated trajectories are highlighted with different colors.

is the weakest. For this reason, further studies to improve the simulation are currently ongoing to have a more precise estimation of the final part of the metastable tails. With the definition of the masks containing the coincidence events for the metastable tails, a scan from 25 to 45 eV has also been done to estimate the energy threshold formation for every metastable tail detected. In Fig. 8 the reader can appreciate the energy formation thresholds for the three main metastable tails (it is important to stretch that the metastable tail for the $H^+ + C_3H^+$ is very weak and thus it results difficult to determine with confidence the exact energy in which it appears, the same counts for the $H_3{}^+ + C_3H^+$ tail, which is also weaker and not reported below).

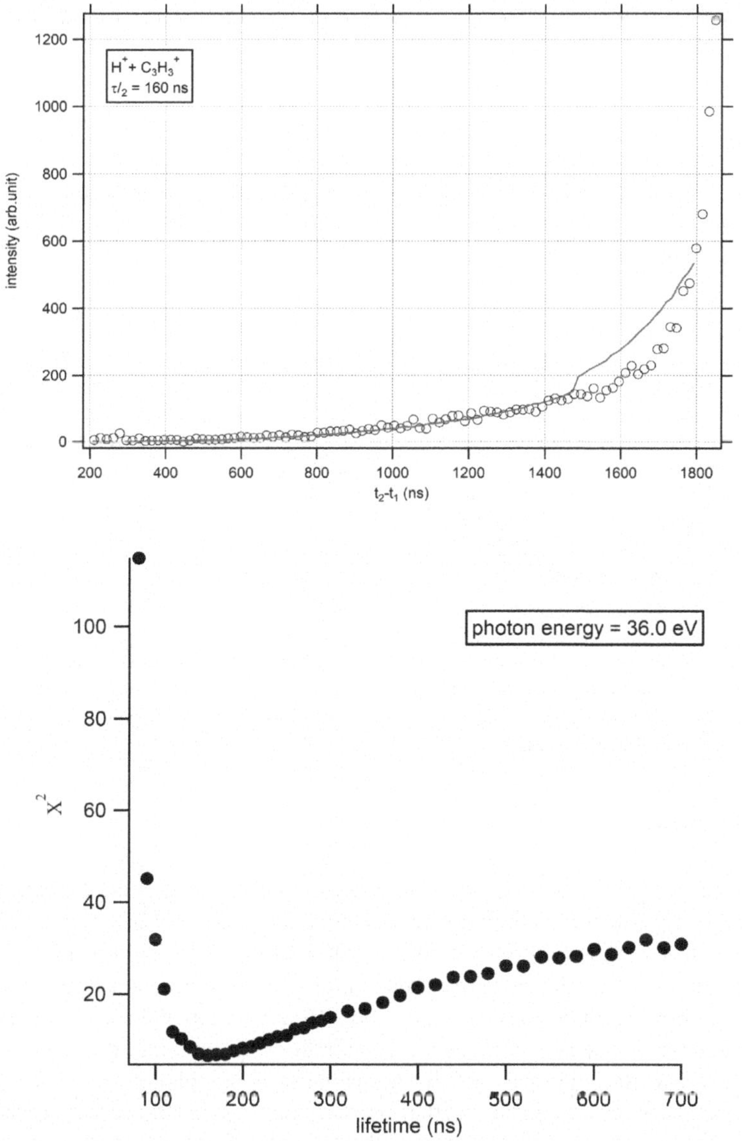

Fig. 7. Analysis of the main metastable tail in the coincidence plot of Fig. 5, in order to derive the lifetime of the dissociation reaction leading to $H^+ + C_3H_3^+$. The upper panel shows the intensity of coincidences along the tail track, as a function of the $t_2 - t_1$ difference (open circles), with the fit (continuous line) by a Monte Carlo trajectory calculation. The lower panel shows the reduced mean square deviation, χ^2, between the simulation and the experimental results, as a function of the lifetime used in the simulation.

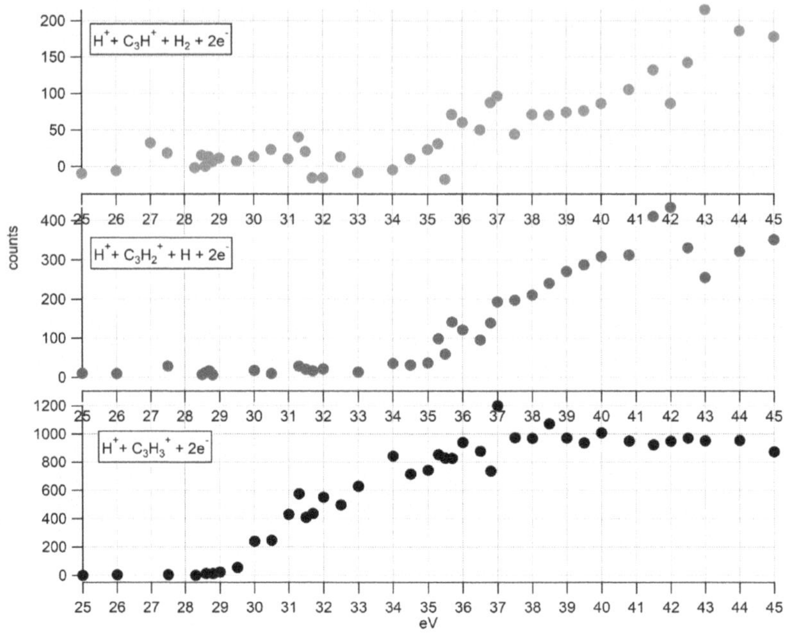

Fig. 8. The energy threshold formation for three of the four metastable tails highlighted, it is quite interesting that the tail referred to the $H^+ + C_3H^+$ ion products seems to start before the $H^+ + C_3H_2^+$ one; this cannot be said with a high level of confidence though, due to the high variability of the data and the weakness of the tail.

4 Conclusions

The experimental procedures we reported describe our apparatus and the data treatment we apply when performing a double photoionization experiment using a synchrotron light source; the example reported summarizes the different steps of the analysis, including the highlights of four different metastable tails. The experimental setup reported permits the measurement of cross sections for the different fragmentation channels together with kinetic energy releases and energy distributions, parameters that can be of great use to understand the chemistry of the ionosphere of our planet and other important bodies like Mars, Venus, or Titan. Further improvements to the data analysis could come from the use of a finer computational method for the analysis of the metastable tail, together with the collection of more statistic to allow a finer calculation of the energy distributions.

Acknowledgments. This work was supported and financed with the "Fondo Ricerca di Base, 2018, dell'Università degli Studi di Perugia" (Project Titled: Indagini teoriche e sperimentali sulla reattività di sistemi di interesse astrochimico). Support from Italian MIUR and University of Perugia (Italy) is acknowledged within the program "Dipartimenti di Eccellenza 2018–2022". The scientific staff of the GASPHASE beamline of ELETTRA Synchrotron Facility of Basovizza (Trieste, Italy) is warmly acknowledged.

References

1. Alagia, M., Balucani, N., Candori, P., Falcinelli, S., et al.: Production of ions at high energy and its role in extraterrestrial environments. Rend. Lincei Sci. Fisiche e Nat. **24**, 53–65 (2013)
2. Pei, L., Carrascosa, E., Yang, N., Falcinelli, S., Farrar, J.M.: Velocity map imaging study of charge-transfer and proton-transfer reactions of CH_3 radicals with $H_3{}^+$. J. Phys. Chem. Lett. **6**(9), 1684–1689 (2015)
3. Candori, P., Falcinelli, S., Pirani, F., Tarantelli, F., Vecchiocattivi, F.: Interaction components in the hydrogen halide dications. Chem. Phys. Lett. **436**(4–6), 322–326 (2007)
4. Stäuber, P., Doty, S.D., Van Dishoeck, E.F., Benz, A.O.: X-ray chemistry in the envelopes around young stellar objects. Astron. Astrophys. **440**(3), 949–966 (2005)
5. Thissen, R., Witasse, O., Dutuit, O., Wedlund, C.S., Gronoff, G., Lilensten, J.: Doubly-charged ions in the planetary ionospheres: a review. Phys. Chem. Chem. Phys. **13**(41), 18264–18287 (2011)
6. Biondini, F., Brunetti, B.G., Candori, P., De Angelis, F., et al.: Penning ionization of N_2O molecules by $He^*(2^{3,1}S)$ and $Ne^*(^3P_{2,0})$ metastable atoms: Theoretical considerations about the intermolecular interactions. J. Chem. Phys. **122**(16), 164308 (2005)
7. Biondini, F., Brunetti, B.G., Candori, P., De Angelis, F., et al.: Penning ionization of N_2O molecules by $He^*(2^{3,1}S)$ and $Ne^*(^3P_{2,0})$ metastable atoms: crossed beam study. J. Chem. Phys. **122**(16), 164307 (2005)
8. Brunetti, B., Candori, P., De Andres, J., Pirani, F., et al.: Dissociative ionization of methyl chloride and methyl bromide by collision with metastable neon atoms. J. Phys. Chem. A **101**(41), 7505–7512 (1997)
9. Bartocci, A., Belpassi, L., Cappelletti, D., Falcinelli, S., et al.: Catching the role of anisotropic electronic distribution and charge transfer in halogen bonded complexes of noble gases. J. Chem. Phys. **142**(18), 184304 (2015)
10. Brunetti, B.G., Candori, P., Falcinelli, S., Pirani, F., Vecchiocattivi, F.: The stereodynamics of the Penning ionization of water by metastable neon atoms. J. Chem. Phys. **139**(16), 164305 (2013)
11. Gronoff, G., et al.: Modelling dications in the diurnal ionosphere of Venus. Astron. Astrophys. **465**(2), 641–645 (2007)
12. Falcinelli, S., et al.: Kinetic energy release in molecular dications fragmentation after VUV and EUV ionization and escape from planetary atmospheres. Plan. Space Sci. **99**, 149–157 (2014)
13. Falcinelli, S., et al.: The escape of O^+ ions from the atmosphere: an explanation of the observed ion density profiles on Mars. Chem. Phys. Lett. **666**, 1–6 (2016)
14. Falcinelli, S., Pirani, F., Vecchiocattivi, F.: The possible role of penning ionization processes in planetary atmospheres. Atmosphere **6**(3), 299–317 (2015)
15. Falcinelli, S.: Mass spectrometric detection of alkaline earth monohalide dications. Mol. Phys. **88**(3), 663–672 (1996)
16. Falcinelli, S., Rosi, M.: Production and characterization of molecular dications: experimental and theoretical efforts. Molecules **25**, 4157 (2020)
17. Eland, J.H.D.: The dynamics of three-body dissociations of dications studied by the triple coincidence technique PEPIPICO. Mol. Phys. **61**(3), 725–745 (1987)
18. Price, S.D., Eland, J.H.D., Fournier, P.G., Fournier, J., Milli´e, P.: Electronic states and decay mechanisms of the N_2O^{2+} dication. J. Chem. Phys. **88**, 1511–1515 (1988)
19. Houfner, S.: Photoelectron Spectroscopy: principles and applications, 3rd edn. Springer-Verlag, Berlin Heidelberg (2013)
20. Oswald, S.: X-ray Photoelectron Spectroscopy in Analysis of Surfaces, Encyclopedia of Analytical Chemistry (2006)

21. Hendon, C.H., Tiana, D., Murray, A.T., Carbery, D.R., Walsh, A.: Helical frontier orbitals of conjugated linear molecules. Chem. Sci. **4**(11), 4278–4284 (2013)
22. Garner, M.H., Hoffmann, R., Rettrup, S., Solomon, G.C.: Coarctate and mobius: the helical orbitals of allene and other cumulenes. ACS Central Sci. **4**(6), 688–700 (2018)
23. Falcinelli, S., et al.: Double photoionization of propylene oxide: a coincidence study of the ejection of a pair of valence-shell electrons. J. Chem. Phys. **148**(11), 114302 (2018)
24. Kaiser, R.I., Stranges, D., Lee, Y.T., Suits, A.G.: Neutral-neutral reactions in the interstellar medium. I. Formation of carbon hydride radicals via reaction of carbon atoms with unsaturated hydrocarbons. Astrophys. J. **477**(2), 982 (1997)
25. Derossi, A., Lama, F., Piacentini, M., Prosperi, T., Zema, N.: High flux and high-resolution beamline for elliptically polarized radiation in the vacuum ultraviolet and soft X-ray regions. Rev. Sci. Instrum. **66**, 1718–1720 (1995)
26. Lavollée, M.: A new detector for measuring three-dimensional momenta of charged particles in coincidence. Rev. Sci. Instrum. **70**(7), 2968–2974 (1999)
27. Alagia, M., et al.: The double photoionization of HCl: an ion–electron coincidence study. J. Chem. Phys. **121**(21), 10508–10512 (2004)
28. Ideböhn, V., Sterling, A.J., Wallner, M., et al.: Phys. Chem. Chem. Phys. **24**, 786–796 (2022)
29. Eland, J.H.D.: Dynamics of fragmentation reactions from peak shapes in multiparticle coincidence experiments. Laser Chem. **11**, 259 (1991)
30. Frasinski, L.J., Stankiewicz, M., Haterly, P.A., Codling, K.: Removal of false coincidences in triple coincidence experiments employing continuous excitation. Meas. Sci. Technol. **3**, 1188 (1992)
31. Falcinelli, S., et al.: Angular distribution of ion products in the double photoionization of propylene oxide. Front. Chem. **7**, 621 (2019)

A Computational Strategy for the Theoretical Investigation of the Reactions Between Atomic Oxygen and Aromatic Compounds

Marzio Rosi[1,2(✉)], Nadia Balucani[3], Piergiorgio Casavecchia[2,3], Luca Mancini[3], Giacomo Pannacci[3], and Gianmarco Vanuzzo[3]

[1] Dipartimento di Ingegneria Civile ed Ambientale, Università of Perugia, 06125 Perugia, Italy
marzio.rosi@unipg.it

[2] Istituto CNR di Scienze e Tecnologie Chimiche "Giulio Natta" (CNR-SCITEC), 06123 Perugia, Italy

[3] Dipartimento di Chimica, Biologia e Biotecnologie, Università of di Perugia, 06123 Perugia, Italy

{nadia.balucani,piergiorgio.casavecchia,luca.mancini, giacomo.pannacci,gianmarco.vanuzzo}@unipg.it

Abstract. The theoretical investigation of large aromatic compounds interacting with atomic oxygen is a computationally very challenging problem for the requirements in terms of computer time and scratch disk space. The use of very accurate methods like the CCSD(T) with very extended basis sets is unfeasible. It is necessary to use less expensive methods. In this article we consider a computational strategy able to provide reasonably accurate results with affordable computing costs.

Keywords: Computational chemistry · Density Functional · Theory methods · Coupled Clusted calculations · Complete basis set limit · Potential energy surface

1 Introduction

Aromatic compounds are very common in a plethora of gaseous environments, ranging from combustion systems to planetary atmospheres and the interstellar medium. Their reactions with atomic oxygen can bring to the degradation of the aromatic ring, as already observed in the first detailed studies of the O(^{3}P) reaction with benzene, pyridine and toluene by some of the present authors [1–4]. More specifically, we took advantage of the versatility of the crossed molecular beam technique with mass spectrometric detection and time-of-flight analysis empowered by the soft ionization approach to measure the experimental branching fractions of the products and infer their reaction mechanism. The interpretation of the experimental results was assisted by electronic structure calculations of the relevant potential energy surface (PES) and kinetics calculations. The results have been surprising and interesting at the same time. Briefly, notable differences have been observed when comparing the reactions with benzene and pyridine

O. Gervasi et al. (Eds.): ICCSA 2024 Workshops, LNCS 14818, pp. 71–83, 2024.
https://doi.org/10.1007/978-3-031-65273-8_5

(even though they are isoelectronic species for which a similar reaction mechanism was expected) or when comparing the reactions of benzene and toluene (even though the methyl substitution was not expected to affect the reaction mechanism). In particular, the ring contraction channel accounts for ca. 30% of the global yield in the case of the reaction with benzene, 98% in the case of the reaction with pyridine and ca. 10% in the case of the reaction with toluene. In other words, nitrogen in the aromatic ring makes the ring structure more "fragile", while the presence of the methyl group "protects" the ring. The comparison of the experimental branching fractions with the predictions of state-of-the-art quantum calculations proved that the observed differences are due to the different extent of intersystem crossing (ISC) to the underlying singlet PES that, in turn, depends on the coupling terms between the triplet and singlet PESs. Therefore, only an experimental or theoretical characterization of the system will allow us to establish the reaction mechanism and the extent of the ring-contraction channel for any reaction of atomic oxygen with other aromatic compounds.

In our laboratory we are now pursuing a systematic investigation of the reactions of atomic oxygen with other aromatics. In particular, we are interested in characterizing the reaction mechanism for functionalized aromatic compounds, like phenol, anisole, aniline, styrene, etc. to clarify whether the substituent groups are protecting or making more fragile the aromatic ring. We are also interested in the reactions of atomic oxygen with polycyclic aromatic compounds. This knowledge will help to address several issues in applied fields, like for instance the resistance of space technology aromatic polymers or the oxidation of organic compounds in the atmosphere. As already seen in the case of benzene, pyridine and toluene, it is imperative to accompany the experimental work with theoretical predictions. However, theoretical methods become very computationally expensive when the size of the investigated systems increases and their application becomes soon unfeasible. For large systems it is necessary to use less expensive methods. It is necessary, therefore, to check the accuracy of the results and to find a reasonable compromise between the accuracy of the results and the computational cost of the method. The extent of ISC can instead be evaluated by comparing the predictions of the branching fractions in the adiabatic singlet and triplet PESs with the experimental values, as already done in the case of the reactions $O(^3P)$ + HCCCN [5] and $O(^3P)$ + C_2H_3CN [6].

In this contribution we will present a computational strategy to study large systems as those relative to the interaction of atomic oxygen with polycyclic aromatic hydrocarbons or functionalized aromatics and, as examples, we will present the study of the interaction of $O(^3P)$ with toluene and ethylbenzene.

2 Computational Details

The optimization of the stationary points of the Potential Energy Surface (PES) of $O(^3P)$ interacting with toluene and ethylbenzene has been performed at Density Functional Theory (DFT) level of calculation using the ωB97X-D [7, 8] functional, and the 6–311 + G(d,p) basis set [9, 10]. In order to assign the saddle points to the relative reactants and products we performed intrinsic reaction coordinate (IRC) calculations [11, 12]. The energy of all the stationary points of the $O(^3P)$ + toluene system, as well as those relative to the *ipso* approach of the $O(^3P)$ + ethylbenzene system was computed also at the CCSD(T) level [13–15] using the correlation consistent aug-cc-pVTZ basis set [16] and the ωB97X-D/6–311 + G (d, p) optimized geometries. A similar computational scheme has been previously successfully applied [17–30].

For the stationary points of the $O(^3P)$ + toluene PES, more accurate calculations were performed at the CCSD(T) level corrected with a Density Fitted (DF) MP2 extrapolation to the complete basis set (CBS) and with corrections for core electrons excitations. In particular, we compute the energies as:

$E_{CCSD(T)/CBS}$=E(CCSD(T)/aug-cc-pVTZ)+[E(CCSD(T,core)/cc-pVTZ)-E(CCSD(T)/cc-pVTZ)]+[E(DF-MP2/CBS)-E(DF-MP2/aug-cc-pVTZ)]

where E(DF-MP2/CBS) is defined as:

E(DF-MP2/CBS) = E[(DF-MP2)/aug-cc-pVQZ)] + 0.5772*[E(DF-MP2/aug-cc-pVQZ)-E(DF-MP2/aug-cc-pVTZ)]

The E(DF-MP2/CBS) extrapolation was performed using Martin's two parameter scheme [31]. These energies will be denoted as CCSD(T)/CBS although they include also the core-valence correlation correction. The accuracy of these calculations should be within ± 5 kJ/mol.

For simplicity, in the following discussion the ωB97X-D/6–311 + G(d,p) calculations will be denoted DFT, the CCSD(T)/aug-cc-pVTZ CC and the CCSD(T)/CBS CBS.

All the energies were corrected to 0 K by adding the zero point energy correction computed using the frequencies evaluated at DFT level.

The DFT and CC calculations were done with Gaussian 09 [32] and the CBS calculations with Molpro [33, 34], while the analysis of the vibrational frequencies was performed using Molekel [35, 36].

3 Results and Discussion

3.1 $O(^3P)$ + toluene

The study of the PES of the interaction of atomic oxygen in its ground state (^3P) and toluene has been performed at the DFT level. The choice of the functional is due to the fact that it allows a good description of weak electrostatic adducts, due to the presence of the dispersion term. Moreover, the ωB97X-D functional provides optimized geometries comparable to those obtained with other functionals, like the B3LYP one which describes very well bonding interactions [37]. The whole PES of $O(^3P)$ + toluene has been reported elsewhere [4]. As can be seen from Fig. 1, where we have reported the PES, the approach of $O(^3P)$ to toluene gives rise to a weakly bound van der Waals adduct. The attack of oxygen to the aromatic ring gives rise to four different isomers, since the oxygen atom can approach the ring in *ipso*, *ortho*, *meta* and *para* sites. In the upper panel of Fig. 1, we have reported the reaction pathway arising from the *ipso* attack of oxygen, while in the lower panel the reaction pathways deriving from the *ortho*, *meta* and *para* approaches of oxygen are reported. The *ipso* approach gives rise to the weakly bound **W1^T** van der Waals adduct, which evolves to the more stable **W2^T** minimum, through the saddle point **TS1^T**. **W2^T** then dissociates to the products **PR1^T**, *i.e.* phenoxy and methyl, overcoming

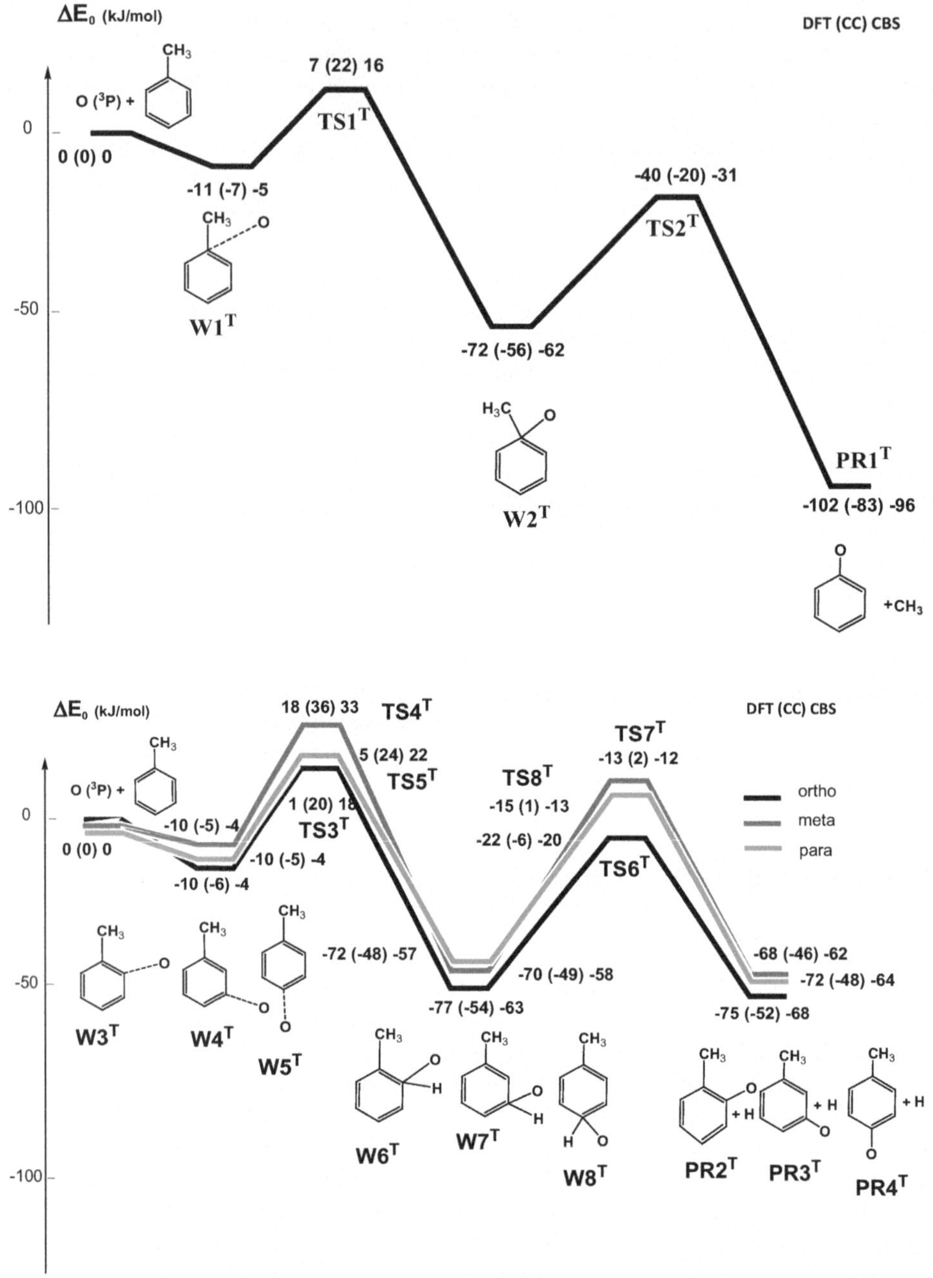

Fig. 1. PES for the O(^{3}P) + toluene reaction. Upper panel refers to the *ipso* attack, while lower panel refers to the *ortho, meta* and *para* attacks. Relative energies (kJ/mol) respect to O(^{3}P) + toluene, computed DFT (in black), CC (in red) and CBS (in blue) level. Zero point energy correction always included.

the saddle point $TS2^T$. The optimized geometries of the minima are shown in Fig. 2, while the optimized structures of the transition states are reported in Fig. 3. From Fig. 2 we can see that the O-C(*ipso*) distance is very long (2.60 Å) corresponding to a weak electrostatic interaction. This distance becomes 1.61 Å in the saddle point and 1.38 Å in $W2^T$ suggesting the presence of a chemical bond between C(*ipso*) and O. A similar scenario happens when the oxygen atom approaches the aromatic ring in *ortho*, *meta* or *para* sites (see lower panel of Fig. 1). First of all, we have the formation of a van der Waals adduct which evolves to a stable minimun where a C-O bond is formed. The approach to the *ortho* site is slightly favored with respect to the *para* and *meta* sites, respectively.

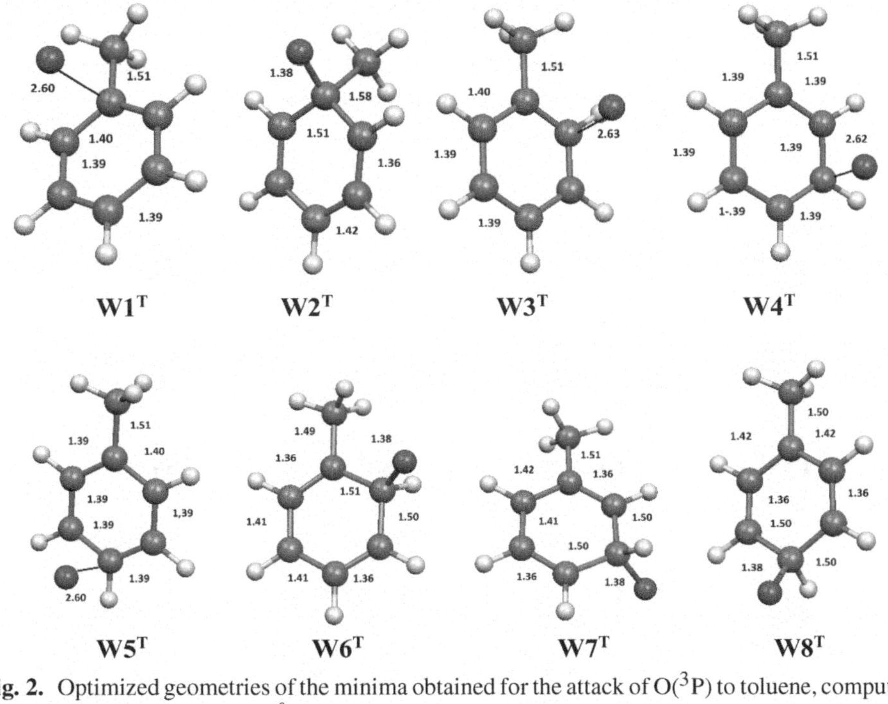

Fig. 2. Optimized geometries of the minima obtained for the attack of $O(^3P)$ to toluene, computed at DFT level. Bond lengths in Å, angles in °. The superscript T means toluene.

In order to check the accuracy of the computed energies, we evaluate the relative energies of the stationary points at the DFT level, and then, using the optimized structures, at the CC and at the CBS levels. The relative energies (kJ/mol) with respect to the energy of $O(^3P)$ + toluene are reported in Fig. 1; the first value is the DFT relative energy, the second is the CC value, while the third is the CBS value. The ΔE_0 and barrier heights (kJ/mol) for all the reaction steps are reported in Table 1. First of all, we can notice that most of the times the CC energies differ significantly with respect to the most accurate CBS energies. In the CBS method we have included with respect to the CC method a core-valence correlation correction and a correction to extrapolate to the complete basis set limit. Since in our systems we have only first and second period elements, the core-valence correlation correction should be almost negligible, the difference between CC and CBS should be ascribed almost entirely to basis set superposition errors. This is a bit surprising since the aug-cc-pVTZ basis set is a relatively extended basis set being of triple-zeta quality with polarization and diffuse functions. However, it is well known that the couple cluster method shows a slow convergence on basis set and in order to obtain accurate results it is necessary to use extended basis set. The values reported in Fig. 1 and Table 1 suggest that the aug-cc-pVTZ basis set is far from convergence on basis set limit. The CC results, although computationally very expensive, therefore, show a not negligible basis set superposition error, The DFT results show surprisingly most of the time a better agreement with the CBS results as we can see form Fig. 1 and Table 1. This result could be due in part to the fact that the DFT methods converge rapidly on basis set and even relatively small basis sets provide converged results with respect to basis set. The basis set superposition error therefore should be lower with DFT methods. The ωB97X-D moreover should consider reasonably well also long range interactions due to the presence of the dispersion term. Therefore the DFT approach seems to be able to provide reasonable results at a very low computational cost.

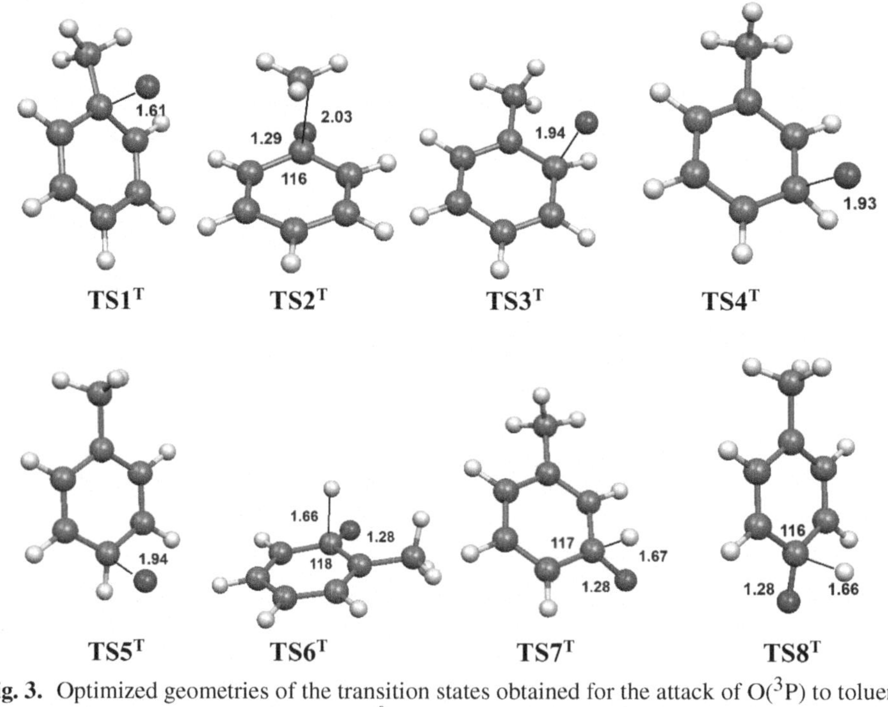

Fig. 3. Optimized geometries of the transition states obtained for the attack of O(^{3}P) to toluene, computed at DFT level. Bond lengths in Å, angles in °. The superscript T means toluene.

3.2 O(^{3}P) + ethylbenzene

The study of the PES of the interaction of atomic oxygen in its ground state (^{3}P) and ethylbenzene has been performed at the DFT level. As can be seen from Fig. 4, where we have reported the PES, the approach of O(^{3}P) to ethylbenzene gives rise to a weakly bound van der Waals adduct. The attack of oxygen to the aromatic ring gives rise to four

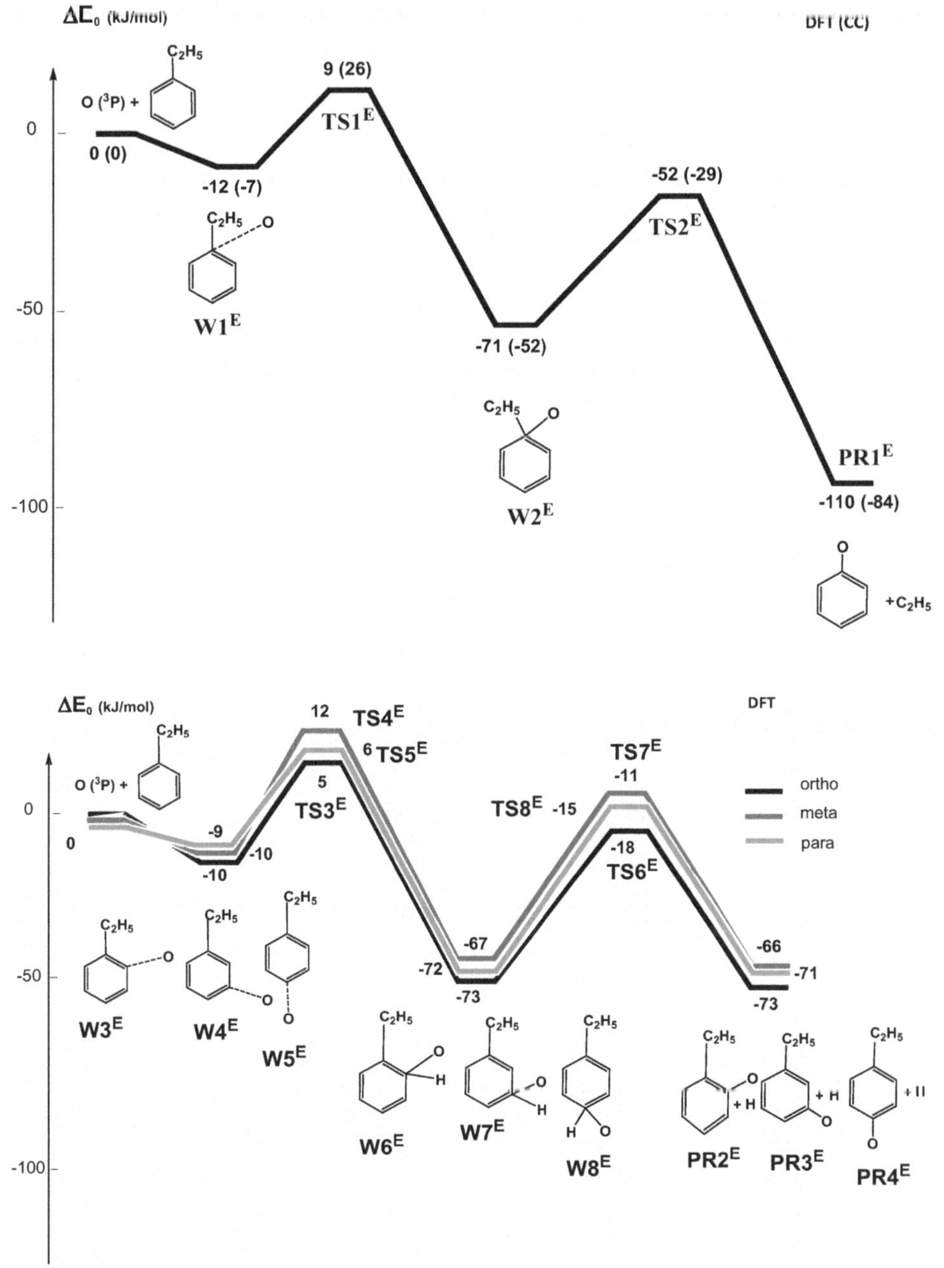

Fig. 4. PES for the O(^{3}P) + ethylbenzene reaction. Upper panel refers to the *ipso* attack, while lower panel refers to the *ortho*, *meta* and *para* attacks. Relative energies (kJ/mol) respect to O(^{3}P) + ethylbenzene, computed at DFT (in black) and CC (in red) level. Zero point energy correction always included.

different isomers, since the oxygen atom can approach the ring in *ipso*, *ortho*, *meta* and *para* sites. In the upper panel of Fig. 4, we have reported the reaction pathway arising from the *ipso* attack of oxygen, while in the lower panel the reaction pathways deriving from the *ortho*, *meta* and *para* approaches of oxygen. The *ipso* approach gives rise to the weakly bound **W1**$^{\mathbf{E}}$ van der Waals adduct, which evolves to the more stable **W2**$^{\mathbf{E}}$ minimum, through the saddle point **TS1**$^{\mathbf{E}}$. **W2**$^{\mathbf{E}}$ then dissociates to the products **PR1**$^{\mathbf{E}}$, *i.e.* phenoxy and ethyl, overcoming the saddle point **TS2**$^{\mathbf{E}}$. The optimized geometries of the minima are shown in Fig. 5, while the optimized structure of the transition states are reported in Fig. 6. From Fig. 5 we can see that the O-C(*ipso*) distance is very long (2.69 Å) corresponding to a weak electrostatic interaction. This distance becomes 1.91 Å in the saddle point and 1.37 Å in **W2**$^{\mathbf{E}}$ suggesting the presence of a chemical bond between C(*ipso*) and O. A similar scenario happens when the oxygen atom approaches the aromatic ring in *ortho*, *meta* or *para* sites (see lower panel of Fig. 4). First of all, we have the formation of a van der Waals adduct which evolves to a stable minimun where a C-O bond is formed. The approach to the *ortho* site is slightly preferred with respect to the *para* and *meta* sites, respectively.

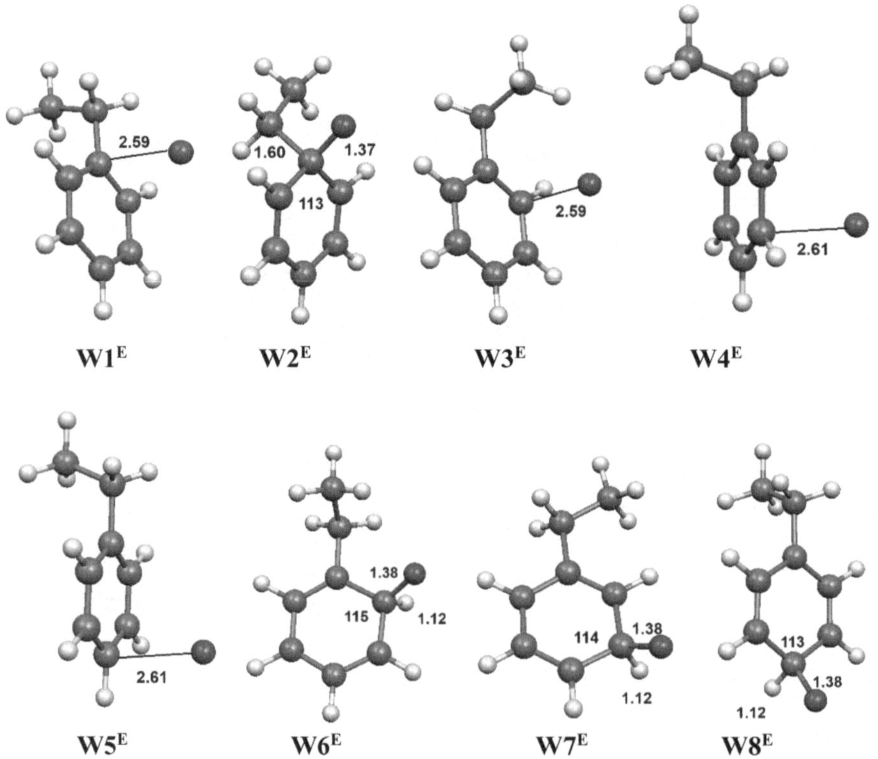

Fig. 5. Optimized geometries of the minima obtained for the attack of O(^{3}P) to ethylbenzene, computed at DFT level. Bond lengths in Å, angles in °. The superscript E means ethylbenzene.

For the reaction pathway relative to the *ipso* approach of the oxygen atom to the ethylbenzene, we have computed the relative energies both at DFT and CC level. The CC calculations were very demanding in terms of computer resources. This effort, however, is probably useless since, due to basis set incompleteness, the CC results are probably less reliable than the DFT values. For this reason for the systems involved in the *ortho*, *meta* and *para* interactions we performed only DFT calculations.

A comparison between the results reported in Fig. 1 and in Fig. 4 suggests that O(^{3}P) + toluene and O(^{3}P) + ethylbenzene systems show a similar behavior as expected for aromatic systems which differ only for the substituent. This comparison however suggests that the DFT method should be adequate also for other aromatic systems interacting with oxygen atoms and, due to its low computational cost, should be feasible also for large polycyclic aromatic compounds.

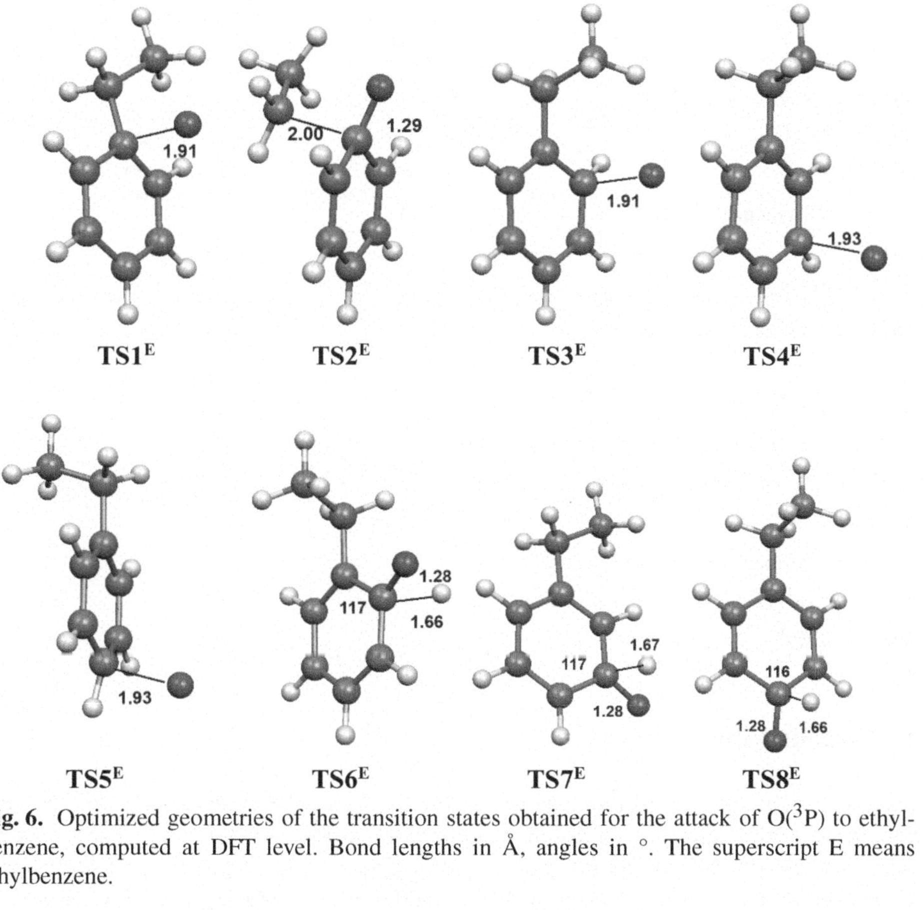

Fig. 6. Optimized geometries of the transition states obtained for the attack of O(^{3}P) to ethylbenzene, computed at DFT level. Bond lengths in Å, angles in °. The superscript E means ethylbenzene.

Table 1. DE_0 and barrier heights (kJ/mol, 0 K) computed at DFT (CC) CBS level for the reaction of O(^{3}P) with toluene or ethylbenzene. All the energies include zero point energy correction. For the labels of the isomers see Figs. 1 and 4.

	$\Delta E^0{}_0$	Barrier height
O(^{3}P) + toluene→W1^T	−11 (−7) −5	
W1^T→W2^T	−61 (−49) −57	18 (29) 21
W2^T→PR1^T	−30 (−27) −34	32 (36) 31
O(^{3}P) + toluene → W3^T	−10 (−6) −4	
W3^T → W6^T	−67 (−48) −59	11 (26) 22
W6^T → PR2^T	2 (2) −5	35.1 (35.1)
O(^{3}P) + toluene → W4^T	−10 (−5) −4	
W4^T → W7^T	−80 (−54) −62	28 (41) 37
W7^T→PR3^T	2 (3) −4	57 (56) 51
O(^{3}P) + toluene → W5^T	−10 (−5) −4	
W5^T → W8^T	−82 (−53) −61	15 (29) 26
W8^T → PR4^T	−2 (0) −7	87 (49) 70
O(^{3}P) + ethylbenzene→W1^E	−12 (−7)	
W1^E→W2^E	−59 (−45)	21 (33)
W2^E→PR1^E	−39 (−32)	19 (23)
O(^{3}P) + ethylbenzene → W3^E	−10	
W3^E → W6^E	−83	15
W6^E → PR2^E	0	55
O(^{3}P) + ethylbenzene → W4^E	−10	
W4^E → W7^E	−77	22
W7^{E}PR3^E	1	56
O(^{3}P) + ethylbenzene → W5^E	−9	
W5^E → W8^E	−63	15
W8^E → PR4^E	1	57

4 Conclusions

The theoretical study at different level of complexity of the interaction of O(^{3}P) with toluene and ethylbenzene suggested a computational strategy for the investigation of large aromatic hydrocarbons interacting with oxygen atoms. The study of O(^{3}P) and toluene has been performed using different methods of increasing complexity, in particular DFT methods with relatively small basis sets, CCSD(T) with a large basis set and CCSD(T) with inclusion of extrapolation to basis set limit. This study suggested that also complex methods like CCSD(T) with large basis sets, as aug-cc-pVTZ, do not provide very accurate results due to basis set superposition errors. It is necessary to include a correction for basis set incompleteness in order to obtain very accurate results. Simpler methods like DFT, converging much faster on basis set, are able to provide acceptable results at a very low computational cost. These methods have been applied to larger systems as O(^{3}P) interacting with ethylbenzene, where CCSD(T)/CBS calculations are unfeasible with the available computational resources. The same methods can be applied to much larger systems with reasonable costs. This is an important result in order to extend our study to complex PAHs.

Acknowledgments. We acknowledge financial support under the National Recovery and Resilience Plan (NRRP), Mission 4, Component 2, Investment 1.1, Call for tender No. 1409 published on 14.9.2022 by the Italian Ministry of University and Research (MUR), funded by the European Union - NextGenerationEU– Project Title P20223H8CK Degradation of space technology polymers by thermospheric oxygen atoms and ions: an exploration of the reaction mechanisms at an atomistic level (ThermOPoly) - CUP J53D23014440001_- Grant Assignment Decree No. 1386 adopted on 01.09.2023 by the Italian Ministry of Ministry of University and Research (MUR).

We acknowledge financial support under the National Recovery and Resilience Plan (NRRP), Mission 4, Component 2, Investment 1.1, Call for tender No. 104 published on 2.2.2022 by the Italian Ministry of University and Research (MUR), funded by the European Union – NextGenerationEU– Project Title 2022JC2Y93 ChemicalOrigins: linking the fossil composition of the Solar System with the chemistry of protoplanetary disks – CUP J53D23001600006 - Grant Assignment Decree No. 962 adopted on 30.06.2023 by the Italian Ministry of Ministry of University and Research (MUR).

References

1. Cavallotti, C., Pratali Maffei, L., Caracciolo, A., Vanuzzo, G., ìBalucani, N., Casavecchia, P.: Theoretical study of the extent of intersystem crossing in the O(^{3}P) + C_6H_6 reaction with experimental validation. J. Phys. Chem. Lett. **11**(22), 9621–9628 (2020)
2. Vanuzzo, G., et al.: Crossed-beam and theoretical studies of the O(^{3}P, ^{1}D) + benzene reactions: primary products, branching fractions, and role of intersystem crossing. J. Phys. Chem. A **125**(38), 8434–8453 (2021)
3. Recio, P., et al.: Intersystem crossing in the entrance channel of the reaction of O(^{3}P) with pyridine. Nat. Chem. **14**(12), 1405–1412 (2022)
4. Balucani, N., et al.: Crossed molecular beam experiments and theoretical simulations on the multichannel reaction of toluene with atomic oxygen. Faraday Discuss. (2024). https://doi.org/10.1039/D3FD00181D
5. Liang, P., et al.: Reactions O(^{3}P,^{1}D) + HCCCN($X^1\Sigma^+$) (Cyanoacetylene): crossed-beam and theoretical studies and implications for the chemistry of extraterrestrial environments. J. Phys. Chem. A **127**, 685–703 (2023)
6. Pannacci, G., et al.: A combined crossed molecular beam and theorerical study of the O (^{3}P,^{1}D)+ acrylonitrile (CH_2CHCN) reactions and implications for combustion and extraterrestrial environments. Phys. Chem. Chem. Phys. **25**(30), 20194–20211 (2023)
7. Chai, J.D., Head-Gordon, M.: Long-range corrected hybrid density functionals with damped atom-atom dispersion corrections. Phys. Chem. Chem. Phys. **10**, 6615–6620 (2008)
8. Chai, J.D., Head-Gordon, M.: Systematic optimization of long-range corrected hybrid density functionals. J. Chem. Phys. **128**, 084106 (2008)

9. Krishnan, R., Binkley, J.S., Seeger, R., Pople, J.A.: Self-consistent molecular orbital methods. XX. A basis set for correlated wave functions. J. Chem. Phys. **72**, 650–654 (1980)
10. Frisch, M.J., Pople, J.A., Binkley, J.S.: Self-consistent molecular orbital methods 28. Supplementary functions for Gaussian basis sets. J. Chem. Phys. **80**, 3265–3269 (1984)
11. Gonzalez, C., Schlegel, H.B.: An improved algorithm for reaction path following. J. Chem. Phys. **90**, 2154–2161 (1989)
12. Gonzalez, C., Schlegel, H.B.: Reaction path following in mass-weighted internal coordinates. J. Phys. Chem. **94**, 5523–5527 (1990)
13. Bartlett, R.J.: Many-body perturbation theory and coupled cluster theory for electron correlation in molecules Ann. Rev. Phys. Chem. **32**, 359–401 (1981)
14. Raghavachari, K., Trucks, G.W., Pople, J.A., Head-Gordon, M.: Quadratic configuration interaction: a general technique for determining electron correlation energies. Chem. Phys. Lett. **157**, 479–483 (1989)
15. Olsen, J., Jorgensen, P., Koch, H., Balkova, A., Bartlett, R.J.: Full configuration–interaction and state of the art correlation calculations on water in a valence double-zeta basis with polarization functions. J. Chem. Phys. **104**, 8007–8015 (1996)
16. Dunning Jr, T.H.: Gaussian basis sets for use in correlated molecular calculations. I.: the atoms boron through neon and hydrogen. J. Chem. Phys. **90**, 1007–1023 (1989)
17. de Petris, G., Cacace, F., Cipollini, R., Cartoni, A., Rosi, M., Troiani, A.: Experimental detection of theoretically predicted N_2CO. Angew. Chem. **117**, 466–469 (2005)
18. De Petris, G., Rosi, M., Troiani, A.: SSOH and HSSO radicals: an experimental and theoretical study of $[S_2OH]^{0/+/-}$ species. J. Phys. Chem. A **111**, 6526–6533 (2007)
19. Bartolomei, M., et al.: The intermolecular potential in NO-N_2 and $(NO\text{-}N_2)^+$ systems: implications for the neutralization of ionic molecular aggregates. PCCP **10**, 5993–6001 (2008)
20. Leonori, F., et al.: Observation of organosulfur products (thiovinoxy, thioketene and thioformyl) in crossed-beam experiments and low temperature rate coefficients for the reaction $S(^1D) + C_2H_4$. PCCP **11**, 4701–4706 (2009)
21. Leonori, F., et al.: Crossed-beam and theoretical studies of the $S(^1D) + C_2H_2$ reaction. J. Phys. Chem. A **113**, 4330–4339 (2009)
22. De Petris, G., Cartoni, A., Rosi, M., Barone, V., Puzzarini, C., Troiani, A.: The proton affinity and gas-phase basicity of sulfur dioxide. PCCP **12**, 112–115 (2011)
23. Berteloite, C., et al.: Low temperature kinetics, crossed beam dynamics and theoretical studies of the reaction $S(^1D) + CH_4$ and low temperature kinetics of $S(^1D) + C_2H_2$. PCCP **13**, 8485–8501 (2011)
24. Rosi, M., Falcinelli, S., Balucani, N., Casavecchia, P., Leonori, F., Skouteris, D.: Theoretical study of reactions relevant for atmospheric models of Titan: Interaction of excited nitrogen atoms with small hydrocarbons. LNCS **7333**, 331–344 (2012)
25. Skouteris, D., Balucani, N., Faginas-Lago, N., Falcinelli, S., Rosi, M.: Dimerization of methanimine and its charged species in the atmosphere of Titan and interstellar/cometary ice analogs. Astron. Astrophys. **584**, A76 (2015)
26. Salvitti, C., Pepi, F., Troiani, A., Rosi, M., de Petris, G.: The Peroxymonocarbonate anion HCO_4^- as an effective oxidant in the gas phase: a mass spectrometric and theoretical study on the reaction with SO_2 molecules, **28**(1), 132 (2023)
27. Troiani, A., Rosi, M., Garzoli, S., Salvitti, C., de Petris, G.: Vanadium Hydroxide cluster ions in the gas phase: bond-forming reactions of doubly-charged negative ions by SO_2-promoted V-O activation. Chem. Eur. J. **23**, 11752–11756 (2017)
28. Salvitti, C., Rosi, M., Pepi, F., Troiani, A., de Petris, G.: Reactivity of transition metal dioxide anions MO_2^- (M = Co, Ni, Cu, Zn) with sulfur dioxide in the gas phase: an experimental and theoretical study. Chem. Phys. Lett. **776**, 138555 (2021). https://doi.org/10.1016/j.cplett.2021.13855533

29. Troiani, A., Rosi, M., Garzoli, S., Salvitti, C., de Petris, G.: Sulphur dioxide cooperation in hydrolysis reactions of vanadium oxide and hydroxide cluster dianions. J. Chem. **42**(6), 4008–4016 (2018)
30. Troiani, A., Rosi, M., Garzoli, S., Salvitti, C., de Petris, G.: Effective redox reactions by chromium oxide anions: sulfur dioxide oxidation in the gas phase. Int. J. Mass Spectrom. **436**, 18–22 (2019)
31. Troiani, A., Rosi, M., Salvitti, C., de Petris, G.: The oxidation of sulfur dioxide by single and double oxygen transfer paths Chem. Phys. Chem. **15**(13), 2723–2731 (2014)
32. Martin, J.M.: Ab initio total atomization energies of small molecules — towards the basis set limit. Chem. Phys. Lett. **259**, 669–678 (1996)
33. Frisch, M.J., et al.: gaussian 09. In: Gaussian. Inc., Wallingford CT, vol. 121, pp. 150–166 (2009)
34. Werner, H.J., Knowles, P.J., Knizia, G., Manby, F.R., Schütz, M.: Molpro: a general-purpose quantum chemistry program package. WIREs Comput. Mol. Sci. **2**, 242–253 (2012)
35. Werner, H.J., et al.: The Molpro quantum chemistry package. J. Chem. Phys. **152**, 144107 (2020)
36. Flükiger, P., Lüthi, H.P., Portmann, S., Weber, J.: MOLEKEL 4.3, Swiss Center for Scientific Computing, Manno. In: An Interactive Molecular Graphics Tool, vol. 54, pp. 766–769, Chimia, Switzerland (2000)
37. nn, M., et al.: A computational study of the reaction between $N(^2D)$ and simple aromatic hydrocarbons. LNCS **14105**, 718–734 (2023)

The Challenge of Characterizing Bimolecular Reactions of Aromatic Compounds via Computational Chemistry. the Case of the N(^{2}D) + Pyridine Reaction and Implications in the Chemistry of the Thermosphere of Titan

Luca Mancini[1(✉)], Marzio Rosi[2,3], and Nadia Balucani[1]

[1] Dipartimento di Chimica, Biologia e Biotecnologie, Università degli Studi di Perugia, 06123 Perugia, Italy
luca.mancini@unipg.it

[2] Dipartimento di Ingegneria Civile ed Ambientale, Università degli Studi di Perugia, 06125 Perugia, Italy

[3] Istituto CNR di Scienze e Tecnologie Chimiche "Giulio Natta" (CNR-SCITEC), 06123 Perugia, Italy

Abstract. Bimolecular reactions involving aromatic compounds are of great interest in many areas of applied science. Given their large weight, an experimental characterization might be difficult using the common experimental methods because of their low vapor pressure. Also the characterization via theoretical methods is difficult, because of the large number of atoms involved and some difficulty in describing delocalization effects with the usual methods for electronic structure calculations. In this contribution, we report a benchmark analysis of the reaction between pyridine and atomic nitrogen in the excited ^{2}D state, which is of relevance in different fields in the chemistry of the thermosphere of Titan. We have used different computational strategies, mainly based on DFT methods, which represent a convenient option in the case of large systems. The use of more accurate methods, such as the coupled-cluster theory, might be difficult to implement considering the significant hardware requirements. The results of this analysis suggest that an acceptable computational strategy is represented by the use of hybrid functionals, such as the B3LYP or M062X, in conjunction with the aug-cc-pVTZ basis set. Implications for the chemistry of the atmosphere of Titan (the largest moon of Saturn) are briefly discussed.

Keywords: Theoretical Chemistry · DFT methods · Chemistry of Planetary Atmospheres

1 Introduction

The computational study of chemical reactions involving aromatic compounds represents a challenge to available theoretical methods. This is particularly true

O. Gervasi et al. (Eds.): ICCSA 2024 Workshops, LNCS 14818, pp. 84–97, 2024.
https://doi.org/10.1007/978-3-031-65273-8_6

if the interest is directed at the reactions of polycyclic aromatic hydrocarbons (PAHs) and their functionalized derivatives given their aromatic nature and the large number of atoms involved. Nevertheless, the great interest in these species has motivated their study with the most advanced methods available (see, for example, the large amount of work carried out by Mebel and his collaborators [1–7]). Among other fields of application, a renewed interest in the chemistry of PAHs and their functionalized derivatives has emerged in the context of astrochemistry and cosmochemistry. Indeed, PAHs and their derivatives have been identified in many extraterrestrial environments, from the interstellar medium [8–10] to comets [11–13] (including 67P/Churyumov-Gerasimenko [14]), carbonaceous chondrites [15], micrometeorites [16] and even in the pristine sample of the carbonaceous asteroid (162173) Ryugu collected by the Hayabusa2 mission [17].

For this reason, in our laboratory we have recently started a systematic investigation of some reactions involving aromatic compounds and atomic nitrogen in its first electronically excited ^2D state [18–22]. These reactions are relevant to the chemistry of the thermosphere of Titan, the massive moon of Saturn, that has long attracted great interest in the study of prebiotic chemistry on a planetary scale [23–26]. We started from the reaction of $N(^2D)$ with benzene (the simplest aromatic compound) as it was suggested as a possible route for the formation of pyridine [18,19], an organic molecule simple enough to form in an abiotic environment but with a notable prebiotic potential given the number of biomolecules characterized by its structure (e.g. niacin, pyridoxine, and nicotinamide adenine dinucleotide). The suggested mechanism, involving a N/CH exchange reaction, was found to be open according to our calculations of the potential energy surface (PES) [19]. However, the competition with other reaction channels makes that channel minor as demonstrated by an experimental study and further theoretical work by us [22] and Chin et al. [27] in which estimates of the product branching fractions (BFs) were obtained via Rice-Ramsperger-Kassel-Marcus (RRKM) theory. Quite surprisingly, the main reaction channel of the $N(^2D)$ + benzene reaction is a ring contraction channel associated with the formation of cyclopentadienyl and HCN while minor contributions come from the channels leading to o-N-cycloheptatriene radical + H, pyrrolyl + acetylene and cyanocyclopentadiene + H [22,27].

The presence of pyridine and pyrimidine in the upper atmosphere of Titan was suggested by the analysis of the Cassini Ion Neutral Mass Spectrometer data, but ground-based attempts to detect it by using ALMA could not confirm their presence and only upper limits of their abundances were derived [24]. To be noted that polycyclic aromatic hydrocarbons (PAHs) and N-containing PAHs have been invoked [28] to explain an unidentified emission band near 3.28 μm analyzed by Dinelli et al. [29]. Furthermore, the Cassini plasma spectrometer ion beam spectrometer data analysis strongly suggests that PAHs and N-containing PAHs account for the ion signal recorded at 170–310 Da [30]. Even if not a single PAH/N-PAH was identified, it seems quite clear that if N-PAHs are present in the thermosphere of Titan, some formation mechanism must be active. $N(^2D)$ reactions with PAHs may be a viable route.

For the above reasons, we have also explored the title reaction that features $N(^2D)$ and pyridine as reactants (incidentally, if the N/CH exchange channel is competitive, pyrimidine could be the formed product). In this case as well, experimental data obtained in crossed molecular beam experiments are available [31] and a comparison between experimental and theoretical BFs is being pursued (Mancini et al., in preparation). Both the cases of $N(^2D)$ + benzene [22] and, in this work, $N(^2D)$ + pyridine have been investigated at a high level of theory. We performed CCSD(T) calculations, extrapolating to complete basis set and including core-valence correlation effects. However, these calculations were extremely costly in terms of CPU time and disk space requirements. Going towards less symmetric and larger systems, it is not possible to perform similar calculations both for the computer time required and the scratch disk space necessary. Therefore, it is necessary to find a compromise between the accuracy of the calculations and their computational costs.

In this contribution, we present a benchmark study of the title reaction only for a portion of the global PES that will be the object of a future paper. In particular, we have considered only one of the possible initial steps of the reaction. Different theoretical methods have been employed to characterize the energetics of the reactants, products, intermediates and transition states along that portion of the PES. The system is much more complex than the reaction with benzene, even though benzene and pyridine are isoelectronic, because of the symmetry loss once N is present in the aromatic ring. The benchmark will be a way to establish which DFT method works better for these type of reactive systems and it will be used to study the reactions of $N(^2D)$ with large PAHs.

2 Theoretical Methods: Electronic Structure Calculations

To assess the best computational strategy, a benchmark study has been performed. In detail, following the same approach already used for the characterization of reactions of $N(^2D)$ with different hydrocarbons [22,32–34], geometry optimization has been performed at the DFT level. Different functionals have been employed, including the B3LYP-D3, ωB97XD, and the M062X, using both the 6–311++G(d,p) Pople's basis set and the correlation consistent valence polarized set aug-cc-pVTZ. Harmonic vibrational frequencies have been calculated using the same level of theory of geometry optimization. The zero point correction obtained has been then used to correct the energies of each stationary point. Finally, IRC calculations [35,36] allowed the identification of the saddle points. The obtained data have been compared with a standard reference, consisting of accurate single-point energy calculations, performed at the coupled-cluster (CCSD(T)) level, corrected with a Density Fitted (DF)MP2 extrapolation to the complete basis set (CBS) and with corrections for core electrons excitations. The scheme of the energy extrapolation is:

$$
\begin{aligned}
E = &E(CCSD(T)/aug-cc-pVTZ)+\\
&[E(CCSD(T,core)/cc-pVTZ) - E(CCSD(T)/cc-pVTZ)]\\
&+ [E(DF-MP2/CBS) - E(DF-MP2/aug-cc-pVTZ)]
\end{aligned}
$$

where E(DF-MP2/CBS) is defined as:

$$
\begin{aligned}
E(DF-MP2/CBS) = &E[(DF-MP2)/aug-cc-pVQZ)]+\\
&0.5772 * [E(DF-MP2/aug-cc-pVQZ)\\
&- E(DF-MP2/aug-cc-pVTZ)]
\end{aligned}
$$

where the E(DF-MP2/CBS) extrapolation has been performed using Martin's two parameter scheme [37]. The results obtained have been corrected considering the zero point energy evaluated at the B3LYP-D3/aug-cc-pVTZ level of theory. All the energy values have been rounded to 1 kJ/mol, to be consistent with the"chemical accuracy" of 1 kcal/mol. The first electronic excited state of atomic nitrogen, N(^{2}D), is difficult to treat through single-reference methods, due to an overlapping between different spin states during the calculations. For this reason, the energy of N(^{2}D) has been estimated by adding the experimental N(^{4}S)-N(^{2}D) separation of 230 kJ/mol [38] to the energy of N(^{4}S) calculated in the present work. The Gaussian09 and Molpro suits of programs have been used to perform electronic structure calculations [39,40].

3 Results

3.1 Potential Energy Surface

The portion of the potential energy surface (PES) resulting from the theoretical investigation is reported in Fig. 1. The energy values reported in the scheme are those evaluated at the CCSD(T)/CBS level of theory.

The initial step of the reaction is represented by the interaction of N(^{2}D) with the π-system of the pyridine molecule. The addition can occur in different positions (ortho, meta, para, and ipso). Here we consider the portion of the PES that is accessed after the addition in the meta/para position with the formation of an intermediate (MIN1) characterized by the presence of a three-member-ring moiety. The process is barrierless and the relative energy of the intermediate is -329 kJ/mol. The breaking of a C-C bond and the "sliding" of the N atom into the ring leads to the formation of a seven-member ring intermediate (MIN2, located 432 kJ/mol below the reactant energy asymptote). A small barrier of 10 kJ/mol must be overcome during the process. Alternatively, a different isomerization process is responsible for the formation of the MIN3 intermediate (located at -527 kJ/mol with respect to the reactants asymptote) via the H-migration and formation of a N-H bond. In this case, the barrier,

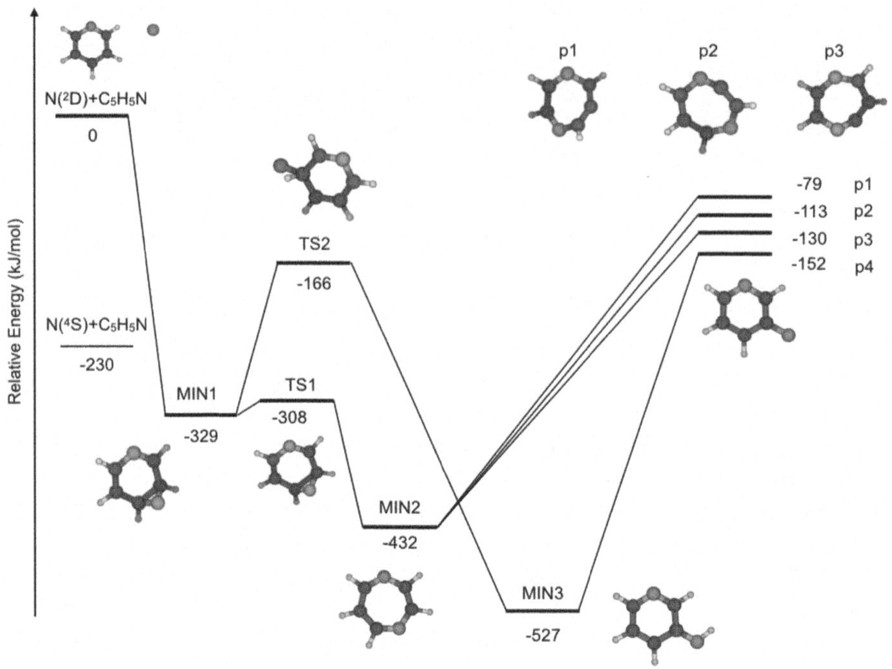

Fig. 1. 2D scheme of the PES obtained for the N(^{2}D)+pyridine reaction.

represented by the TS2 saddle point, is higher (163 kJ/mol) than that associated to the other isomerization process. Therefore, this pathway is expected to be less competitive. Once formed, the MIN2 intermediate can directly decompose, through a H-displacement process. Due to the loss of symmetry with respect to the case of benzene, the H atom can be released from three different positions and three different isomeric products can be formed. The enthalpy change associated with the three product channels (p1, p2 and p3) is -79 kJ/mol, -113 kJ/mol and -130 kJ/mol, respectively. Also the MIN3 intermediate can lose one H atom, leading to the formation of the p4 co-fragment, a radical in a triplet spin state. All H-loss processes are barrierless, that is, the dissociation takes place without the presence of a saddle point in the exit channel.

3.2 Benchmark Analysis

As described above (see Sect. 2), DFT calculations have been performed using different functionals and basis sets to establish the best computational strategy for the analysis of the PES. Both the energies and the geometrical parameters have been compared. In detail, the geometrical analysis has been focused on three stationary points representative of different scenarios, the pyridine molecule (representing a singlet closed-shell molecule) and two open-shell structures: MIN2, being a seven-member ring in the doublet electronic state, and the TS1 saddle point. In Fig. 2 are reported the structures of the stationary points with the labeled bonds object of the study, while in Table 1, the values of the bond lengths (in Å) obtained at different levels of theory are listed.

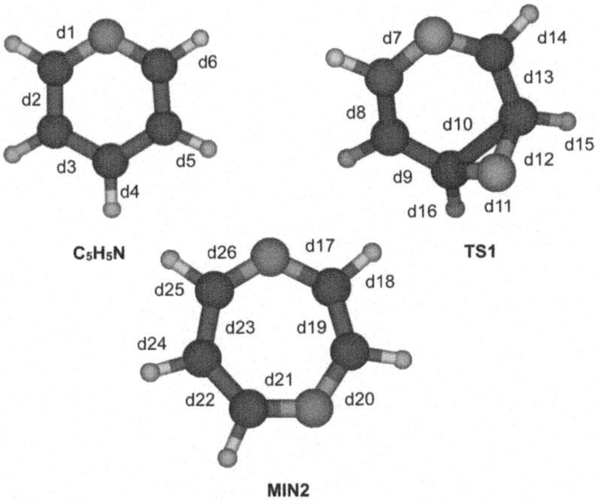

Fig. 2. Structure of the three stationary points used to compare the geometrical parameters, resulting from the theoretical analysis.

As can be seen from the data presented in Table 1, the differences in the values of bond lengths are very small (almost negligible), even considering the use of a more accurate basis set (i.e., the aug-cc-pVTZ). Therefore, the only comparison of the bond distances is not sufficient to select the most appropriate computational strategy. For this reason, the energies of all the stationary points identified have been evaluated. In the first instance, the zero point energy correction, derived from the frequency calculations can be analysed. As can be seen from Table 2, small differences between the various levels of theory are present. The main deviation in the case of the pyridine molecule is 0.0010 Hartree/particles, which corresponds to 2.74 kJ/mol and is lower than the chemical accuracy. One of the main deviations can be appreciated in the case of the MIN1 intermediate, where the major difference is obtained comparing the B3LYP/6–311++G(d,p) and the ωB97XD/aug-cc-pVTZ level, leading to a value of 0.0017 Hartree/particles (equal to 4.59kJ/mol, barely around the limit value of chemical accuracy of 1 kcal/mol).

The relative energy with respect to the reactant asymptote (taken as zero reference) obtained at the various levels of theory is reported in Table 3. Those values reported include the zero-point corrections (evaluated at the same level of theory used for the geometry optimization). The comparison is done with the more accurate CCSD(T)/CBS data. It can be noticed how the use of a bigger basis set, i.e., the aug-cc-pVTZ, leads to values more in line with the complete basis set extrapolation limit. All three functionals provide similar results, with small deviations. The main differences can be appreciated in the case of p4

Table 1. Bond lengths (in Å) evaluated for the main distances in the structures reported in Fig. 2, where *6–311* and *TZ* represent the 6–311++G(d,p) and aug-cc-pVTZ basis sets, respectively.

Dist	B3LYP-D3/ 6–311	B3LYP-D3/ TZ	ωB97XD/ 6–311	ωB97XD/ TZ	M062X/ 6–311	M062X/ TZ
d1	1.337	1.334	1.332	1.330	1.332	1.330
d2	1.395	1.391	1.390	1.386	1.392	1.388
d3	1.393	1.388	1.388	1.385	1.389	1.386
d4	1.084	1.082	1.084	1.082	1.084	1.082
d5	1.084	1.081	1.084	1.081	1.083	1.081
d6	1.086	1.084	1.086	1.084	1.086	1.084
d7	1.354	1.351	1.343	1.341	1.342	1.339
d8	1.377	1.373	1.378	1.375	1.372	1.379
d9	1.421	1.417	1.408	1.404	1.404	1.400
d10	1.792	1.785	1.824	1.820	1.835	1.832
d11	1.415	1.412	1.404	1.402	1.407	1.406
d12	1.419	1.416	1.411	1.409	1.413	1.412
d13	1.423	1.420	1.409	1.405	1.406	1.402
d14	1.088	1.086	1.088	1.085	1.087	1.085
d15	1.090	1.087	1.089	1.086	1.088	1.086
d16	1.089	1.087	1.089	1.086	1.088	1.086
d17	1.379	1.376	1.376	1.373	1.376	1.373
d18	1.087	1.085	1.087	1.084	1.086	1.085
d19	1.371	1.367	1.365	1.361	1.367	1.363
d20	1.379	1.376	1.376	1.373	1.376	1.373
d21	1.314	1.312	1.310	1.307	1.309	1.307
d22	1.413	1.409	1.409	1.405	1.411	1.407
d23	1.413	1.409	1.409	1.405	1.411	1.407
d24	1.085	1.082	1.085	1.082	1.083	1.081
d25	1.091	1.088	1.091	1.088	1.090	1.088
d26	1.314	1.312	1.310	1.307	1.309	1.307

and MIN3. The reason for such a difference is due to correlation effects in the case of the presence of a multi-reference character of the stationary points, as can be seen from the value of the T1 diagnostic parameter of 0.028 and 0.027, respectively. As reported in previous studies [41,42], a value of T1 higher than 0.02 leads to a strong multi-reference character, which can be properly taken into account by using different methods (including CASSCF and/or CASPT2).

Table 2. Zero-point energy correction (in Hartree/particle) calculated for the main stationary points in the PES at different levels of theory. *6–311* and *TZ* represent the 6–311++G(d,p) and aug-cc-pVTZ basis sets, respectively.

Point	B3LYP-D3/ 6–311	B3LYP-D3/ TZ	ωB97XD/ 6–311	ωB97XD/ TZ	M062X/ 6–311	M062X/ TZ
C_5H_5N	0.088376	0.088596	0.089275	0.089421	0.089223	0.089338
MIN1	0.090302	0.090424	0.091870	0.092049	0.091831	0.091940
MIN2	0.090208	0.0900470	0.091194	0.091412	0.090666	0.090967
MIN3	0.091410	0.091642	0.092312	0.092569	0.092189	0.092424
TS1	0.088964	0.089087	0.090248	0.090333	0.090185	0.0900221
TS2	0.084813	0.084910	0.086491	0.087686	0.085602	0.085733
p1	0.079424	0.079542	0.080665	0.080719	0.080751	0.080833
p2	0.079390	0.079507	0.080570	0.080646	0.080734	0.080760
p3	0.079188	0.079291	0.080403	0.080467	0.0880411	0.080452
p4	0.078815	0.079076	0.079593	0.079853	0.079735	0.079981

Table 3. Relative energies (in kJ/mol) obtained for the main stationary points in the PES at different levels of theory. B3 = B3LYP-D3 6–311 = 6–311++G(d,p) TZ = aug-cc-pVTZ ωB = ωB97XD M06 = M062X

Point	B3/ 6–311	B3/ TZ	ωB/ 6–311	ωB/ TZ	M06/ 6–311	M06/ TZ	CCSD(T)/ CBS
MIN1	-335	-332	-331	-335	-330	-337	-329
MIN2	-460	-454	-436	-442	-424	-432	-432
MIN3	-549	-542	-531	-536	-517	-524	-527
TS1	-328	-322	-313	-319	-309	-316	-308
TS2	-169	-163	-117	-122	-138	-154	-166
p1	-117	-109	-94	-101	-100	-109	-113
p2	-137	-129	-114	-122	-122	-132	-130
p3	-79	-73	-56	-62	-60	-67	-79
p4	-184	-180	-170	-172	-156	-161	-152

In general, to characterize the reaction mechanism and predict the product branching fractions, it is important to derive the minimum energy path for all competing reaction channels. The presence of a series of energy barriers can promote or hinder certain reaction paths, leading to an appreciable effect on the final reaction micro-mechanism. In Table 4 are reported the enthalpy changes and the barrier heights (in kJ/mol) for the different reaction paths, leading to the various product channels identified in the PES.

As can be noticed, the B3LYP functional (together with the M062X one), in conjunction with the aug-cc-pVTZ basis set, provides results in line with the

Table 4. Enthalpy changes (ΔH_0^0) and Barrier Heights, reported in kJ/mol, for the reaction pathways in the PES, evaluated at different levels of theory. B3 = B3LYP-D3 TZ = aug-cc-pVTZ ωB = ωB97XD M06 = M062X CBS = CCSD(T)/CBS

Path	B3/TZ		ω B/TZ		M06/TZ		CBS	
	ΔH_0^0	Barrier	ΔH_0^0	Barrier	ΔH_0^0	Barrier	ΔH_0^0	Barrier
C_5H_5N →MIN1	-332	-	-337	-	-337	-	-329	-
MIN1 → MIN2	-122	10	-105	18	-95	21	-103	21
MIN1 → MIN3	-211	169	-199	215	-187	192	-198	163
MIN2 →p1	381	-	380	-	365	-	353	-
MIN2 →p2	345	-	341	-	323	-	319	-
MIN2 →p3	325	-	320	-	300	-	302	-
MIN3 → p4	363	-	364	-	363	-	375	-

accurate CCSD(T)/CBS reference. It represents also a useful strategy for the optimization of the stationary points in terms of CPU time. In general the differences in the values of energy are not expected to contribute significantly, also in the case of MIN3 and TS2 (which shows a strong multi-reference character), since the relative heights and the paths probability are basically unchanged. This is at variance with the case of the reactions involving atomic oxygen, where the presence of open-shell species with unpaired electrons (both in the singlet and in the triplet state) can alter the energy (and, consequently, the morphology) of the PES [43–45]. In some cases, indeed the strong multi-reference character of some stationary points identified in the PES required the refinement of the energy of minima and transition states at a higher level of theory (such as the CASPT2).

4 Conclusions

The present contribution reports a benchmark study on the reaction between atomic nitrogen, in its first electronically excited ^{2}D state, and the pyridine molecule when considering the addition of N in the meta/para positions. A more complete work is currently underway (Mancini et al., in preparation). The theoretical branching fractions of the products will be obtained considering all the possible addition sites and a comparison with the experimental results by [31] will be carried out. In this work, different DFT methods have been considered and compared with the predictions of the more accurate CCSD(T) theory with the extrapolation on the complete basis set limit. In general, one of the best choices for the estimate of the energies is the use of correlation methods (such as CCSD(T)) but in the cases in which the dimensions of the system are too large, it is difficult to use because of the high computational cost. All the methods used here provide similar results in the case of geometry optimization. The more accurate aug-cc-pVTZ basis set can be considered a better alternative to the 6–311++G(d,p) Pople's basis set since it provides, as expected, more accurate results. A more general conclusion is that the energy values determined at the DFT level can be used to represent the competing reaction channels in the potential energy surface, providing that the energy values associated with the transition states and reaction intermediates homogeneously scale. In this way, albeit differing with respect to the reference value, the relative probability of the occurrence of each channel remains unchanged. The data obtained at the DFT level can be considered a good starting point for a kinetic analysis, which allows us to derive the product branching fractions. Among the plethora of available DFT functionals, the B3LYP appears to be a good compromise between chemical accuracy and computational cost for the analysis of the PES.

After this benchmark work, we will apply the B3LYP method to the theoretical characterization of reactions involving $N(^2D)$ (and also other atomic and radical species) with larger aromatic species, like naphthalene and indene (which are supposed to be intermediate steps toward the formation of more complex PAHs starting from benzene), as well as N-PAHs, like quinoline and 1,3,5 triazine, or other aromatic species that have been speculated to be present in the upper atmosphere of Titan [28,30]. For some of those systems, the comparison with the experimental results obtained by the crossed molecular beam method will allow us to verify if the level of calculations is appropriate. If necessary, we will resort to the higher-level calculations. This combined experimental and theoretical work is necessary because the crossed molecular beam method nicely reproduces the low-pressure environment of the upper atmosphere of Titan, but not its low temperature. Therefore, the theoretical predictions, once validated by the comparison with the experimental results at high collision energies, will be extended to the temperatures relevant to the very cold moon of Saturn as was done in previous work by us [46–50]. Those results will contribute to a thorough understanding of the rich atmospheric chemistry of Titan [24].

Acknowledgements. This work was supported by the Italian Space Agency (Bando ASI Prot. n. DC-DSR-UVS-2022-231, Grant no. 2023-10-U.0 "Modeling Chemical Complexity: all'Origine di questa e di altre Vite per una visione aggiornata delle missioni spaziali (MIGLIORA)" Codice Unico di Progetto (CUP) F83C23000800005). M. Rosi also acknowledge financial support under the National Recovery and Resilience Plan (NRRP), Mission 4, Component 2, Investment 1.1, Call for tender No. 104 published on 2.2.2022 by the Italian Ministry of University and Research (MUR), funded by the European Union - NextGenerationEU- Project Title 2022JC2Y93 ChemicalOrigins: linking the fossil composition of the Solar System with the chemistry of protoplanetary disks - CUP J53D23001600006 - Grant Assignment Decree No. 962 adopted on 30.06.2023 by the Italian Ministry of Ministry of University and Research (MUR).

References

1. Kaiser, R.I., Parker, D.S., Mebel, A.M.: Reaction dynamics in astrochemistry: low-temperature pathways to polycyclic aromatic hydrocarbons in the interstellar medium. Annu. Rev. Phys. Chem. **66**, 43–67 (2015)
2. Parker, D.S., et al.: Low temperature formation of naphthalene and its role in the synthesis of PAHs (polycyclic aromatic hydrocarbons) in the interstellar medium. Proc. Natl. Acad. Sci. **109**(1), 53–58 (2012)
3. Yang, T., et al.: Hydrogen-abstraction/acetylene-addition exposed. Angew. Chem. Int. Ed. **55**(48), 14983–14987 (2016)
4. Zhao, L., et al.: Pyrene synthesis in circumstellar envelopes and its role in the formation of 2D nanostructures. Nature Astron. **2**(5), 413–419 (2018)
5. Kaiser, R., Parker, D., Zhang, F., Landera, A., Kislov, V., Mebel, A.: PAH formation under single collision conditions: reaction of phenyl radical and 1, 3-butadiene to form 1, 4-dihydronaphthalene. J. Phys. Chem. A **116**(17), 4248–4258 (2012)
6. Kislov, V., Sadovnikov, A., Mebel, A.: Formation mechanism of polycyclic aromatic hydrocarbons beyond the second aromatic ring. J. Phys. Chem. A **117**(23), 4794–4816 (2013)
7. Frenklach, M., Mebel, A.M.: On the mechanism of soot nucleation. Phys. Chem. Chem. Phys. **22**(9), 5314–5331 (2020)
8. Allamandola, L., Tielens, A., Barker, J.: Polycyclic aromatic hydrocarbons and the unidentified infrared emission bands-auto exhaust along the Milky Way. Astrophys. J. Part 2-Lett. Editor (ISSN 0004-637X), vol. 290, March 1, 1985, p. L25–L28. **290**, L25–L28 (1985)
9. Allamandola, L., Tielens, A., Barker, J.: Interstellar polycyclic aromatic hydrocarbons-The infrared emission bands, the excitation/emission mechanism, and the astrophysical implications. Astrophys. J. Suppl. Ser. (ISSN 0067–0049), vol. 71, Dec. 1989, p. 733–775. Res. support. NASA DOE NSF. **71**, 733–775 (1989)
10. Tielens, A.G.: Interstellar polycyclic aromatic hydrocarbon molecules. Annu. Rev. Astron. Astrophys. **46**, 289–337 (2008)
11. Moreels, G., Clairemidi, J., Hermine, P., Brechignac, P., Rousselot, P.: Detection of a polycyclic aromatic molecule in comet P/Halley. Astron. Astrophys. (ISSN 0004-6361), vol. 282, no. 2, p. 643–656 **282**, 643–656 (1994)
12. McKay, A.J., Roth, N.X.: Organic matter in cometary environments. Life **11**(1), 37 (2021)
13. Clemett, S.J., Sandford, S.A., Nakamura-Messenger, K., Horz, F., McKAY, D.S.: Complex aromatic hydrocarbons in stardust samples collected from comet 81P/wild 2. Meteorit. Planet. Sci. **45**(5), 701–722 (2010). https://doi.org/10.1111/j.1945-5100.2010.01062.x, https://onlinelibrary.wiley.com/doi/abs/10.1111/j.1945-5100.2010.01062.x
14. Hänni, N., et al.: Identification and characterization of a new ensemble of cometary organic molecules. Nat. Commun. **13**(1), 3639 (2022)
15. Spencer, M.K., Hammond, M.R., Zare, R.N.: Laser mass spectrometric detection of extraterrestrial aromatic molecules: mini-review and examination of pulsed heating effects. Proc. Natl. Acad. Sci. **105**(47), 18096–18101 (2008)
16. Maurette, M.: Carbonaceous micrometeorites and the origin of life. Orig. Life Evol. Biosph. **28**(4), 385–412 (1998)
17. Naraoka, H., et al.: Soluble organic molecules in samples of the carbonaceous asteroid (162173) Ryugu. Science **379**(6634), eabn9033 (2023)

18. Balucani, N., Pacifici, L., Skouteris, D., Caracciolo, A., Casavecchia, P., Rosi, M.: A theoretical investigation of the reaction N(2D) + C6H6 and implications for the upper atmosphere of titan. In: Gervasi, O., et al. Computational Science and Its Applications – ICCSA 2018. ICCSA 2018. LNCS, vol. 10961. Springer, Cham (2018). https://doi.org/10.1007/978-3-319-95165-2_53
19. Balucani, N., Pacifici, L., Skouteris, D., Caracciolo, A., Casavecchia, P., Falcinelli, S., Rosi, M.: A computational study of the reaction n (2 d)+ c 6 h 6 leading to pyridine and phenylnitrene. In: Computational Science and Its Applications–ICCSA 2019: 19th International Conference, Saint Petersburg, Russia, July 1–4, 2019, Proceedings, Part III 19. pp. 316–324. Springer (2019)
20. Rosi, M., et al.: A computational study on the insertion of N (2 D) into a C-H or C-C bond: the reactions of N (2 D) with benzene and toluene and their implications on the chemistry of titan. In: Computational Science and Its Applications–ICCSA 2020: 20th International Conference, Cagliari, Italy, July 1–4, 2020, Proceedings, Part III 20, pp. 744–755. Springer (2020). https://doi.org/10.1007/978-3-030-58808-3_54
21. Rosi, M., et al.: A Computational study of the reaction between N(2D) and simple aromatic hydrocarbons. In: Gervasi, O., et al. (eds.) Computational Science and Its Applications – ICCSA 2023 Workshops: Athens, Greece, July 3–6, 2023, Proceedings, Part II, pp. 718–734. Springer Nature Switzerland, Cham (2023). https://doi.org/10.1007/978-3-031-37108-0_46
22. Balucani, N., et al: An experimental and theoretical investigation of the N(2D) + C6H6 (benzene) reaction with implications for the photochemical models of Titan . Faraday Discuss. **245**, 327–351 (2023). https://doi.org/10.1039/D3FD00057E
23. Sagan, C., Thompson, W.R., Khare, B.N.: Titan: a laboratory for prebiological organic chemistry. Acc. Chem. Res. **25**(7), 286–292 (1992). https://doi.org/10.1021/ar00019a003
24. Nixon, C.A.: The Composition and chemistry of Titan's atmosphere. ACS Earth Space Chem. **8**(3), 406–456 (2024). https://doi.org/10.1021/acsearthspacechem.2c00041
25. Vuitton, V., Dutuit, O., Smith, M.A., Balucani, N.: Chemistry of titan's atmosphere. In: Müller-Wodarg, I., Griffith, C.A., Lellouch, E., Cravens, T.E. (eds.) Titan: Interior, Surface, Atmosphere, and Space Environment, pp. 224–284. Cambridge University Press (2014). https://doi.org/10.1017/CBO9780511667398.010
26. Balucani, N.: Elementary reactions of N atoms with hydrocarbons: first steps towards the formation of prebiotic N-containing molecules in planetary atmospheres. Chem. Soc. Rev. **41**(16), 5473 (2012). https://doi.org/10.1039/c2cs35113g
27. Chin, C.-H., Zhu, T., Zhang, J.Z.H.: Cyclopentadienyl radical formation from the reaction of excited nitrogen atoms with benzene: a theoretical study. Phys. Chem. Chem. Phys. **23**(21), 12408–12420 (2021). https://doi.org/10.1039/D1CP00133G
28. López-Puertas, M., et al.: Large abundances of polycyclic aromatic hydrocarbons in titan's upper atmosphere. Astrophys. J. **770**(2), 132 (2013)
29. Dinelli, B., et al.: An unidentified emission in titan's upper atmosphere. Geophys. Res. Lett. **40**(8), 1489–1493 (2013)
30. Haythornthwaite, R.P., et al.: Heavy positive ion groups in titan's ionosphere from cassini plasma spectrometer IBS observations. Planet. Sci. J. **2**(1), 26 (2021)
31. Recio, P., et al.: A crossed molecular beam investigation of the N (2D)+ pyridine reaction and implications for prebiotic chemistry. Chem. Phys. Lett. **779**, 138852 (2021)

32. Vanuzzo, G., et al.: Reaction N (2D)+ CH2CCH2 (Allene): an experimental and theoretical investigation and implications for the photochemical models of titan. ACS Earth Space Chem. **6**(10), 2305–2321 (2022)
33. Mancini, L., et al.: The reaction N (2D)+ CH3CCH (methylacetylene): a combined crossed molecular beams and theoretical investigation and implications for the atmosphere of titan. J. Phys. Chem. A **125**(40), 8846–8859 (2021)
34. Vanuzzo, G., et al.: The N(2D)+ CH2CHCN (vinyl cyanide) reaction: a combined crossed molecular beam and theoretical study and implications for the atmosphere of titan. J. Phys. Chem. A **126**(36), 6110–6123 (2022)
35. Gonzalez, C., Schlegel, H.B.: An improved algorithm for reaction path following. J. Chem. Phys. **90**(4), 2154–2161 (1989). https://doi.org/10.1063/1.456010
36. Gonzalez, C., Schlegel, H.B.: Reaction path following in mass-weighted internal coordinates. J. Phys. Chem. **94**(14), 5523–5527 (1990). https://doi.org/10.1021/j100377a021
37. Martin, J.M.: Ab initio total atomization energies of small molecules-towards the basis set limit. Chem. Phys. Lett. **259**(5–6), 669–678 (1996)
38. Moore, C.E.: Atomic Energy Levels. National Bureau of Standards, US Department of Commerce (1949)
39. Frisch, M., et al.: Gaussian 09, Revision A. 02, 2009, Gaussian. Inc., Wallingford CT (2009)
40. Werner, H.J., et al.: The molpro quantum chemistry package. J. Chem. Phys. **152**(14), 144107 (2020)
41. Leonori, F., et al.: Experimental and theoretical studies on the dynamics of the O(3P)+ propene reaction: primary products, branching ratios, and role of intersystem crossing. J. Phys. Chem. C **119**(26), 14632–14652 (2015)
42. Gimondi, I., Cavallotti, C., Vanuzzo, G., Balucani, N., Casavecchia, P.: Reaction dynamics of O(3P)+ propyne: Ii. primary products, branching ratios, and role of intersystem crossing from ab initio coupled triplet/singlet potential energy surfaces and statistical calculations. J. Phys. Chem. A **120**(27), 4619–4633 (2016)
43. Recio, P., et al.: Intersystem crossing in the entrance channel of the reaction of O(3P) with pyridine. Nat. Chem. **14**(12), 1405–1412 (2022)
44. Balucani, N., et al.: Crossed molecular beam experiments and theoretical simulations on the multichannel reaction of toluene with atomic oxygen. Faraday Discuss. (2024)
45. Pannacci, G., et al.: A combined crossed molecular beam and theorerical study of the O (3 P, 1 D)+ acrylonitrile (CH 2 CHCN) reactions and implications for combustion and extraterrestrial environments. Phys. Chem. Chem. Phys. **25**(30), 20194–20211 (2023)
46. Balucani, N., et al.: Combined crossed molecular beam and theoretical studies of the N (2D)+ CH4 reaction and implications for atmospheric models of titan. J. Phys. Chem. A **113**(42), 11138–11152 (2009)
47. Balucani, N., et al.: Formation of nitriles and imines in the atmosphere of titan: combined crossed-beam and theoretical studies on the reaction dynamics of excited nitrogen atoms N(2D) with ethane. Faraday Discuss. **147**, 189–216 (2010)
48. Balucani, N., et al.: Combined crossed beam and theoretical studies of the N(2D)+ C2H4 reaction and implications for atmospheric models of titan. J. Phys. Chem. A **116**(43), 10467–10479 (2012)

49. Liang, P., et al.: Combined crossed molecular beams and computational study on the N(2D)+ HCCCN (x1σ+) reaction and implications for extra-terrestrial environments. Mol. Phys. **120**(1–2), e1948126 (2022)
50. Mancini, L., et al.: A computational characterization of the reaction mechanisms for the reactions N(2D)+ CH3CN and hc3n and implications for the nitrogen-rich organic chemistry of titan. Comput. Theor. Chem. **1229**, 114341 (2023)

Econometric and Multidimensional Evaluation in Urban Environment (EMEUE 2024)

SDGs Localization: A New Tool for a Comprehensive Assessment of Neighbourhood Sustainability

Valeria Saiu(✉), Ivan Blečić, Italo Meloni, and Francesco Piras

Department of Environmental and Civil Engineering and Architecture, University of Cagliari, Cagliari, Italy
v.saiu@unica.it

Abstract. The localization of the Sustainable Development Goals (SDGs) is a crucial factor for achieving all 17 goals outlined in the UN 2030 Agenda, particularly at the urban level. While SDG 11 specifically focuses on cities and communities, it is important to note that all SDGs encompass an "urban component". However, there are limited tools available that offer a comprehensive evaluation in this regard. This paper introduces the methodology employed to create a new tool for localizing SDGs at the urban and neighbourhood level, aimed at supporting the Regional Sustainable Development Strategy (Sardegna2030) of the Sardinia Region in Italy. The evaluation framework, indeed, is designed to become a centrepiece in the institutionalisation of sustainability assessment across the public administrations, and to promote sustainability through urban projects and policies. This work raises critical questions about the current practices and suggests how SDGs-oriented urban strategies and projects can offer a comprehensive a common framework of principles and criteria against which different interventions can be assessed and justified. The chapter concludes with a discussion of the potential more general relevance of such an approach, and some suggestions for future research and practice.

Keywords: 2030 Agenda · SDGs Localisation · Urban Sustainability · Evaluation Tools · Neighbourhood Sustainability Assessment

1 Introduction

The United Nations' 2030 Agenda for Sustainable Development, approved in September 2015, defined an action plan for People, Planet, Prosperity, Peace, and Partnership (the so-called "5P's"), with the aim of accelerating ecological transition without leaving anyone behind. In this context, the Sustainable Development Goals (SDGs) constitute the comprehensive operational framework of the 2030 Agenda. Their 17 goals and 169 targets seek to integrate environmental and social objectives into a single strategic vision, which had previously been predominantly addressed separately.

For the goals and targets of the 2030 Agenda to become operational, they need to be "localized". This involves adapting global goals to the specific reality of each

O. Gervasi et al. (Eds.): ICCSA 2024 Workshops, LNCS 14818, pp. 101–112, 2024.
https://doi.org/10.1007/978-3-031-65273-8_7

local context, with a commitment from local authorities and all stakeholders to define significant strategies and priority actions aimed at extending benefits at the grassroots level [1–3].

Local goals and specific indicators are translated by "Voluntary Local Reviews" (VLRs) at different scales (national, regional and municipal), providing a useful support for assessing and monitoring current situation and the effectiveness of various policies, projects and interventions [4–7].

In Italy, the National Strategy for Sustainable Development, approved by the Italian Government in 2017 for the period 2017–2030, is the primary document supporting the process of SDGs implementation and monitoring across the territory through specific regional and sub-regional strategies [8, 9]. Among these local strategies, the Sardinian "Regional Sustainable Development Strategy" (Sardinia2030 Strategy) came into in force in 2021 [10, 11] and provides a Positioning Report that analyses the degree of SDGs achievement, comparing with other Italian and European regions.

In national and regional strategies, urban settlements are recognized as pivotal for attaining various sustainability goals. As UN-Habitat highlighted, the 23% of the 234 SDGs indicators include an "urban component" [12]. As a result, urban decisions that transform built, natural, social and economic environments have a significant impact on the attainment of multiple goals [13]. Thus, effective plans and urban design strategies require integrated site-specific decision support tools to make better choices in relation to different aspects of sustainability.

However, there are limited tools available to assist public administrations in SDGs-oriented decision-making. The current data, indicators, and evaluation models primarily focus on the municipal level as a whole. This macro-urban approach tends to flatten information through average values that are insufficient in describing specific situations at the neighbourhood or suburban scale (the urban micro-scale), which is the level where ordinary city transformation processes occur [14–16].

In this paper, we present a new tool specifically designed to facilitate the "localization" of SDGs at the neighbourhood level. The tool has been developed as part of the research project titled "Evaluation of Urban Sustainability through Mobility and Capability Analysis" (Urban SOSLab), funded by the Italian Ministry of Environment and Energy Security to support the development of the Sardinia2030 Strategy. Furthermore, the part of the instrument focused on mobility is under development through the project "GLOSSA: GLOcal" knowledge System for Sustainability Assessment" and within the Sustainable Mobility National Research Centre (MOST), funded within the framework of the Italian National Recovery and Resilience Plan (PNRR). The project is carried out by an multidisciplinary research team within the Department of Civil and Environmental Engineering and Architecture at the University of Cagliari (Italy).

This paper provides a comprehensive overview of the structure of the "Urban SOSLab" evaluation framework, offering insights into its intricacies and components. It delves into the methodology behind the selection of indicators, shedding light on the rationale and criteria employed. Furthermore, the paper explores the key findings derived from the application of the proposed evaluation tool and discusses the broader policy implications of the proposed tool, considering its potential impact on urban governance, decision-making processes, and the pursuit of sustainable urban development goals.

2 The SDGs Assessment Tool: Methodology

The development of the assessment tool involved four phases each characterized by specific objectives, methodologies, and outcomes (Fig. 1).

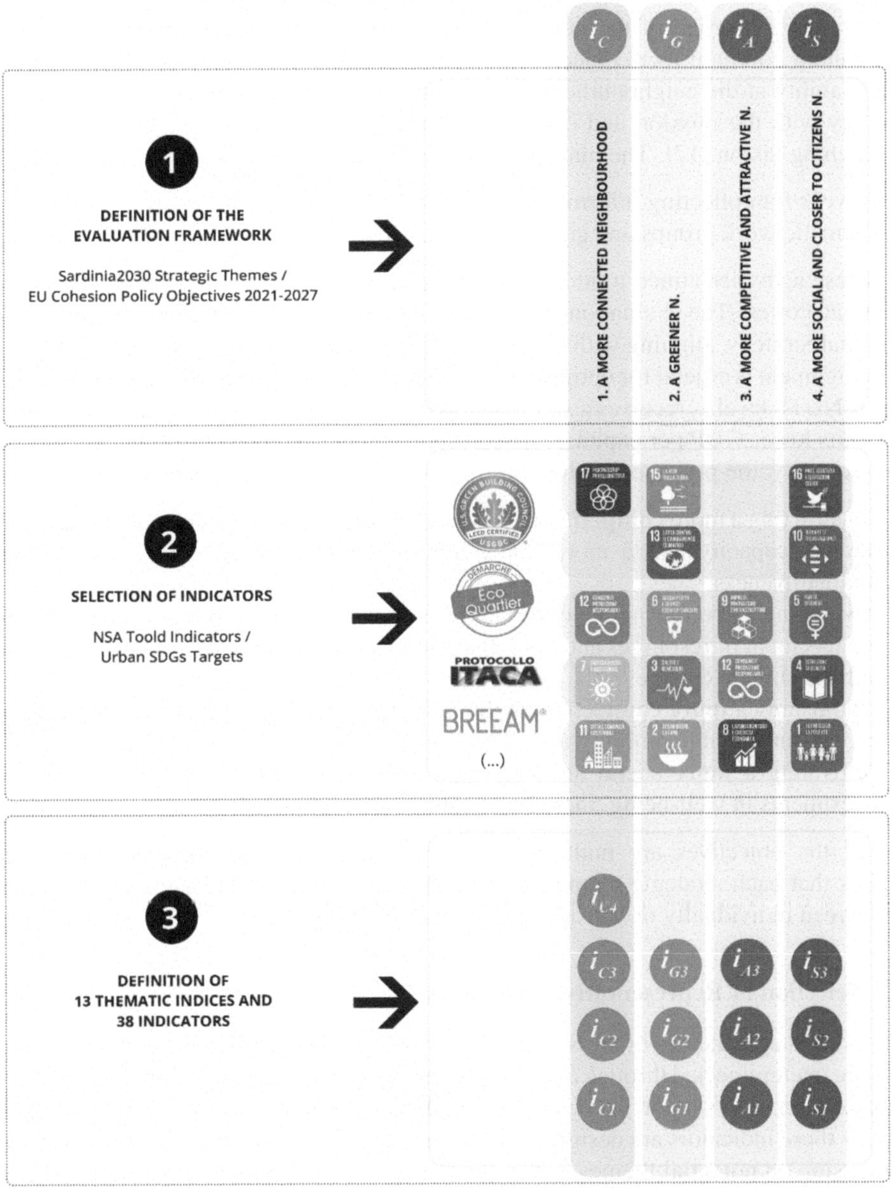

Fig. 1. The proposed Evaluation Framework: steps for the definition and structure.

In the next subsections we describe these phases in detail to provide readers a comprehensive understanding of the iterative and systematic approach undertaken to create the evaluation tool.

2.1 Strategy and Policy: Build a Common Vision

The primary objective of the initial phase was to establish the evaluation framework by defining common local objectives, ensuring a comprehensive assessment of urban sustainability at the neighbourhood scale. In this context, we aligned the Sardinia2030 Strategy with the mission and objectives of the Urban SOSLab to establish a unified overarching vision [17]. The main activities undertaken included:

- Surveys for collecting information.
- Thematic work groups and collective discussion.

These activities aimed to identify relevant evaluation issues and define key assessment categories. The evaluation framework maintains the five strategic themes of the Regional Strategy, aligning with the 2021–2027 EU cohesion policy objectives [10, 11]. This alignment is crucial for optimizing resources, given that the Sardinia Region, classified as NUTS level 1, receives a significant portion of funds for growth and employment due to its lower GDP per capita, which currently stands below 75% of the EU average. Consequently, the proposed evaluation framework encompasses goals related to:

1. Constructing a more **competitive** and **smarter** neighbourhood with enhanced administrative capacity and a more innovation-oriented production system for greater competitiveness.
2. Developing a **greener** neighbourhood focused on the conservation, management, and enhancement of resources and land, fostering an ecological and resilient transition to address climate change.
3. Establishing a **more connected** neighbourhood by promoting inclusive accessibility through sustainable mobility and digital accessibility.
4. Cultivating a **more social** and **closer to citizens** neighbourhood by encouraging investments in well-being, including health, social inclusion, equality, and culture.

All the objectives are multidimensional, synergistic, and complementary. This implies that each strategic theme and goal should be interrelated with the others and not be read individually or considered as self-supporting.

2.2 Selection of Representative Indicators

The five strategic themes of the Sardinia2030 Strategy, described in the previous subsection, are interconnected through a matrix presenting a specific set of 102 indicators. These indicators serve to support and guide the implementation of actions and policies. However, as these indicators are designed for the urban macro-scale, they may only be useful in focusing on potential themes and evaluation elements. For our neighbourhood-scale evaluation, we require indicators at a smaller scale.

Therefore, to establish specific indicators at the proposed level for the five strategic themes, we conducted an analysis and selection of evaluation and certification tools and systems. These tools were chosen from among the most significant ones currently employed internationally. We considered evaluation systems at both the city-wide and neighbourhood levels to formulate a list that is as representative as possible of the diverse objectives and thematic indicators in use.

Over the past few decades, there has been a noticeable increase in both academic and practical interest surrounding sustainability assessment tools, certification systems, and indicator frameworks. This trend reflects a tangible commitment to embracing multi-criteria analysis and sustainable accounting methodologies to gauge the efficacy of urban interventions.

These evolving patterns not only highlight a growing engagement with comprehensive evaluative approaches but also signify a maturing recognition within academic circles of the pivotal role these instruments play. They serve as essential tools for not only discerning but also enhancing various dimensions of sustainability outcomes stemming from neighbourhood level interventions in urban environments. This heightened scholarly and practical attention underscores the growing imperative to employ robust evaluative mechanisms to navigate the complexities inherent in fostering sustainable urban development [15, 18–21].

Proposing varied sets of criteria and indicators, these tools aim to measure, monitor, and evaluate the sustainability of existing urban areas, neighbourhoods, and planned interventions across multiple dimensions. These include environmental aspects, such as the relevance of green space provision, resource management (including water, energy, and waste), and transportation usage patterns.

Social considerations encompass factors such as housing affordability, the provision of services and public spaces, as well as levels of community participation and involvement. Additionally, economic dimensions involve evaluating economic activities and their impact on job creation within these neighbourhoods. This holistic approach underscores the interconnectedness of environmental, social, and economic factors in shaping sustainable urban development [15, 22, 23].

Regarding the urban scale, the primary point of reference includes the studies and tools developed by UN-Habitat for the implementation of the New Urban Agenda (NUA), such as the "City Prosperity Index" (CPI). The CPI serves as a comprehensive global index founded on a robust statistical approach that integrates multiple indicators to measure various dimensions of prosperity and sustainability. Recently, it underwent an update to incorporate all indicators of Goal 11 and the SDGs with an urban component. In its latest iteration from 2017, the CPI evolved into the City Prosperity Initiative-Perception Index (CPI-PI), also known as "CPI-SDGs", aligning itself with the urban sustainable development goals associated with the SDGs and the NUA.

Simultaneously, a selection of relevant assessment and certification tools for urban sustainability at the neighbourhood scale, known as "Neighbourhood Sustainability Assessment Tools" (NSA Tools), were also analyzed. These tools can be considered a significant reference for shaping a monitoring platform for Goal 11 and urban targets at the micro-urban scale [16]. Among these: LEED for Cities and Communities (US Green Building Council 2019), GBC neighbourhoods (Green Building Council Italia 2015), BREEAM Communities (Building Research Establishment Global Ltd.

2017), HQE2R, STAR Community Rating System (STAR Communities 2015), ÈCO-QUARTIERS (Ministère de la Transition Écologique 2021), ITACA Protocol (Institute for Innovation and Transparency in Government Procurement and Environmental Compatibility 2020) and others [15, 18–21, 24].

The correlations between UN sustainability goals and NSA (Neighbourhood Sustainability Assessment) tools has been emphasized in recent studies conducted by researchers [16, 25, 26] and private organizations dedicated to advancing the achievement of SDGs through their tools. Notably, tools such as BREEAM and DGNB have been highlighted for providing a correlation matrix that illustrates the alignment between global goals, indicators, and the criteria used for evaluating buildings or neighbourhoods. Moreover, the DGNB System District Criteria version 2020 offers an additional incentive known as the "Agenda 2030 bonuses" in nine criteria. These additional points are awarded to projects that make a significant contribution to urban climate, biodiversity, life-cycle assessment, energy, and mobility infrastructure [27–29].

In our study, we have curated a selection of indicators tailored to our specific case study. This process involved a comprehensive analysis of the environmental, social, and economic challenges within the Sardinia Regional Strategy (see Sect. 2.1) coupled with an extensive review of academic literature and best practices in urban sustainability assessment (see Sect. 2.2). Each indicator was chosen based on its relevance to the unique characteristics of the study area and its ability to provide meaningful insights into sustainability at the neighbourhood scale.

2.3 Identification of SDGs "Urban Targets"

Concurrently with the preceding step, an examination is conducted on the 230 SDGs indicators that hold relevance for an urban project. Beyond Goal 11, a comprehensive scrutiny of all SDGs is undertaken to pinpoint additional objectives linked to urban planning and design interventions. Employing this methodology, a distinct set of "urban targets" and indicators is identified, as outlined in the work by Saiu and Blečić [26]. In this way, the set identified in Sect. 2.2. is further refined and adapted to identify indicators aligned with the SDGs.

2.4 Definition of Local Indicators

To make the assessment tool effective within the local context of Sardinian urban areas, an analysis is conducted on open-access databases and information sources, at the national, regional and municipal level [27]. This examination aims to ascertain which indicators can be calculated and update periodically. The intended evaluation tool is designed for user-friendliness, targeting public administrations and local stakeholders. Consequently, it necessitates straightforward low-cost and easy management procedures.

The final set of indicators is achieved by applying the evaluation framework to a designated case study. This experimentation serves as a crucial test of the tool's validity and reliability in practice. Indicators are calculated through an automated procedure integrated into a Geographic Information System (GIS) platform that can generate georeferencing for thematic layers, facilitating overlay and map calculations. This process provides a comprehensive view of the neighbourhood's diverse performances across various thematic areas of evaluation, in relation to different SDGs.

The resulting visualization highlights priorities and proposes actions for improvement. The set of indicators and their descriptions is provided in Table 1.

Table 1. Evaluation framework: themes and indicators

Strategic Theme	Elements of evaluation / Goals	N. of Indices	N. of Indicators
A more competitive and atractive neighbourhood	social diversity, universal accessibility of urban space, support for vulnerable social categories, degree of resident participation, and women's proactivity	3	11
A greener neighbourhood	environmental quality concerning health and safety, assessed through the presence of urban greenery, the area's carbon dioxide (CO2) sequestration capacity, albedo effect, and permeability of public and publicly used surfaces, with the possibility of extending the calculation to private surface	3	7
A more connected and accessible neighbourhood	urban connectivity and permeability, vehicular, pedestrian, and cycling mobility, as well as public transportation	4	14
A more social and closer to citizens neighbourhood	functional diversity, variety and accessibility of different services and activities within the neighbourhood	3	8
	TOTAL	**13**	**38**

Figure 2 shows an exemplary calculation summary of indicators, which in this specific case illustrates the indicator "1.2.1 – Road surface dedicated to pedestrian pathways" one of the 11 indicators included within the strategic theme "A more connected and accessible neighbourhood".

The summary is divided into two parts: the side column contains the indicator's name, description, formula, and the data required for calculation, followed by the indicator value. On the left side, you can view the map of the area under examination, where colours represent the different values assumed by the indicator in various parts of the study area.

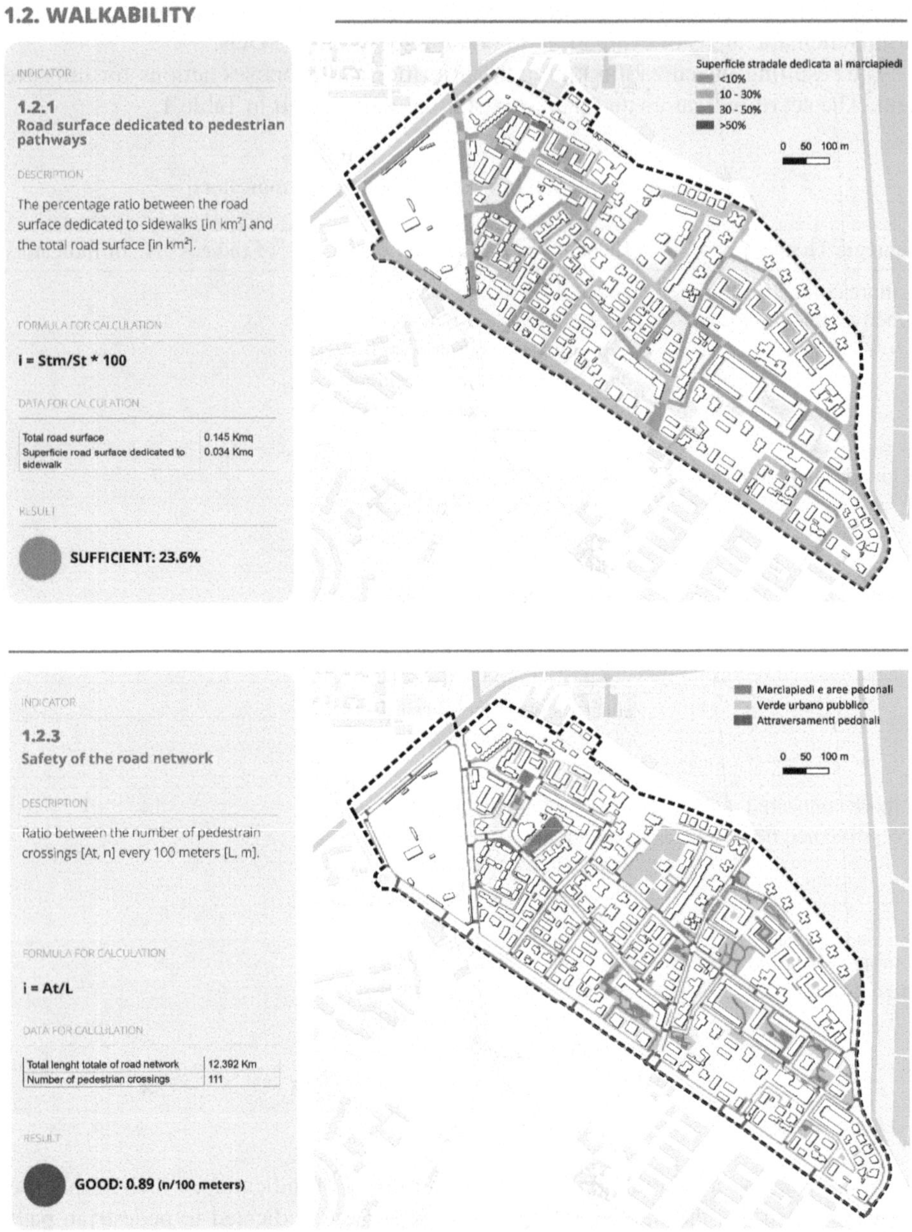

Fig. 2. An exemplary calculation summary prepared to evaluate two of selected indicators.

3 Conclusions

The methodology delineated in this paper presents a robust approach to assessing and localizing Sustainable Development Goals (SDGs) at the neighbourhood level. By aligning with the objectives of the Sardinia2030 Strategy and employing a comprehensive evaluation framework, this tool stands as a cornerstone for advancing sustainable urban development initiatives. Its systematic approach not only addresses the complexities of urban sustainability but also offers a practical means to translate global SDGs into actionable strategies at the local level.

Through meticulous selection and adaptation of indicators, this methodology ensures a nuanced understanding of neighbourhood-level sustainability, taking into account the unique context and challenges of each locality. By focusing on urban targets across various SDGs, it enables a holistic evaluation that encompasses environmental, social, and economic dimensions of sustainability. This multi-dimensional perspective is crucial for fostering resilient and inclusive communities in the face of contemporary urban challenges.

Furthermore, the methodology's integration of data from open-access databases and utilization of Geographic Information System (GIS) technology enhances its analytical capabilities, allowing for the comprehensive assessment of neighbourhood performance across diverse thematic areas. This data-driven approach not only facilitates evidence-based decision-making but also enables stakeholders to track progress over time and identify areas for improvement.

The user-friendly design of the tool ensures accessibility for public administrations and local stakeholders, democratizing the process of urban planning and decision-making. By providing a common framework and language for discussing sustainability goals, the methodology fosters collaboration and consensus-building among diverse stakeholders, thereby enhancing the likelihood of successful implementation.

In conclusion, this methodology offers more than just a technical tool; it represents a paradigm shift towards a more integrated and participatory approach to urban sustainability. Its potential applications extend beyond the specific context of Sardinia, serving as a valuable resource for urban planners, policymakers, and researchers worldwide.

Furthermore, by providing actionable insights and tangible outcomes, this methodology has the potential to inform real-world urban development projects and policy interventions. It equips decision-makers with evidence-based information to guide strategic investments and interventions, ultimately contributing to the creation of healthier, more resilient, and equitable cities for future generations.

We have begun to explore the potential for scaling up our methodological approach into a more generalized framework. This endeavour entails a careful examination of the outcomes derived from our specific case study, assessing the transferability and adaptability of our methodology to diverse urban contexts. Our aim is to develop a flexible methodology that can be readily applied across various urban settings to assess sustainability and inform planning and development decisions for creating healthier, more resilient, and equitable cities for future generations.

Acknowledgement. This research was founded by the Italian Ministry of Ecological Transition (now Ministry of Environment and Energy Security) within the project titled "SOSLabs. Laboratori di ricerca-azione per la Sostenibilità urbana: Lab2 Urban Sustainability" for the period 2020–2022.

The portion of this study focusing on objective 11 of the 2030 Sustainable Development Goals "Sustainable Cities and Communities", is currently undergoing expansion through the project "A GLOcal knowledge System for Sustainability Assessment of urban projects (GLOSSA)", a Research Project of National Relevance (PRIN) funded by the Ministry of University and Research (MUR) (CUP F53D23005650001). This project received financial support by the Italian Ministry of University and Research (MUR) under the National Recovery and Resilience Plan (NRRP), Mission 4, Component 2, Investment 1.1, Call for tender No.104 published on 2.2.2022, funded by the European Union – NextGenerationEU.

Additionally, the segment of this study focusing on mobility is currently being expanded within the MOST – Sustainable Mobility National Research Centre, which received funding from the European Union Next-Generation EU (PNRR – Italian National Recovery and Resilience Plan – Mission 4 Component 2, Investment 1.4—D.D. 1033 17 June 2022, CN00000023).

References

1. United Nations Development Group (UNDP): Localizing the post-2015 development agenda: dialogues on implementation, New York (2014)
2. SDSN -Sustainable Development Solutions Network: Getting started with the SDGs in cities: a Guide for Stakeholders. https://resources.unsdsn.org/getting-started-with-the-sdgs-in-cities. Accessed 24 Jan 2024
3. Fisher, A., Fukuda-Parr, S.: Introduction—data, knowledge, politics and localizing the SDGs. J. Hum. Dev. Capab. **20**(4), 375–385 (2019). https://doi.org/10.1080/19452829.2019.1669144
4. Khussamov, R., Galiy, E., Anisimov, E., Ershova, L., Nemkov, D.: National strategies for sustainable development G-7: trends 2010–2020. In: E3S Web of Conferences. EDP Sciences, Les Ulis, France (2020)
5. Ortiz-Moya, F., Saraff Marcos, E., Kataoka, Y., Fujino, J.: State of the voluntary local reviews 2021: from reporting to action. Institute for Global Environmental Strategies (2021)
6. Narang Suri, S., Miraglia, M., Ferrannini, A.: Voluntary local reviews as drivers for SDG localisation and sustainable human development. J. Hum. Dev. Capab. **22**, 725–736 (2021). https://doi.org/10.1080/19452829.2021.1986689
7. Siragusa, A., Stamos, I., Bertozzi, C., et al.: European Handbook for SDG Voluntary Local Reviews, 2022nd edn. Publications Office of the European Union, European Commission, Joint Research Centre (2022)
8. Cavalli, L., Farnia, L., Vergalli, S., Lizzi, G., Romani, I.G., Alibegovic, M.: Knowing the present for a sustainable future: the SDGs Index for the provinces and metropolitan cities of Italy (2020)
9. Saiu, V., Blečić, I.: SDGs implementation in Italy: a comparative assessment of subnational strategies for sustainable development. In: Gervasi, O., Murgante, B., Misra, S., Ana, M.A., Rocha, C., Garau, C. (eds.) Computational Science and Its Applications – ICCSA 2022 Workshops: Malaga, Spain, July 4–7, 2022, Proceedings, Part III, pp. 627–638. Springer International Publishing, Cham (2022). https://doi.org/10.1007/978-3-031-10545-6_42
10. Regione Autonoma della Sardegna, Sardegna2030: La Strategia della Regione Sardegna per lo Sviluppo Sostenibile. https://delibere.regione.sardegna.it/protected/57127/0/def/ref/DBR57095/. Accessed 4 Oct 2021
11. Regione Autonoma della Sardegna, Sardegna2030. http://www.regione.sardegna.it/j/v/2847?s=1&v=9&c=94636&na=1&n=4&nodesc=1&ph=1&disp=2. Accessed 30 Jan 2022

12. UN-Habitat: Sustainable Development Goals Cities. https://unhabitat.org/programme/sustainable-development-goals-cities. Accessed 26 Jan 2022
13. Saiu, V.: Evaluating outwards regeneration effects (OREs) in neighborhood-based projects: a reversal of perspective and the proposal for a new tool. Sustainability **12**, 10559 (2020). https://doi.org/10.3390/su122410559
14. Sala Benites, H., Osmond, P., Rossi, A.M.G.: Developing low-carbon communities with LEED-ND and climate tools and policies in São Paulo, Brazil. J. Urban Plan. Dev. **146** (2020). https://doi.org/10.1061/(ASCE)UP.1943-5444.0000545
15. Sharifi, A., Dawodu, A., Cheshmehzangi, A.: Neighborhood sustainability assessment tools: a review of success factors. J. Clean. Prod. **293**, 125912 (2021). https://doi.org/10.1016/j.jclepro.2021.125912
16. Saiu, V., Blečić, I., Meloni, I.: Making sustainability development goals (SDGs) operational at suburban level: potentials and limitations of neighbourhood sustainability assessment tools. Environ. Impact Assess. Rev. **96**, 15 (2022). https://doi.org/10.1016/j.eiar.2022.106845
17. Saiu, V., Blečić, I., Cocco, G., Meloni, I.: Urban sustainability and SDGs implementation between regional strategy and local practice: case of Sardinia. In: Filho, W.L., Dinis, M.A.P., Moggi, S., Price, E., Hope, A. (eds.) SDGs in the European Region, pp. 1–32. Springer International Publishing, Cham (2022). https://doi.org/10.1007/978-3-030-91261-1_64-1
18. Feleki, E., Vlachokostas, C., Moussiopoulos, N.: Characterisation of sustainability in urban areas: an analysis of assessment tools with emphasis on European cities. Sustain. Cities Soc. **43**, 563–577 (2018). https://doi.org/10.1016/j.scs.2018.08.025
19. Kaur, H., Garg, P.: Urban sustainability assessment tools: a review. J. Clean. Prod. **210**, 146–158 (2019). https://doi.org/10.1016/j.jclepro.2018.11.009
20. Sharifi, A., Murayama, A.: Neighborhood sustainability assessment in action: cross-evaluation of three assessment systems and their cases from the US, the UK, and Japan. Build. Environ. **72**, 243–258 (2014). https://doi.org/10.1016/j.buildenv.2013.11.006
21. Sharifi, A., Murayama, A.: A critical review of seven selected neighborhood sustainability assessment tools. Environ. Impact Assess. Rev. **38**, 73–87 (2013). https://doi.org/10.1016/j.eiar.2012.06.006
22. Berardi, U.: Sustainability assessment of urban communities through rating systems. Environ. Dev. Sustain. **15**, 1573–1591 (2013). https://doi.org/10.1007/s10668-013-9462-0
23. Merino-Saum, A., Halla, P., Superti, V., Boesch, A., Binder, C.R.: Indicators for urban sustainability: Key lessons from a systematic analysis of 67 measurement initiatives. Ecol. Indic. **119**, 106879 (2020). https://doi.org/10.1016/j.ecolind.2020.106879
24. Saiu, V.: The three pitfalls of sustainable city: a conceptual framework for evaluating the theory-practice gap. Sustainability **9**, 2311 (2017). https://doi.org/10.3390/su9122311
25. Wen, B., et al.: The role and contribution of green buildings on sustainable development goals. Build. Environ. **185**, 107091 (2020). https://doi.org/10.1016/j.buildenv.2020.107091
26. Saiu, V., Blečić, I., Cocco, G., Meloni, I.: Urban Sustainability and SDGs Implementation between Regional Strategy and Local Practice: Case of Sardinia. In: Leal Filho, W., Dinis, M.A.P., Moggi, S., Price, E., Hope, A. (eds) SDGs in the European Region. Implementing the UN Sustainable Development Goals – Regional Perspectives. Springer, Cham. https://doi.org/10.1007/978-3-031-17461-2_64
27. DGNB: Building for a better world. How buildings contribute to the UN sustainable development goals (SDGs) (2020). https://www.dgnb.de/de/verein/publikationen/bestellung/downloads/DGNB_Report_Building_for_a_better_world.pdf
28. DGNB: DGNB System District Criteria Set. Version 2020 (2020)

29. DGNB: Building for a better world: new DGNB report on the sustainable development goals. https://www.dgnb.de/de/aktuell/pressemitteilungen/2021/dgnb-report-sdgs. Accessed 3 Mar 2022
30. Saiu, V., Blečić, I., Meloni, I., Piras, F., Scappini, B.: Towards a SDGs Based Neighborhood Sustainability Evaluation Framework: A Tool for Assessing Sustainability at the Urban Micro-Scale. In: Abastante, F., et al. Urban Regeneration through Valuation Systems for Innovation. Green Energy and Technology. Springer, Cham, pp. 195–215 (2022). https://doi.org/10.1007/978-3-031-12814-1_12

Relationship Between Real Estate Indices and Inflation Trends: An Application to the European Residential Property Market

Marco Locurcio[1], Pierluigi Morano[1], Francesco Tajani[2](✉), Francesco Sica[2], and Tiziana Acquafredda[1]

[1] Polytechnic University of Bari, 70126 Bari, Italy
[2] "Sapienza" University of Rome, 00196 Rome, Italy
francesco.tajani@uniroma1.it

Abstract. The present paper aims to analyse the European residential real estate market through a set of indicators elaborated by Eurostat. By means of the use of correlation analysis, the comparison of several housing market indicators related to European nations enables the creation of a synthetic and unambiguous representation of the residential property market trend, as well as the comprehension of the dynamics that are seen within each nation and of the relationships that exist between the indicators with nations. The use of deflated indicators and the analysis of inflation trends has allowed us to highlight the peculiarities of the Italian residential property market and its ability to act as a safe-haven asset.

Keywords: Inflation · Housing prices · Real estate portfolio

1 Introduction

The role of the real estate investment in the context of the global economy is a controversial issue. The idea that real estate allows to protect capital from the depreciation due to inflation is widespread among investors, but the validity of this assumption is often called in questions especially during periods of uncertainty.

In Italy in January 2024, the estimated National consumer price index for the Whole Community (NIC) gross of tobacco [1], and the consumer price index for Manual and Non-manual workers' Families (FOI) [2] both register an increase of 0.8% on a yearly basis. In December 2023 the producer prices of construction for "Residential and non-residential buildings" increased by 0.4% on a yearly basis, while industrial producer prices fall by 16.0% on a yearly basis [3].

The uncertainty deriving from the rise of inflation rate and construction prices is also fuelled by the energetic crisis and the tensions in the current geopolitical backdrop. In such unpredictable circumstances it is questionable whether the real estate is a good hedge against inflation and, thus, an advantageous asset class to invest in, with an appropriate return with respect to the risks involved.

O. Gervasi et al. (Eds.): ICCSA 2024 Workshops, LNCS 14818, pp. 113–125, 2024.
https://doi.org/10.1007/978-3-031-65273-8_8

2 Aim

The objective of the present research is to determine whether investing in the residential real estate is an appropriate choice in order to hedge the capital against inflation, namely if the residential property investments reliably perform as shield against the loss of the nominal value of households' savings due to inflation.

Such issue can be addressed by exploring how house prices behave in relation to the fluctuations in the level of inflation. Practically, the aim of this work is to find the relationship between the characteristics of the residential real estate market and the trend in consumer prices inflation in the European economic context, in order to determine to what extent the real estate can be considered as a safe haven.

In the next sections of this work, a portion of the scientific literature on the dynamics of residential markets and their relationship to inflation and other socioeconomic indicators is reviewed. The next paragraph focuses on methodological suggestions and data sources. The last phase involves putting the case study into practice. This is done by conducting two correlation analyses, each from a different perspective, using data related to the European Union. After looking over the empirical findings, some recommendations for residential real estate investment guidelines are made.

3 Literature Review

The scholarly research on the influence of rising prices for goods and services on price increases for homes frequently seeks to determine if a residential property is an investment capable of protecting investors' equity from the negative repercussions of inflationary processes. It is feasible to claim that research on this subject does not always yield comparable outcomes based on the sample of papers examined for this study's scope.

In general, the majority of the analyzed papers find an existing correlation between house prices dynamics and inflation. For example, Goodhart and Hofmann [4] report that residential properties inflation is one of the three most influential factor in the prediction of future consumer price inflation, while Tsatsaronis and Zhu [5] detect that inflation is the most influential determinant of the changes in housing price, in particular when inflation rate is high.

In the most part of the studies in which a correlation is identified, this turns out to be positive [6]. Specifically, several studies [7–10] agree on the asymmetry of such positive relation, with a bigger impact of inflation on housing prices than that of housing prices on inflation, and outline that residential properties investments effectively hedge inflation, as also reported by Wu and Tidwell [11], Christou et al. [12], Tsatsaronis and Zhu [5].

The empirical literature on this subject generally suggests that the residential property presumably has inflation-hedging characteristics. Nonetheless there is no absolute agreement on this assumption and, more in general, on the existence and the typology of relation between inflation and house prices trends.

Various deductions can be drawn from the same empirical analysis depending on how certain parameters are changed, such as the geographic or temporal focus on which housing market data is observed, as well as the varied data sources and methodology that are utilised. Wu and Tidwell [11] highlight that the inflation-hedging efficacy depends on the analyzed area of the real estate market. Tsatsaronis and Zhu [5] remark that this efficacy is bigger during the high-inflation periods. Azmi et al. [13] outline that the residential property can generally be considered as a partially inflation-hedging asset, but the extent of this characteristic depends on the specific property market (geographical collocation and market behavior) and on the specific asset typology. Lee [14] finds out that investing in small and medium size real estate assets during high inflation periods contributes to the preservation of purchasing power. Instead, during the low inflation and deflation periods, investment on residential properties is not an advisable choice. Nneji et al. [15] claim that movements in the inflation rate notably impact on house prices in the steady-state and boom regimes only, while a disconnection between them is reported during the housing busts. Clapp and Giaccotto [16] distinguish among expected and unexpected inflation, finding that the former is negatively related while the latter is positively correlated with housing prices.

The study carried out by Demary [17] demonstrates that changes in house prices can be used as an indicator of forthcoming inflation, in accordance with Goodhart and Hofmann [4]. But, in contrast with the other studies mentioned above, this study finds that a negative relationship exists between these two elements. In fact, as a consequence of an inflationary shock the house prices show a gradually decreasing trend, which can be due to the rise in interest rates. This result in contrast with the majority of the literature can be attributed to the fact that a simplified methodology was employed. Demary [17] also shows that the volatility of housing prices is higher than the volatility of cyclical components of inflation, in agreement with the results obtained by Tang et al. [18] which show a higher House Price Index (HPI) aggregate growth rate compared to that of inflation. This finding makes the real estate investment profitable in periods of rapid market growth. Nevertheless, Tang et al. [18] do not detect any significant long-run transmission between housing prices and actual, expected or unexpected inflation, demonstrating that housing price cannot hedge inflation in the Chinese market. The authors attribute the reason for the discrepancy between their findings and those of previous Chinese housing market studies to the use of different data sources, different methodologies and different composition of the employed indicator. Wolski [19] achieves comparable outcomes in a different real estate market. For the Polish real estate market, there exists no short-term relationships between residential property prices and inflation indexes. The author explains this lack of correlation with the fact that the analyzed market is shallow, and that the research period is short and only includes a little period of elevated inflation.

4 Methodology

The eventual existence of a relationship between the house prices and consumer prices inflation can be investigated through a correlation analysis, that eventually reveals also what kind of relationship it is in terms of type and strength. Indeed, based on the correlation coefficient it is possible to deduce if two variables have a linear association, if this possible relationship is positive or negative, and if it is weak, moderate or strong.

Specifically, two different correlation analyses will be carried out: (i) the first one will be performed between the various selected variables, for each country, and (ii) the second one will be implemented between the different countries, for each variable. Consequently, two different groups of correlation matrices will be produced: the first group investigates different real estate phenomena (represented by diverse variables) within each single country; meanwhile each matrix belonging to the second group analyzes the possible relationship existing for a given phenomenon among different countries.

The characteristics of the housing market and the trend of consumer prices in the European Union area can be observed through the indicators and variables provided by the Eurostat database [20]. The consistency among the national data collection and elaboration is a significant benefit of using this sort of information source.

For each of the following indicators annual data were chosen, rather than monthly and quarterly ones, in order to observe the evolution of values from a broader point of view. Moreover, the average indices were always preferred over the rates of change. For consistency and unity of perspective, all the annual average indices picked up were referred to 2015 as the base year, namely the year in which each index value is conventionally set to equal 100. The gathered information covers the years 2010–2022, which is a little period of time but guarantees the ongoing existence of the historical series, what has also been used as a criterion for choosing the indicators and the member states to look at.. As regards the geographical coverage of the investigation, national data were collected for 24 member countries of the European Union equipped with available and complete time series of the indicators deemed consistent with the scope of the work.

The Eurostat "Detailed dataset" describes 9 different statistical themes, among which the "Economy and finance" section collects the majority of indicators concerning the subject matter of this work. In fact, in the sub-category "Prices" it is possible to find the House Price Index (HPI), the Owner-Occupied House Price Index (OOHPI) and the Harmonised Index of Consumer Prices (HICP). Meanwhile the deflated version of the HPI is included in the "EU policies" dataset.

Firstly, the HPI is important because it captures the changes in transaction prices of all types of dwellings, both new and existing, purchased by households from any sector and for any purpose (e.g. owner-occupancy, rent, investment etc.). This index measures inflation in the housing market. The HPI is available with respect to purchases of newly built dwellings and purchases of existing dwellings separately, but also with regards to all purchases. The inflation-adjusted version of the latter is the deflated HPI, which is therefore an interesting indicator to be considered.

Another indicator considered in the empirical analysis is OOHPI, which is more inclusive than HPI from the point of view of the encompassed expenditures and narrower than HPI as regards the purpose of the purchases. While the prices considered for the calculation of the HPI only take into account the purchase price of a property excluding additional expenses which arise for ownership and maintenance purposes (e.g. fees, commissions, major repairs etc.), the OOHPI incorporates all of these costs. Moreover, OOHPI covers the owner-occupancy as the only dwelling acquisition purpose: this indicator solely refers to the dwellings which are owned and occupied by the same household. To be more specific, the OOHPI encompasses the changes in prices incurred by households (i) for the "acquisition of dwellings" that were previously outside the household sector, for the purpose of owner-occupancy, and (ii) for the "ownership of dwellings", which are consequent to the occupancy of the property. The "existing dwellings new to households" and "self-build dwellings and major renovations" sub-components are excluded because data are not available for every considered European country. The sub-component called "other services related to the acquisition of dwellings" is also excluded because it regards some factors which are not relevant for the scope of this study, such as property transfer tax, notary fees and real estate agent commissions. The second component, related to the ownership, is investigated only with regards to the "major repair and maintenance" sub-component, because it has a more significant weight on the total. In fact, the weight associated with this sub-component is 187.2 in EU, compared with 24.0 corresponding to "insurance connected with dwellings" and 8.8 related to "other services related to ownership of dwellings". Moreover, the home insurance market is not widespread in Italy and in many other European countries, which causes a low incidence of insurance costs for residential property on the total costs incurred by households and consequent a low relevance of such indicator. [21].

Lastly, in this analysis the HICP is considered in terms of overall index representing all-items inflation. As regards housing, the HICP solely includes actual rents, incurred by tenants, and light maintenance costs, incurred by renters and owner-occupiers [22].

It is possible, but in this case it is not necessary, to separately consider the HICP related to Housing, which is one of the 12 division according to the European Classification of Individual Consumption according to Purpose (ECOICOP) and includes (i) actual rentals for housing, (ii) imputed rentals for housing, (iii) maintenance and repair of the dwelling, (iv) water supply and miscellaneous services relating to the dwelling, (v) electricity, gas and other fuels. The individual investigation of this portion of expenditures, which is exclusively related to the dwelling's ownership, is not necessary to accomplish the scope of this analysis because the focus is mainly on transaction prices for the dwelling's acquisition.

As mentioned above, also the evolution of the costs of the production factors (mostly labor, material and energy), and consequently the evolution of the prices proposed by the building constructor to the client, have a strong influence on house prices. Eurostat describes these two aspects providing respectively the Construction Cost Indicator (CCI) and the Construction Producer Price Index (CPPI), but in this analysis they are excluded because they are both solely focused on the economic activity of new residential buildings' construction. Moreover, complete CCI time series are not available for some of the 24 analysed countries, and the HICP is representative of these aspects because it reflects the price development of factors of production such as materials and energy.

Table 1 captures the selected Eurostat variables and their definitions.

Table 1. List of the considered Eurostat indicators and the associated description.

N.	Eurostat indicator	Description
1	House Price Index - HPI	Index which measures the movements in prices of all dwellings bought by households, independently of their previous owners and the purpose of the purchase. The land component is included

(*continued*)

Table 1. (*continued*)

N.	Eurostat indicator	Description
2	HPI, purchases of newly built dwellings	HPI limited to purchases of newly built dwellings
3	HPI, purchases of existing dwellings	HPI limited to purchases of existing dwellings
4	Deflated HPI	It is the ratio between the HPI and the national accounts deflator for private final consumption expenditure. Thus, it quantifies inflation in the residential property market with respect to that in the private final consumption expenditure of households and non-profit institutions serving households (NPISHs)
5	Owner-Occupied Housing Price Index - OOHPI	Index measuring changes in (i) prices incurred by owner-occupiers (households living in a dwelling that they own) for the initial acquisition of dwellings that are new to household sector, and (ii) in other costs related to the ownership of dwellings
6	OOHPI, acquisition of dwellings	Index measuring changes in expenditures incurred by households for the initial acquisition of dwellings that are new to household sector (new dwellings; existing dwellings new to households) and other services related to the acquisitions of dwellings
7	OOHPI, new dwellings	It reflects changes in prices incurred by own-occupies for new dwellings (newly built dwellings' acquisition; expenditures for self-built dwellings and for major renovations)
8	OOHPI, purchases of newly built dwellings	It reflects changes in prices for newly built dwellings purchased by households for their own use. Similar to "HPI, purchases of newly built dwellings" but: (i) purchases for purposes different from owner-occupancy are included in HPI and not in OOHPI; (ii) in the OOHPI the land value is excluded from weights but included in prices, while in the HPI the land value is included in both weights and prices
9	OOHPI, major repair and maintenance	It is a sub-category of OOHPI related to the ownership of dwellings. It reflects the change in the owner-occupier's expenditure for the major repair and maintenance of the dwellings they own and in which they live
10	Harmonised Index of Consumer Prices - HICP	HICP provides a standardized and comparable measure of inflation for countries and country groups. It measures the movements over time in prices of consumer goods and services acquired by households

5 Case Study

The two correlation analyses, which were previously introduced, are performed on the data for the ten selected Eurostat indicators for the 2010–2022 period with regards to the 24 European Member States for which complete series are available (Belgium, Bulgaria, Czechia, Denmark, Germany, Estonia, Ireland, Spain, France, Croatia, Italy, Cyprus, Lithuania, Luxembourg, Hungary, Malta, Netherlands, Austria, Poland, Portugal, Romania, Slovenia, Slovakia, Sweden). In addition to the data for individual countries, the analysis was also performed with regards to the EA-20 and EU-27 aggregates, which are the aggregate indices covering respectively the countries of the Euro Area (20, from 2023) and all the European Union countries (27, since 2020) [23].

5.1 Correlation Analysis for Indicators Comparison

The first correlation analysis allows to investigate the trends of different real estate phenomena, represented by the selected indicators variables, within each single country.

The obtained results are similar when such analysis is performed on the data for the vast majority of the 24 EU countries taken individually, for the 20 EA aggregate and for the 27 EU aggregate (see Table 2): the existence of a strong positive correlation is detected among the variables describing real estate market and inflation dynamics. A different behaviour can be observed only in a few countries: the results of the first correlation analysis for Italy (see Table 3), Cyprus and Romania show inhomogeneity in the obtained correlation coefficients, with some indicators being positively correlated, some others being negatively correlated, and with a wide and varied range of coefficients.

Specifically, by looking at the correlation coefficients among Italian indicators it is possible to notice that in the Italian housing market the HPI for all dwellings has a strong positive correlation (0.99) with the HPI for existing dwellings, while it is weakly positively correlated (0.23) with the HPI for new dwellings. Consistently, the deflated HPI turn out to have a strong positive correlation (0.97) with the HPI for existing dwellings and a weak negative relationship (–0.12) with the HPI for new dwellings. This is confirmed by the fact that outcomes of the correlation analysis between the total HPI and each other indicator are similar to those obtained analysing the correlation between the existing dwellings HPI and the remaining indicators, in both cases contrasting with European results. Opposite results instead stand out from the correlation analysis between the new dwellings HPI and the other indicators, which generally show the existence of strong positive relationships, in compliance with the results observed at the European level.

The behavioural similarity reported for the market of existing dwellings and for the overall national housing market may be a consequence of the preponderance of the existing properties with respect to the entirety of Italian residential real estate assets. However, this consideration is valid for the majority of EU Member States, as proven by 2023 data about weights of newly built and existing dwellings in total dwellings [24], therefore it is not a satisfactory explanation of the difference between the behaviour of the Italian real estate market and that of the majority of European countries. Another possible explanation for this finding is the different effect exerted by location on the price formation for new and existing buildings. In fact, existing buildings are typically located in urban areas which are well equipped with road infrastructure and served by public transport, activities, and essential services. Differently, new buildings are mainly located in peripheral areas which do not dispose of such facilities. The differences in location, and specifically in the characteristics of the urban area, strongly influences house prices. In the Italian housing market, which is subdivided into OMI zones reflecting the characteristics of different locations [25], there is a wide gap among house prices in central and in peripherical areas, with rising house prices by switching from no-urban areas to urban ones, and furtherly increasing for metropolitan zones. This can depend on the contrast between the high demand for housing services and the low elasticity of housing supply, on many social and economic indicators [26], but also on the quality and on the typology of the urban area. In fact, d'Acci [27] has found that in Italy a 1% increase in the quality of the urban area in which the property is located results in a 0.58% increase in its value; in addition, Casolaro and Fabrizi [28] demonstrated that the correlation between the quality of housing and house prices turns out to be stronger in the municipalities outside of the metropolitan areas.

The difference in the correlation coefficients measured among each indicator and, respectively, the HPI for new dwellings and the HPI for existing ones is consistent with the nearly absent correlation between these two HPIs (0.10). In fact, the existing and new housing markets behave differently in relation to inflationary dynamics: there is a strong positive relationship between the HPI for new dwellings and the HICP (0.71), while there is a moderate negative relationship (–0.60) between the HPI for existing dwellings and the HICP. On the one hand, the reason of this finding can be associated to the close link between the prices of the factors of production and the newly built houses: in high inflation periods the prices of materials and energy increase, leading to higher construction costs incurred by building contractors and consequently to higher prices for newly built dwellings. On the other hand, high inflation causes a reduction in consumers' purchasing power and, hence, the prices of existing houses are forced to contract.

The results of the first correlation analysis for Italian data (i) between the OOHPI categories and the HPI for new dwellings, (ii) among the several categories of the OOHPI, and (iii) between the OOHPI categories and HICP, are consistent with those obtained at the European level. Moderate and strong relationships (from 0.60 to 1) exist among the different categories of OOHPI and also between such categories and the HPI for new dwellings, while a strong positive correlation (from 0.71 to 0.92) is captured between the OOHPI and the HICP. This finding indicates that, when focusing exclusively on the retail residential real estate market, concerning the purchases of the "first home" and leaving aside the acquisitions for investment purposes, the Italian market follows the prevailing European trend. Differently, the relationships existing between the OOHPI categories and, respectively, the total HPI, the HPI for existing dwellings and the deflated HPI, range from a weak positive correlation (0.23) to a strong negative one (–0.85). Outcomes similar to those obtained for Italy also result, even if in a less marked way, when performing the first correlation analysis on data from Cyprus and Romania.

Table 2. First correlation analysis among the ten selected Eurostat indicators (numbered according to Table 1), with reference to European Union - 27 countries (from 2020).

EU-27	1	2	3	4	5	6	7	8	9	10
1	1.00									
2	1.00	1.00								
3	1.00	1.00	1.00							
4	0.98	0.98	0.98	1.00						
5	0.99	0.99	0.99	0.94	1.00					
6	0.99	0.99	0.99	0.94	1.00	1.00				
7	0.99	0.99	0.99	0.94	1.00	1.00	1.00			
8	1.00	1.00	1.00	0.98	0.98	0.99	0.99	1.00		
9	0.96	0.96	0.96	0.89	0.99	0.99	0.99	0.96	1.00	
10	0.92	0.92	0.92	0.83	0.97	0.97	0.97	0.92	0.92	1.00

Table 3. First correlation analysis among the ten selected Eurostat indicators (numbered according to Table 1), with reference to Italy.

Italy	1	2	3	4	5	6	7	8	9	10
1	1.00									
2	0.23	1.00								
3	0.99	0.10	1.00							
4	0.94	–0.12	0.97	1.00						
5	–0.22	0.89	–0.34	–0.53	1.00					
6	–0.03	0.96	–0.16	–0.36	0.98	1.00				
7	–0.31	0.85	–0.43	–0.62	0.99	0.96	1.00			
8	0.23	1.00	0.10	–0.12	0.89	0.96	0.85	1.00		
9	–0.63	0.60	–0.72	–0.85	0.89	0.79	0.93	0.60	1.00	
10	–0.50	0.71	–0.60	–0.76	0.92	0.85	0.96	0.71	0.96	1.00

5.2 Correlation Analysis for Geographical Comparison

The second correlation analysis allows the comparison between different countries of the same phenomenon among those described by the previously introduced indicators. Such analysis highlights that the Italian HPI (especially with regards to existing dwellings) and its deflated version are negatively correlated with those of almost all the European countries, which are instead characterised by similar behaviours, hence strongly positively correlated.

The only large European country which is not perfectly aligned to the European average is Spain, while Cyprus and Romania turn out to be quite uncorrelated to the rest of Europe. The graph in Fig. 1 presents the comparison of the main indicators, namely the HPI, the deflated HPI and the OOHPI, with regards to Italy, Europe and other Member States such as Germany, Spain and France. It is interesting to outline that Italy follows a pattern which is symmetrical to the European average, and also a more limited range of fluctuations appear in the period 2010–2022.

Specifically, a closer inspection of the total HPI graph reveals different trends in the European countries: a linear growth in Germany; in France, in compliance with the average European behaviour, a constant trend in the period 2010–2014 and a growing trend from 2014 onwards; for Spain, a sharp decrease until 2013 and a sharp turnaround from 2013 onwards; in Italy, a sinusoidal trend with temporally shorter cycles but shorter excursions, and a slower recovery of HPI. The same dynamics can be seen with reference to the deflated HPI, but with less pronounced growth trends especially from 2021 because of the strong inflation observed from such year onwards.

Finally, for all the in countries with the exception of Italy, the graphical representation of total OOHPI shows results similar to those shown for HPI, although with minor variations; in Italy there is a total OOHPI less variable in the period analysed; differently, in Italy the total OOHPI was less variable in the analysed period.

5.3 Conclusions for Investment Portfolio

As highlighted in the previous analyses, the purchase of a residential property with the aim of including it in the investment portfolio can generally be an advantageous choice for the investor [29]. Some of the possible benefits that this kind of investment can provide are increased income flow and protection against inflation.

However, the extent to which this choice is convenient depends on various factors such as the type of property, thus the chosen sub-sector of the residential real estate market, and the current market conditions, related to the geographical area in which the property is located and the characteristics of the specific local market.

Thus, the first way to protect the capital through the diversification towards risk is to differentiate the real estate portfolio from a typological point of view. A distinctive feature of the Italian housing market, which can assist Italian investors in the selection of the more convenient property type, is the weak correlation among the existing dwellings market and the new dwellings one. This finding allows investors to minimize the risks that they face when they build a portfolio of real estate assets, even when it is uniform in terms of residential use, by diversifying it across new and existing properties.

Another way to minimize the risks linked to the real estate investment regards the geographical diversification of the portfolio on a European scale. Investing in the real estate markets of different nations which are correlated by a strong positive relationship, as most European countries are, may not be safe. Conversely, the negative correlation existing between Italy and the majority of other European Member States with regards to the HPI makes the investment in the Italian housing market highly attractive. The discrepancy between the behaviour of the Italian market and that of other European countries, indeed, is a beneficial characteristic when it comes to diversifying the portfolio with the goal of consequently minimising risks.

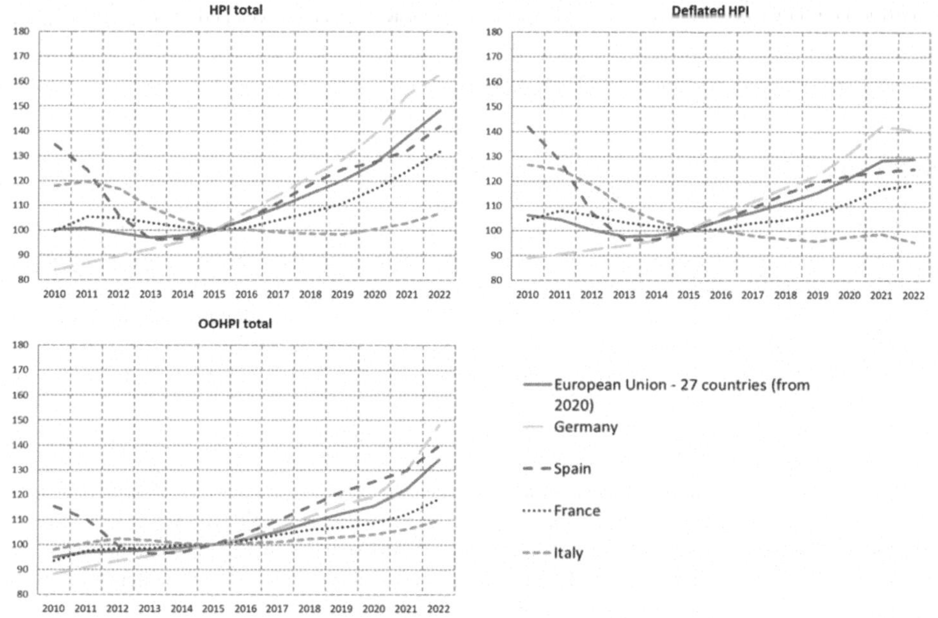

Fig. 1. Graphical comparison of the HPI total, the deflated HPI and the OOHPI total from five geographical points of view: EU-27 aggregate, Germany, Spain, France, Italy.

6 Conclusions

The analysis of the European residential property market, through the main indicators developed by Eurostat, showed a homogeneous behaviour of most countries and, in general, a high correlation between house prices and inflation.

Italian housing market's behaviour turns out to differ, at least in part, from the average European one, probably due to its distinctive features: the fact that households play a predominant role in the market, and the specificities of the existing properties (included in the corresponding HPI sub-category), whose price formation is strongly influenced by the location component.

Such uniqueness, which is also found in other European countries to a lesser extent (Cyprus and Romania), enables the derivation of valuable insights that can aid in asset allocation. The inclusion, within a residential real estate portfolio, of a share of apartments located in Italy, better if located in the consolidated urban territory (thus included within the HPI for existing dwellings), allows geographical differentiation of investment risk and provides protection against inflationary pressures, which are advantages that other European countries do not offer.

A possible progression of this work, which could be developed in future studies, is to analyse the relationship between the residential properties market, the inflationary dynamics, and other socio-demographical variables. Factors such as population and income could show interesting correlations with the indicators already involved in the present study, giving rise to further considerations which can have implications on asset allocation strategies useful for investors.

Note: This study has been developed within the current research P.R.I.N. Project 2022: "INSPIRE—Improving Nature-Smart Policies through Innovative Resilient Evaluations", Grant number: 2022J7RWNF.

References

1. Istat - Prezzi al consumo. https://www.istat.it/it/archivio/293510. Accessed 05 March 2024
2. Istat - Indice dei prezzi al consumo per le rivalutazioni monetarie. https://www.istat.it/it/archivio/30440. Accessed 27 Feb 2024
3. Istat - prezzi alla produzione dell'industria e delle costruzioni. https://www.istat.it/it/archivio/293447. Accessed 27 Feb 2024
4. Goodhart, C., Hofmann, B.: Do asset prices help to predict consumer price inflation? Manch. Sch. **68**, 122–140 (2000)
5. Tsatsaronis, K., Zhu, H.: What drives housing price dynamics: cross-country evidence. BIS Q. Rev. (2004)
6. Yu, H., Huang, Y.: Regional heterogeneity and the trans-regional interaction of housing prices and inflation: evidence from China's 35 major cities. Urban Stud. **53**(16), 3472–3492 (2016)
7. Anari, A., Kolari, J.: House prices and inflation. Real Estate Econ. **30**(1), 67–84 (2002)
8. Kuang, W., Liu, P.: Inflation and house prices: theory and evidence from 35 major cities in china. Int. Real Estate Rev. **18**(2) (2015)
9. Aqsha, N.S., Masih, M.: Is residential property the ultimate hedge against inflation? new evidence from Malaysia based on ARDL and nonlinear ARDL. MPRA Paper No. 91508, (2018)
10. Ting, H., Huangjin, L.: Dynamic relationship between real estate prices and inflation rate. In: IEEE International Conference on Granular Computing (GrC), pp. 153–156. IEEE, China (December 2013)
11. Wu, Y., Tidwell, A.: Inflation-hedging properties of regional Chinese real estate market: evidence from 35 cities in China. Appl. Econ. **47**(60), 6580–6598 (2015)
12. Christou, C., Gupta, R., Nyakabawo, W., Wohar, M.E.: Do house prices hedge inflation in the US? a quantile cointegration approach. Int. Rev. Econ. Financ. **54**, 15–26 (2018)
13. Azmi, A.S.M., et al.: Inflation hedging characteristics of Malaysian residential properties. In: International Real Estate Research Symposium (IRERS), Kuala Lumpur, Malaysia (2010)
14. Lee, K.N.H.: Inflation and residential property markets: a bounds testing approach. Int. J. Trade Econ. Financ. **3**(3), 183 (2012)
15. Nneji, O., Brooks, C., Ward, C.W.: House price dynamics and their reaction to macroeconomic changes. Econ. Model. **32**, 172–178 (2013)
16. Clapp, J.M., Giaccotto, C.: The influence of economic variables on local house price dynamics. J. Urban Econ. **36**(2), 161–183 (1994)
17. Demary, M.: The link between output, inflation, monetary policy and housing price dynamics. MPRA Paper No. 15978, (2009)
18. Tang, J., Ye, K., Qian, Y.: Rethinking the relationship between housing prices and inflation: new evidence from 29 large cities in China. Int. J. Strateg. Prop. Manag. **23**(3), 142–155 (2019)
19. Wolski, R.: Residential real estate as a potential hedge of capital against inflation. Real Estate Manag. Valuation **31**(1), 36–42 (2023)
20. Eurostat database. https://ec.europa.eu/eurostat/data/database. Accessed 16 Feb 2024
21. Housing price statistics - owner-occupied housing price index. https://ec.europa.eu/eurostat/statistics-explained/index.php?title=Housing_price_statistics_-_owner-occupied_housing_price_index#Weights_for_the_calculation_of_OOHPI. Accessed 27 Feb 2024

22. Cournède, B.: House prices and inflation in the Euro area. In: OECD Economics Department Working Papers No. 450. OECD Publishing (2005)
23. Eurostat. https://ec.europa.eu/eurostat/statistics-explained/index.php?title=Tutorial:Country_codes_and_protocol_order#EU_and_euro_area_aggregates. Accessed 27 Feb 2024
24. Housing price statistics – house price index. https://ec.europa.eu/eurostat/statistics-explained/index.php?title=Housing_price_statistics_-_house_price_index#Dynamics_in_the_housing_market:_uses_of_the_house_price_index_and_policy_implications. Accessed 27 Feb 2024
25. Lisi, G., Iacobini, M.: Measuring the effect of location on house prices in italy. Reg. Econ. Dev. Res. 54–62 (2020)
26. Manzoli, E., Mocetti, S.: The house price gradient: evidence from Italian cities. Ital. Econ. J **5**, 281–305 (2019)
27. d'Acci, L.: Quality of urban area, distance from city centre, and housing value. case study on real estate values in Turin. Cities **91**, 71–92 (2019)
28. Casolaro, L., Fabrizi, C.: House prices in local markets in Italy: dynamics, levels and the role of urban agglomerations. Bank Italy Occas. Paper No. 456, (2018)
29. Del Giudice, V., De Paola, P., Torrieri, F., Nijkamp, P.J., Shapira, A.: Real estate investment choices and decision support systems. Sustainability **11**(11), 3110 (2019)

An Evaluation Method of the Building Decarbonization Process at the International Level

Francesco Tajani[1], Francesco Sica[1], Pierluigi Morano[2], Marco Locurcio[2], and Antonella Roma[1(✉)]

[1] Department of Architecture and Design, "Sapienza" University of Rome, 00196 Rome, Italy
antonella.roma@uniroma1.it

[2] Department of Civil, Environmental, Land, Building Engineering and Chemistry, Polytechnic University of Bari, 70125 Bari, Italy

Abstract. The anthropogenic greenhouse effect results from human activity that causes greenhouse gas (GHG) emissions, the effects of which are now directly observable in terms of pollution and global warming. Among the negative consequences of rising global temperatures are the advancing processes of desertification, rising water levels, and biodiversity loss. The seriousness of the problem is underlined by the many plans and policies being implemented by governments internationally. The paper analyses the real estate sector's responsibility for GHG emissions. The existing building stock, built before introducing energy efficiency measures, needs to be energetically renovated to meet the current requirements and performance standards that governments now demand.

The paper presents an analysis of forecast data from the Carbon Risk Real Estate Monitor (CRREM) project, an EU-funded initiative to provide information on the decarbonization path to be followed by the existing building stock in 30 European countries. The study's contribution is the construction of a Building Decarbonisation Performance Index (I_{BDP}). The index was constructed using forecast data provided by the CRREM project, which were processed by implementing the operational logic of multi-criteria techniques. I_{BDP} was designed with the objective of monitoring and assessing the degree of alignment between European countries and the decarbonization paths set by EU policies.

Keywords: Decarbonization pathway · CRREM · Real Estate

1 Introduction

Climate change is one of the most pressing issues that humanity has ever faced. The current uncontrolled use of natural land, the appropriation and consequent degradation of ecosystem services, and the progressive loss of biodiversity are issues of global magnitude and unprecedented in human history [1]. The impacts on ecosystems, economic sectors, health, and human well-being are severe. Climate change is closely linked to GHG emissions, which are produced and emitted by human activities. The consequences

O. Gervasi et al. (Eds.): ICCSA 2024 Workshops, LNCS 14818, pp. 126–142, 2024.
https://doi.org/10.1007/978-3-031-65273-8_9

and risks associated with GHG emissions can now be seen in the desertification phenomena affecting the globe, acidification of the oceans, loss of vitality and resilience of ecosystems, and the progressive rise in temperatures [2]. The relevance and urgency of this issue is demonstrated by the introduction of climate change mitigation and adaptation laws and policies that have been adopted in worldwide countries [3–5].

In September 2015, 193 Member States of the United Nations (UN) signed the 2030 Agenda [6], a program of action comprising 17 Sustainable Development Goals (SDGs) to be achieved by 2030. The SDGs cover a range of common issues of global concern, including the fight against climate change[1].

In December 2015, during the 21st Conference of the Parties (COP21) of the United Nations Framework Convention on Climate Change (UNFCCC)[2] [7], a legally binding international treaty on climate change was adopted: the Paris Agreement. The goal of the agreement is to keep the global average temperature increase well below 2 °C above pre-industrial levels (1850) and to limit the increase to 1.5 °C. The Paris Agreement is based on a five-year cycle of climate action to be taken by countries. Specifically, from 2020 onwards, each country will submit its national climate action plans, called National Determining Contributions (NDCs), in which it will communicate the strategies it intends to implement to reduce GHG emissions. In this context of global legislation and policies, the European Commission presented the 'Green Deal', a package of policy initiatives aimed at making Europe a climate-neutral continent by 2050 [8]. With the passing of the EU Climate Change Act[3], the Green Deal commitment has become a binding obligation; specifically, the legislation requires EU countries to reduce their GHG emissions by 55% (compared to 1990 levels) by 2030 to achieve climate neutrality by 2050.

In this context, an analysis of the sectors that contribute the most to GHG emissions is required. This study focuses on the building and construction sector, which is a significant contributor to energy consumption and GHG emissions [9]. According to the European Environment Agency[4] [10], the building sector accounts for 40% of EU energy consumption and 36% of annual GHG emissions from the EU energy sector. Similarly, the 2022 Global Status Report for Buildings and Construction[5] (Buildings-GSR) [11] shows a negative trend in the sector's decarbonization process, with construction activity returning to pre-pandemic levels in 2021 (Covid-19, 2020). This means that the gap between the required decarbonization path and the actual energy consumed and emissions generated becomes increasingly large [12].

---

[1] Goal 13 of Agenda 2030 is to "Promote action, at all levels, to combat climate change.".

[2] The UN organization is charged with supporting the global response to the threat of climate change.

[3] Regulation (EU) 2021/1119 of the European Parliament and of the Council of 30 June 2021 establishing a framework for achieving climate neutrality and amending Regulation (EC) N. 401/2009 and Regulation (EU) 2018/1999 ('European Climate Regulation').

[4] The European Environment Agency (EEA) is an agency of the European Union that delivers knowledge and data to support Europe's environmental and climate goals.

[5] The Global Status Report for Buildings and Construction (Buildings-GSR) is a publication of the Global Alliance for Buildings and Construction (Global ABC), the leading global platform for all building stakeholders committed to the vision of a zero-emission system.

The criticality of the building stock in terms of GHG emissions forces a reflection on the existing building stock [13]. Buildings have a long lifecycle, and it is important to remember that focusing on energy efficiency in the design of new buildings will have an effect that is not immediately visible.

Approximately 80–90% of the existing building stock, for which no sustainability certification guidelines have been established, will still be in existence in 2050 [9], which is why it is necessary to accelerate energy renovation.

Consumers and investors are increasingly interested in sustainable products, and governments are moving toward stricter policies on the energy efficiency of building stock. From this perspective, a building that does not meet regulatory requirements and market expectations gradually loses value [14]. In this context, several certifications that monitor and certify the sustainability of building stock (GRESB[6], LEED[7], BREEAM[8]) have become widespread [15–17].

Among the various monitoring projects active at the international level, the EU-funded CRREM project [18] occupies a prominent position. The main objectives of the CREEM project are to reduce the risk factors associated with premature obsolescence and potential depreciation of the existing building stock, issues that will arise because of legislation and market demand for more sustainable products. CREEM is a support tool for companies operating in the real estate sector to face these risks and promote energy efficiency to avoid the 'stranding' of the many assets that will not meet future energy performance standards.

This paper addresses the issues previously outlined. The European Union has set forth stringent parameters to be met by all member states. These include the European Commission's goal to reduce GHG emissions by 55% by the year 2030 with the ultimate objective of achieving climate neutrality by the year 2050. All economic sectors must contribute to the reduction of GHG emissions. Achieving climate neutrality entails reducing GHG emissions to the greatest extent possible and offsetting any remaining emissions. In this context, the European Union introduced the CRREM project, which provides forecasting data to guide the decarbonization path that the real estate sector must take, to meet the targets set by the EU. The CRREM project provides indications of the rate of GHG emissions expressed in $kgCO2e/m^2/yr$ for each year from 2020 to 2050. These indications are intended to guide 10 classes of real estate use in their efforts to achieve climate neutrality in 2050.

This paper contributes to the construction of the I_{BDP}, which was developed from the processing of forecast data contained in the CRREM project. The data provided by CRREM was processed using the logic of multi-criteria techniques to structure an index that would synthesize the CRREM data and provide an overview of the energy performance of the building stock in 30 European countries. The index developed is intended to identify which European countries are furthest from and closest to the net zero emissions target to be reached by 2050. The development of this index is a valuable support for government decision-making. It allows for the identification of countries

[6] Global Real Estate Sustainability Benchmark (GRESB).

[7] Leadership in Energy and Environmental Design (LEED).

[8] Building Research Establishment Environmental Assessment Method (BREEAM).

where more stringent policies are needed to facilitate the decarbonization of the real estate sector, thereby incentivizing renovation activities.

The second phase of the research involved structuring an etiological analysis. Specifically, the authors identified a set of synthetic indicators, which were derived from statistics portals that were active at the European level (Eurostat). The indicators, such as national population numbers and population density, were correlated with I_{BDP} values. The objective of this correlation process was to determine whether there was a relationship between the indicators and the I_{BDP}. In conclusion, the necessary path of decarbonization of the real estate sector to be undertaken and realized by European nations can be influenced by different parameters and aspects that this paper attempts to identify through etiological analysis.

2 Material and Method

The decarbonization trajectory of fossil fuel-reliant economic systems is being meticulously monitored by analytical systems designed to track the advancement of distinct sectors. This attention is necessary not only to limit pollution rates but also because of the expected economic impacts. The current GHG emission rates of most economic sectors, if continued over their lifetimes, would be incompatible with the 1.5–2 °C degree temperature rise limit. Failing to adhere to the decarbonization trajectory prescribed by international law would result in a loss of asset value, a phenomenon commonly referred to as a "stranding asset". The phenomenon of stranding assets can manifest itself in many ways [19]. These include:

i) the premature decommissioning of assets,
ii) a progressively reduced utilization of assets, and
iii) ever-increasing costs of asset efficiency.

In the forthcoming period, properties with suboptimal energy performance will be subjected to a process of depreciation, thereby increasing the risk of premature economic obsolescence. The increasing development of European regulations on sustainable finance will encourage economic actors to invest their resources in assets that comply with regulations and market demands. In this context, the importance of energy certification is becoming increasingly apparent. Although certified buildings are more expensive to construct, they have a higher return on investment. The majority of certifications to date have been for new buildings. However, what about existing buildings?

The CRREM project offers a comprehensive understanding of the future of real estate portfolios and the decarbonization trajectory that must be pursued, enabling investors to evaluate not only the current emissions of their assets but also the prospective risk of stranded assets [20].

2.1 Carbon Risk Real Estate Monitor (CRREM)

The CRREM initiative represents the most significant global initiative to date in the field of operational decarbonization of real estate assets. The initiative provides incentives and accelerates energy retrofit initiatives in real estate portfolios by communicating the financial risks associated with poor building energy performance. The primary objective is to reduce the downside risk of properties that will not meet future energy performance standards and for which energy upgrades will become financially unaffordable. The CRREM toolkit generated a decarbonization pathway forecast for 10 asset classes (office, retail high street, retail shopping center, retail warehouse, industrial distribution warehouse cooled, hotel, healthcare, lodging leisure & recreation, residential multi-family) across 30[9] European countries.

In particular, the CRREM tool translated the aspiration to limit global temperature rise by 2050 into different country- and property-specific trajectories. The project employed a scientific approach to allocate the remaining global anthropogenic emissions budget, following the Paris Agreement targets, by country. The CRREM tool is presented in XLS format and requires the input of a set of information that defines the energy efficiency status of an individual asset or portfolio of assets. This enables the identification of assets that are at risk of stranding. Furthermore, the tool enables the measurement of the impact of asset refurbishment on the entire property portfolio, including the progress resulting from the refurbishment of a single asset.

However, in the absence of precise data on individual buildings, the project provides the "*Decarbonization Target Tool*", which enables the assessment of decarbonization and energy reduction pathways without the need to enter precise data on building portfolios. The initial baseline year is 2020, at which point the tool provides data on GHG emissions (in kgCO2e/m^2/yr) and energy use intensity (in kWh/m^2/yr) derived from market averages. Having established the baseline and the target of a global temperature rise of no more than 1.5 °C, the tool presents annual emission and energy use reduction targets from 2020 to 2050, building a pathway to near-zero emissions.

2.2 Method

The objective of this paper is to carry out a comparative analysis of the decarbonization pathways of the building stock of 30 European countries, the same as those considered by the CRREM project. The objective of this comparative analysis is to identify which countries are furthest away from the climate neutrality targets set by European legislation and which countries are most aligned with European targets in terms of GHG emissions. In particular, this analysis is necessary to identify which countries require the implementation of more stringent and meticulous policies that will incentivize and accelerate the renovation of the nation's housing stock.

It should be noted that data made available through the CRREM project, and in particular within the worksheet designated as the "Decarbonisation Target Tool", are disaggregated. This means that they do not provide an overview of the overall situation in terms of GHG emissions for the Country in question. The CRREM dataset provides the GHG emission figure for each year from 2020 to 2050 (thus 31 GHG emission rates in

[9] Austria, Belgium, Bulgaria, Croatia, Cyprus, Czech Republic, Denmark, Estonia, Finland, France, Germany, Greece, Hungary, Ireland, Italy, Latvia, Lithuania, Luxembourg, Malta, Netherlands, Norway, Poland, Portugal, Romania, Slovakia, Slovenia, Spain, Sweden, Switzerland and United Kingdom.

$kgCO2e/m^2/yr$) for 10 classes of real estate use (*office, retail high street, retail shopping center, retail warehouse, industrial distribution cooled warehouse, hotel, healthcare, lodging leisure & recreation, residential multi-family)* for each of the 30 European nations examined. The data presented in this way offers insights that can inform the decarbonization strategy to be pursued for each class of real estate use within each nation. The decarbonization pathway was developed by CRREM through an analytical approach, utilizing 2020 GHG emissions data as the foundation. Nevertheless, the disaggregated presentation of the data precludes certain considerations, such as: i) Which real estate use class is the most responsible in terms of GHG emissions? ii) Which European nation is furthest from the targets set by the European Commission in EU policies?

In order to address these questions an analysis of the data made available by the CRREM toolkit was conducted. In particular, the data available in the Decarbonization Target Tool worksheet were extracted and processed using the operational logic of the pairwise comparison. The logic of the pairwise comparison constitutes the basis for the implementation of several multicriteria techniques, among which the Analytic Hierarchy Process (AHP) represents one of the most widely applied. By combining qualitative and quantitative analysis [21, 22], the AHP is a multicriteria technique used by researchers and decision-makers in many fields to select the best alternative, guide resource allocation decisions, and optimize decision-making processes [23–25]. The versatility of the technique allows it to be applied to different topics, and there is a growing body of literature on using AHP to evaluate decarbonization processes in economic systems [21, 26, 27].

In this study, the logic of pairwise comparison was employed and applied to the dataset extracted from the CRREM toolkit. In particular, the initial database is formed by 310 data points on GHG emissions expressed in $kgCO2e/m^2/year$ for each of the 30 Nations. The operational logic of the pairwise comparison has permitted the evaluation of the 10 classes of real estate use within each nation.

The objective of this initial operation was to identify which class of real estate use was more responsible in terms of GHG emissions within each nation. This process has also allowed for the development of a hierarchization of the classes of real estate use, enabling the creation of a ranking of the sectors from the most to the least energy-efficient.

The initial phase of the research involved the pairwise comparison of the different sectors of real estate use. This was followed by a further phase of aggregation of the data, which enabled the researchers to obtain an overall indicator of the energy performance of each nation. The ultimate objective of this research was to construct I_{BDP}, a descriptive index of the performance of the 30 European nations in terms of GHG emissions.

The operational methodology that has led to the construction of the I_{BDP} consists of 3 steps, as described below.

3 Operational Methodology and Results

3.1 Pairwise Comparison

The first phase of the methodology involved the construction of pairwise comparison matrices. The operational logic of the pairwise comparison is to compare the GHG emissions of each of the 10 real estate use classes within the same nation. As described above, CRREM provides an indication of GHG emissions for each real estate use sector for each year from 2020 to 2050. Therefore, the dataset considered contains 31 data (one for each year) for each of the use sectors. To compare the sectors, 31 matrices (one for each year from 2020 to 2050) of order 10×10 (since 10 are the classes of use to be compared) were constructed for each of the 30 European countries included in the CRREM. This operation resulted in the construction of a total of 930 comparison matrices.

Figure 1 below shows a randomly selected example of a constructed comparison matrix. Specifically, the matrix below in Fig. 1 contains data (GHG emissions) extracted from the CRREM related to Belgium for the year 2020 for all 10 sectors of real estate use. The logic used to construct the matrix is the same as that used for all the other Nations.

2020 Belgium	Office	Retail, High Street	Retail, Shopping Centre	Retail, Warehouse	Industrial, Distribution Warehouse Cooled	Industrial, Distribution Warehouse	Hotel	Healthcare	Lodging, Leisure & Recreation	Residential Multi-Family
Office	1	0,807554835	0,965025874	1,183854261	1,345678676	3,523231442	0,954653394	0,755344068	0,756300058	1,191205851
Retail, High Street	1,238306003	1	1,194997332	1,465973838	1,666361982	4,362838644	1,182153029	0,935347094	0,936530902	1,475077356
Retail, Shopping Centre	1,036241646	0,836821952	1	1,226759088	1,394448286	3,65091915	0,989251605	0,782718981	0,783709617	1,234377113
Retail, Warehouse	0,844698569	0,682140414	0,815155974	1	1,136692852	2,976068557	0,806394356	0,638038053	0,638845577	1,006209878
Industrial, Distribution Warehouse Cooled	0,743119452	0,600109707	0,717129498	0,87974513	1	2,618181818	0,709421507	0,56131087	0,562021285	0,88520824
Industrial, Distribution Warehouse	0,283830346	0,229208568	0,273903628	0,336013765	0,381944444	1	0,270959603	0,214389568	0,214660907	0,338100369
Hotel	1,047500597	0,845914172	1,010865178	1,240088044	1,409599216	3,690587037	1	0,791223362	0,792224762	1,24778884
Healthcare	1,323899984	1,069121834	1,277597739	1,567304637	1,781543978	4,664406051	1,263865614	1	1,001265635	1,577037408
Lodging, Leisure & Recreation	1,322226528	1,067770426	1,27598281	1,565323509	1,779292043	4,658510077	1,262268043	0,998735965	1	1,575043977
Residential Multi-Family	0,839485467	0,677930548	0,810125196	0,993828447	1,129677691	2,957701591	0,80141765	0,634100367	0,634902907	1

Fig. 1. Pairwise comparison matrix of real estate sectors for the year 2020 for the country Belgium

The ratio of GHG emissions of the two sectors being compared is displayed in each cell of the pair comparison matrix (see Fig. 1). The values obtained within each row have been added together, resulting in the generation of 10 scores (one for each row) associated with the sectors of real estate use (see Table 1). These scores are intended to describe the energy performance in terms of GHG of the real estate class for the specific year considered in the matrix.

The illustrative example provided concerning the processing of data for the year 2020 in Belgium offers a preliminary insight. In Belgium in 2020, the sector with the greatest responsibility for GHG emissions is Healthcare. Conversely, the most efficient sector is Industrial distribution warehouses.

3.2 Identifying the Real Estate Sector Class Most Responsible for GHG Emissions for Each Nation

The second step aims to understand, for each of the 30 countries, which real estate sector class is most responsible for GHG emissions. This was done by adding together

Table 1. Index Belgium, describing GHG emissions performance by real estate class sector.

Belgium (2020)	
Real Estate sector classes	Score of the sector
Industrial distribution warehouse	3,543011
Industrial distribution warehouse cooled	9,276247
Residential multi-family	10,479169
Retail warehouse	10,544244
Office	12,482848
Retail shopping center	12,935247
Hotel	13,075791
Retail high street	15,457586
Lodging, leisure & recreation	16,505153
Healthcare	16,526042

the 31 scores (relative to the year) for each country for the same real estate category, as determined in the previous phase.

For example, considering the *Office* sector for Belgium, after obtaining the score of the office sector for each year from 2020 to 2050, these 31 values have been added. By repeating the same operation for each of the 10 real estate sector classes, it has been possible to obtain the first partial index for each country. This partial index makes it possible to understand which sector requires greater commitment in terms of policies to be implemented to achieve the net zero emissions target. At this stage, the scores of the property sector classes have been normalized. This means that the values have been normalized to values between 0 and 1, making them easier to interpret. The higher scores on the index correspond to higher GHG emissions from the property sector and therefore poorer performance on a decarbonisation pathway. The sectors with lower scores are better performing and therefore closer to the net zero target.

Taking Belgium as an example, Table 2 shows the calculated values in ascending order for each of the real estate sectors included in the analysis. It can be seen that for Belgium, the sector most responsible for GHG emissions is the *healthcare* sector. In contrast, the sector closest to the decarbonization target is the *industrial distribution warehouse.*

The second step of the methodology showed that the *healthcare* sector has the highest GHG emissions for 46% of the countries[10]. However, for 93.5% of the nations, i.e., almost the entire sample, *industrial distribution warehouse* is the most efficient sector.

[10] Latvia, Belgium, Lithuania, Croatia, Estonia, Romania, Slovakia, Slovenia, Spain, Hungary, Ireland, Switzerland, Italy and the United Kingdom.

Table 2. Index Belgium, describing GHG emissions performance by real estate class sector.

Belgium	
Real Estate sector classes	Score of the sector
Industrial distribution warehouse	0.0320
Industrial distribution warehouse cooled	0.063
Residential multi-family	0.072
Retail warehouse	0.090
Office	0.106
Hotel	0.111
Retail shopping center	0.113
Retail high street	0.133
Lodging, leisure & recreation	0.137
Healthcare	0.139

3.3 Construction of the I_{BDP}

The aim of the third step of the proposed methodology is to develop the I_{BDP}. The second step has led to the construction of a national scale index consisting of 10 values, each corresponding to a real estate class sector. In the third step, the 10 scores of each real estate use class have been added together for each nation. This operation carried out similarly for each of the 30 nations in the sample, has led to the construction of an overall index consisting of 30 values, each describing the energy performance of the buildings in each nation. As in the previous phase, the index values have been normalized after summation to obtain values between 0 and 1. Similar to the second phase, lower I_{BDP} scores indicate that a country is closer to its 2050 decarbonization targets; conversely, higher scores indicate that countries need to adopt more stringent and urgent policies to meet the Paris Agreement targets. Table 3 shows the I_{BDP} scores calculated for the 30 countries considered, in ascending order.

The I_{BDP} has been then georeferenced to allow a geographical comparison of the results obtained (see Fig. 2).

Based on the I_{BDP}, it can be seen that Denmark has the best energy performance, with a lower GHG emission rate, and therefore is more in line with the targets defined in the Paris Agreement. In contrast, Cyprus, Malta, Sweden, and Spain have the highest GHG emissions. Therefore, in the context of the decarbonization path, which is mandatory at the global level, it is highlighted that it is necessary to apply more stringent policies to achieve the net zero emissions target in these countries.

Table 3. I_{BDP} calculated for 30 European nations

Nations	I_{BDP}
Denmark	0.031007
Slovenia	0.032291
Latvia	0.032356
Lithuania	0.032374
Estonia	0.032382
Ireland	0.032561
Slovakia	0.032587
Poland	0.032679
Hungary	0.032704
Belgium	0.032790
Romania	0.032800
Luxembourg	0.032815
Norway	0.032855
Bulgaria	0.032902
Germany	0.032953
Austria	0.032026
Croatia	0.032986
Finland	0.033039
Switzerland	0,033103
Netherlands	0.033655
United Kingdom	0.033715
Czech Republic	0.034008
Portugal	0.034181
Greece	0.034190
France	0.034348
Italy	0.034689
Spain	0.035229
Sweden	0.035238
Malta	0.035720
Cyprus	0.035815

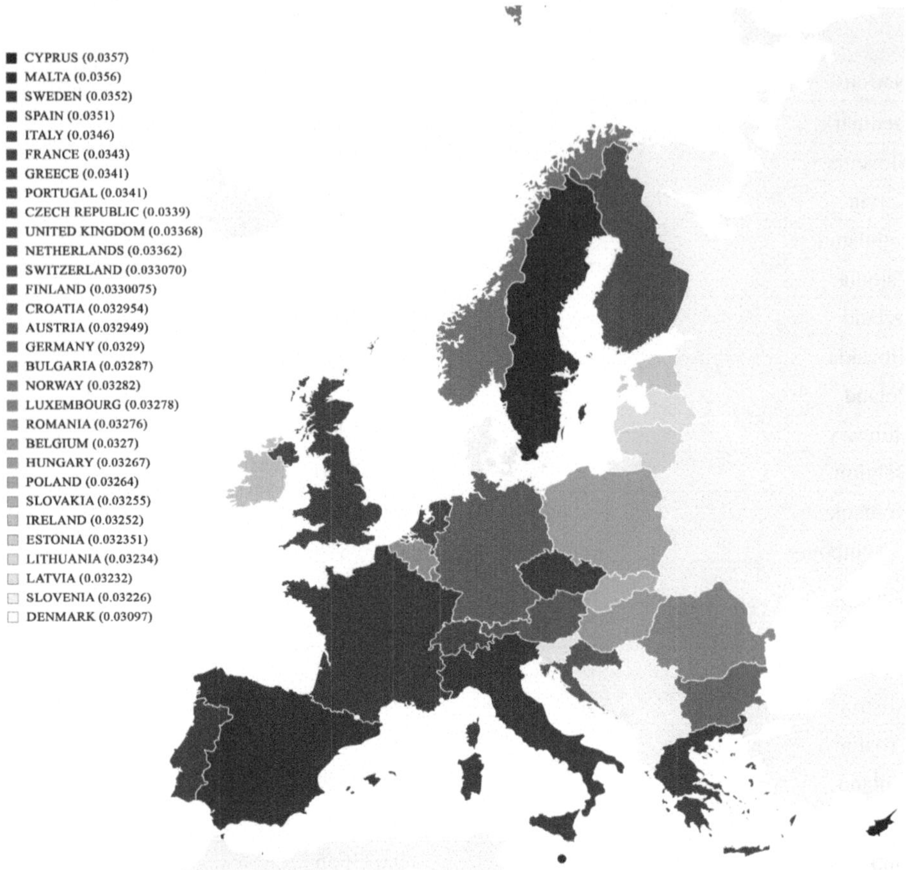

Fig. 2. Georeferenced I_{BDP} representing the building energy performance of 30 European countries.

4 GHG Emission Driver

The I_{BDP} aims to provide a qualitative framework that outlines European nations that deserve greater attention based on their performance in terms of GHG emissions associated with the building stock. However, this process assesses the location of emissions, and further consideration needs to be given to the processes and factors that determine a country's greater or lesser adherence to the required decarbonization pathway [28].

For instance, a nation's geographical location and level of urbanization influence its energy needs for heating, cooling, and lighting [29]. In 2019, an analysis of GHG emissions in 28 EU countries revealed that the energy sector is responsible for 77.1% of GHG emissions, making it the primary driver of this environmental problem [30]. Consequently, international policies have introduced more stringent measures to facilitate the transition to renewable energy sources. This is because countries that rely heavily on fossil fuels, such as coal, oil, and natural gas, tend to have higher GHG emission rates.

Similarly, it is necessary to consider emissions associated with industrial production. Industrial activities represent a significant contributor to global greenhouse gas emissions, largely due to the dependence on energy-intensive energy sources, the mismanagement of industrial waste, and the use of chemicals. Government policies, incentives, and cooperation between the public and private sectors play a pivotal role in the promotion of more sustainable industrial processes. Nevertheless, if industrial production were to be regarded as the primary contributor to GHG emissions in isolation, a false assumption might be made, that countries where industry plays a dominant role are the furthest from the net zero emissions goal. In this context, it is important to consider the role of consumption in addition to industrial activity. Countries with a larger population and less independence in terms of industrial production tend to have higher rates of consumption and imports, which have related consequences for the environment [31]. A comprehensive analysis of the factors believed to be responsible for greenhouse gas emissions would enable the assessment of all the parameters and aspects that need to be improved in order to ensure sustainable development.

4.1 Etiological Analysis

It is crucial to analyze the factors that may contribute to an increase in greenhouse gas emissions at the national level to identify those aspects that require attention and to determine the appropriate course of action.

In this context, an etiological analysis was developed in the present research to explore the correlation between socioeconomic variables selected by the authors GHG emissions of the analyzed nations. In particular, data from the main public statistics portals was extracted and subjected to correlation analysis to ascertain the relationship between the aggregated data and the I_{BDP} index.

The selection of indicators for correlation analysis with I_{BDP} was based on general data. For instance, academic literature indicates that countries with a higher level of industrialization and greater productivity tend to have a higher rate of pollution associated with GHG emissions. In order to investigate this issue further, it was deemed appropriate to investigate the possible relationship between developed I_{BDP} and GDP associated with the nations under study. Similarly, an investigation into the correlation between the rate of renewable energy sources utilized within a nation and its I_{BDP} is being conducted to elucidate to what extent the utilization of alternative energy sources constitutes a strategic solution to the global warming phenomenon. The correlation analysis between the structured IBDP and general aggregate indicators of this type represents the foundation for future developments in this research. The identification of macro-themes at the national level that are associated with a nation's GHG emissions allows for the development of a framework. Specifically, the future developments of the research will entail a deeper and more specific analysis of the indicators obtained in the first phase, to create a dataset of potential problems and issues that policy-makers should consider because of decarbonization pathways.

Specifically, the etiological analysis conducted considered 8 socio-economic variables to analyze their correlation with the I_{BDP}:

1. *Population*; Data have been extracted from the Eurostat database[11] [32]. Specifically, for 29 of the Nations analyzed, the latest available number of population for 2021 has been used. It should be noted that it has not been possible to correlate this variable with I_{BDP} index for the United Kingdom.
2. *Share of imports (%) of mineral fuels, lubricants, and related materials.* Data have been extracted from the Eurostat database. Specifically, for 27 of the Nations analyzed, the latest available percentage share data for 2022 have been used. It has not been possible to correlate this information with the I_{BDP} index for Norway, Switzerland, and the United Kingdom.
3. *Share of exports (%) of mineral fuels, lubricants, and related materials.* Data have been extracted from the Eurostat database. Specifically, for 27 of the Nations analyzed, the latest available percentage share data for 2022 were used. It has not been possible to correlate this information with the I_{BDP} index for Norway, Switzerland, and the United Kingdom.
4. *GDP*. Data have been extracted from the Trading Economics[12] [33] database. Specifically, for all 30 Nations analyzed, the latest available data on GDP expressed in USD, for 2022 have been used.
5. *Deaths*. Data have been extracted from the Eurostat database. Specifically, for 25 of the Nations analyzed, the latest available number of deaths data for 2021 has been used. It has not been possible to correlate this information with the I_{BDP} index for Cyprus, Italy, Luxembourg, Slovenia, and the United Kingdom.
6. *Population density*. Data have been extracted from the Eurostat database. Specifically, for 29 of the Nations analyzed, the latest available data on population density expressed in person/km^2 for 2022 have been used. It has not been possible to correlate this information with the I_{BDP} index for the United Kingdom.
7. *Soil sealing*. Data have been extracted from the Eurostat database. Specifically, for 29 of the Nations analyzed, the latest available data on km^2 of land sealed with impermeable materials due to urban development and construction for 2018 has been used. It has not been possible to correlate this information with the I_{BDP} index for the United Kingdom.
8. *Renewable energy*. Data have been extracted from the Eurostat database. Specifically, for 28 of the Nations analyzed, the latest available data on the share of renewable energy consumption in gross final energy consumption for 2022 have been used. It has not been possible to correlate this information with the I_{BDP} index for the United Kingdom and Switzerland.

All numerical data of the variables were normalized to eliminate the unit of measurement and to obtain values between 0 and 1, which could be easily compared with the I_{BDP}. A correlation analysis was conducted between the identified variables and the I_{BDP}, employing scatter plots and subsequently constructing a correlation matrix.

[11] Eurostat is the statistical office of the European Union.

[12] Trading Economics provides its users with accurate information for 196 countries, including historical data and forecasts for more than 20 million economic indicators, exchange rates, stock market indexes, government bond yields and commodity prices.

Figure 3[13] illustrates one of the eight scatter diagrams produced as an illustrative example of the analysis performed. The diagrams produced enabled an initial series of reflections to be formulated, which were then expanded upon with the results derived from the correlation matrix.

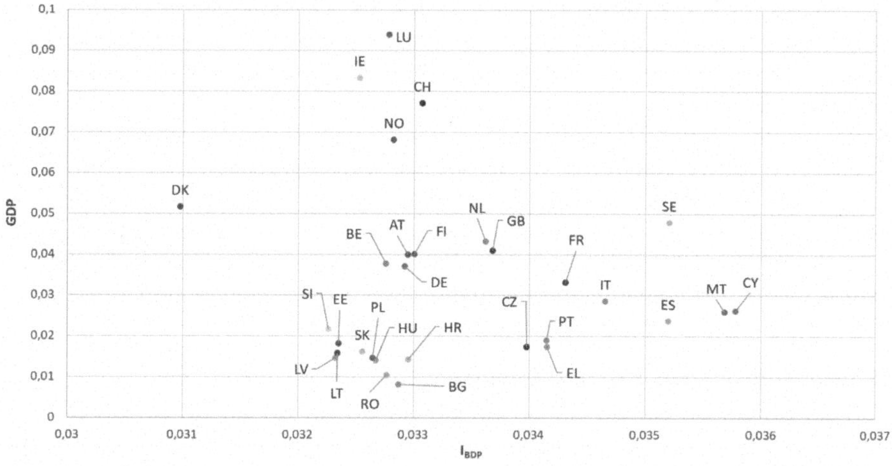

Fig. 3. Scatter plots for analysis of the correlation between I_{BDP} and GDP

Correlation analysis allows to assess the strength and direction of the relationship between two quantitative variables, i.e., the extent to which variations in one variable are associated with variations in another variable.

The scatterplot is a useful tool for correlation analysis because it provides an initial view of the possible relationship between two variables and helps to identify the possible presence of outliers that may alter the correlation coefficient. Several conclusions can be drawn from the analysis of the generated scatterplots.

There is no strong correlation between I_{BDP} and the selected demographic and socio-economic variables, but there is clustering around I_{BDP} values between 0.032 and 0.034.

In addition, in some cases outliers appear in the graph: i) Germany for the correlation between I_{BDP} and population; ii) Germany for the correlation between I_{BDP} and deaths; iii) Netherlands for the correlation between I_{BDP} and fossil fuel export rate; iv) Malta for the correlation between I_{BDP} and population density.

From the scatterplots, the correlation matrix allows the relationship between I_{BDP} and the selected variables to be made explicit. The correlation analysis shows a moderate positive correlation between GHG emissions and population density, i.e., more densely populated countries have higher GHG emissions. All other selected socio-economic and

[13] Austria (AT), Belgium (BE), Bulgaria (BG), Croatia (HR), Cyprus (CY), Czech Republic (CZ), Denmark(DK), Estonia (EE), Finland (FI), France (FR), Germany (DE), Greece (EL), Hungary (HU), Ireland (IE), Italy(IT), Latvia(LV), Lithuania (LT), Luxembourg (LU), Malta(MT), Netherlands (NL), Norway (NO), Poland (PL), Portugal (PT), Romania(RO), Slovakia (SK), Slovenia (SI), Spain (ES), Sweden (SE), Switzerland (CH), United Kingdom (GB).

demographic variables show a weak positive correlation (I_{BDP}-population; I_{BDP}-fuels import; I_{BDP}-fuels export; I_{BDP}-decreases; I_{BDP}-sealing).

Two variables are an exception: GDP and renewable energy, where the correlation is weak but negative. It is clear, therefore, that lower GHG emissions are associated with countries that make greater use of renewable energy sources, and this is reflected in the analyses carried out by the European Union, for which the energy sector is one of the main contributors to GHG emissions. For this reason, the energy renovation of the European building stock is one of the main tasks of sustainability policies implemented by governments.

However, the correlation matrix has shown a weak negative correlation between GDP and I_{BDP}. Gross domestic product measures the aggregate level of production of a state, i.e. all final goods and services produced by a state in a given year. The correlation analysis shows that countries with lower GDP have higher GHG emissions. This result highlights that countries with a lower level of economic prosperity have the most difficulty in implementing the decarbonization pathway and require greater effort.

5 Conclusions and Future Insights

Anthropogenic activities are the main source of GHG emissions, which are the main cause of climate change. The effects of climate change are now visible worldwide through phenomena such as desertification, melting glaciers, uncontrolled fires, and loss of biodiversity. In this context, policies have been developed at the global level to encourage and commit all nations to sustainable development by implementing measures to mitigate and adapt to climate change. The European Union has been one of the main proponents, adopting urgent legislation forcing member states to embark on decarbonization pathways, especially for those sectors deemed most responsible for GHG emissions. The EU is aiming for complete decarbonization by 2050 to meet the target set by the United Nations in the Paris Agreement. This research aims to analyze the contribution of the EU building stock in this context. Buildings, which are energy intensive and have a long life cycle, are responsible for a high proportion of GHG emissions and increasingly stringent policies need to be implemented to encourage their energy renovation. The EU-funded CRREM project analyses the decarbonization pathway that 10 classes of the building sector in 30 European countries need to implement to meet international zero net emissions targets.

This paper has developed the I_{BDP}, an index based on data provided by CRREM. The index has been developed using the logic of multicriteria techniques, to understand which of the 30 countries included in the CRREM analysis are furthest away from the decarbonization targets to be achieved. The result has shown that Denmark is the most efficient nation, in contrast to Cyprus, Malta, Sweden, Spain, and Italy, which are furthest away from their decarbonization targets. This analysis, based on the developed index, serves as a starting point for governments to signal and highlight which countries need more urgent legislation to meet the targets set by the Paris Agreement.

Another important issue to investigate is related to the factors that affect GHG emissions in a country. A first approach to this analysis has been developed in this study by considering some aggregate indicators at the national level, reflecting the socio-economic and demographic factors of the nation. In order to understand the possible correlation between these variables and the developed index, scatter plots and a correlation matrix have been developed. The analysis has shown that there is a weak positive correlation with population density and a weak negative correlation with GDP. This result points out that more attention should be paid to countries with lower levels of economic prosperity, as it is precisely these countries that need to make greater efforts and investments in sustainable development.

This research is a starting point for the extension of the analysis to other variables to analyze their possible correlation with GHG emissions. Another aspect to highlight is the scatter plots obtained. The presence of densities corresponding to concentrated values in the index (0.032–0.034) requires further investigation using a cluster analysis of the countries for which the same range of values is recorded.

References

1. The Intergovernmental Panel on Climate Change (IPCC) homepage. https://www.ipcc.ch/srccl/. Accessed 15 Feb 2024
2. Malhi, Y., et al.: Climate change and ecosystems: threats, opportunities and solutions. Philos. Trans. R. Soc. B **375**(1794) (2020)
3. Eskander, S.M., Fankhauser, S.: Reduction in greenhouse gas emissions from national climate legislation. Nat. Clim. Chang. **10**(8), 750–756 (2020)
4. Sica, F., Tajani, F., Sáez-Pérez, M.P., Marín-Nicolás, J.: Taxonomy and Indicators for ESG Investments. Sustainability **15**(22), 15979 (2023)
5. De Paola, P., Tajani, F., Locurcio, M., Di Liddo, F.: Sustainable real estate and resilient cities: management. Assess. Innov. Sustain. **14**(15), 8984 (2022)
6. United Nation, Sustainable Development Goals (SDGs) homepage. https://unric.org/it/agenda-2030/. Accessed 10 Feb 2024
7. United Nations Climate Change homepage. https://unfccc.int/. Accessed 03 Feb 2024
8. Fetting, C: The European green deal. ESDN report, 53 (2020)
9. Basbagill, J., Flager, F., Lepech, M., Fischer, M.: Application of life-cycle assessment to early stage building design for reduced embodied environmental impacts. Build. Environ. **60**, 81–92 (2013)
10. European Enviroment Agency homepage. https://www.eea.europa.eu/en. Accessed 27 Jan 2024
11. UN environment programme homepage. https://www.unep.org/resources/publication/2022-global-status-report-buildings-and-construction. Accessed 28 Feb 2024
12. Röck, M., et al.: Embodied GHG emissions of buildings–the hidden challenge for effective climate change mitigation. Appl. Energy **258**, 114107 (2020)
13. Sáez-Pérez, M.P., García Ruiz, L.M., Tajani, F.: Assessment of the thermal properties of buildings in Eastern Almería (Spain) during the summer in a Mediterranean climate. Sustainability **16**(2), 746 (2024)
14. Hirsch, J., Spanner, M., Bienert, S.: The carbon risk real estate monitor—developing a framework for science-based decarbonizing and reducing stranding risks within the commercial real estate sector. J. Sustain. Real Estate **11**(1), 174–190 (2019)
15. GRESB homepage. https://www.gresb.com/nl-en/about-us/. Accessed 20 Jan 2024
16. Guide to LEED certification homepage. https://www.usgbc.org/guide-LEED-certification. Accessed 20 Feb 2024
17. BREEAM homepage. https://bregroup.com/products/breeam/. Accessed 12 Feb 20245

18. About CREEM homepage. https://www.crrem.eu/about-crrem/. Accessed 27 Feb 2024
19. Cahen-Fourot, L., Campiglio, E., Godin, A., Kemp-Benedict, E., Trsek, S.: Capital stranding cascades: the impact of decarbonisation on productive asset utilisation. Energy Econ. **103**, 105581 (2021)
20. Spanner, M.M., Wein, J.: Carbon risk real estate monitor: making decarbonisation in the real estate sector measurable. J. Euro. Real Estate Res. **13**(3), 277–299 (2020)
21. You, Y., Yi, L.: Energy industry Carbon neutrality transition path: corpus-based AHP-DEMATEL system modelling. Energy Rep. **8**, 25–39 (2022)
22. Locurcio, M., Tajani, F., Morano, P., Anelli, D., Manganelli, B.: Credit risk management of property investments through multi-criteria indicators. Risks **9**(6), 106 (2021)
23. Vaidya, O.S., Kumar, S.: Analytic hierarchy process: an overview of applications. Eur. J. Oper. Res. **169**(1), 1–29 (2006)
24. Guarini, M.R., D'Addabbo, N., Morano, P., Tajani, F.: Multi-criteria analysis in compound decision processes: the AHP and the architectural competition for the chamber of deputies in Rome (Italy). Buildings **7**(2), 38 (2017)
25. Tajani, F., Anelli, D., Di Liddo, F., Morano, P.: An innovative methodology for the assessment of the social discount rate: an application to the European states for ensuring the goals of equitable growth. Smart Sustain. Built Environ. (2023)
26. Hottenroth, H., et al.: Beyond climate change. multi-attribute decision making for a sustainability assessment of energy system transformation pathways. Renew. Sustain. Energy Rev. **156**, 111996 (2022)
27. Papapostolou, A., Karakosta, C., Kourti, K.A., Doukas, H., Psarras, J.: Supporting Europe's energy policy towards a decarbonised energy system: a comparative assessment. Sustainability **11**(15), 4010 (2019)
28. Bastianoni, S., Pulselli, F.M., Tiezzi, E.: The problem of assigning responsibility for greenhouse gas emissions. Ecol. Econ. **49**(3), 253–257 (2004)
29. Dodman, D.: Forces driving urban greenhouse gas emissions. Curr. Opinion Environ. Sustain. **3**(3), 121–125 (2011)
30. European Parliament: EU greenhouse gas emissions by country and sector: Infographic (2018)
31. Satterthwaite, D.: The implications of population growth and urbanization for climate change. Environ. Urban. **21**(2), 545–567 (2009)
32. Eurostat homepage. https://ec.europa.eu/eurostat/web/main/home. Accessed 1 Mar 2024
33. Trading Economics homepage. https://tradingeconomics.com/. Accessed 04 Mar 2024

Applications and Economic Profiles of the Turnkey Contract Business Model

Edda Donati(✉), Sergio Copiello, and Pietro Bonifaci

Department of Architecture and Arts, University IUAV of Venice, Santa Croce 191, 30135 Venice, Italy
{edonati,sergio.copiello,pbonifaci}@iuav.it

Abstract. The real estate sector, particularly in Europe, plays a significant role in energy consumption, with older and post-World War II buildings contributing most to air emissions. To address this, the European Union has implemented directives since 2002, emphasizing energy efficiency in both existing and new buildings. In response, European countries' governments have introduced incentives to encourage efficiency in existing facilities and the construction of new, high-efficiency buildings. This has led to the development of innovative and sustainable business models, notably the Turnkey Contract model, originating from the United States and exclusively used for new constructions.

This article comprehensively analyses the Turnkey Contract business model, focusing on 11 European case studies from Belgium, Germany, the Netherlands, and Spain. The study employs Multiple Correspondence Analysis (MCA) and K-modes cluster analysis to identify economic profiles associated with this model. The results reveal two distinct clusters: one with private ownership, smaller-scale buildings for private use, and a construction company as the main contractor; the other with public ownership, larger-scale buildings for public use, and an architecture/engineering firm as the main contractor.

The analysis also highlights commonalities between the two clusters identified, such as a project duration of 1–3 years, energy class A properties, and subsidized interventions like insulation and renewable energy installations. The study provides insights into how ownership structure and intended use influence the variations in the Turnkey Contract business model, offering a foundational exploration for future research to consider economic, financial, and cultural specificities across European countries.

Keywords: economic profiles · Turnkey Contract · sustainable business model

1 Introduction

Since the 1990s, it has been recognized that the majority of greenhouse gas emissions are due to human activity, especially in the real estate sector. Research shows that old, inefficient buildings and properties constructed after the Second World War are responsible for the majority of these emissions [1, 2]. The European real estate sector reflects these characteristics perfectly. Not only is it inefficient and highly dependent on fossil

O. Gervasi et al. (Eds.): ICCSA 2024 Workshops, LNCS 14818, pp. 143–161, 2024.
https://doi.org/10.1007/978-3-031-65273-8_10

fuels for energy production, but it also accounts for around 38% of energy consumption in the EU. This result is due to the long-life cycle of existing properties, the construction of a limited number of new efficient buildings and an initial lack of 2 confidence in the increase in property value resulting from the efficiency measures promoted and the associated price premiums and cost premiums [3–6].

To overcome this problem, the European Union has promoted several directives since 2002 aimed at increasing the efficiency of existing installations or the construction of new buildings such as near-zero energy (NZeB) ones [7, 8]. Several amendments have been made to the two main directives (Directive 2002/91/EC and Directive 2010/31/EU) aimed at increasing the efficiency of existing buildings and the construction of new buildings with high energy efficiency. Those that focus mainly on the construction and efficiency of new buildings are Directive 2012/27/EU, Directive 2018/844 and the Energy Efficiency Directive (EU/2023/1791) [9–11], the latter, which is recently adopted, made the efficiency of buildings a fundamental principle of EU energy policy, giving it legal value for the first time [11].

In response to these directives, various European governments have issued several incentives over the years to promote the efficiency of existing facilities and the construction of new high-efficiency buildings [12]. This has also led to the development and simultaneous introduction of innovative and sustainable business models (SBMs) for the efficiency of existing facilities and the construction of new real estate assets [13–17].

In contrast to traditional ones, these SBMs aim to produce a service or a product that can be used by as many people as possible, rather than incentivizing the profit that results from its introduction. They are also characterized by the presence of numerous actors within the long production chain who share data and various information throughout the life of the project, which helps to reduce waste [15, 17–21]. Finally, unlike a traditional model, the revenue of a SBM must be calculated in the short and long term due to the extension of the production chain [15, 22–24].

This article represents an important contribution resulting from broader research dedicated to the analysis of different SBMs [25]. In this article, special emphasis is placed on examining and analyzing the SBM of the Turnkey Contract. This model, which originated in the United States, is used exclusively for the construction of new buildings and requires the presence of a main contractor who takes care of the design, construction and building of the real estate asset on behalf of the client. The article is organized as follows: Sect. 2 presents a literature review on the Turnkey Contract, describing its structure, its functioning and the different actors that interact and are involved within the model. Section 3 the method of analysis used, Sect. 4 presents the data analyzed, Sect. 5 the results and, finally, Sect. 6 the conclusions.

2 Literature Review

The Turnkey Contract business model originated in the United States within the construction and oil and gas industries before expanding to various sectors and other countries [26].

A first description of this model can be found in the works of Freeman [27]. He defines the Turnkey Contract as a contract whereby a single contractor secures and establishes all premises, facilities, tools, materials, and operating people required to complete a project and make it operationally ready. The consumer only needs to turn the key to use the new facility fully and effectively, hence the name [28].

This business model envisages five different actors: the customer, the contractor (who is determined through a call for tenders), the project manager (who is appointed by the contractor), the project coordinator and the subcontractors (Fig. 1). Before the commencement of works, the contractor warrants that he has the necessary skills and organization to fulfil the client's needs and complete the works within the agreed deadlines [29]. When participating in the tender, the contractor must submit the project design. He must also submit a bid package in which he demonstrates his marketing, management, and coordination skills. Finally, he must submit a list of suppliers and subcontractors who can support him in fulfilling his contractual obligations.

Once awarded the contract, the contractor is under the obligation to appoint a project manager who will be responsible for the organization and the execution of the project. The project manager must have adequate managerial and technical training and be fully informed about the contractor's intervention methodology. Finally, he must be able to translate and interpret the client's requests and assign the most suitable staff depending on the phases of the project [26, 30–32].

Further to the above, the project manager is also obliged to duly inform the client on the state progress of the works and provide updates about the search and signing of contracts with subcontractors, should the contractor not be able to carry out some of the works under the main contract. Also, he must ensure that works are carried out pursuant to the deadlines agreed upon by the parties [26, 30–32].

For his part, the project manager may require that a project coordinator is appointed in order to attend to the definition of the different design phases and the setting of the budget, as well as to provide quality control, production and timely delivery of the project [29, 32–34].

The execution of works may however run into some setbacks, should the customer request any variation, or the delivery of the works be non-timely [28, 29, 35, 36].

As for the first case, usually, the contract entitles the customer to demand that variations be made when the project was not fully defined at the beginning. As a result, the construction company is under the obligation to perform the requested additional works, provided the additional price is paid by the customer. As far as the non-timely delivery is concerned, failure to execute the works under the agreed deadlines may be due to impediments which are related to both the customer and the contractor. Such situations are often expressly addressed by the parties in the contract, which may or may not allow for a time extension should execution by the contracting company be non-timely [26, 33–36].

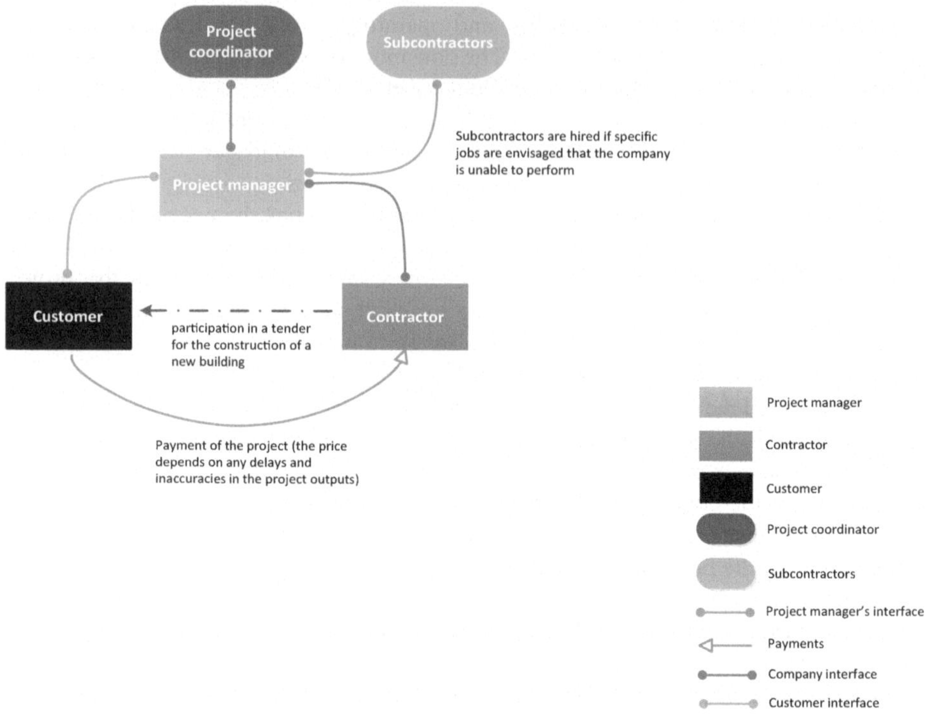

Fig. 1. Scheme of the Turnkey Contract business model

3 Method

3.1 Multiple Correspondence Analysis (MCA)

Multiple Correspondence Analysis (MCA) was chosen to analyze the relationship between two or more categorical variables in a multidimensional table.

This approach is an extension of Correspondence Analysis (CA), a statistical method for examining and visualizing relationships between the rows and columns of a contingency table using a map. In the data set, each point is placed in a multidimensional space and has a specific weight and the chi-square distance between the points can be calculated. The latter is an important tool for evaluating and visually representing relationships between categories of categorical variables, allowing patterns, associations and structures to be recognized in contingency data. The chi-square distance is calculated based on the chi-square statistic, which measures the discrepancy between the observed frequencies and the expected frequencies in the contingency table and reflects how much the observed frequencies differ from those that would be expected if the two variables were independent [37–39].

The graphical representation and analysis of the correlations and distances between the different categories of the considered variables form the basis for interpreting the results (e.g. the variable categories that are close to each other in two-dimensional space are more strongly associated, while those that are far away are less correlated) [37].

The qualitative characteristics of each point determine its position within the diagram, and the chi-square distance measures the distance between them. Let us define:

- $x_i = [x_1, x_2 \ldots x_n]$ a vector representing one set of coordinates or characteristics for a data point in a multidimensional space. Each element $(x_1, x_2 \ldots x_n)$ corresponds to a specific feature or dimension;
- $y_i = [y_1, y_2 \ldots y_n]$ a vector representing a second set of coordinates or characteristics for another data point in the same multidimensional space. Similar to x_i, , each element $(y_1, y_2 \ldots y_n)$ corresponds to a feature or dimension.
- $c_i = [c_1, c_2 \ldots c_n]$ a vector of ratios, where c_i is the ratio of the total frequencies in the columns of a contingency table to the overall sum of frequencies in the table. These ratios are associated with each dimension or feature in the vectors x and y.

The equation for determining the Chi-square distance for the vectors x_i and y_i given their vector of ratios c_i is as follows:

$$\text{Chi} - \text{square distance} = \sqrt{\frac{(x_1 - y_1)^2}{c_1} + \frac{(x_2 - y_2)^2}{c_2} + \ldots + \frac{(x_n - y_n)^2}{c_n}} \tag{1}$$

To reflect the difference between the groups under consideration, the Pearson chi-square statistic is calculated in the table and the result is divided by the sample size n. This value, often referred to as inertia, represents the total variance in CA and MCA. A higher inertia indicates that the observed data points are more dispersed or varied in the multidimensional space, suggesting a richer structure or more pronounced relationships between categories [37–39].

After collecting the data through interviews, questionnaires or other survey methods that reflect the characteristics or preferences in the different categories of interest, these are represented in the vectors x_i and y_i. These vectors are then used to create a map that clearly highlights the appropriate scale for the categories and the differences between each group. The results can be represented by a symmetrical or asymmetrical map. In a symmetrical map, the positions of points/categories on the map reflect the overall structure of relationships in the data without emphasizing specific variables. It provides a global view of the associations between categories, irrespective of the variable they belong to. An asymmetrical map emphasizes the relationships between categories within specific variables. It allows for a more detailed exploration of the associations within each variable individually, revealing how categories within one variable relate to each other [37, 39].

3.2 Cluster Analysis: K-modes Algorithm

Once the initial exploratory phase of data analysis is complete, cluster analysis is proceeded with to determine the results. Clustering analysis processes the data by generating associations between data points based on the characteristics. Each cluster is determined by a center called the centroid (a point in the feature space that mediates the distances between all the data belonging to the cluster) and by a distance (generally, Euclidean distance is used for clustering). Data belonging to the same region (and therefore close to the centroid of the cluster) can be defined as similar, while data belonging to different regions (with large distances compared to the analyzed centroid) can be defined as dissimilar. The calculation of the centroid of a cluster in the k-means analysis is given by the squared Euclidean distance between two objects X and Y in a two-dimensional space:

$$d(X, Y) = \sum_{i=1}^{n} (x_i - y_i)^2 \tag{2}$$

where

- n is the number of attributes (features) in each object X and Y;
- x_i and y_i are the values of the i-th attribute in objects X and Y, respectively;
- $(x_i - y_i)^2$ calculates the squared difference between the i-th features of X and Y;
- denotes the summation over all dimensions, adding up the squared differences for each dimension.

The algorithm aims to minimize the sum of these squared distances within clusters, leading to well-defined cluster centroids and compact clusters.

In this article, it is decided to use the k-modes algorithm, an extension of the k-means algorithm, for the data analysis (for a detailed analysis of cluster analysis with the k-means algorithm, please refer to the following literature [40–45]). It is decided to use this algorithm because the dataset consists only of categorical data for which there is no a priori definable sorting criterion and which have no geometric properties [46, 47]. To calculate the distance (d_1) in the k-modes cluster analysis, it is determined by the number of deviations between the attributes of two objects X and Y and results from the following formula:

$$d_1(X, Y) = \sum_{i=1}^{n} \delta(x_i, y_i) \tag{3}$$

where

- n, x_i and y_i have the same meanings in the formula (2);
- $\delta(x_i, y_i)$ is a function that equals 0 if $x_i = y_i$ (attributes are equal) and 1 if $x_i \neq y_i$ (attributes are different).

So, $d_1(X, Y)$ is calculated by summing up the values of $\delta(x_i, y_i)$ for each attribute. The function δ essentially counts the number of attributes that differ between X and Y.

The model, defined as the maximum relative frequency for each attribute, acts as a representative vector of a group of objects (Q_l) in the k-modes analysis and minimizes the sum of the distances of all objects in a cluster $d(X_{l,i}, Q_l)$ from the mode itself [46, 48, 49]. The distance calculation with the k-modes algorithm results from the following minimization formula:

$$D(X_l Q_l) = \sum_{i=1}^{n_l} d(X_{l,i}, Q_l) \tag{4}$$

where

- $X_l = \{X_{l,1}, \ldots, X_{l,n}\}$ corresponds to the set of objects in cluster l;
- Q_l represents the mode (representative vector) for cluster l;

- $d(X_{l,i}, Q_l)$ is the distance between is the distance between the object X_l and the mode Q_l;
- n_l is the number of objects in the cluster l.

The MCA is performed using the statistical software Stata, while the cluster analysis with the k-modes algorithm is performed using the programming language Python, as Stata only displayed the cluster analysis using the k-means algorithm and the k-median.

4 Data

The data analysis includes 11 European case studies of buildings built from scratch using the Turnkey Contract business model in specific countries. The properties are located respectively in Belgium (CSEU01), Germany (CSEU02, CSEU03, CSEU04 and CSEU05), the Netherlands (CSEU06 and CSEU07) and Spain (CSEU07, CSEU08, CSEU09, CSEU10 and CSEU11). They were built between 2015 and 2020 and are both publicly and privately owned (Table 1 and Fig. 2).

Table 1. List and description of the 11 case studies

Code	Case study	Country	Ownership	Construction year
CSEU01	Center Perex, Namur	Belgium	Public	2017
CSEU02	Passivhaus Parklogen München-Schwabing, Munich	Germany	Private	2017
CSEU03	Project NEWTON house 1, Berlin Adlershof	Germany	Private	2018
CSEU04	Wood hybrid construction: 132 new residential units in passive house standard, Wuppertal	Germany	Public	2020
CSEU05	Wood hybrid construction: 258 new living spaces in passive house standard, Bochum	Germany	Public	2020
CSEU06	City Hall, Venlo	Netherlands	Public	2016
CSEU07	City Hall, Almelo	Netherlands	Public	2015
CSEU08	Entrepatios Las Carolinas: ecological cohousing, Madrid	Spain	Private	2020
CSEU09	Villa Eulietta, Donostia-San Sebastian	Spain	Private	2017
CSEU10	Cooperative of 74 dwellings in the Park of the Offices, La Coruña	Spain	Private	2017
CSEU11	29Soto, Pamplona	Spain	Private	2017

Once the case studies had been localized, the process of defining the basic variables for the subsequent analyses began. The variables considered cover several aspects, including the actors involved in the Turnkey Contract model, their roles, the investment provided and the physical characteristics of the building being constructed, as well as the works carried out as part of the project (Table 2 and Table 3).

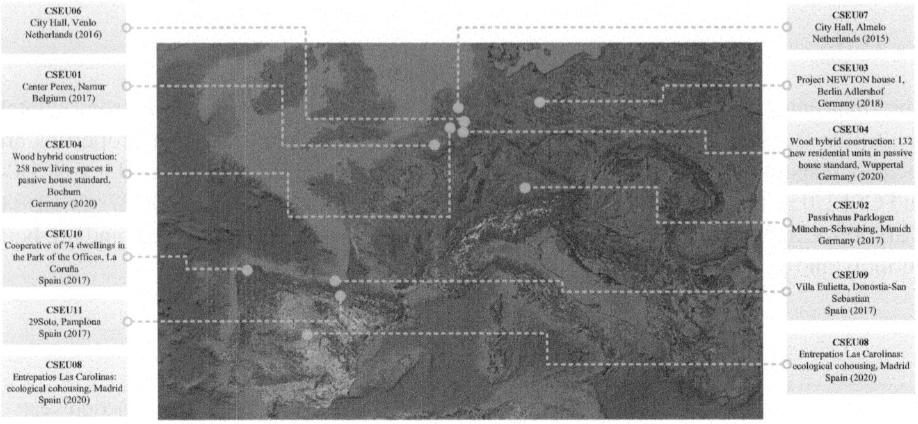

Fig. 2. Geolocalization of the 11 case studies

Table 2. Variables description

Code	Variable	Description
OW	Ownership	The ownership of the property to be built
CST	Customer	Subject to whom the construction of the property is directed. He can provide initial results for the realization of the project, but the design and execution phase is the exclusive task of the contractor with whom the contract has been concluded
C	Contractor	The main contractor is responsible for the design, construction and inspection of the property. He may also use subcontractors to carry out certain work that he cannot do himself
SubC	Subcontractor	Existence or non-existence of subcontractors within the business model who were responsible for carrying out certain work that the main contractor could not carry out itself

(continued)

Table 2. (*continued*)

Code	Variable	Description
IC	Investment cost (€)	Total cost of constructing the property
PP	Project period	Time spent on the construction of the property (excluding the planning phase)
NFA	Net floor area (m^2)	Net floor area of the built property
Use	Intended use	Intended use of the built property. It can be intended for private purposes (apartment or office) or for public purposes (office)
EC	Energy class	The energy class achieved after the construction of the property
EPI	Energy performance index after the efficiency measures (kWh/m^2 year)	The energy performance index obtained after the construction of the property
Insul	Insulation	Promotion of works carried out as part of the project that have favored the insulation of the building envelope or even the installation of highly energy-efficient devices
Syst	Systems	Support for works carried out as part of the project that have improved the building's heating, cooling and ventilation system
Renew	Renewable energy	Promotion of works within the project that favor the installation of renewable energy systems: photovoltaic and solar thermal panels

Table 3. List of variables and their values

Code	Variables	Values
Ow	Ownership	1:Private; 2:Public
CST	Customer	1:Natural person; 2:Real estate company; 3:Capital company; 4:Construction company; 5:Public financial institution
C	Contractor	1:Architecture/engineering studio; 2:Construction Company
SubC	Subcontractor	1:Only 1 subcontractor; 2:Between 2 and 3 subcontractors; 3:More than 3 subcontractors; 4:No subcontractors
IC	Investment cost (€)	0: N/A; 1: €509.000 – €3.000.000; 2: €3.071.611 – €9.509.725; 3: €16.000.000 – €19.130.000; 4: €35.316.112 – €58.765.578

(*continued*)

Table 3. (*continued*)

Code	Variables	Values
PP	Project period	0: N/A; 1: Between 1 and 3 years; 2: Between 4 and 10 years
NFA	Net floor area (m^2)	1: 162 m^2–1.732 m^2; 2: 2.012 m^2– 7.884 m^2; 3: 8.400 m^2– 9.681 m^2; 4: More than 10.000 m^2
Use	Intended use	1: Private homes and offices; 2: Public offices
EC	Energy class certificate	1: A; 2: B; 3: D
EPI	Energy performance index after the efficiency measures (kWh/m^2 year)	
	A	1: 1,00–17,00 kWh/m^2 year; 2: 22,00–33,30 kWh/m^2 year; 3: 39,00–50,00 kWh/m^2 year
	B	4: 80,00–90,00 kWh/m^2 year
	D	5: 155,00–158,33 kWh/m^2 year
Insul	Insulation	1: Walls; 2: Walls and windows
Syst	Systems	1:Heating system, cooling system and ventilation system; 2:Heating system and ventilation system; 3:Heating system and cooling system
Renew	Renewable energy	1: Installation of photovoltaic and solar thermal panels; 2:Installation of photovoltaic panels; 3:Installation of solar thermal panels; 4:Anything

5 Results and Discussion

5.1 Multiple Correspondence Analysis (MCA)

Due to the large number of variables, it has been decided to perform an initial MCA analysis by first considering all the variables related to the business model of the Turnkey Contract and then the variables related to the energy aspects of the buildings.

The variables of the business model concern the ownership of the property to be built (OW), the client (CST), the main contractor (C), the existence of subcontractors (SubC), the investment costs (IC), the construction period (PP), the intended use and the net floor area (Use, NFA). Variables related to the energy aspect of the building include the energy class (EC), the energy performance index (EPI) and the work on the envelopes, systems and renewable energy sources (Insul, Syst, Renew).

Figure 3 shows the MCA analysis for the variables representing the characteristics of the Turnkey Contract business model. The first aspect that can be seen from the graph is how the y-axis contributes to the formation of 2 clusters within the dataset: one on the right and one on the left.

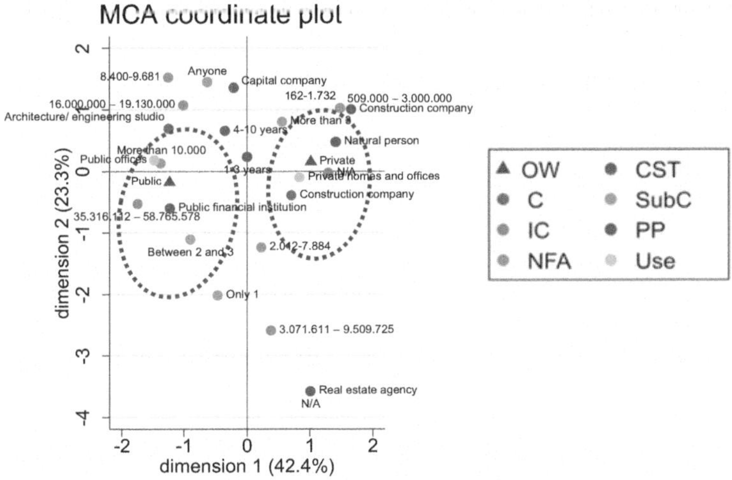

Fig. 3. MCA among the variables: Nature of customer (NCST), Customer (CST), Contractor (C), Subcontractor (SubC), Intervention cost (IC), Project period (PP), Net floor area (NFA) and Intended use (Use). The percentages given as Dimension 1 (42.4%) and Dimension 2 (23.3%) refer to the proportion of variation explained by the two main dimensions of the map. These percentages indicate how effective the main dimensions of the map are in capturing the variation in the original data (higher percentages may explain a greater proportion of the variation in the data).

The cluster on the right side of the y-axis is characterized by private ownership and a construction company as the main contractor within the business model, as well as an intended use associated with private homes and offices. Furthermore, these elements seem to be closely related to having a natural person as a client and having more than 3 subcontractors within the project. Finally, although to a lesser extent, these variables are associated with a net built-up area of between 162 and 1.732 m^2 and an investment cost of between €509,000 and €3,000,000.

The cluster to the left of the y-axis is characterized by public ownership, the identification of a public institution as the main client and an architecture/engineering studio as the main contractor. Other characteristic elements in this cluster are the presence of 2 or 3 subcontractors within the model. Other variables related to those previously described are an investment cost between €16,000,000 and €19,300,000, a project duration between 4 and 10 years and the construction of real estate for public and institutional purposes with a net floor area between 35.316 and 58.765 m^2.

Figure 4 shows the MCA analysis for the variables associated with the physical and energy characteristics of the finished property. It can be seen from the graph that energy class A is the most strongly associated compared to the other two classes present in the dataset (B and D). Indeed, an energy performance index (EPI) between 1.00 and 17.00 kWh/m^2 year and between 22,00 e 33,30 kWh/m^2 year are closely associated with this energy class.

When exploring the other considered variables, those most related to class A concern the insulation of the walls, all the building systems (heating, cooling and ventilation) and the one-off installation of photovoltaic and solar thermal panels. Also significant, although to a lesser extent than those previously described, are the project works that only involve the building's heating and ventilation system, as well as the simultaneous installation of photovoltaic and solar thermal panels and, finally, an index of energy performance (EPI) between 39.00 and 50.00 kWh/m^2 year.

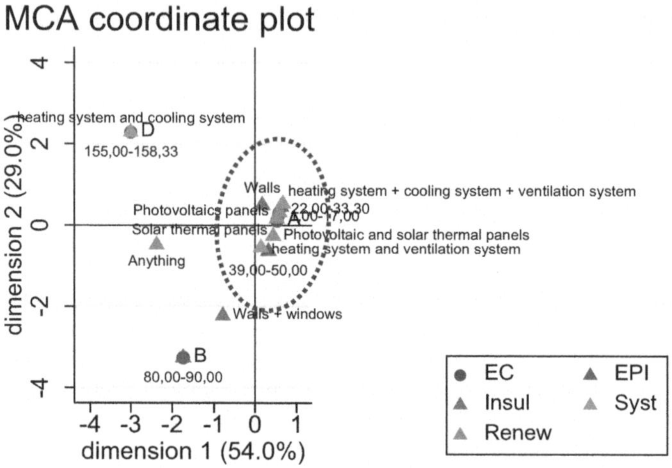

Fig. 4. MCA among the variables: Energy Class (EC), Energy performance index (EPI), Insulation (Insul), Systems (Syst) e Renewable energy (Renew). The percentages given as 'Dimension 1 (54.0%) and Dimension 2 (29.0%)' refer to the proportion of variation explained by the two main dimensions of the map. These percentages indicate how effective the main dimensions of the map are in capturing the variation in the original data.

5.2 Cluster Analysis: K-modes Algorithm

The cluster analysis with the k-modes algorithm begins with an implementation of the "elbow method" to determine the optimal number of clusters (k) for the analyzed data set. In this method, the clustering algorithm is run for a set of cluster values and the explained variation is plotted as a function of the number of clusters. The "elbow" in the graph represents the point at which adding additional clusters does not significantly improve the explained variation. Therefore k = 2 turns out to be the optimal number of clusters to analyze the dataset, since upon reaching k = 3 numerous similarities emerge between 2 out of 3 clusters (Fig. 5).

After establishing k = 2, the identification and subdivision of the different case studies based on their characteristics are conducted, with each case study belonging to a cluster being assigned a value of 1 (Table 4 and Fig. 6).

Upon completion of this process, the definition of the characteristics of each cluster and, consequently, each economic profile starts (see Table 5). The k-modes analysis performed on the dataset shows that the two identified clusters (economic profiles) differ from each other in almost all variables, with the exception of the energy class of the built property (EC), its energy performance index (EPI), the project duration (PP) and, finally, the project works carried out (Insul, Syst and Renew) (Table 4). The two clusters therefore have similar economic profiles in terms of the project work carried out, but differ completely in the structure of the business model.

Although both apply the Turnkey Contract business model for the construction of the property, the first profile (cluster = 0) is characterized by private ownership, a natural person as the customer and a construction company as contractor, as well as the presence of more than 3 subcontractors. Other characteristic and exclusive elements of this profile are the investment costs between €509,000 - €3,000,000, a constructed property with a net floor area between 162–1.732 m^2 and finally the private use of the property which can be traced back to private homes and offices.

The second profile (cluster = 1) has, as part of its Turnkey Contract business model, a public property, a client associated with a public financial institution, a main contractor associated with an architecture/engineering firm and the presence of 2 or 3 subcontractors working with the main contractor in the construction of the property. Other exclusive elements of this profile are investment costs between €16,000,000 and €19,130,000, a net floor area of the built property of more than 10.000 m^2 and finally, the intended public use of the property which can be traced back to public offices or institutions.

The common characteristics of both economic profiles instead concern the duration of the project implementation, the energy characteristics of the property and the interventions promoted. The properties of both economic profiles have a project period between 1 and 3 years, a final energy class of the property of A and an energy performance index between 1,00–17,00 kWh/m^2 year. Finally, in both clusters (economic profiles), the insulation of the envelope, the installation of a heating and ventilation system and photovoltaic systems are subsidized as part of the new construction project (Fig. 6).

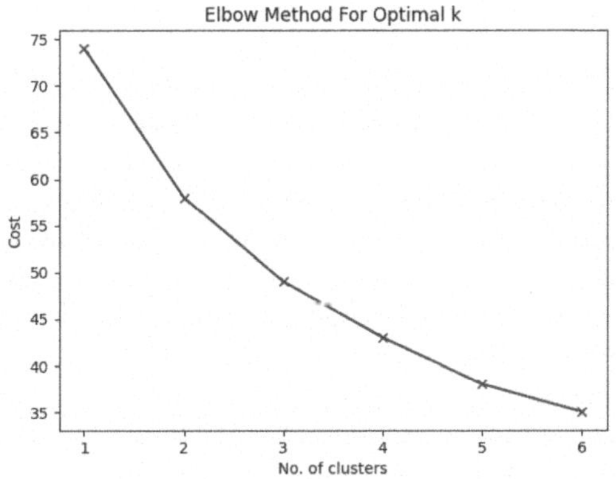

Fig. 5. "elbow method" graphic

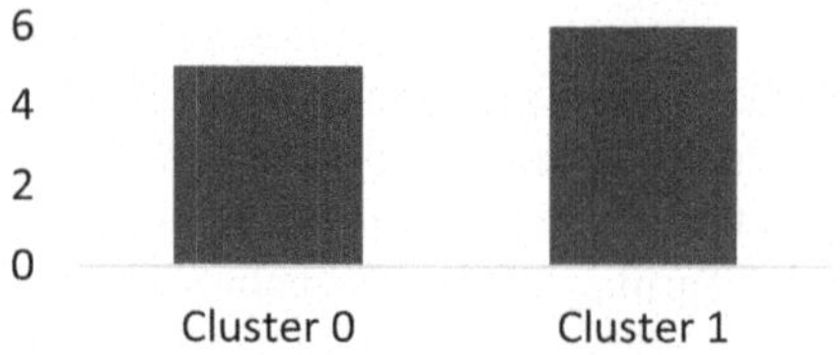

Fig. 6. Histogram k-modes analysis

Table 4. Cluster assignment for each case study

Code	Cluster 0	Cluster 1
CSEU01	0	1
CSEU02	1	0
CSEU03	1	0
CSEU04	0	1
CSEU05	0	1
CSEU06	0	1
CSEU07	0	1
CSEU08	1	0
CSEU09	1	0
CSEU10	0	1
CSEU11	1	0

5.3 Discussion

The analysis carried out using Multiple Correspondence Analysis (MCA) and the K-modes algorithm provides a detailed overview of the characteristics of the Turnkey Contract business model and the energy aspects of the buildings.

Cluster 0 is characterized by private properties, with an individual customer and a construction company as the main contractor, as well as the presence of more than 3 subcontractors. This profile is associated with lower investment costs and smaller properties that are mainly used as private residences or offices.

Cluster 1, on the other hand, shows a tendency towards public properties, with public institutions as customers and architecture/engineering firms as main contractors, together with 2 or 3 subcontractors. This profile is associated with significantly higher investment costs and larger properties intended for public use, such as government offices or institutions.

These differences have a significant impact on both the real estate market and urban planning. It may be more favorable for private investors to focus on smaller projects suitable for private residential or commercial use, while the public sector is more likely to invest in larger projects for public services or infrastructure.

Table 5. Cluster properties description

Variables	Cluster 0	Cluster 1
Ownership (OW)	1	2
Customer (CST)	1	5
Contractor (C)	2	1
Subcontractor (SubC)	3	2
Investment cost (IC)	1	3
Project Period (PP)	1	1
Net floor area (NFA)	1	4
Intended use (Use)	1	2
Energy certification (EC)	1	1
Energy performance index (EPI)	1	1
Insulation (Insul)	1	1
System (Syst)	2	2
Renewable sources (Renew)	2	2

The similarities between the two profiles, such as the energy class A, the energy performance index, and the specifications of the subsidized buildings, indicate that energy efficiency and sustainable technologies are priority aspects for both sectors. This suggests a general trend towards more sustainable and environmentally friendly building practices, regardless of differences in the business model or end use of the properties.

6 Conclusions

The research aims to deepen the knowledge of the sustainable Turnkey Contract business model by analyzing a dataset of 11 case studies. The aim is to identify the economic profiles associated with this model. After defining the variables, the MCA highlights two different groups in relation to the business model, one with private ownership and one with public ownership, and instead a single group for the variables associated with the energy characteristics of the property. This shows that energy class A is the most significant within the dataset and that the latter is closely correlated with an EPI between 1.00 and 17.00 kWh/m^2 year and between 22.00 and 33.30 kWh/m^2 year. Among the variables representing the work carried out as part of the project, those most strongly associated with class A concern the insulation of the walls and all the building's systems (heating, cooling and ventilation).

The case studies are then divided into clusters using the k-modes analysis. The number of clusters is determined using the "elbow method", which suggests an optimal number of clusters of 2. Both clusters are given energy class A properties with similar characteristics from an energy point of view (EPI between 1.00 and 17.00 kWh/m^2 year), the project works carried out and the project period (between 1 and 3 years) but they are distinguished by their intended use: private homes and offices in cluster 0 and public offices in cluster 1.

The variables that differentiated the two economic profiles are mainly related to the structure of the Turnkey Contract business model. Cluster 0 is associated with a private owner, a natural person as the customer, a construction company as the contractor, the presence of more than 3 subcontractors within the business model and the construction of a property for private use, while Cluster 1 is associated with a public owner, a public financial institution as the client, an architecture/engineering firm as the contractor, the presence of 2 or 3 subcontractors within the business model and the construction of a building for public use.

The results of the study are of great importance for various areas such as real estate, urban planning, sustainability, and finance. Examining the different economic profiles associated with Turnkey Contract projects sheds light on market demand and urban redevelopment. In addition, the increasing focus on energy efficiency reflects a growing commitment to sustainable approaches. Furthermore, the distinction between private and public ownership provides valuable insights for refining financing and investment strategies aligned with risk and return considerations. Further exploration of factors such as cultural influences and regional differences could provide deeper insights into the adoption of Turnkey Contract models.

Continued research in this area promises to enrich the relevance and timeliness of the topic and promote a more holistic understanding of sustainable urban redevelopment practices.

References

1. Mironiuc, M., Ionaşcu, E., Huian, M.C., Ţaran, A.: Reflecting the sustainability dimensions on the residential real estate prices. Sustainability **13**(5) (2021). https://doi.org/10.3390/su13052963
2. Vanags, J., Butane, I.: Major aspects of development of sustainable investment environment in real estate industry. Procedia Eng. **57**, 1223–1229 (2013). https://doi.org/10.1016/J.PROENG.2013.04.154
3. Hirsch, J., Spanner, M., Bienert, S.: The carbon risk real estate monitor - developing a framework for science-based decarbonizing and reducing stranding risks within the commercial real estate sector. J. Sustain. Real Estate **11**(1), 174–190 (2019). https://doi.org/10.22300/1949-8276.11.1.174
4. Tan, Y., Liu, G., Zhang, Y., Shuai, C., Shen, G.Q.: Green retrofit of aged residential buildings in Hong Kong: a preliminary study. Build. Environ. **143**, 89–98 (2018). https://doi.org/10.1016/J.BUILDENV.2018.06.058
5. Copiello, S., et al.: Building industry and energy efficiency: a review of three major issues at stake. In: Gervasi, O., Murgante, B., Misra, S., Garau, C., Blečić, I., Taniar, D., Apduhan, B.O., Rocha, A.M.A.C., Tarantino, E., Torre, C.M. (eds.) ICCSA 2021. LNCS, vol. 12954, pp. 226–240. Springer, Cham (2021). https://doi.org/10.1007/978-3-030-86979-3_17
6. Copiello, S., Donati, E.: Is investing in energy efficiency worth it? evidence for substantial price premiums but limited profitability in the housing sector. Energy Build **251**, 111371 (2021). https://doi.org/10.1016/j.enbuild.2021.111371
7. DIRECTIVE 2002/91/EC OF THE EUROPEAN PARLIAMENT AND OF THE COUNCIL of 16 December 2002 on the energy performance of buildings

8. DIRECTIVE 2010/91/EU of 10 December 2010 amending Council Directive 91/414/EEC to include as active substance and amending Decision 2008/934/EC, EFSA Journal, vol. 8, no. 5. Wiley-Blackwell Publishing Ltd, May 01 (2010). https://doi.org/10.2903/j.efsa.2010.1592
9. DIRECTIVE 2012/27/EU OF THE EUROPEAN PARLIAMENT AND OF THE COUNCIL of 25 October 2012 on energy efficiency, amending Directives 2009/125/EC and 2010/30/EU and repealing Directives 2004/8/EC and 2006/32/EC (2012)
10. Directive (EU) 2018/844 of the European Parliament and of the Council of 30 May 2018 amending Directive 2010/31/EU on the energy performance of buildings and Directive 2012/27/EU on energy efficiency (Text with EEA relevance)
11. I (Legislative acts) DIRECTIVES DIRECTIVE (EU) 2023/1791 OF THE EUROPEAN PARLIAMENT AND OF THE COUNCIL of 13 September 2023 on energy efficiency and amending Regulation (EU) 2023/955 (recast) (Text with EEA relevance)
12. Kuzemko, C.: Re-scaling IPE: local government, sustainable energy and change. Rev. Int. Polit. Econ. **26**(1), 80–103 (2019). https://doi.org/10.1080/09692290.2018.1527239
13. Bocken, N., Short, S., Rana, P., Evans, S.: A value mapping tool for sustainable business modelling. Corp. Gov. Int. J. Bus. Soc. **13**(5), 482–497 (2013). https://doi.org/10.1108/CG-06-2013-0078
14. Munaro, M.R., do C. D. Freitas, M., Tavares, S.F., Bragança, L.: Circular business models: current state and framework to achieve sustainable buildings. J. Constr Eng. Manag. **147**(12), 04021164 (2021). https://doi.org/10.1061/(ASCE)CO.1943-7862.0002184
15. Morioka, S.N., Evans, S., de Carvalho, M.M.: Sustainable business model innovation: exploring evidences in sustainability reporting. Procedia CIRP **40**, 659–667 (2016). https://doi.org/10.1016/j.procir.2016.01.151
16. Boons, F., Lüdeke-Freund, F.: Business models for sustainable innovation: state-of-the-art and steps towards a research agenda. J. Clean. Prod. **45**, 9–19 (2013). https://doi.org/10.1016/j.jclepro.2012.07.007
17. Zhao, X., Pan, W., Lu, W.: Business model innovation for delivering zero carbon buildings. Sustain. Cities Soc. **27**, 253–262 (2016). https://doi.org/10.1016/j.scs.2016.03.013
18. Stubbs, W., Cocklin, C.: Conceptualizing a 'sustainability business model.' Organ. Environ. **21**(2), 103–127 (2008). https://doi.org/10.1177/1086026608318042
19. Xu, A.: Turunen, Mäntymäki, and Heikkilä, platform-based business models: insights from an emerging Ai-enabled smart building ecosystem. Electronics (Basel) **8**(10), 1150 (2019). https://doi.org/10.3390/electronics8101150
20. Ensign, P., Roy, S., Brzustowski, T.: Decisions by key office building stakeholders to build or retrofit green in toronto's urban core. Sustainability **13**(12), 6969 (2021). https://doi.org/10.3390/su13126969
21. Leising, E., Quist, J., Bocken, N.: Circular economy in the building sector: three cases and a collaboration tool. J. Clean. Prod. **176**, 976–989 (2018). https://doi.org/10.1016/j.jclepro.2017.12.010
22. Antonini, E., Longo, D., Gianfrate, V., Copiello, S.: Challenges for public-private partnerships in improving energy efficiency of building sector. Int. J. Housing Sci. Appl. **40** (2016)
23. Sovacool, B.K., Turnheim, B., Martiskainen, M., Brown, D., Kivimaa, P.: Guides or gatekeepers? incumbent-oriented transition intermediaries in a low-carbon era. Energy Res. Soc. Sci. **66**, 101490 (2020). https://doi.org/10.1016/j.erss.2020.101490
24. Tronchin, L., Manfren, M., Nastasi, B.: Energy efficiency, demand side management and energy storage technologies – a critical analysis of possible paths of integration in the built environment. Renew. Sustain. Energy Rev. **95**, 341–353 (2018). https://doi.org/10.1016/j.rser.2018.06.060

25. Donati, E., Copiello, S.: The one-stop shop business model for improving building energy efficiency: analysis and applications. In: Lecture Notes in Computer Science (including subseries Lecture Notes in Artificial Intelligence and Lecture Notes in Bioinformatics), Springer Science and Business Media Deutschland GmbH, pp. 422–439 (2023). https://doi.org/10.1007/978-3-031-37111-0_30
26. Schneider, M.E.: Turnkey Contracts - Concepts, liabilities, claims. In: The international Construction law Review, in 4, vol. 3. LLYOD'S OF LONDON PRESS LTD (1986). https://www.lalive.law/wp-content/uploads/2019/10/mes_turnkey_contracts.pdf
27. Ralston, A., Reilly, E.D.: Encyclopedia of Computer Science and Engineering, 2nd edn. Van Nostrand Reinhold Co, New York (1983)
28. Merna, A., Smith, N.: Project managers and the use of Turnkey contracts. Int. J. Project Manage. **8**(3), 183–189 (1990). https://doi.org/10.1016/0263-7863(90)90021-3
29. Doe, J., Nitek, D., Minamikata, N., Smith Freehills, H.: Construction Arbitration and Turnkey Projects, Construction Arbitration and Turnkey Projects (2021). https://globalarbitrationreview.com/guide/the-guide-construction-arbitration/fourth-edition/article/construction-arbitration-and-turnkey-projects
30. Encinas, F., Marmolejo-Duarte, C., Sánchez de la Flor, F., Aguirre, C.: Does energy efficiency matter to real estate-consumers? survey evidence on willingness to pay from a cost-optimal analysis in the context of a developing country. Energy Sustain. Dev. **45**, 110–123 (2018). https://doi.org/10.1016/J.ESD.2018.05.008
31. Christopher Gordon, B.M.: Choosing Appropriate Construction Contracting Method
32. Kumar Guha, S.: Turnkey Construction Contracts in Construction and Infrastructure Projects: An Introductory Overview **11**(2) (2020). http://financial-dictionary.thefreedictionary.com/Turnkey+Construction+Contract
33. Olsson, S., Malmqvist, T., Glaumann, M.: Managing sustainability aspects in renovation processes: Interview study and outline of a process model. Sustainability (Switzerland) **7**(6), 6336–6352 (2015). https://doi.org/10.3390/su7066336
34. Rodney, M., Mshali, M., Aldridge, M.D.: Public Infrastructure Financing Using EPC (Engineer-Procure-Construct)/Turnkey Contracts
35. Solimene, F.: Use of FIDIC forms in the oil and gas construction sector and possible amendments to the yellow and silver books. J. World Energy Law Bus. **7**(6), 558–571 (2014). https://doi.org/10.1093/jwelb/jwu035
36. Guha, S.K.: Turnkey construction contracts in construction and infrastructure projects: an introductory overview. Indian J. Law Justice **11**(2), 211–224 (2020). https://www.scopus.com/inward/record.uri?eid=2-s2.0-85169149600&partnerID=40&md5=22e282abd7f9e3386e7b88d449053cf8
37. Greenacre, M.J.: Correspondence analysis. Wiley Interdiscip. Rev. Comput. Stat. **2**(5), 613–619 (2010). https://doi.org/10.1002/wics.114
38. Moschidis, O.E.: A different approach to multiple correspondence analysis (MCA) than that of specific MCA. Mathématiques et Sci. Humaines **186**, 77–88 (2009). https://doi.org/10.4000/msh.11091
39. Abdi, H., Valentin, D.: Multiple Correspondence Analysis. http://www.utd.edu/
40. Ahmed, M., Seraj, R., Islam, S.M.S.: The k-means algorithm: a comprehensive survey and performance evaluation. Electronics (Basel) **9**(8) (2020). https://doi.org/10.3390/electronics9081295
41. Ding, C., He, X.: K-means clustering via principal component analysis. In: Proceedings of the Twenty-First International Conference on Machine Learning, in ICML 2004. New York, NY, USA: Association for Computing Machinery, p. 29 (2004). https://doi.org/10.1145/1015330.1015408

42. Na, S., Xumin, L., Yong, G.: Research on k-means clustering algorithm: an improved k-means clustering algorithm. In: 2010 Third International Symposium on Intelligent Information Technology and Security Informatics, pp. 63–67 (2010).https://doi.org/10.1109/IITSI.2010.74
43. Kanungo, T., Mount, D.M., Netanyahu, N.S., Piatko, C.D., Silverman, R., Wu, A.Y.: An efficient k-means clustering algorithm: analysis and implementation. IEEE Trans. Pattern Anal. Mach. Intell. **24**(7), 881–892 (2002). https://doi.org/10.1109/TPAMI.2002.1017616
44. Wu, J., Xiong, H., Chen, J.: Adapting the right measures for K-Means Clustering. In: Proceedings of the 15th ACM SIGKDD International Conference on Knowledge Discovery and Data Mining, in KDD 2009. New York, NY, USA: Association for Computing Machinery, pp. 877–886 (2009). https://doi.org/10.1145/1557019.1557115
45. Morissette, L., Chartier, S.: The k-means clustering technique: General considerations and implementation in Mathematica. Tutor Quant Methods Psychol. **9**(1), 15–24 (2013). https://doi.org/10.20982/tqmp.09.1.p015
46. Chaturvedi, A., Green, P.E., Caroll, J.D.: K-modes Clustering. J. Classif. **18**(1), 35–55 (2001). https://doi.org/10.1007/s00357-001-0004-3
47. Dorman, K.S., Maitra, R.: An efficient k-modes algorithm for clustering categorical datasets. Stat. Anal. Data Min. **15**(1), 83–97 (2022). https://doi.org/10.1002/sam.11546
48. Carreira-Perpiñán, M.A., Wang, W.: The K-modes algorithm for clustering (2013). http://arxiv.org/abs/1304.6478
49. Goyal, M., Aggarwal, S.: A review on K-Mode clustering algorithm. Int. J. Adv. Res. Comput. Sci. **8**(7). https://doi.org/10.26483/ijarcs.v8i7.4301

Comparative Analysis of Economic Evaluation Techniques for Healthcare Initiatives

Pierluigi Morano[1(✉)], Maria Rosaria Guarini[2], Antonella Roma[2], Felicia Di Liddo[1], and Marco Locurcio[1]

[1] Department of Civil, Environmental, Land, Building Engineering and Chemistry, Polytechnic University of Bari, 70125 Bari, Italy
marco.locurcio@poliba.it

[2] Department of Architecture and Design, "Sapienza" University of Rome, 00196 Rome, Italy

Abstract. The COVID-19 pandemic of 2020 has highlighted the vulnerability of global health systems and the need for resilient health infrastructures capable of meeting current and emerging challenges. In response, governments around the world are reassessing their health systems and recognizing the urgency of change. This paper provides an overview of economic evaluation techniques used to assess the feasibility of health interventions, particularly relevant in the current context of increased funding for health systems strengthening. Economic evaluation techniques such as Cost-Effectiveness analysis (CEA), Cost-Utility analysis (CUA), Least-Cost Analysis (LCA), and Cost-Benefit Analysis (CBA) are reviewed, each with its limitations and challenges, especially when applied to the complex health sector. The paper highlights the need for critical evaluation of methodologies and suggests future research directions aimed at overcoming these limitations, possibly through integrated approaches or additional quality assessment techniques such as Multi-Criteria Analysis (MCA).

Keywords: Economic Appraisal · Healthcare interventions · evaluation · founding sources · decision-making tools

1 Introduction

In March 2020, the World Health Organization (WHO) declared a global state of emergency due to the spread of the Coronavirus (COVID-19). Covid-19 was declared a pandemic and the rate of infections, hospitalizations, and deaths worldwide reached levels that health systems were struggling to withstand and absorb.

Health is a fundamental right that institutions must guarantee to their communities, and therefore they must have strong and resilient health systems capable of meeting the needs of their populations. The COVID-19 emergency caused a major public health and economic crisis worldwide [1] and highlighted the vulnerability of health systems unable to cope with an epidemiological crisis of this magnitude [2].

Medical facilities were unprepared and inadequately equipped to deal with a public health emergency [3] demonstrating the failure of the hospital-centric model of health

O. Gervasi et al. (Eds.): ICCSA 2024 Workshops, LNCS 14818, pp. 162–176, 2024.
https://doi.org/10.1007/978-3-031-65273-8_11

systems. Due to a lack of technology, training, health workers, and hospital beds, overburdened hospitals were unable to cope with large numbers of patients in a limited time [4]. This crisis exposed the structural dysfunction of health systems around the world, particularly in developing countries [5].

In this context, governments have recognized the urgent need for effective transformation. The health needs of populations are changing and health systems are facing new challenges due to an aging population, the increasing prevalence of chronic diseases, the risk of epidemics, etc. Within this framework, the concept of resilient health systems has emerged, which must be able to cope with health emergencies while continuing to provide care for the population [6].

The concept of resilience is applied in many fields (environmental policy, engineering, infrastructure) and is used in government policy to describe an essential quality of systems to meet today's challenges. This concept has been associated with the health sector in recent years, particularly since 2014, when the WHO called for action to build stronger health systems in the wake of the Ebola outbreak in Africa, [7].

A strong health system is rooted and integrated at the local level [8]; in particular, the health model needs to shift from the centrality of hospitals to an extensive territorial network [9] within which the community can meet its own care needs.

The ambition to build solid health systems requires a range of effective investments, and one constraint is the limited financial resources available [10]. Indeed, the health sector is characterized by ever-increasing costs due to several factors, including i) an aging population, which increases the proportion of the population requiring care; ii) the increasing prevalence of chronic diseases; iii) price inflation of health resources; and iv) technological advances, which are available at higher costs [11].

In response to the problem of limited financial resources, national administrations are seeking tools to help them select interventions that offer the best value for money, maximizing benefits to the population while containing health expenditure [12]. Economic evaluation, for example, supports the decision-making process of those involved in investment decisions, but its application to health investments is a controversial and debated area.

2 Aim

Economic evaluation is the most widely used tool for assessing plans and projects that have an impact on the community. The main purpose of economic evaluation is to compare the costs and benefits associated with a plan or project in order to identify the best option from a range of competing alternatives [13, 14]. These evaluations are essential in all contexts, particularly in the health sector, where the increasing pressure on available budgets, combined with the need to reorganize healthcare models, requires analyses to identify the best allocation of limited financial resources [15].

Economic evaluations operate in contexts of uncertainty, are predictive, and have methodological and ethical limitations that are widely discussed in academic research. These limitations are most evident when economic evaluations are applied to the health sector [16]. The difficulty relates to the effects generated by the implementation of a health intervention. The construction of a health facility, the introduction of innovative technology, or a prevention campaign, are activities that have an impact on a more or less large catchment area or territory and whose beneficial effects are visible over a relatively long time.

This paper aims to provide a framework for the main techniques used to assess the feasibility of a health intervention, given the topicality of the issue, as evidenced by the large number of funding sources. The aim is to outline their main characteristics, differences, and inherent methodological limitations. This process will make it possible to establish a framework that can serve as a starting point for the development of alternative economic evaluation models, or models that combine the techniques currently used, in order to make a significant contribution to the topic.

Section 3 of this document provides an overview of the funding and investment sources currently (2024) being promoted by the European Union (EU) and the Italian government, with a particular focus on the National Recovery and Resilience Plan (NRRP), one of the objectives of which is to rethink and make more efficient the articulation of health services across the territory. Section 4 describes the main economic evaluation techniques used to assess the feasibility of health interventions, highlighting their characteristics and limitations. Section 5 presents the conclusions and future developments of this research.

3 European Financial Framework in Healthcare

Economic evaluation in health is important, particularly in the current context of increased focus on health by European and national institutions, which have allocated significant funding in response to the challenges of the COVID-19 pandemic and the need to strengthen health systems. In the context of such funding, economic evaluation has a key role in ensuring that resources are allocated effectively and efficiently. A summary description of the main European and Italian national funding programs aimed at strengthening the health sector is provided below to illustrate the centrality of this issue.

3.1 European Union Strategies

The critical issues and gaps identified in health systems worldwide during the COVID-19 pandemic were highlighted by the EU, which implemented a series of actions and funding plans aimed at Member States to address this issue. EU Member States are responsible for organizing and providing medical care to their citizens, so the actions implemented by the EU aim to complement national plans to join forces and work towards a stronger Health Union.

EU4Health[1] [17] is one of the EU's response to the pandemic crisis. It is an unprecedented financial support to Member States, worth €5.1 billion for the period 2021–2027. The program is based on the idea that health is an investment. The aim is to have a healthy and active population that is more productive and has a better quality of life. EU4Health thus complements Member States' health policies and has four general objectives: i)

[1] Established by Regulation (EU) 2021/522 of the European Parliament and of the Council of 24 March 2021 establishing a Programme for the Union's action in the field of health ('EU4Health Programme') for the period 2021–2027, and repealing Regulation (EU) No 282/2014.

improving and promoting health, ii) protecting people, iii) access to essential medicines, medical devices, and products, and iv) strengthening health systems.

EU4Health is complemented by the *Cohesion Policy* [18], the EU's principal investment policy. Since the establishment of the European Community, now the European Union, there have been significant disparities between Member States and within the regions of Member States. In this context, cohesion policy aims to promote the harmonious development of all Member States by strengthening economic, social, and territorial cohesion.

The European Commission has developed the Cohesion Policy for the period 2021–2027, which is based on five key policy objectives[2]. These objectives are to be achieved through strategic actions implemented by Member States. EU cohesion policy is implemented through the main Structural Funds[3].

The Cohesion Policy developed by the EU, includes a series of strategic actions to designed to reinforce the health services provided by Member States within their respective territories. The legislation establishing the common rules for the utilization of Structural Funds is Regulation (EU) 2021/1060 of the European Parliament and of the Council of 24 June 2010[4] [19].

Member States are permitted to utilize the Structural Funds for the implementation of a diverse range of interventions. In accordance with Regulation 2021/1060 [19], the EU develops a list of "*enabling conditions*" to "*ensure the necessary conditions for the effective and efficient use of Union support granted through the Funds*". Each of the EU's enabling conditions is linked to specific objectives, and it is the responsibility of Member States to develop programs and policies that meet the EU's enabling conditions in order to access funding.

With regard to the health sector, the EU identifies enabling condition *4.6* which is defined as a: *'strategic policy framework for health and long-term care'*. This condition is based on the assumption that health crises, aging populations, and technological progress present challenges to the robustness of health systems.

For this reason, European support, through the Structural Funds, is intended to complement the financial resources of each Member State in order to ensure the long-term, efficient, accessible, and sustainable provision of health care for its citizens.

---

[2] The 5 policy objectives of UE Cohesion Policy: i) a more competitive and smarter Europe, ii) a greener and low-carbon transition towards a net zero carbon economy, iii) a more connected Europe by improving mobility, iv) a more social and inclusive Europe, v) a Europe closer to its citizens by promoting the sustainable and integrated development of all types of territories.

[3] UE Structural funds: i) European Regional Development Fund (ERDF), ii) European Social Fund Plus (ESF +), iii) Cohesion Fund (CF), and iv) Just Transition Fund (JTF).

[4] Regulation (EU) 2021/1060 of the European Parliament and of the Council of 24 June 2021 laying down common provisions on the European Regional Development Fund, the European Social Fund Plus, the Cohesion Fund, the Just Transition Fund, and the European Maritime, Fisheries and Aquaculture Fund and financial rules for those and for the Asylum, Migration and Integration Fund, the Internal Security Fund and the Instrument for Financial Support for Border Management and Visa Policy.

The long-term budget developed by the European Union (Cohesion Policy 2021–2027) is complemented by an additional and important temporary investment plan promoted by the EU to revive Europe and repair the damage caused by the Covid-19 pandemic: the *Next Generation EU* [20]. The Next Generation EU comprises a number of funds[5], the largest of which is the *Recovery and Resilience Facility* (RRF) [21]. A significant part of these funds will be allocated to the restructuring of EU Member States' health systems.

The RRF, established by Regulation (EU) 2021/241[6] [22], has areas of intervention structured around six key pillars[7]. One of these is area e) which concerns *health and economic, social, and institutional resilience*. This area includes measures to strengthen crisis response and preparedness.

The relevance and importance of the matter is demonstrated by the series of EU-funded grants which, in the context of the post-pandemic challenges, seek to address the structural difficulties that currently characterize health care provision.

3.2 NRRP in Italy, a Focus on Mission 6: Health

The NRRP is a strategic action plan launched by the Italian government within the framework of the RRF established by the EU. The NRRP was approved by the European Commission on 22 April 2021, and is structured around three strategic axes that are shared at the European level: i) digitalization and innovation, ii) environmental transition, and iii) social inclusion. In terms of the plan's structure, the Italian government, in accordance with European guidelines, has divided NRRP into six missions, which are further divided into a total of 16 components [23]. The missions are designated as follow:

- Mission 1: Digitization, innovation, competitiveness, culture and tourism
- Mission 2: Green revolution and ecological transition
- Mission 3: Infrastructure for sustainable mobility
- Mission 4: Education and research
- Mission 5: Cohesion and Inclusion
- Mission 6: Health

One of the purposes of this document is to present a synthesis of the principal characteristics of Mission 6, which includes a series of plans and projects designed to reinforce the national health system and health infrastructure.

[5] Funds under the Next Generation EU investment plan: i) Recovery Assistance for Cohesion and the Territories of Europe (REACT-EU), ii) Horizon Europe, iii) InvestEU, iv) The European Agricultural Fund for Rural Development, v) The Just Transition Fund (JTF), vi) RescEU, and vii) Recovery and Resilience Facility.

[6] Regulation (EU) 2021/241 of the European Parliament and of the Council of 12 February 2021 establishing the Recovery and Resilience Facility.

[7] The scope of application of the Facility shall refer to policy areas of European relevance structured in six pillars: a) green transition, b) digital transformation, c) smart, sustainable and inclusive growth, including economic cohesion, jobs, productivity, competitiveness, research, development and innovation, and a well-functioning internal market with strong SME, d) social and territorial cohesion, e) health, and economic, social and institutional resilience, with the aim of, inter alia, increasing crisis preparedness and crisis response capacity and f) policies for the next generation, children and the youth, such as education and skills

As previously stated, the global pandemic of 2020 demonstrated the universal value of health and highlighted critical structural issues that could be amplified in the future due to the growing demand for care associated with demographic (aging population), epidemiological, and social trends. The NRRP's criticality framework identifies several key areas for improvement, including i) significant territorial disparities in the provision of health services, ii) inadequate integration between hospital services, territorial services, and social services, iii) long waiting times for treatment, and iv) poor ability to achieve synergies in the definition of strategies to respond to environmental, climate and health risks.

Mission 6 of the NRRP is divided into two components [24] defined as follows.

- M6C1- Proximity networks, facilities, and telemedicine for territorial health care; the interventions that form part of this component aim to strengthen the services provided in the territory by reinforcing and creating structures and territorial garrisons (community homes, community hospitals). Another objective is to strengthen home care, develop telemedicine, and improve integration with all social and health services.
- M6C2- Innovation, Research and Digitization of the National Health Service: The actions included in this component will enable the renewal and modernization of existing technological and digital facilities, the completion and dissemination of the Electronic Health Record (EHR), and the improvement of the delivery and monitoring capacity of the Essential Levels of Care (ELC) through more effective information systems. Significant resources will also be devoted to scientific research and promoting technology transfer, as well as strengthening the skills and human capital of the NHS, including through improved staff training.

In particular, the M6C1 component is designed to enhance the capacity of the National Health System (NHS). Italy is currently experiencing an aging population and a prevalence of chronic diseases, with 40% of the population affected. In order to reinforce the healthcare infrastructure across the country, the NRRP plans to establish, by 2026, 1288 Community Homes, which will function as a multi-city center. These centers will be staffed by doctors from various specialties, ensuring that the community is adequately served. Furthermore, the construction of 381 community hospitals across the country is planned, with the objective of alleviating the burden on existing hospitals. The Community Hospitals will be configured as a network of assistance, with the objective of facilitating the admission and treatment of patients requiring low/medium intensity interventions. Finally, it is planned to activate 602 Territorial Operational Centers with the intention of increasing the volume of home-based services, thus promoting the diffusion of telemedicine.

The M6C2 component will fund two distinct lines of action. The first line of action is technological and digital modernization, which will entail the purchase of new equipment, the creation of an electronic health file, and the structural modernization of hospital facilities to bring them into line with current seismic regulations. The second line of action is concerned with training, scientific research, and technology transfer. It is proposed that support for biomedical research be increased and health workers be trained.

4 Economic Appraisal in Healthcare

The framework of funding sources activated at the European and Italian national level is useful in understanding the topicality of the issue. In the context of this funding, economic evaluation plays a pivotal role in ensuring that resources are allocated in an optimal manner. It aims to identify the most promising strategies for improving the health of the population and maximizing the impact of funding in the health sector.

Economic evaluations can be employed to assess health infrastructure projects, such as new hospitals or clinics, as well as complex subsystems, such as emergency departments, but also health interventions, such as emergency departments. They can also be used to evaluate health interventions, such as prevention programs or the use of alternative treatments.

It is important to note that economic evaluation models are essential for many of the funding sources announced by the EU under the Cohesion Policy 2021–2027. In accordance with Regulation (EU) 2021/1060 [19], the European Commission requires managing authorities to ensure adequate value for money in the selection of operations to be financed, and Member States are responsible for this. The EU's RRP is designed to provide support to Member States in the implementation of reforms and investments. Member States are responsible for structuring their intervention plans and for making an appropriate selection and prioritization of projects based on a set of criteria, including the results of the Economic Appraisal (EA). The EA is a methodological process through which the value generated by a project for all stakeholders can be understood and determined. This process allows for the assessment of whether society will benefit from the investment.

In this context, the Directorate-General for Regional and Urban Policy[8], with the support of JASPERS[9], has developed an *Economic Evaluation Vademecum (EAV)* [25] as a tool to support the implementation of economic evaluations to be carried out for the funding sources made available to the EU in the period 2021–2027. The EAV summarizes good practices for carrying out economic evaluations based on established national and international experience and is also consistent with the EA approach followed by the European Investment Bank. The EAV is complementary to the European Commission's Guide to Cost-Benefit Analysis (CBA) of investment projects for the period 2014–2020 [26].

The European Commission provides financial support for a diverse range of projects, including infrastructure, energy efficiency, healthcare, and research and innovation. The specific approach employed in each sector is dependent on the nature of the project requiring an EA. CBA represents one of the principal techniques in the context of EA. In numerous EU countries, it is the principal instrument employed to identify the optimal design choices that maximize the community's welfare while respecting financial constraints. Nevertheless, the application of CBA is often impractical for smaller projects due to the significant time and financial resources required. In this context, decision-makers

---

[8] European Commission department responsible for EU policy on regions and cities.

[9] Join Assistance to Support Projects in European Regions.

may opt to employ alternative techniques, including least-cost analysis (LCA), cost-effectiveness analysis (CEA), cost-benefit analysis (CBA) and multi-criteria analysis (MCA).

The European Commission proposes, by way of example but not exhaustively, the most appropriate EA techniques to be used, diversified according to the project area. The in-depth analysis of the EAV and the scientific literature has revealed that the most prevalent economic evaluation techniques employed in the health sector are: *i) Least-Cost Analysis (LCA), ii) Cost-Effectiveness Analysis (CEA), iii) Cost-Utility Analysis (CUA), iv) Cost-Benefit Analysis.*

This research aligns with the aforementioned context, and an analysis of the scientific literature in this field has enabled an understanding of the mechanisms for implementing economic evaluation techniques in the health sector. A critical reading of this literature has been attempted.

4.1 Comparative Analysis of the Main Used Health Economic Evaluation Techniques

The selection of the most appropriate economic evaluation technique is contingent upon the purpose and subject of the analysis. CEA, CUA, and LCA are mainly employed in the assessment of health programs, such as the comparison of therapeutic alternatives, prevention plans, and chronic disease screening [27–29]. The use of CBA in the assessment of the feasibility of health programs is marginal [30–32]. The EAV indicates that CBA is the most appropriate tool for evaluating health infrastructure projects, such as the construction of a new hospital or outpatient centers.

It is a common practice for economic evaluators to evaluate the financial implications of a project or program. However, there is considerable variation in the way in which the benefits of such initiatives are quantified [33]. The only exception is LCA, which is the simplest approach. LCA is used to compare two health interventions, typically health programs, whose beneficial effects are known and are equivalent. According to this principle, since the benefits are known and not decisive, only the costs are analyzed in order to select the low-priced alternative [34].

The quantification of benefits in EA is a topic that has been the subject of much debate. In CEA, costs are quantified in monetary units, while benefits are quantified in natural units. Some examples of natural units used to quantify the effectiveness of an intervention are Years of Life Gained (LYG) or Years of Life Lost (YLL) due to mortality. The distinguishing feature of CEA from CUA is the parameter used to quantify the benefit. In particular, the researchers considered that considering solely the number of Years of Life Gained (LYG) was reductive, given that undergoing a life-saving procedure could have consequences such as disability. For this reason, the Quality-Adjusted Life Year (QALY) indicator was introduced. The QALY is a metric that combines "*the effects of health interventions on mortality and morbidity into a single index*", thus providing a "*common currency*" that allows for comparisons between different disease areas [35]. The calculation of a QALY is a complex process. The objective is to relate the number of years spent in a state of health to the quality of that state, which is derived from Health-Related Quality of Life (HRQoL). Patients are required to rate their health status on a scale of 0 (death) to 1 (excellent health) in order to obtain the requisite data for the calculation of a complete indicator. This indicator is derived by multiplying the rating by a given quantity (LYG).

To summarize, CEA and CUA are used to compare therapeutic alternatives, procedures, technologies, etc., to determine which alternative provides the greatest health benefit for the cost incurred. This is achieved by calculating the incremental cost and the incremental effectiveness of each alternative in comparison to the next most effective alternative. These calculations lead to the final measure of CEA and CUA, which is the Incremental Cost/Effectiveness Ratio (ICER) or Incremental Cost/Utility Ratio (ICUR) [36, 37].

The primary distinction between CBA and the EA previously described is the unit of measurement employed to quantify the benefits, which in this case is a monetary unit (pounds sterling, dollars, euros). As all the costs and benefits of the project alternatives to be compared are expressed in monetary terms, CBA allows for a more accurate comparison, even between different sectors of the economy, in order to select the most optimal allocation of the available financial resources [38].

The primary challenge in applying CBA to complex contexts, such as health interventions, is the monetization of intangible benefits. The academic literature notes that the monetization of benefits following the implementation of a health intervention is a widely criticized area due to the subjectivity inherent in such valuations. The monetization of health benefits is a challenging aspect of CBA. Approaches used in CBA for monetizing health benefits include i) the human capital approach [39], ii) the revealed preference approach, iii) the Contingent Valuation (CV) approach [40, 41].

The human capital method is based on the principle that a health intervention is an investment because time spent in good health is linked to increased productivity in the labor market. By employing monetary values, such as market wage levels, and associating them with time spent in good health, it is possible to monetize the benefits associated with the project being evaluated. On the other hand, the Revealed Preference (RP) Method analyses the choices made by the consumer. In particular, an example of the RP method is the hedonic wage approach, which analyses the relationship between the health risks associated with a chosen hazardous job and the salary required to accept the job [42, 43]. This type of report therefore highlights the preferences of individuals who accept risks in exchange for money. Finally, the most widely used method for quantifying benefits in monetary terms is the CV, a technique of stated preferences based on the construction of a questionnaire describing a hypothetical market in which (non-market) goods are traded. The monetary valuation of benefits in the CV is measured by the maximum willingness-to-pay (WTP), i.e. the maximum amount of money that users would pay to gain more years of life or a better quality of life by undergoing a health project.

As described above, it can be said that the economic evaluation techniques used in the health field differ mainly in the method of quantifying the benefits associated with an intervention, the outputs of the analysis, and the choice of one or the other depends on the subject of the analysis, i.e. the qualification of the health intervention to be evaluated. The following table (see Table 1) provides a summary of the principal similarities and differences between the various economic evaluation techniques employed to assess the feasibility of interventions in the health sector. of simplification.

Table 1. Comparative analysis of the health economic evaluation techniques

	LCA	CEA	CUA	CBA
Project type	Disease prevention/ treatment programmes/ new technology			Healthcare infrastructure
Costs associated with an intervention	Monetary units ($, €..)			
Benefits associated with an intervention	–	Natural units (YLG, YLL)	QALY	Monetary units ($, €..)
Final feasibility indicator	–	ICER	ICUR	Net Present Value (NPV); Internal Rate of Return (IRR)

4.2 Common Pitfalls of Health Economic Appraisal

It is important to note that each of the EA techniques described has methodological limitations, which become particularly apparent when applied to controversial areas such as health. Critics argue that CEA and CUA fail to capture all the characteristics of the costs and benefits associated with a health intervention, mainly because of a lack of data. Many CEAs and CUAs use decision-analytic models, such as Markov models, to estimate costs and effectiveness. These models incorporate assumptions based on published data and/or expert opinion. If the elements of the model are not accurate, the conclusions may be erroneous. Consequently, the analyses carried out by different evaluators may lead to different conclusions.

A further criticism raised by academic literature is the concept of QALY used as a measure of benefit in CUA. The ethical issues associated with the QALY indicator are among the most frequently cited concerns. The assumption that individuals can assign a value to the quality of their own lives presupposes the existence of life and consciousness, which precludes the consideration of interventions aimed at, for example, fetuses or brain-damaged patients [44]. Furthermore, the QALY metric has been criticized for its inability to measure improvements in the form of minor but clinically significant benefits. This is exemplified by the case of cancer patients who may benefit from treatment even if their life expectancy remains low [45].

The limitations of the indicators employed to quantify benefits in the CUA and CEA are that the calculation method is discriminatory against groups of patients who receive less benefit. In the case of patients receiving palliative care or the elderly, the calculation of QALYs is significantly influenced by the number of years of life associated with these user groups. A similar phenomenon can be observed in patients with disabilities or non-self-sufficiency. While these individuals may experience improvements in their health through specific interventions, they are unable to claim a high level of health status due to the permanent condition that affects them [46].

A further step in the analysis is to move from the indicators used in the CEA and CUA to the final parameter of the analysis (ICER/ICUR). This allows for the identification of intrinsic criticalities. The ICER or ICUR is employed by decision-makers as a means of prioritizing the allocation of available financial resources. This report provides information on the incremental cost per unit of health benefit gained when comparing two competing alternatives. In order to ascertain the acceptability of the ratio value, it is necessary to compare it with a pre-defined threshold value. In the literature, the threshold is derived from the outcome of some already widely accepted and reimbursed medical and therapeutic strategies. For instance, in the United Kingdom and the United States, values between 50,000 and 75,000 dollars/pounds are considered an acceptable proportion of life-saving expenditure [47]. The establishment of a universally accepted threshold is considered unethical, and no state has set a universally accepted threshold. This naturally leads to subjectivity and arbitrariness in the assessment [48].

However, CBA has many limitations that make it difficult to apply. When applied to the evaluation of a health project, these include the upper limit and the difficulty of quantifying non-market goods and intangible benefits in monetary terms. The assignment of a monetary value to benefits is controversial, and the ethical validity of this issue is debated, as human life, death, quality of life and illness are immeasurable goods. Consequently, the CBAs conducted and reported in the literature frequently result in a mere comparison of costs, with the quantification of benefits being overlooked [49–52].

Furthermore, the absence of active global guidelines frequently results in the exclusion of costs and benefits from the analysis that could be of fundamental importance to the evaluation. For instance, the analysis employs values such as i) hospitalization rates, ii) costs of care, iii) disease incidence rates, iv) mortality rates, and so on. A variety of sources are available for the collection of this information, resulting in inconsistency between the inputs used by evaluators, who may arbitrarily select which information to utilize [53–55].

As mentioned above, CBA uses different methods to assign a monetary value to intangible benefits. The CV, although the most widely used technique, is controversial because of some fundamental implications. There is a possibility that respondents to the questionnaire may not accurately reflect the true value of their WTP for a health benefit, as they may be inclined to report exceedingly high amounts if they are required to assign a value to their health [51]. At the same time, the declared WTP reflects a distortion of the economic capacity of the unit, leading to income discrimination. Units with greater economic capacity are more likely to declare higher amounts, which results in a bias in favor of these units. In contrast, units with lower economic capacity are less likely to declare a WTP [56].

Similarly, the human capital approach also presents certain complications. The monetary value associated with a person's illness is estimated based on lost productivity (absence from work) using wage rates. Consequently, this approach fails to account for the value of lives of individuals who are not employed, whether as unemployed or retired. Furthermore, the labor market is known to be an imperfect market. Indeed, wages reflect inequalities, with lower earnings associated with gender or ethnicity [52].

Concerning the quantified benefits of the revealed preferences approach, the estimated values vary considerably according to the type of employment considered in the analysis, and there is controversy as to which types of workers are more at risk than others.

Based on the critical issues identified in the analysis of the literature, there are therefore many problems that need to be investigated further in order to identify solutions that are more effective in the economic assessment of uncertain contexts.

5 Conclusion and Future Insights

The challenges that have arisen in the 2020 global health crisis, which has affected health systems and care provision worldwide, must be addressed. A multitude of regulatory frameworks and sources of funding are available at the European and Italian national levels to address these issues. It is therefore of the utmost importance that these sources of funding, which are available for a limited period (2021–2027) and are limited in number, are used in the most effective manner possible to achieve the greatest possible impact. In this context, economic evaluations are of significant importance as decision-making tools, as they assist in defining the optimal allocation of resources. However, when applied to challenging contexts such as health, economic evaluations demonstrate more clearly the limitations that are characteristic of such contexts.

This research has sought to provide a comprehensive overview of the most commonly employed methods in the health sector, with a view to identifying their inherent limitations. The framework will facilitate further development. The future development of the research anticipates the implementation of an analysis of the limitations of the EA, to propose approaches to overcome them by building a support model for the authorities in the definition of intervention programs.

The authors will investigate the possibility of constructing an integrated approach by combining the techniques employed thus far in the health field, or by combining the methodologies outlined in this document with additional quality assessment techniques, such as multi-criteria analyses, in order to optimize the potential of the model.

Note. The research has been developed within the project "MISTRAL - a toolkit for dynaMic health Impact analysiS to predicT disability-Related costs in the Ag"-HORIZON-HLTH-2022-ENVHLTH04 - Grant Agreement Project n. 101095119 of the Polytechnic of Bari (Italy).

References

1. Hartley, D.M., Perencevich, E.N.: Public health interventions for COVID-19: emerging evidence and implications for an evolving public health crisis. JAMA **323**(19), 1908–1909 (2020)
2. Malik, M.A.: Fragility and challenges of health systems in pandemic: lessons from India's second wave of coronavirus disease 2019 (COVID-19). Global Health J. **6**(1), 44–49 (2022)
3. Filip, R., Gheorghita Puscaselu, R., Anchidin-Norocel, L., Dimian, M., Savage, W.K.: Global challenges to public health care systems during the COVID-19 pandemic: a review of pandemic measures and problems. J. Personalized Med. **12**(8), 1295 (2022)
4. Capolongo, S., Gola, M., Brambilla, A., Morganti, A., Mosca, E.I., Barach, P.: COVID-19 and healthcare facilities: a decalogue of design strategies for resilient hospitals. Acta Bio. Medica: Atenei Parmensis **91**(9-S), 50–60 (2020)
5. Barach, P., et al.: Disruption of healthcare: Will the COVID pandemic worsen non-COVID outcomes and disease outbreaks? Prog. Pediatr. Cardiol. **59**, 101254 (2020)
6. Haldane, V., et al.: Health systems resilience in managing the COVID-19 pandemic: lessons from 28 countries. Nat. Med. **27**(6), 964–980 (2021)
7. Turenne, C.P., Gautier, L., Degroote, S., Guillard, E., Chabrol, F., Ridde, V.: Conceptual analysis of health systems resilience: a scoping review. Soc. Sci. Med. **232**, 168–180 (2019)

8. El Bcheraoui, C., Weishaar, H., Pozo-Martin, F., Hanefeld, J.: Assessing COVID-19 through the lens of health systems' preparedness: time for a change. Glob. Health **16**, 1–5 (2020)
9. Gola, M., Fior, M., Arruzzoli, S., Galuzzi, P., Capolongo, S., Buffoli, M.: A research method for locating community healthcare facilities in Italy: how to guarantee healthcare for all. J. Integr. Care **34**(1), 98–114 (2024)
10. Fried, L.P.: The need to invest in a public health system for older adults and longer lives, fit for the next pandemic. Front. Public Health **9**, 682949 (2021)
11. Simoens, S.: Health economic assessment: a methodological primer. Int. J. Environ. Res. Public Health **6**(12), 2950–2966 (2009)
12. Turner, H.C., et al.: An introduction to the main types of economic evaluations used for informing priority setting and resource allocation in healthcare: key features, uses, and limitations. Front. Public Health **9**, 722927 (2021)
13. Luz, A., Santatiwongchai, B., Pattanaphesaj, J., Teerawattananon, Y.: Identifying priority technical and context-specific issues in improving the conduct, reporting and use of health economic evaluation in low-and middle-income countries. Health Res. Policy Syst. **16**, 1–12 (2018)
14. Goodacre, S., McCabe, C.: An introduction to economic evaluation. Emerg. Med. J. EMJ **19**(3), 198 (2002)
15. Eddama, O., Coast, J.: A systematic review of the use of economic evaluation in local decision-making. Health Policy **86**(2–3), 129–141 (2008)
16. Locurcio, M., Morano, P., Tajani, F., Di Liddo, F., Bortone, I.: The cost-benefit analysis for the validation of next generation EU investments: an application to the healthcare sector. In: International Conference on Computational Science and Its Applications, Cham: Springer International Publishing, pp. 27–44 (2022). https://doi.org/10.1007/978-3-031-10548-7_3
17. Public Health, EU4Health. https://health.ec.europa.eu/funding/eu4health-programme-2021-2027-vision-healthier-european-union_en. Accessed 02 Mar 2024
18. Cohesion Policy. https://ec.europa.eu/regional_policy/2021-2027_en. Accessed 05 Mar 2024
19. Regulation (EU) 2021/1060 of the European Parliament. https://eur-lex.europa.eu/legal-content/EN/TXT/?uri=CELEX%3A32021R1060. Accessed 10 Mar 2024
20. NextGenerationEU homepage. https://next-generation-eu.europa.eu/index_it. Accessed 01 Mar 2024
21. The Recovery and Resilience Facility homepage. https://commission.europa.eu/business-economy-euro/economic-recovery/recovery-and-resilience-facility_en. Accessed 20 Mar 2024
22. Regulation (EU) 2021/241 of the European Parliament of the Council of 12 February 2021. https://eur-lex.europa.eu/legal-content/EN/TXT/?uri=CELEX%3A32021R0241. Accessed 15 Mar 2024
23. Italia Domani NRRP homepage. https://www.italiadomani.gov.it/content/sogei-ng/it/it/home.html. Accessed 20 Mar 2024
24. NRRP – Health. https://www.pnrr.salute.gov.it/portale/pnrrsalute/dettaglioContenutiPNRRSalute.jsp?lingua=italiano&id=5833&area=PNRR-Salute&menu=missionesalute. Accessed 25 Mar 2024
25. Economic Appraisal Vademecum. https://ec.europa.eu/regional_policy/en/information/publications/guides/2021/economic-appraisal-vademecum-2021-2027-general-principles-and-sector-applications. Accessed 26 Mar 2024
26. Guide to Cost-Benefit Analysis of Investment Projects for Cohesion Policy 2014–2020. https://ec.europa.eu/regional_policy/en/information/publications/guides/2014/guide-to-cost-benefit-analysis-of-investment-projects-for-cohesion-policy-2014-2020. Accessed 21 Mar 2024
27. Boger, P.C., Turner, D., Roderick, P., Patel, P.: A UK-based cost–utility analysis of radiofrequency ablation or oesophagectomy for the management of high-grade dysplasia in Barrett's oesophagus. Aliment. Pharmacol. Ther. **32**(11–12), 1332–1342 (2010)

28. Kwon, J.S., Carey, M.S., Goldie, S.J., Kim, J.J.: Cost-effectiveness analysis of treatment strategies for Stage I and II endometrial cancer. J. Obstet. Gynaecol. Can. **29**(2), 131–139 (2007)
29. Whitmore, R.G., Thawani, J.P., Grady, M.S., Levine, J.M., Sanborn, M.R., Stein, S.C.: Is aggressive treatment of traumatic brain injury cost-effective? J. Neurosurg. **116**(5), 1106–1113 (2012)
30. Battistoni, G., Genco, M., Marsilio, M., Pancotti, C., Rossi, S., Vignetti, S.: Cost–benefit analysis of applied research infrastructure. Evidence from health care. Technol. Forecasting Soc. Change **112** 79–91 (2016)
31. Langabeer, J.R., et al.: Cost–benefit analysis of telehealth in pre-hospital care. J. Telemed. Telecare **23**(8), 747–751 (2017)
32. de Sonneville-Koedoot, C., Bouwmans, C., Franken, M.C., Stolk, E.: Economic evaluation of stuttering treatment in preschool children: the RESTART-study. J. Commun. Disord. **58**, 106–118 (2015)
33. Hoomans, T., Severens, J.L.: Economic evaluation of implementation strategies in health care. Implement. Sci. **9**, 1–6 (2014)
34. Burns, C.L., Wishart, L.R., Kularatna, S., Ward, E.C.: Knowing the costs of change: an introduction to health economic analyses and considerations for their use in implementation research. Speech Lang. Hearing **23**(1), 30–36 (2020)
35. Whitehead, S.J., Ali, S.: Health outcomes in economic evaluation: the QALY and utilities. Br. Med. Bull. **96**(1), 5–21 (2010)
36. Gray, A.M., Clarke, P.M., Wolstenholme, J.L., Wordsworth, S.: Applied Methods of Cost-effectiveness Analysis in Healthcare, vol. 3. OUP Oxford (2010)
37. Karlsson, G., Johannesson, M.: The decision rules of cost-effectiveness analysis. Pharmacoeconomics **9**(2), 113–120 (1996)
38. Drummond, M.F., Sculpher, M.J., Claxton, K., Stoddart, G.L., Torrance, G.W.: Methods for the Economic Evaluation of Health Care Programmes. Oxford University Press (2015)
39. Grossman, M.: The human capital model. In: Handbook of Health Economics, vol. 1, pp. 347–408. Elsevier (2000)
40. Bloom, D., Canning, D.: Health as human capital and its impact on economic performance. the Geneva papers on risk and insurance. Issues Pract. **28**(2), 304–315 (2003)
41. Haefeli, M., Elfering, A., McIntosh, E., Gray, A., Sukthankar, A., Boos, N.: A cost-benefit analysis using contingent valuation techniques: a feasibility study in spinal surgery. Value Health **11**(4), 575–588 (2008)
42. Park, M., Jit, M., Wu, J.T.: Cost-benefit analysis of vaccination: a comparative analysis of eight approaches for valuing changes to mortality and morbidity risks. BMC Med. **16**, 1–11 (2018)
43. Cole, M.A., Elliott, R.J., Lindley, J.K.: Dirty money: is there a wage premium for working in a pollution-intensive industry? J. Risk Uncertain. **39**, 161–180 (2009)
44. Ng, Y.K.: Quality-adjusted life years (Qualys) versus willingness to pay in matters of life and death. Int. J. Soc. Econ. **25**(6/7/8), 1178–1188 (1998)
45. Pettitt, D.A.R.S., et al.: The limitations of QALY: a literature review. J. Stem Cell Res. Therapy **6**(4) (2016)
46. Petrou, S., Renton, A.: The QALY: a guide for the public health physician. Public Health **107**, 327–336 (1993)
47. Woods, B., Revill, P., Sculpher, M., Claxton, K.: Country-level cost-effectiveness thresholds: initial estimates and the need for further research. Value Health **19**(8), 929–935 (2016)
48. Eichler, H.G., Kong, S.X., Gerth, W.C., Mavros, P., Jönsson, B.: Use of cost-effectiveness analysis in health-care resource allocation decision-making: how are cost-effectiveness thresholds expected to emerge? Value Health **7**(5), 518–528 (2004)

49. Culyer, A.J., Chalkidou, K.: Economic evaluation for health investments en route to universal health coverage: cost-benefit analysis or cost-effectiveness analysis? Value Health **22**(1), 99–103 (2019)
50. Johannesson, M., Jönsson, B.: Economic evaluation in health care: is there a role for cost-benefit analysis? Health Policy **17**(1), 1–23 (1991)
51. Hansson, S.O.: Philosophical problems in cost–benefit analysis. Econ. Philos. **23**(2), 163–183 (2007)
52. Zarnke, K.B., Levine, M.A., O'Brien, B.J.: Cost-benefit analyses in the health-care literature: don't judge a study by its label. J. Clin. Epidemiol. Label **50**(7), 813–822 (1997)
53. Joglekar, P.N.: Cost-benefit studies of health care programs: choosing methods for desired results. Eval. Health Prof. **7**(3), 285–303 (1984)
54. Dunlop, D.W.: Benefit-cost analysis: a review of its applicability in policy analysis for delivering health services. Soc. Sci. Med. **9**(3), 133–139 (1975)
55. Perkins, C., et al.: Cost–benefit analysis: methodological challenges of evaluating large-scale public health interventions and a worked example of the costs and benefits of part-night lighting. In: What is the Effect of Reduced Street Lighting on Crime and Road Traffic Injuries at Night? A Mixed-Methods Study. NIHR Journals Library (2015)
56. Smith, R.D.: Construction of the contingent valuation market in health care: a critical assessment. Health Econ. **12**(8), 609–628 (2003)

Healthcare Investments Valuation. A Case Study of a Property in Disuse to Be Enhanced in a Nursing Home

Francesco Tajani[1], Spartaco Paris[2], Pierluigi Morano[3], Giuseppe Cerullo[1](✉), and Melania Arenas Morente[4]

[1] Department of Architecture and Design, "Sapienza" University of Rome, 00196 Rome, Italy
giuseppe.cerullo@uniroma1.it

[2] Department of Structural and Geotechnical Engineering, "Sapienza" University of Rome, 00196 Rome, Italy

[3] Department of Civil, Environmental, Land, Building Engineering and Chemistry (DICATECh), Polytechnic University of Bari, 70126 Bari, Italy

[4] School of Engineering, University of Basilicata, 85100 Potenza, Italy

Abstract. Italy is increasingly becoming an aging nation. According to the national Institute of Statistics, nearly one in four individuals is over the age of 65, and by 2050, it is estimated that this demographic trend could constitute 35% of the population. This aging also brings with it an increase in the number of people who are not self-sufficient. Currently, there are approximately 4 million individuals in Italy with a severe reduction in autonomy in daily activities, and of these, 92% live at home. The lack of care provision for an aging population has been considered a significant opportunity by the real estate market. The sector of Residential Healthcare Facilities (RHFs), in fact, is emerging as one of the most promising, and it is also beginning to attract increasing interest from investors in Italy seeking attractive returns. This research aims to analyze the main aspects related to the RHF sector. In this sense, an economic evaluation case study is presented, concerning the enhancement of a nursing home located in the Friuli Venezia Giulia region (Italy).

Keywords: Healthcare · DCFA · Health Facilities · Property Investment · Nursing Home

1 Introduction

The improvements in living conditions, due to discoveries in the scientific-health field, the increase in life expectancy, which has grown at different rates depending on the regions of the world, and the decline in the birth rate, together with the role played by the change in the value system in the last decades of the last century, have led to a gradual and progressive aging of the population. This process presents different intensities depending on the regions of the world, being particularly advanced in Europe and America [1]. In 2022, worldwide, there were 771 million people aged 65 or older,

O. Gervasi et al. (Eds.): ICCSA 2024 Workshops, LNCS 14818, pp. 177–191, 2024.
https://doi.org/10.1007/978-3-031-65273-8_12

representing about 10% of the world's population. This percentage is set to more than double in the next eighty years, when those over 65 will constitute 24% of the world's population [2]. In Italy, the life expectancy in 1880 was 35.4 years. However, over the course of our country's history, life expectancy has seen a significant increase: in 1900 it rose to 42.8 years [3] to reach 83 years for men and 86 years for women today [4]. Italy ranks among the top countries for life expectancy at birth, but it is at the bottom of the list for its ability to provide long-term care for its elderly.

This situation will have a significant impact on social protection policies, which will have to address the needs of an increasing number of elderly people. This issue has been further highlighted by the emergency related to the spread of the Covid-19 epidemic, which has highlighted some critical issues that characterize the Italian socio-health sector, in particular residential facilities for the elderly [5]. Although the Covid-19 pandemic has played a role in emphasizing the critical elements of the sector, it has allowed for a collective debate on services for the elderly, especially residential ones, and on the necessary changes and investments [6–8].

In this context, the sector of Residential Healthcare Facilities (RHFs) is emerging as one of the most promising and, also in Italy, is attracting more and more interest from investors looking for interesting returns and countercyclical operations.

2 Aim

This research aims to analyse the main aspects related to the RHF sector. The focus is represented by the observation of the demographic change that will result in a decrease of 9.1 million people under the age of 65 and a parallel increase of 4.6 million people over the age of 65 [9]. On the other hand, the lack of care provision for an aging population has been considered a great opportunity by the real estate market. In this sense, an economic evaluation case study is presented, in order to highlight the most significant aspects regarding the management of RHF initiatives by relating cost items and revenues, focusing on a business plan that is decisive in the return on investment for the private operator.

The work is articulated as follows. In Sect. 3, an overview of the RHFs is presented, examining the demand and the supply, the regional fragmentation that impacts the costs and the tariffs in the RHFs, and the possible future scenarios. In Sect. 4, a case study of a disused building located in Friuli Venezia Giulia (Italy) is illustrated, and a financial analysis of the investment for its enhancement is developed. Finally, the conclusions are discussed in Sect. 5.

3 Framework

3.1 Statistics

Italy is currently, one of the countries with the highest longevity in the EU and the world [10]. Despite the tragic number of deaths due to the Covid-19 pandemic [1], the average age of the Italian population has been characterized by an increase, rising from 45.7 years at the beginning of 2020 to 46.4 at the start of 2023 [2].

A significant statistic pertains particularly to the condition of the elderly who are not self-sufficient, i.e. people aged at least 65 years who are unable to independently carry out daily life activities and without assistance. These individuals' number is approximately 2.9 million, and it has strongly increased in recent decades and is set to rise further in the coming years [3]. In 2050, 1/3 Italians are expected to be over 65 years old. Socio-economic developments (e.g. the decrease in multi-generational family units and an increase in female participation in the labor market) have contributed to reducing the availability of family caregivers to provide care and assistance to the elderly.

Therefore, RHFs have been subject to a rapid, demand-driven expansion in the last two decades [4]. In particular, RHFs are facilities that provide non-self-sufficient individuals, both elderly and otherwise, who have outcomes of physical, psychological, sensory, or mixed pathologies that cannot be treated at home, with a medium level of medical, nursing, and rehabilitative care. This is accompanied by a "high" level of custodial and hotel assistance, modulated based on the care model adopted by the autonomous regions and provinces [5]. RHF is an extra hospital facility of the National Health Service, that is part of the network of first-level territorial services [6].

In 2021, there were 325,032 elderly individuals over 65 who had at least one episode of stay in a RHF during the year, an amount that increased to 362,249 in 2022. In the same year, 2.58% of the elderly aged 65 or more had one or more admissions to a RHF [7]. Despite Italy showing an upward trend in the number of elderly people accommodated in RHFs, the current availability of 4,629 facilities distributed across the entire national territory [8] is not adequate to satisfy the emerging demand.

Currently in Italy, there are 18.6 beds available in RHFs for every 1,000 older residents. This is significantly below the OECD average of 43.8 [4]. At the territorial level, the availability is greater in the North-East with 9.9 beds per 1,000 residents, while in the South it is just over three beds per 1,000 residents and covers only 10% of the total beds [9].

3.2 Tariffs

In Italy, the RHFs represent an extremely fragmented reality with strong regional heterogeneity. There are vast differences from numerous perspectives, not only in terms of the services offered but especially in terms of regional regulation. The proportions allocated to regional health funds also vary, as do the tariffs [10, 11].

Indeed, the national regulations assign to the regions the responsibility of establishing the division of social and health costs for health and social care services aimed at non-self-sufficient or disabled subjects. Furthermore, the regions are tasked with determining the criteria for user participation in the costs of services [12]. This diversity, in a sense, is natural, considering the initial heterogeneity of the different Regions that have regulated socio-health residential services. These services may seem homogeneous, but within them, they present requirements, target ranges of recipients, and a mix of professionals that are significantly different [13].

The financing of RHFs is based on a principle of "co-payment" among the National Health Service, the user, and the social services of the Municipalities [14]. The distribution of the tariffs among these entities is influenced by both national regulations on Essential Levels of Assistance (LEA) and regional and municipal policies. National directives, as established by the decree on LEA [15], suggest an equitable distribution of the tariffs, with a 50% division between the social and health sectors and the remaining share is burdened by the private user.

As for the costs borne by the user, in all Italian regions, these are associated with the hotel and care component of the service. The social share of the tariff of RHFs includes the so-called hotel expenses, which generally include food, cleaning of premises, laundry, meal assistance, and may also include other services. A precise and nationally shared definition of the social/hotel services of RHFs has never been made. The "hotel" quota borne by the assisted person varies greatly [12].

3.3 Market

According to recent research by BNP Paribas Asset Management [16], changes in global demographics have led European investors to healthcare opportunities that are "significantly attractive". As a result, new opportunities also arise and investing in sectors that can meet these demographic challenges provide long-term opportunities. [17]. Healthcare is always seen by developers and investors as a valuable alternative asset class, with clinics, nursing homes, and assisted homecare reported among the European and Italian top trends in real estate investments; therefore, a high level of quality is requested [18, 19]. From 2015 to Q3 2021, Italy's alternative investments hierarchy saw infrastructure (40%) leading, followed by nursing homes (23%), and hospitals/clinics (15%). This situation highlights their growing significance in Italy's investment landscape [20].

The RHF sector in Italy is relatively new, with significant investment volumes only starting to be registered in 2016. In the first three quarters of 2022, healthcare investment activity in Italy reached approximately 0.3€ billion, half of which was invested in the elderly care sector, whereas the remainder was allocated to hospitals [21]. It should also be noted that, as these are often new developments, it is much easier to integrate ESG principles [22–24] into these types of assets quickly and simply. Indeed, they allow for investments with a significant social impact, as they are primarily assets linked to social and demographic trends [25].

Even if decreasing, the level of yield observed in Italy is still attractive for European investors in comparison with other European mature markets. For secondary assets, yields could rise to 6.00%−8.50% depending on the property and location [26]. The market, characterized by the presence of a multitude of local managers, has been affected by the arrival on the national territory of professional operators (Korian, Kos Care, etc.), who through a consolidation process carried out through acquisitions, a process still ongoing, has led to a concentration of the market [27]. The consolidation of the market, the arrival of new international operators, and the numerous developing opportunities make this segment worthy of being analyzed, understood, and evaluated.

4 Case Study

The case study of present research concerns an office building currently in disuse [28, 29], located in the central area of the province of Udine (Northern Italy). The property, about 4,000 m^2, has been built in the late 1980s. It comprises two above-ground levels and a subterranean floor. The vertical circulation within the building is facilitated by two staircases, both internal and external, in addition to elevators (Fig. 1).

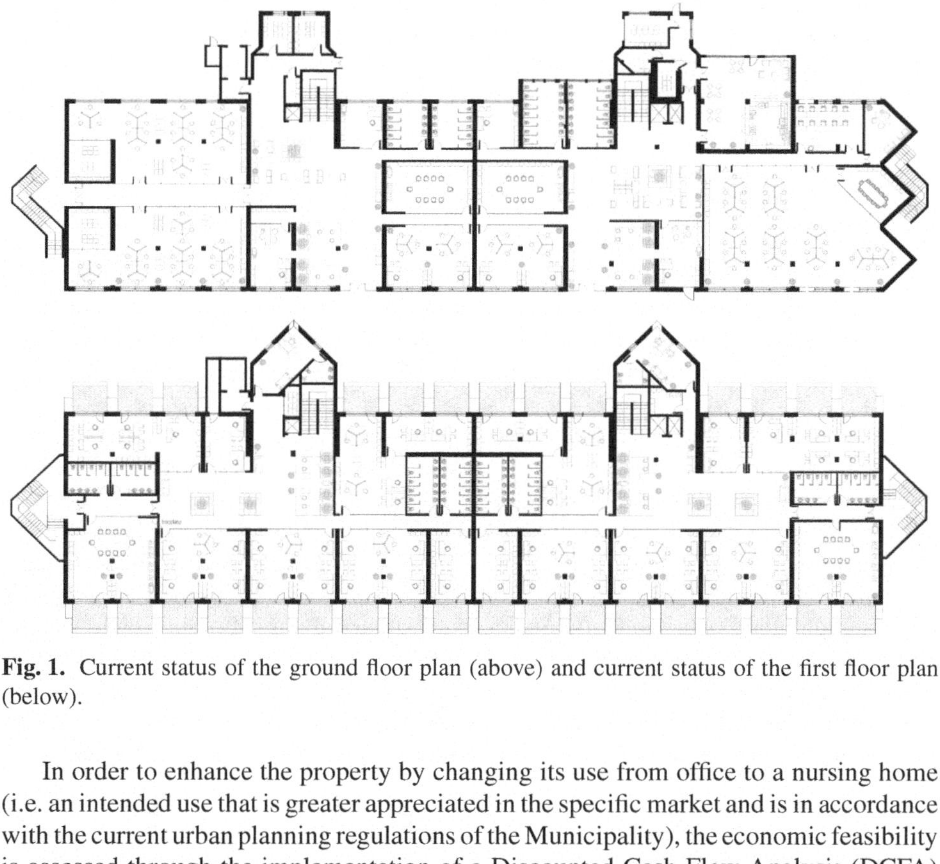

Fig. 1. Current status of the ground floor plan (above) and current status of the first floor plan (below).

In order to enhance the property by changing its use from office to a nursing home (i.e. an intended use that is greater appreciated in the specific market and is in accordance with the current urban planning regulations of the Municipality), the economic feasibility is assessed through the implementation of a Discounted Cash Flow Analysis (DCFA) technique [30, 31]. The year 2024 is the reference date of the valuation. The performance indicators of the evaluation have been obtained by assessing a series of critical factors, to be considered for the activation of a RHF, which can be traced back to the location of the intervention, market analysis and management, economic and financial aspects.

4.1 Market Analysis

Demand. The population referred to December 31, 2021, the latest available data, in Friuli Venezia Giulia region amounts to 1,194,647 residents and, compared to 2011, decreases by 2.1%. Between 2020 and 2021, the birth rate slightly decreased (from 6.2 to 6.1 per thousand), with a decrease of 165 births in absolute value. The population of Friuli Venezia Giulia region, in 2021, has been characterized by a significantly older structure compared to the total of the country. The average age, 48.1 years, is increasing compared to 2020 (47.9), and higher than the 46.2 years of the national average. The aging index also grows, going from 227.1 in 2020 to 231.8 in 2021 [32].

Supply. Residential structures, which include RHFs, are widely spread in Lombardy and Piedmont regions, which represent the first two ones in terms of numbers. However, they are less numerous in Friuli Venezia Giulia region, where there are 198 structures between RHFs and other residential structures, with a total of 9,806 beds. Focusing on the activities carried out by these residential structures in the field of elderly care, we can identify the specific activities of the RHFs. At the regional level, in Friuli Venezia Giulia region, there are 93 nursing homes for the elderly, all attributable to the RHFs [33].

4.2 Design

The general strategy of the intervention on the existing building supports the transition from old to new, emphasizing internal transformations. The primary aim is to create new distributive spaces while avoiding extensive interventions on the structural skeleton. Inside the new building, 122 beds are arranged in 7 single rooms and 115 double rooms (Fig. 2).

The condition of the elderly determines a conception of spaces dedicated to them that cannot disregard the respect of norms designed for people with disabilities. Consequently, in RHFs, architectural barriers are be abolished, and the available space is made fully usable by all guests of the residences [34]. The basic unit of the nursing homes is the module or core, composed of 20–25 places for the elderly who are not self-sufficient. The total reception capacity can vary from 20 to 120 places [6]. In their entirety, the solutions configure, on a residential scale, an articulated living fabric in which, next to the rooms, there are areas dedicated to social relations, connections between the various areas to facilitate the mobility of those present, and spaces reserved for occupational activities and laboratory work to actively engage guests, according to their experiences and attitudes [34]. The use of materials compliant with safety standards is mandatory [35].

4.3 Valuation

The evaluation of the enhancement investment on the case study property has been developed using the DCFA technique. This monetary evaluation does not consider the intangible social impacts of the project, therefore the perspective is strictly private, meaning that for the investor the optimal behaviour is to seek the maximum earning from the investment.

The analysis period has been determined as a balance between the specific factors of the property and the ability to predict market fluctuations. In this case, the DCFA has been carried out over an analysis period, by considering a lease temporal horizon equal to eighteen years (i.e. an ordinary 9 + 9 contract), as provided by the Italian Law No. 392/1978 for hospitality structures. The first two years of the analysis period have been allocated for the renovation of the entire property.

Due to the current inflation rate trend, both for revenues and costs, a decreasing rate has been adopted, equal to 2.4% in the second year, and 2% in the remaining years (the latter is the target value in the EURO zone in the medium-long term).

Fig. 2. Design of the ground floor plan (above) and design of the first floor plan (below).

Costs. The costs have been divided into three main categories: rental fees, operating costs and transformation costs. It should be outlined that in the first year, the expenses are related to transformation costs. The rental fees start in the second year, whereas the operating costs enter in the full operation condition in the third year (50% in the second year).

Rental Fees: Ordinarly, for RHFs the lease fees currently paid vary in a range between 5,000€ and 9,500€ per bed. In the specific case study, a lease fee of about 5,000€ per bed has been assumed. Taking into account that the nursing home activity will start at the second year of the analysis period, a reduction equal to the 50% of the lease fee has been considered for the first year. In Table 1 the temporal distribution of the rental fees is represented.

Operating costs: The literature agrees that [36], in order to interpret the dynamics and determinants of operating costs, it is necessary to classify them into two main categories: the hotel service and the socio-health service.

– The hotel service includes several cost categories (Table 2):

  - Laundry and ironing: Since the facility is equipped with an internal laundry, it is necessary to employ at least 4 workers, for a total annual cost of 107,016.21€ (i.e. 3% of the total operating costs).

Table 1. Account of expenses incurred for rental fees.

Periods	1	2	3	17	18
Total Rent [€/year]	300,625.75	601,251.50	613,276.53	809,205.35	825,389.46
% on revenues	–	24%	21%	12%	12%
Unit Rent [€/bed]	–	4,928.29	5,026.86	6,632.83	6,765.49

- Cleaning and sanitization of premises: This cost can vary significantly based on various factors, such as size, complexity of required services, and cleaning frequency. In this case, the presence of 9 staff members divided into different areas with daily frequency has been considered, resulting in an annual cost of 186,105€. Additionally, the cleaning of common areas, entrusted to an external company at a monthly cost of 3,000€ for 3 interventions per week, is accounted for. The total cost is 222,105€ per year (i.e. 7% of the total operating costs).
- Dining services: Meal management is entrusted to an external company, but as per regional regulations, the facility must be equipped with a kitchen, thus requiring at least 4 staff members, resulting in a total annual cost of 118,300€. In addition to this cost, the daily cost of meals is taken into account, which amounts to 11.83€. The total cost of the dining service amounts to 636,454.00€ annually (i.e. 20% of the total operating costs).
- Receptionist: To manage reception activities, 3 employees are expected to be hired, each working 8 h per day to cover the entire day. The total cost amounts to 91,260.00€ per year (i.e. 3% of the total operating costs).
- Gardening: For the maintenance of the green, internal management is assumed through the hiring of a gardener, the cost is 22,164.74€ per year (i.e. 1% of the total operating costs).

– For socio-healthcare costs (Table 3), the main component is represented by staff dedicated to this specific typology of assistance. Human resources constitute the most significant cost element in RHFs. In this case study, they are thus divided as follows:

- General Practitioner: Considering an average work week of 38 h for each individual doctor, it is necessary to have 4 doctors in the facility to cover the total hours. Therefore, the total annual cost amounts to 224,669.95€ (i.e. 7% of the total operating costs).
- Nurse: Considering a total of 36,254.40 work hours annually to be allocated for the total number of patients, it is hypothesized that 18 nurses are employed in the facility. Hence, the total annual cost amounts to 587,248.77€ (i.e. 18% of the total operating costs).

- Socio-Health Operator: The employment of approximately 40 socio-health operators in the facility is considered, for a total cost of 1,151,375.09€ (i.e. 36% of the total operating costs).
- Physiotherapists: Considering the number of users, it is deemed appropriate to employ 4 active physiotherapists in the facility. The total annual cost amounts to 180,492€ (i.e. 6% of the total operating costs).

Table 2. Account of expenses incurred for operating costs.

Periods	1	2	3	17	18
Laundry [€]	–	64,209.72	107,016.21	141,205.61	144,029.73
Cleaning [€]	–	133,263.00	222,105.00	293,062.83	298,924.09
Gardening [€]	–	22,164.74	22,608.03	29,830.82	30,427.44
Receptionist [€]	–	30,420.00	91,260.00	120,415.63	122,823.94
Dining service cost [€]	–	310,892.40	518,154.00	683,693.20	697,367.06
Kitchen employee [€]	–	70,980.00	118,300.00	156,094.34	159,216.22
Total dining [€]	–	381,872.40	636,454.00	839,787.54	856,583.29
Total hotel service [€]	–	631,929.86	1,079,443.24	1,424,302.43	1,452,788.48
% of total costs	–	11%	28%	28%	28%

Table 3. Account of expenses incurred for socio-healthcare costs.

Periods	1	2	3	17	18
General Practitioner [€]	–	134,801.97	224,669.95	296,447.23	302,376.17
Nurse [€]	–	352,349.26	587,248.77	774,862.28	790,359.53
Socio-Health Operator [€]	–	690,825.06	1,151,375.10	1,519,214.99	1,549,599.29
Physiotherapists [€]	–	108,295.20	180,492.00	238,155.36	242,918.47
Total socio-healthcare [€]	–	1,286,271.49	2,143,785.82	2,828,679.86	2,885,253.46
% of total costs	–	22%	56%	56%	56%

Transformation Costs: The expenditures necessary for transitioning from the office intended use to a nursing home are assessed (Table 4). The cost breakdown is determined as follows:

- Construction costs: Borrowing the data of the DEI price list [37], a parametric cost of 740.00€/m^2 has been estimated. This value has undergone adjustments. The total construction costs amount at 4,531,646.18€ and are referred to the first and the second years of the analysis period.

- Urbanization charges: These costs represent 10% of the sum of hard costs and security and coordination costs. The total amount is 475,823€ (i.e. 10% of the construction costs).
- Design and construction management: This item includes surveys, verifications, design, work management, etc., amounting to a total of 570,987€ (i.e. 12.6% of the construction costs).
- Social security charges: These amount to approximately 624,280€ (i.e. 13.8% of the construction costs).
- FF&E Acquisition: This is equal to 154€/m^2 for the furnishings for the patient rooms and health services, while a cost of 142€/m^2 for the collective services. The total cost amounts to 824,916.12€.
- Photovoltaic System: This is equal to 350,000€.

The transformation costs for the purposes of building renovation in the first two years amount to a total of 7,604,234.45€.

Table 4. Account of expenses incurred for transformation costs.

Periods	1	2
Hard and soft cost [€]	3,857,591.00	2,571,727.33
	60%	40%
FF&E Acquisition [€]	329,966.45	494,949.67
	40%	60%
Photovoltaic System [€]	210,000.00	140,000.00
	60%	40%
Total transformation costs [€]	4,397,557.45	3,206,677.01

Revenues. They are primarily constituted by the tariffs of long-term users (Table 5). Additional revenues are derived from ancillary structures and complementary services, which represent a marginal percentage, approximately 2.5% of the total revenue in the full operation of the nursing home.

The rates, divided by bed type, have been determined by examining the private share that is published for competitors, having comparable characteristics and located near the asset under examination. Figure 3 shows a competitive set composed of selected comparable structures has been identified.

The analysis of the tariff's private share published for the competitors has led to the determination of an average daily rate for beds in double rooms equal to 89.73€ per day. For beds in single rooms, the average daily rate is 93.57€.

Starting from the fifth year, in which the expected occupancy (equal to 90%) remains constant, it is planned to evaluate the regional tariff as revenue following the accreditation process[1].

With regard to the Friuli Venezia Giulia region, the D.G.R. of April 11, 2013, no. 650 approved the procedure for the authorization and accreditation of public and private RHFs. The requirements of the RHFs were subsequently integrated with the DGR of May 13, 2016, no. 817, currently in force. The D.G.R. no. 2088 of October 26, 2017 defined the regional accreditation program for public RHFs and specified the applicable requirements and procedure.

- The Friuli Venezia Giulia region determines the relative rates as follows:
- Basic care level: equal to 111€ per day of hospitalization;
- Average care level: equal to 134€ per day of hospitalization;

Rehabilitative care level: equal to 143€ per day of hospitalization.

Assuming a diversified user base with different levels of assistance, an average tariff equal to 129.33€ has been determined.

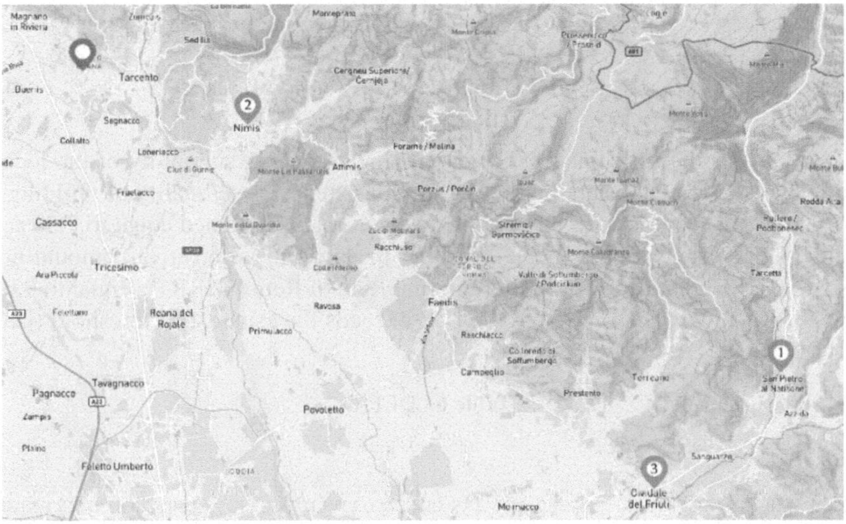

Fig. 3. Localization of the subject (in red) and the identified comparables (in blue).

The discount rate has been determined by borrowing the logic of the Weighted Average Cost of Capital (WACC): a financial structure composed of 50% of equity means and 50% of third-party means is hypothesized; taking into account Treasury bonds

[1] The accreditation processes in the healthcare sector were initiated starting from the '90s, so currently all Regions have defined the structural standards and regulated the organization of the RHFs. In the social assistance sector, the accreditation process for social-assistance structures has been introduced by law no. 328/2000 [41].

Table 5. Revenues.

Periods	1	2	3	17	18
Beds	0	122	122	122	122
Single room	–	7	7	7	7
Double room	–	115	115	115	115
Occupancy (%)	–	60%	70%	90%	90%
No. of beds	–	73	85	110	110
Rate-Tariff Euro/year	RATE			TARIFF	
Bed in a single room [€]	–	34,151.83	34,834.87	59,869.47	61,066.86
Bed in a double room [€]	–	32,752.67	33,407.72	59,869.47	61,066.86
Annual revenue					
Single room [€]	–	143,437.70	170,690.86	377,177.65	384,721.20
Double room [€]	–	2,259,934.00	2,689,321.46	6,196,489.90	6,320,419.70
Total bed revenues [€]	–	2,403,371.70	2,860,012.32	6,573,667.55	6,705,140.90
Other revenues [€]	–	60,084.29	71,500.31	164,341.69	167,628.52
% other revenues	–	2,50%	2,50%	2,50%	2,50%

rates and the specific investment risk (geographical context, suitability of the location to the intended use, real estate market trend; consistency, type/quality of building and maintenance, vacancy risk, solvency), the discount rate is assumed equal to 7.50%.

The Net Present Value (NPV) of the investment is positive, amounting to 1,068,344.07€, and the corresponding Internal Rate of Return (IRR) is equal to 9.00%. As the NPV is higher than zero and the IRR exceeds the established threshold (i.e. the

Table 6. DCFA.

Periods		1	2	3	17	18
Total revenues [€]		0,00	2,463,455.99	2,931,512.63	6,738,009.24	6,872,769.42
Total costs [€]		4,698,183.20	5,726,129.86	3,836,505.59	5,062,187.65	5,163,431.40
Cash flow [€]		−4,698,183.2	−3,262,673.8	−904,992.96	1,675,821.59	1,709,338.02
Discount rate	7,50%					
Discounted cash flow [€]		−4,370,402.9	−2,823,298.1	−728,483.6	490,099.08	465,024.24
NPV [€]		**1,068,344.07**				
IRR [%]		**9,0%**				

discount rate used for the implementation of the DCFA), the project is financially viable (Table 6).

5 Conclusions

Over the past few decades, the issue of care for the elderly who are not self-sufficient has gained increasing relevance, primarily due to the aging demographic trend of the Italian population. In 2022, in Italy the number of elderly significantly exceeds that of children, with a ratio of over five elderly for every child under six years old. This transformation is evident when compared to 1971, when for every 100 young people under the age of 15, there were only 46 people over the age of 65, whereas today this ratio has risen to 193 [38].

To address this anticipated demographic growth in the coming years, a substantial increase in the number of available beds in residential facilities for the elderly will be necessary. Currently, in Italy, only 1.7% of the elderly population can access such services, a coverage rates much lower than the 4–5% recommended by the European Union. To reach a coverage rate of 5%, it would be necessary to add about 600,000 beds by 2035, doubling the current offer of RHFs [39].

In response to this need, over the past ten years, the healthcare real estate sector in Europe has been characterized by a significant interest from institutional investors [40], transforming from a niche market to a strategic one. In particular, Italy represents one of the countries with the greatest opportunities for investment growth, due to the relevant gap between existing supply and demand, with only 4,000 RHFs and 200,000 beds in the country.

The attractiveness of investments in the sector is amplified by the fact that the return on RHFs is not strongly affected by their geographical location, but rather by the proportion of places agreed with the local health system. However, a regulatory harmonization among the different Italian regions is fundamental, as currently there are divergent rules regarding accreditations, agreements, and prices. This regulatory uniformity could prevent further geographical imbalances between the North and the South. Another positive factor is that this asset class guarantees long-term cash flows and is less influenced by economic fluctuations, as the socio-healthcare activity has a temporal constancy in the need for services from its users, independent of economic cycles, ensuring a low investment risk and providing a positive social impact.

The future of this class of real estate activities will certainly be characterized by strong investment growth, which will have to meet the growing demand linked to the aging of the population.

Note: The current study has been developed within the current research "Resilience and sustainability of urban regeneration initiatives. Methods for assessing the effectiveness of investments on the built environment", Call for Research Projects 2022, Sapienza University of Rome.

References

1. Eurostat Statistics Explained. https://ec.europa.eu/eurostat/statistics-explained/index.php?title=Ageing_Europe_-_statistics_on_population_developments#Older_people_.E2.80.94_global_developments. Accessed 25 Jan 2024
2. Il Sole 24 Ore. https://www.infodata.ilsole24ore.com/2023/08/11/nel-2100-il-24-della-popolazione-mondiale-avra-piu-di-65-anni-e-in-italia. Accessed 11 Feb 2024
3. Livi Bacci, M.: La popolazione nella storia d'Europa, Laterza (1998)
4. Buccianti, C., Semboloni, M.: L'invecchiamento della popolazione italiana e i suoi effetti in ambito economico. Rivista Italiana di Economia Demografia e Statistica, **24**(2), (2020)
5. William, G., David, S., Nicholas, B.: The coronavirus and the risks to the elderly in long-term care. J. Aging Soc. Policy 1–6 (2020)
6. Hartley, D., Perencevich, E.: Public health interventions for COVID-19: emerging evidence and implications for an evolving public health crisis. JAMA **323**(19), 1908–1909 (2020)
7. Malik, M.A.: Fragility and challenges of health systems in pandemic: lessons from India's second wave of coronavirus disease 2019. Glob. Health J. **6**(1), 44–49 (2022)
8. Del Giudice, V., De Paola, P., Del Giudice, P.F.: COVID-19 infects real estate markets: short and mid-run effects on housing prices in campania region (Italy). Soc. Sci. **9**(7), 114 (2020)
9. Censis, 57° Rapporto sulla situazione sociale del Paese/2023, Franco Angeli (2023)
10. European Observatory on Health Systems and Policies: Italy: Country Health Profile 2023. State of Health in the EU, Brussels (2023)
11. Fried, L.P.: The need to invest in a public health system for older adults and longer lives, fit for the next pandemic. Front. Public Health **9**, 682949 (2021)
12. ISTAT, rapporto annuale 2023. https://www.istat.it/it/archivio/286191. Accessed 23 Jan 2024
13. Longo, F., Cinelli, G.: Un Servizio nazionale per gli anziani non autosufficienti. Manag.ed Economia Sanitaria **118**(2), 155–179 (2021)
14. De Belvis, A.G., Meregaglia, M., Morsella, A., Adduci, A.: Italy health system review 2022. Health Syst. Transit. **24**(4) (2022)
15. D.P.R. 14 gennaio 1997. https://www.gazzettaufficiale.it/eli/id/1997/02/20/097A1165/sg. Accessed 14 Feb 2024
16. Linee guida del Ministero della Sanità n. 1 del gennaio 1994 (1994)
17. Vetrano, D.L.: Trend di fragilità e long-term care in Italia. Italia longeva per l'invecchiamento e la longevità attiva (2023)
18. Garante nazionale per la geolocalizzazione delle strutture sociosanitaria assistenziali. https://www.garantenazionaleprivatiliberta.it/gnpl/it/dettaglio_contenuto.page?contentId=CNG7869&modelId=10021i. Accessed 20 Feb 2024
19. ISTAT. https://www.istat.it/it/archivio/290577. Accessed 2 Feb 2024
20. Commissione straordinaria per la tutela e la promozione dei diritti umani, RSA_file_definitivo.docx.pdf (senato.it). Accessed 7 Feb 2024
21. Fosti, G., Notarnicola, E., Perobelli, E.: Il personale come fattore critico di qualità per il settore Long Term Care (2023)
22. Pesaresi, F.: La ripartizione della spesa nelle RSA. Tendenze nuove (2012)
23. Notarnicola, E., Perobelli, E.: Sistemi socio-sanitari per gli anziani non autosufficienti: quali premesse per l'implementazione della riforma nazionale anziani? I luoghi della cura (2023)
24. Brizioli, E., Masera, F.: I sistemi di tariffazione. In: L'assistenza agli anziani non autosufficienti in Italia, Maggioli (2011)
25. Decreto del Presidente del Consiglio dei Ministri 29 novembre 2001: Definizione dei livelli essenziali di assistenza (G.U. Serie Generale , n. 33 del 08 febbraio 2002)
26. BNP PARIBAS REAL ESTATE, The investment case for european healthcare real estate (2019)

27. Savills: UK & European. Care Homes 2022. The Countercyclical Asset Class? (2022)
28. Bachmann, M., et al.: The increasing investment of real estate in the health system - a comparison between the USA and Europe. Healthcare **9**, 1633 (2021)
29. PwC: Emerging Trends in Real Estate. (2019)
30. CBRE Research: Italian alternative real estate investments (2021)
31. Jones Lang LaSalle S.p.A.: Real Estate Healthcare snapshot. Milan (2022)
32. Boffo, R., Patalano, R.: ESG Investing: Practices. Progress and Challenges. OECD, Paris (2020)
33. Calabrò, F., Iannone, L., Pellicanò, R.: The historical and environmental heritage for the attractiveness of cities. the case of the umbertine forts of pentimele in Reggio Calabria, Italy. In: New Metropolitan Perspectives. NMP 2020. Smart Innovation, Systems and Technologies, vol. 178, pp. 1990–2004 (2021)
34. De Paola, P., Tajani, F., Locurcio, M.: Sustainable real estate: management. Assess. Innov. Sustain.y **13**, 4676 (2021)
35. Tajani, F., Morano, P.: Evaluation of the economic sustainability of the projects in social housing. In: Computational Science and Its Applications – ICCSA 2014, pp. 135–147 (2014). https://doi.org/10.1007/978-3-319-09150-1_11
36. Cushman & Wakefield: Marketbeat Nursing Homes in Europe (2022)
37. Panu, M.: Healthcare: asset in crescita o segmento di «nicchia»?. Il Sole 24 Ore (2020)
38. Morano, P., Tajani, F., Anelli, D.: Urban planning decisions: an evaluation support model for natural soil surface saving policies and the enhancement of properties in disuse. Prop. Manag. **38**(5), 699–723 (2020)
39. Morano, P., Tajani, F., Di Liddo, F., Amoruso, P.: The public role for the effectiveness of the territorial enhancement initiatives: a case study on the redevelopment of a building in disuse in an Italian small town. Buildings **11**(3) (2021)
40. Tajani, F., Manganelli, B., Cerullo, G., Morano, P., Morente, M.A.: The student housing as a catalyst for virtuous processes of "win-win" revitalization of property assets in disuse. In: Gervasi, O., et al. Computational Science and Its Applications – ICCSA 2023 Workshops. ICCSA 2023. Lecture Notes in Computer Science, vol. 14109, pp. 387–400. Springer, Cham (2023). https://doi.org/10.1007/978-3-031-37120-2_25
41. Calabrò, F.: Integrated programming for the enhancement of minor historical centres. The SOSTEC model for the verification of the economic feasibility for the enhancement of unused public buildings. ARCHISTOR, **17**(3), 1509–1523 (2020)
42. Istat. https://www.istat.it/it/archivio/288397. Accessed 13 Jan 2024
43. Nomisma: Supporto alla due diligence di un portafoglio immobiliare in ambito Healthcare (2023)
44. Pesaresi, F.: L'evoluzione normativa nazionale e la classificazione delle regioni. In: Residenze sanitarie per anziani, Il Mulino, Bologna (2002)
45. D.P.C.M. 22 dicembre 1989. https://www.gazzettaufficiale.it/eli/id/1990/01/03/089A5965/sg. Accessed 20 Feb 2024
46. Pesaresi, F.: I costi delle RSA. Welfare Oggi, 77–83 (2013)
47. DEI tipografia del Genio civile: Prezzi Tipologie Edilizie. DEI, Milan (2019)
48. Il Sole 24 Ore. https://www.ilsole24ore.com/art/i-dati-istat-italia-cinque-anziani-ogni-bambino-2050-rischio-collasso-pensioni-AFIUVu6B. Accessed 18 Feb 2024
49. Cavestri, L.: Residenze per anziani, scatta l'interesse ma pesano le diverse regole regionali e i costi. Il Sole 24 Ore (2023)
50. Newell, G., Marzuki, J.: The increasing importance of the alternate real estate sectors for institutional investors. J. Gen. Manag. **49**(1), 46–64 (2023)
51. Masera, F., Chiatti, C., Rocchetti, C.: Gli standard di qualità e di personale nelle residenze. L'assistenza agli anziani non autosufficienti in Italia (2011)

Port Cities and Evaluation: A Literature Review to Explore Their Interplay in Planning

Giuseppe Ciciriello[1](✉), Sabrina Sacco[1], Carmelo Maria Torre[2], and Maria Cerreta[1]

[1] Department of Architecture, University of Naples Federico II, Naples, Italy
{giuseppe.ciciriello,sabrina.sacco,maria.cerreta}@unina.it
[2] Department of Civil, Environmental, Land, Construction Engineering and Chemistry, Polytechnic University of Bari, Bari, Italy
carmelomaria.torre@poliba.it

Abstract. Port cities constitute complex urban systems in which human settlement, production activities, and coastal resource uses intersect. Sustainable planning in these cities is critical due to the intricate relationship and interdependence between environmental phenomena and socio-economic processes. This literature review investigates the role of evaluation in the study of port cities by cross-checking the concepts of "evaluation" and "port cities". The Scopus database was queried to obtain publication data; then VOSviewer software was used for bibliometric analysis of the results. The analysis of the scientific literature identifies the themes, scale of application, key methodologies and evaluation issues found. Afterwards, the classification of scientific literature is differentiated by scale of application, topic, method, and evaluative problem. The results show how research tends to distinguish and separate the evaluation of the economic performance of the port, seen as transportation and logistics infrastructure, from the assessment of socio-environmental processes, referring to the urban waterfront and coast. This separation is reflected in the identified research gaps and suggests that the application of evaluation to the transformative processes of port cities must therefore promote the integration of the different dimensions involved, aiming for the recognition of the complexity of the values involved and pursuing Sustainable Development Goals.

Keywords: Port city · Evaluation models · Sustainable Development Goals

1 Introduction

Port cities represent a particular application category of urban and regional planning, configuring a complex interplay between human settlements, productive activities, and the use of coastal resources.

The concept of a port city encapsulates a delicate balance between physical, symbolic, economic, and cultural cohesion, resulting in a complex organism shaped by the intensity of the relationship between land and sea territories [1].

O. Gervasi et al. (Eds.): ICCSA 2024 Workshops, LNCS 14818, pp. 192–209, 2024.
https://doi.org/10.1007/978-3-031-65273-8_13

Port cities have evolved symbiotically, with infrastructure development closely intertwined with settlement expansion, embodying an urban synecdoche [2]. However, technological advancements have led to the specialisation of port infrastructure, exacerbating tensions between productive activities and urban centres and necessitating administrative autonomy and separation between port and city institutions. That phenomenon has led to consider port and city systems as separate entities [3].

As noted by Zheng et al., [4] a port city comprises two distinct systems. The first is the port system, marked by logistics, imports, exports, and large-scale infrastructure projects propelled by technology, efficiency, and competitive forces. The second system encompasses its interaction with the city, characterised by social, ecological, and economic diversity, facilitating the development of urban and regional infrastructure.

Ports act as catalysts for the economic, social and cultural prosperity of cities stimulating employment generation, facilitating trade, and supporting industries and tourism, thus enhancing the economic vitality of port cities [5, 6].

Despite their economic benefits, port operations have a huge environmental impact on cities through air and noise pollution, soil contamination, habitat destruction, and traffic congestion. These issues are leading civil society to keep even more attention to social and environmental issues concerning port activities [4].

Port cities, being complex urban environments, require careful planning and management due to their significant economic, social, and environmental impacts.

To interpret and understand intervention at various levels of transformation and development of port cities the application of evaluation models could play a crucial role in port cities by helping to assess, monitor and guide their sustainable development [7].

Port cities present themselves as pivotal centres for the pursuit and implementation of Sustainable Development Goals (SDGs) [8], both through shared, inclusive, and regenerative planning policies (goals 10 and 11); as well as in the implementation of circular economic models capable of generating wealth and investments at the territorial level (goals 8 and 12); and finally, preserving and restoring marine and terrestrial ecosystems, countering the effects of climate change (goals 13, 14, and 15).

Adopting sustainable development strategies for port city ecosystems means focusing on reducing negative impacts and fostering cooperation among different stakeholders.

Port cities, being complex urban environments, require careful planning and management due to their significant economic, social, and environmental impacts [9].

By assessing the effectiveness of policies, projects, and interventions, evaluation methods help ensure that port cities meet the goals of sustainability [10]. Furthermore, evaluation methodologies enable the various stakeholders to make informed choices, understanding the consequences of various development decisions on the city's infrastructure, economy, environment, and society.

The research intends to investigate what is the state of the art in the application of evaluation models to the analysis of processes concerning port cities and what recurring and effective methods can be deduced for the study of transformations of port cities. In order to seek answers to these questions, it was decided to conduct a review of the scientific landscape, investigating the Scopus database, and crossing the concept of evaluation with that of port cities.

The contribution is divided into three parts: in Sect. 2 the methodology and the functional tools for bibliometric analysis are presented; in Sect. 3 the results were elaborated and refined through bibliometric maps and defined criteria, identifying the main thematic groups in which research concerning evaluation applied to the city-port context can be attributed. Finally, Sect. 4 provides a discussion of the results and the conclusions.

2 Material and Methods

In order to answer the main research question, as presented above, it was decided to conduct a literature review, questioning the Scopus database [11–13], using a two-step advanced search (Fig. 1).

The search query is the combination of the terms "evaluation" or "assessment" and "port city", focusing on results that contain the word 'port', and its variants, within the title, abstract, or as a keyword.

The subject fields explored include Environmental Science, Engineering, Social Sciences, Computer Science, Econometric Sciences, and Decision Sciences.

The query results in a large number of documents, about 3262, to be exported in CSV format.

This first set of results has therefore been analysed using the VOSviewer software [14], through a co-occurrence analysis of authors' and index keywords, aimed at the realisation of a bibliometric map [15]. The Network Visualization of the co-occurrence map consists of a cloud-like layout, in which the nodes are filled with the keywords that recur at least 20 times in the searched documents, and whose dimension is directly related to the number of times they are found; the links between the words represent the strength of the bond between two keywords, i.e. the number of times they are co-occurred. Within the map, it is possible to identify the division of keywords into different coloured clusters, which represent special groups of elements with a higher degree of interconnection strength, which allow for understanding thematic and qualitative information [16].

The findings obtained from the bibliometric map, which will be explained hereafter, have led to the decision to narrow the field of investigation by refining the search query.

The terms "port plan", "waterfront", and "maritime planning" were added to the previous string, as well as "analysis", "method", "assessment", "evaluation", "scenario", and "process", searching the database for documents that use or contain these keywords in their title and abstract. Moreover, subjects have been narrowed down, focusing on the topics that carry the most relevant number of documents and are most closely related to the research field: Environmental Sciences, Engineering, Social Sciences, and Computer Science. To further refine the results, only journal articles and conference papers published in English have been selected, targeting only the 2010–2024 publication interval, as it contains the most representative number of articles.

A total of 185 results have been obtained and analysed iteratively through the VOSviewer software; a second bibliometric keyword co-occurrence map was then built, with a minimum recurrence threshold of 5 times, highlighting further significant clusters.

Finally, the resulting 185 publications of the refined search have been examined in detail and selected according to the following criteria: source availability in the associated databases; helpfulness to the research question; consistency with the clusters identified in the related bibliometric map, and the possibility of deriving an evaluation method. The step-by-step filtering, yielded a result of 39 articles, which were then assembled into a classification table. The chosen articles were therefore organised according to four main criteria: the scale of application of the proposed research, whether it deals exclusively with port processes, port-city interaction, or those affecting the regional scale; the topic mainly covered; the evaluation method in detail explained and employed; finally, the main evaluation problem addressed.

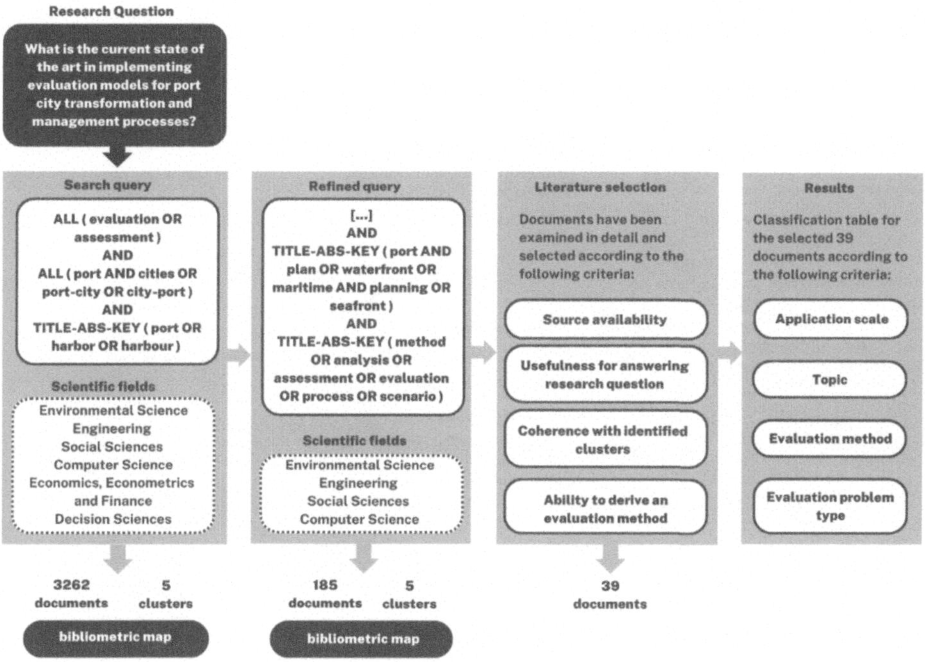

Fig. 1. Methodological approach scheme.

3 Results

The first phase of the research yielded 3262 results, considering the intersection between the concepts of evaluation and port cities, exploring Social Sciences, but also Engineering, Environmental Science, Decision Theory, Economics, and Computer Science. This group of outcomes presents a rather broad and varied distribution, providing information of a global nature regarding publications dealing with the evaluation of port city processes.

The Network Visualisation of the co-occurrence bibliometric map (Fig. 2) displays how the most recurrent keywords are organised around 5 main clusters, arranging themselves according to the relative correlation between them, and therefore according to a common semantic value. The clusters on the right (marked in blue and green, respectively) represent scientific contributions concerning air pollution, caused using fossil fuels for ship operation; closely related to these are, of course, those covering environmental monitoring, as well as water and coastal sediment pollution. Among them, it is interesting to note the presence of a small cluster of research dealing with the evaluation of health risks associated with exposure to environmental pollutants. The remaining groups (in orange and yellow) centre around the themes of the port, both as a transport infrastructure and as an urban issue; but also, around the assessment of risks and impacts due to climate change and extreme natural events.

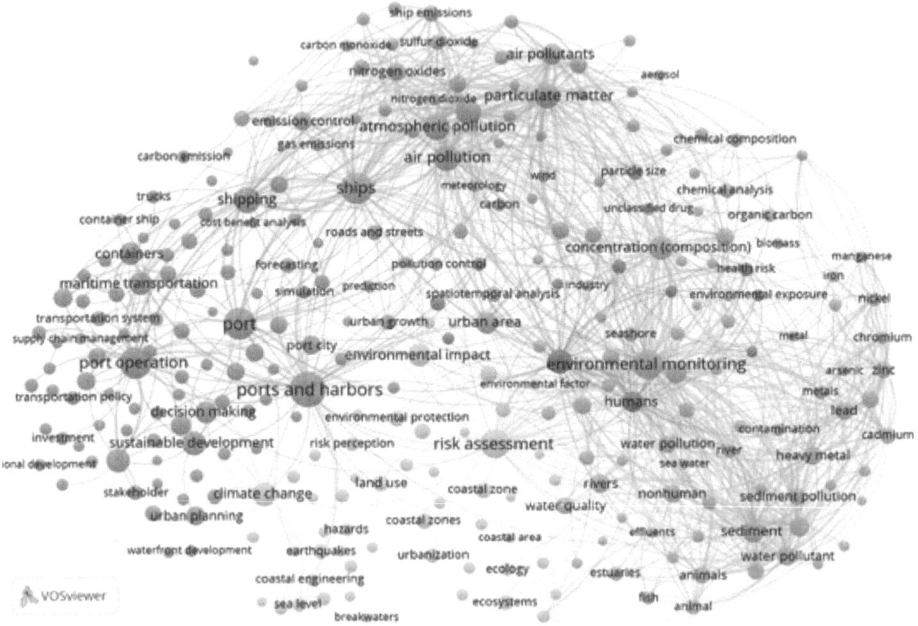

Fig. 2. VOSviewer network visualisation of co-occurrence map based on 3262 documents.

In the second phase, search terms were narrowed down in order to arrive at more detailed and relevant information from a research demand perspective. Therefore, only the most relevant disciplinary fields were considered, and the research period was narrowed down between 2010 and 2024 since it contains, with a good approximation, the most significant number of publications. The 185 results obtained were analysed in a second bibliometric co-occurrence map (Fig. 3), leading to the identification of five thematic clusters.

The keyword cloud presents a layout in which two main groupings are recognisable: in the right half (marked in dark and light blue) are the keywords related to logistics, supply chains, and analogously transportation development and port operation; in the left half, instead, there are two related clusters (in shaded green), namely urban planning and sustainable development, alongside environmental planning and monitoring. The sharp polarisation of the graph shows how the scientific literature tends to treat the topics of "city" and "port" separately and not in an integrated manner, assessing only the operational aspects of economic performance on the one hand, or dealing with urban transformation and environmental issues on the other. Furthermore, the occurrence of terms such as "risk assessment" and "decision support system" in the operational port clusters might indicate particular publications dealing with risk management in the port domain; and, again, it can be seen that the logic of sustainability and ecology are more integrated in urban planning. In the middle-low part of the cloud, a last small cluster can be distinguished, represented by the keywords "decision making", "fuzzy mathematics", "analytic hierarchy process", which indicates how the field of evaluation is equally distant and still rather marginal, in terms of publications, compared to the other clusters.

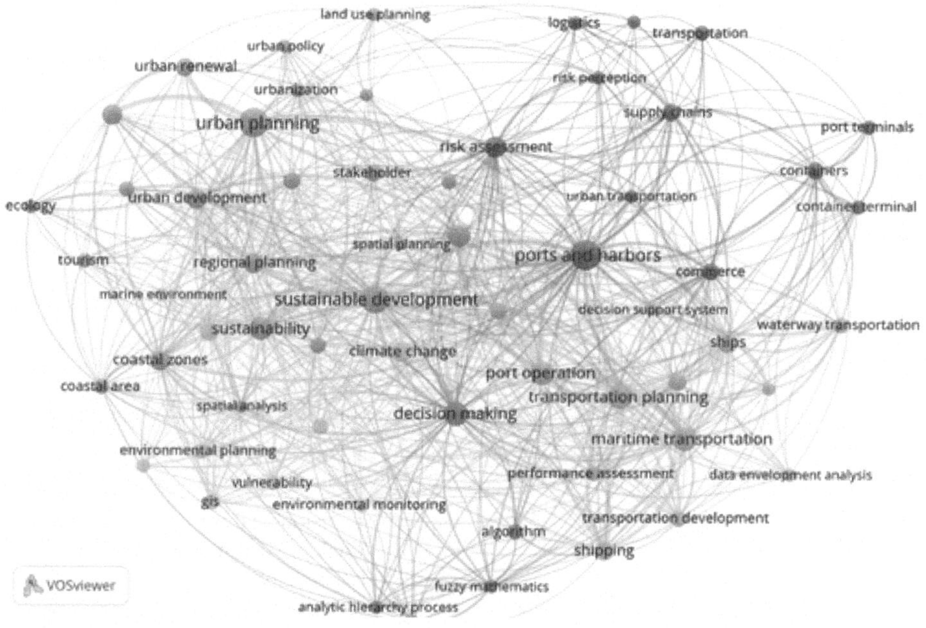

Fig. 3. VOSviewer network visualisation of co-occurrence map based on 185 documents.

Subsequently, publications were manually filtered based on resource availability, relevance to the research question, coherence with the aforementioned clusters, and the ability to clearly derive an evaluation method.

The 39 selected documents were classified into a classification table (Table 1). By aiming to understand potential trends or common practices in publications, the following table presents an organisation of the examined literature by linking specific evaluation methods to particular themes and their corresponding spatial scope.

Besides authors, title and year of publication, four classification categories were identified: the application scale of the analysed process, the main theme investigated, the method used to address the specific evaluation problem.

City-Port interaction, refers to the dynamic and multidimensional relationship and interdependence between City and Port.

Table 1. Classification table of the analysed articles.

Application scale	Topic	Year	Author(s)	Evaluation method	Evaluation problem type
City-Port interaction	Heritage	2022	Daldanise, G. et al.	Preference Ranking Organization Method for Enrichment Evaluation (PROMETHEE) + Geometrical Analysis for Interactive Aid (GAIA)	Preferable scenario selection
		2013	De Rosa, F. et al.	Comparative method using sustainability indicators	Indicator selection
	Natural landscape	2022	Martin, R. et al.	Delphi Survey + Analytic Hierarchy Process (AHP)	Ranking of alternatives + Indicator selection
		2019	Martin, R. et al.	Delphi Survey	Indicator selection
	Port Planning	2012	Cerreta, M. et al.	Integrated Spatial Assessment (ISA)	Impact assessment
	Risk management	2024	Chondrogianni, D. V. et al	Method applying Bioclimatic Index	Ranking of alternatives
		2022	Chondrogianni, D. V. et al.	Method applying Bioclimatic Index	Ranking of alternatives
		2022	Guan, C. et al.	Decision-Making Trial and Evaluation Laboratory (DEMATEL) - Interpretive Structural Modeling (ISM) - and Social Network Analysis (SNA)	Performance levels
		2021	Saleh, D. et al.	Spatial Analysis + Analytic Hierarchy Process (AHP)	Performance levels + Indicator selection

(*continued*)

Table 1. (*continued*)

Application scale	Topic	Year	Author(s)	Evaluation method	Evaluation problem type
	Urban Planning	2024	Fan, M. et al.	Method applying Space Syntax Theory's indicators	Performance levels
		2020	Andrade, M. J. et al.	Qualitative and quantitative indicator-based method	Performance levels
		2020	Cerreta, M. et al.	Sustainability indicators (SI) + Role-Playing Method (RPM) + Preference Ranking Organization Method for Enrichment Evaluation (PROMETHEE)	Preferable scenario selection
		2020	Cerreta, M. et al.	Method applying Survey + Likert scale evaluation	Indicator selection
		2020	Jardas, M. et al.	PROMETHEE	Preferable scenario selection
		2019	Schipper, C	Comparative method applying Green Port Policy (GPP) + Sustainable Development Goals (SDGs) framework	Indicator selection
		2019	Ignaccolo, M. et al.	Method applying Level of Service (LOS) and Utility	Performance levels
		2017	Schipper, C	Sustainable Integrated Condition Index (SICI)	Ranking of alternatives + Indicator selection

(*continued*)

Table 1. (*continued*)

Application scale	Topic	Year	Author(s)	Evaluation method	Evaluation problem type
		2013	Nicolini, E. et al.	Community Impact Evaluation (CIE) - Novel Approach to Imprecise Assessment and Decision Environments (NAIADE)	Ranking of alternatives
		2010	Cuturi, C	Method applying Sustainability indicators	Performance levels
Port	Port externalities	2022	Tovar, B. et al.	Data Envelopment Analysis (DEA)	Impact assessment + Performance levels
		2020	Viana, M. et al.	Health impact assessment (HIA)	Impact assessment + Performance levels
		2017	Papaefthimiou, S et al.	Quantitative and cost indicator-based method	Impact assessment + Performance levels
	Port Planning	2021	Majidi, A. et al.	Stepwise weight-assessment ratio analysis (SWARA) + Measurement of alternatives and ranking according to compromise solution (MARCOS) + Combined compromise solution (CoCoSo) + Technique for order preference by similarity to ideal solution (TOPSIS)	Ranking of alternatives

(*continued*)

Table 1. (*continued*)

Application scale	Topic	Year	Author(s)	Evaluation method	Evaluation problem type
		2020	Ignaccolo, M. et al.	Comparative method applying ASI (Avoid/Reduce, Shift/Maintain, Improve) analysis + ASI/Sustainability matrix	Indicator selection
		2020	Molavi, A. et. al.	Comparative method applying Smart Port Index (SPI) indicators	Indicator selection + Performance levels
		2019	Hill, R. J	Multi-criteria analysis (MCA)	Preferable scenario selection
	Risk management	2021	Dhanak, M. et. al.	Hybrid multi-modal simulation model	Performance levels
		2018	Yang, Z. et al.	Cost Benefit analysis based on Evidential Reasoning (ER)	Performance levels + Feasibility Evaluation
	Transportation and logistics	2023	Celik, M. S	Semi-structured interview + SWOT (Strenght, Weakness, Opportunity, Threat) + TOWS (Threat, Opportunity, Weakness, Strenght) + Analytic Network Process (ANP)	Preferable scenario selection
		2021	Sirotić, M. et al.	Survey + SWOT (Strenght, Weakness, Opportunity, Threat) + Gap analysis	Preferable scenario selection
		2020	Mou, N. et al.	Fuzzy Analytic Hierarchy Process (FAHP) + Entropy model	Ranking of alternatives
		2020	Feng, H. et al.	Analytic Hierarchy Process (AHP)	Ranking of alternatives

(*continued*)

Table 1. (*continued*)

Application scale	Topic	Year	Author(s)	Evaluation method	Evaluation problem type
Regional	Heritage	2016	Pourebrahim, S. et al.	Fuzzy Technique for Order Preference by Similarity to Ideal Solution (TOPSIS) + Fuzzy Analytic Hierarchy Process (FAHP)	Ranking of alternatives
	Port externalities	2024	Madhumitha, R. et al.	Spatial Analysis + Analytic Hierarchy Process (AHP)	Impact assessment
	Risk management	2023	Hemmerling, S. A. et al	Social Return on Investment (SROI)	Preferable scenario selection + Impact assessment
		2022	Amoura, R. et al.	Spatial analysis using Geographical Information System (GIS)	Performance levels
		2016	Zanetti, V. B. et al.	Method applying Socio-Environmental Vulnerability Index for Coastal Areas (SEVICA)	Performance levels
		2013	Messner, S. et al.	Spatial analysis using Geographical Information System (GIS) + Cost Benefit analysis based on Net Ecosystem Services Analysis (NESA)	Performance levels + Feasibility Evaluation
	Transportation and logistics	2023	Fisher-Holloway, B. et al.	Spatial analysis using Geographical Information System (GIS)	Resource Allocation

Research into the city-port dynamic represents a focal point for numerous studies exploring decision-making processes and transformative endeavours within urban waterfronts. These investigations also delve into evaluative inquiries regarding the impacts of both natural phenomena and human activities. They share a holistic interest in integrating infrastructure, settlements, and the surrounding environment.

Daldanise et al. [17] delve into the potential for implementing planning and regeneration strategies for waterfronts, centred on heritage re-use. Their research proposes the definition of a decision support system employing multicriteria evaluation, showcasing the transformative potential of cultural-led urban development. Meanwhile, De Rosa et al. [18], conducting a comparative research of adequate indicators, advocate for the Historic Urban Landscape (HUL) framework as a guiding light for the regeneration of historical urban waterfronts, underscoring its symbiotic relationship with the layered urban fabric.

Martin et al. [19, 20] conduct interviews to stakeholder to derive soft indicators facilitating a nuanced understanding of the interplay between natural landscapes and tourist port infrastructure impact. Their prioritisation of integration actions underscores the need for holistic planning approaches.

By proposing an innovative Integrated Spatial Assessment protocol, Cerreta et al. [21] advocate for a comprehensive evaluation framework that incorporates environmental, social and economic indicators alongside spatial analysis and decision-making processes.

Strategies for adapting to extreme weather phenomena are analysed by Chodrogianni et al. [22, 23], through a quantitative ranking of various waterfront planning solutions based on bioclimatic performance, and by Guan et al. [24], who propose a methodology for analysing and evaluating natural and infrastructural risk factors, aimed at supporting the development of plans for the safety of port settlements. Saleh et al. [25] use remote sensing techniques for the identification of geomorphological risk areas, together with the evaluation of the most vulnerable anthropized areas, based on specific land use.

The topic of urban planning is particularly explored: Fan et al. [26] and Ignaccolo et al. [27] use the calculation of various indices to quantify the degree of pedestrian accessibility of coastal settlements with the waterfront and port area; Andrade et al. [28] and Cuturi [29] address comparative studies of different port cities in search of scalable indicators useful for replicating successful strategies. Schipper [30, 31] proposes a set of indicators to assess the sustainability of port city planning, correlated with the SDGs. Through the application of a specific index (Sustainable Integrated Condition Index) for the evaluation of various case studies, it is shown that alternatives where economy, society, and environment are more integrated are those with the best performance. Nicolini et al. [32] also resort to a set of multidimensional indicators, applying it to the ranking of strategic design alternatives for the port city of Palermo. Jardas et al. [33] use the PROMETHEE multicriteria evaluation method to assess the preferable development scenario of a city on the Croatian coast, among various possible strategies. Finally, Cerreta et al. [34, 35] construct a set of quantitative and qualitative indicators for the regeneration of Naples port city system. By applying the PROMETHEE multicriteria method, the preferred transformative scenario was selected.

Port, refers to its infrastructure aspect, the management and planning processes of port functions, as well as the external effects they can produce.

Papaefthimiou et al. [36] tackle the issue of externalities, particularly focusing on measuring the pollution stemming from port activities on the surrounding environment. They devise quantitative indicators aimed at economically assessing the impact of pollutants, drawing insights from operational data across a range of ports. Moreover, Tovar et al. [37] employ a mathematical approach to compute the social and environmental external costs attributed to ship pollution in several Spanish cities. Instead, Viana et al. [38] undertake a Health Impact Assessment, establishing a direct correlation between ship pollutant emissions and mortality and disease rates in diverse case studies.

Port planning is approached by selecting appropriate indicators, exemplified by Ignaccolo et al. [39], who introduce a methodology linking suitable quantitative measures with diverse qualitative strategies for sustainable development. Similarly, Molavi et al. [40] devise indicators influenced by the Smart Port Model, covering environmental, energy, operational and security aspects of the port. Majidi et al. [41] employ the TOPSIS multicriteria method to assess sustainability performance across various ports in Iran. Hill [42] adopts this multicriteria approach also for his research, evaluating the preferred scenario within specific port development projects and emphasising the necessity of a decision support system to manage conflicting interests among stakeholders.

Dhanak et al. [43], using a multi-modal simulation model, predict the effects of extreme weather events on the operability of transport chains, evaluating the different levels of adaptation and infrastructural resilience of the port. Yang et al. [44] apply a Cost-Benefit Analysis to assess the feasibility of climate change adaptation interventions in relation to the magnitude of the risks encountered.

Using a qualitative methodology based soft data collected through surveys, Sirotić et al. [45] propose a strategic perspective for the sustainable development of transport in the Adriatic Sea; Celik [46] adopts the Analytic Network Process (ANP) to identify the optimal alternative among various development proposals in the context of trade between the Mediterranean and China. In the Chinese context, Feng et al. [47] employ the Analytic Hierarchy Process (AHP) multicriteria method to classify the most efficient ports in terms of logistics and transportation within a specific region. Similarly, Mou et al. [48] utilise Fuzzy-AHP method for assessing the economic development potential of ports in the Yangtze Delta context.

Regional, refers to a territorial and broader dimension in which urban and port dynamics influence a larger spatial context.

Pourebrahim et al. [49] apply the TOPSIS method to various alternative scenarios aimed at conserving cultural and natural heritage, endangered by anthropogenic phenomena. Subsequently, the Fuzzy-AHP method is used to establish a hierarchy of priorities among different conservation areas.

Madhumitha et al. [50] assess the impacts generated by infrastructure and port activities on the coast and hinterland, using satellite observation data. Additionally, they adopt the AHP method to determine the extent of impacts across different coastal settlements, employing specific socio-environmental indicators.

The use of spatial analysis methods and Geographic Information System (GIS) techniques is particularly significant for risk management. Amoura et al. [51] employ remote sensing technologies and mathematical methods to assess and predict the extent of coastal flooding due to future sea level rise, mapping the at-risk areas; similarly, Zanetti et al. [52] proposes a combined index to jointly assess vulnerability to coastal and inland hydrological hazards. Messner et al. [53] develop a prediction framework to evaluate sea level rise in port areas and the associated hydraulic risk, additionally proposing a methodology to measure the economic impact of these phenomena and support the planning of adaptation interventions.

Hemmerling et al. [54] present a participatory approach for assessing the social impact of potential coastal restoring interventions, applying the Social Return on Investment (SROI) framework. The results demonstrate a convergence between social and environmental evaluations.

Fisher-Holloway et al. [55] utilise spatial analysis techniques such as Cluster analysis and Kernel density analysis within GIS to pinpoint the location of logistics services. This approach supports decision-making processes within port-hinterland transportation systems.

4 Discussion and Conclusions

After reviewing the vast literature dealing with the application of evaluation methods in the port city domain, it is evident the extent of the investigated field. Studies address a spectrum of scales, from the circumscribed area of port operations to the broader territorial context. This inquiry makes clear how entangled are the social, economic and environmental values involved in port cities.

Review of the research literature reveals a differentiation in the proposed evaluation methodologies: some studies use quantitative approaches, exploiting mathematical models to quantify the performance levels of particular targets, or to express values in econometric terms; others rely on the soft data gathered from qualitative approaches as common in social survey techniques; and others mix qualitative and quantitative indicators to capture the complexity of port city ecosystems.

In addition, the methods examined differ considerably depending on the topic: urban and port planning assumes evaluation models in order to make the best choices, based on objectively defined criteria, in support of policy-making; risk management is concerned with predicting potential effects due to more or less predictable extrinsic factors, while port emissions assessment focuses instead on the social and environmental effects already produced; logistics and transportation development are analyzed to maximize the efficiency in resource use and resulting economic performance provided; landscape and heritage assessment, by contrast, invokes the need for tools to recognize alternate values, such as environmental, social and cultural. In this sense, a research gap arises, namely the involvement of stakeholders and communities in transformative processes. While some of the research discussed explores the interplay between stakeholder interests and the proper methods of evaluation and quantification, exhibiting interesting outcomes, and succeeding in simultaneously capturing social and economic preferences; other studies fail in this task, coming up with solutions that only partially reflect the complexity of port cities.

Given the deep entanglement of economic vitality, community and natural environment, port cities require rigorous evaluative approaches to capture the interdependence among all the factors and values recalled. Examining the landscape of academic publications, this literature review offers a systematization of evaluative methods and case studies for obtaining planning solutions in several topics.

Author Contributions. Conceptualization, G.C. and S.S.; methodology, G.C. and S.S.; validation, M.C. and C.M.T.; formal analysis, G.C.; investigation, G.C.; writing—original draft preparation, G.C. and S.S.; writing—review and editing, G.C. and S.S.; visualisation, G.C.; supervision, M.C. and C.M.T. All authors have read and agreed to the published version of the manuscript.

Disclosure of Interests. The authors have no competing interests to declare that are relevant to the content of this article.

References

1. Moretti, B.: Beyond the Port City. De Gruyter (2021). https://doi.org/10.1515/9783868599503
2. Bruttomesso, R.: Città-porto: Mappe per nuove rotte urbane, Grande Sud, Palermo, Mediterraneo. Marsilio (2006)
3. Hesse, M.: Approaching the relational nature of the Port-City interface in Europe: ties and tensions between seaports and the urban. Tijdschr. voor Econ. en Soc. Geogr. **109**, 210–223 (2018). https://doi.org/10.1111/tesg.12282
4. Zheng, Y., Zhao, J., Shao, G.: Port city sustainability: a review of its research trends. Sustainability **12**, 8355 (2020). https://doi.org/10.3390/su12208355
5. Merk, O., Dang, T.-T.: The effectiveness of port-city policies (2013). https://doi.org/10.1787/5k3ttg8zn1zt-en
6. Couling, N., Hein, C.: The urbanisation of the sea : from concepts and analysis to design(2020).https://doi.org/10.7480/isbn.9789462085930
7. Merk, O.: The competitiveness of global port-cities: synthesis report (2013). https://doi.org/10.1787/5k40hdhp6t8s-en
8. THE 17 GOALS. Sustainable Development Homepage. https://sdgs.un.org/goals. Accessed 07 May 2024
9. AIVP Homepage. https://www.aivp.org/en/%0A. Accessed 11 Jan 2021
10. Cerreta, M., Cannatella, D., Poli, G., Sposito, S.: Climate change and transformability scenario evaluation for venice (Italy) port-city through ANP method. In: Gervasi, O., et al. (eds.) ICCSA 2015. LNCS, vol. 9158, pp. 50–63. Springer, Cham (2015). https://doi.org/10.1007/978-3-319-21410-8_4
11. Scopus Homepage. https://www.scopus.com/home.uri. Accessed 6 May 2024
12. Cerreta, M., La Rocca, L.: Urban regeneration processes and social impact: a literature review to explore the role of evaluation. In: Gervasi, O., et al. (eds.) ICCSA 2021. LNCS, vol. 12954, pp. 167–182. Springer, Cham (2021). https://doi.org/10.1007/978-3-030-86979-3_13
13. Sacco, S., Di Martino, F., Cerreta, M.: Smart circular cities and stakeholders engagement: a literature review to explore the role of artificial intelligence. In: Gervasi, O., et al. Computational Science and Its Applications – ICCSA 2023 Workshops. ICCSA 2023. Lecture Notes in Computer Science, vol. 14108, pp. 239−258. Springer, Cham (2023).https://doi.org/10.1007/978-3-031-37117-2_18
14. van Eck, N.J., Waltman, L.: Visualizing bibliometric networks. In: Ding, Y., Rousseau, R., Wolfram, D. (eds.) Measuring Scholarly Impact, pp. 285–320. Springer, Cham (2014). https://doi.org/10.1007/978-3-319-10377-8_13
15. Radhakrishnan, S., Erbis, S., Isaacs, J.A., Kamarthi, S.: Novel keyword co-occurrence network-based methods to foster systematic reviews of scientific literature. PLoS ONE **12**, e0172778 (2017). https://doi.org/10.1371/journal.pone.0172778
16. van Eck, N.J., Waltman, L.: VOSviewer manual. Manual for VOSviewer version 1.6.20 software documentation. Univeristeit Leiden (2023)

17. Daldanise, G., Clemente, M.: Port cities creative heritage enhancement (PCCHE) scenario approach: culture and creativity for sustainable development of naples port. Sustainability **14** (2022). https://doi.org/10.3390/su14148603
18. De Rosa, F., Di Palma, M.: Historic urban landscape approach and port cities regeneration: naples between identity and outlook. Sustainability **5**, 4268–4287 (2013). https://doi.org/10.3390/su5104268
19. Martín, R., Yepes, V.: The concept of landscape within marinas: Basis for consideration in the management. Ocean Coast. Manag. **179** (2019). https://doi.org/10.1016/j.ocecoaman.2019.104815
20. Martín, R., Yepes, V.: Assessing the relationship between landscape and management within marinas: the managers' perception. Land **11** (2022). https://doi.org/10.3390/land11070961
21. Cerreta, M., De Toro, P.: Strategic environmental assessment of port plans in Italy: experiences, approaches, tools. Sustainability **4**, 2888–2921 (2012). https://doi.org/10.3390/su4112888
22. Chondrogianni, D. V, Stephanedes, Y.J.: Evaluation of urban planning methods toward bioclimatic and resilient urban spaces. Environ. Plan. B Urban Anal. City Sci. **49**, 1354–1370 (2022). https://doi.org/10.1177/23998083211063220
23. Chondrogianni, D. V, Stephanedes, Y.J.: Assessment of the bioclimatic index for resilient urban spaces in Mediterranean cities. Environ. Plan. B Urban Anal. City Sci. **51**, 234–258 (2024). https://doi.org/10.1177/23998083231175894
24. Guan, C., Dong, D., Shen, F., Gao, X., Chen, L.: Hierarchical structure model of safety risk factors in new coastal towns: a systematic analysis using the DEMATEL-ISM-SNA method. Int. J. Environ. Res. Public Health **19** (2022). https://doi.org/10.3390/ijerph191710496
25. Saleh, D., Kawy, A.A. El: Problematic soil risk assessment approach for sustainable spatial suitability of urban land use in new cities by using GIS and remote sensing: a case study of new east port said city. Int. J. Sustain. Dev. Plan. **16**, 615–628 (2021). https://doi.org/10.18280/ijsdp.160402
26. Fan, M., Marzbali, M.H., Abdullah, A., Tilaki, M.J.M.: Using a space syntax approach to enhance pedestrians' accessibility and safety in the historic city of george town, Penang. Urban Sci. **8** (2024). https://doi.org/10.3390/urbansci8010006
27. Ignaccolo, M., Inturri, G., Cocuzza, E., Giuffrida, N., Torrisi, V.: Framework for the evaluation of the quality of pedestrian routes for the sustainability of port–city shared areas. In: WIT Transactions on the Built Environment, pp. 11–22 (2019). https://doi.org/10.2495/CC190021
28. Andrade, M.J., Costa, J.P., López, J.B.: 3DPortCityMeasure: methodology for the comparative study of good practices in port-city integration. Sustainability **12** (2020). https://doi.org/10.3390/su12030880
29. Cuturi, C.: The revival of urban waterfront areas: evaluation of British and Irish experiences. Int. J. Sustain. Dev. **13**, 122–148 (2010). https://doi.org/10.1504/IJSD.2010.035104
30. Schipper, C.A., Vreugdenhil, H., de Jong, M.P.C.: A sustainability assessment of ports and port-city plans: comparing ambitions with achievements. Transp. Res. Part D Transp. Environ. **57**, 84–111 (2017). https://doi.org/10.1016/j.trd.2017.08.017
31. Schipper, C.: Understanding the sustainable development goal approach for ports of the future. In: Proceedings of the IAHR World Congress, pp. 3195–3204 (2019). https://doi.org/10.3850/38WC092019-1913
32. Nicolini, E., Pinto, M.R.: Strategic vision of a Euro-Mediterranean port city: a case study of palermo. Sustain. **5**, 3941–3959 (2013). https://doi.org/10.3390/su5093941
33. Jardas, M., Schiozzi, D., Smojver, Ž.: Application of multi-criteria analysis of determining sea port development models in the spatial concept of a town, based on the example of the town of rovinj. Teh. Vjesn. **27**, 297–307 (2020). https://doi.org/10.17559/TV-20181116194755

34. Cerreta, M., Muccio, E., Poli, G., Regalbuto, S., Romano, F.: City-port circular model: towards a methodological framework for indicators selection. In: Gervasi, O., et al. Computational Science and Its Applications – ICCSA 2020. ICCSA 2020. Lecture Notes in Computer Science, vol. 12251, pp. 855–868 . Springer, Cham (2020). https://doi.org/10.1007/978-3-030-58808-3_61
35. Cerreta, M., Giovene di Girasole, E., Poli, G., Regalbuto, S.: Operationalizing the circular city model for Naples' city-port: a hybrid development strategy. Sustainability **12**(7), 2927 (2020). https://doi.org/10.3390/su12072927
36. Papaefthimiou, S., Sitzimis, I., Andriosopoulos, K.: A methodological approach for environmental characterization of ports. Marit. Policy Manag. **44**, 81–93 (2017). https://doi.org/10.1080/03088839.2016.1224943
37. Tovar, B., Wall, A.: The external costs of port activity for port cities: an environmental efficiency analysis of Spanish ports. Int. J. Sustain. Transp. **16**, 820–832 (2022). https://doi.org/10.1080/15568318.2021.1943074
38. Viana, M., et al.: Estimated health impacts from maritime transport in the Mediterranean region and benefits from the use of cleaner fuels. Environ. Int. **138**, (2020). https://doi.org/10.1016/j.envint.2020.105670
39. Ignaccolo, M., Inturri, G., Giuffrida, N., Torrisi, V.: A sustainable framework for the analysis of port systems. Eur. Transp. - Trasp. Eur. 1–19 (2020)
40. Molavi, A., Lim, G.J., Race, B.: A framework for building a smart port and smart port index. Int. J. Sustain. Transp. **14**, 686–700 (2020). https://doi.org/10.1080/15568318.2019.1610919
41. Majidi, A., Mirzapour Al-E-Hashem, S.M.J., Zolfani, S.H.: Sustainability ranking of the Iranian major ports by using MCDM methods. Mathematics **9**, (2021). https://doi.org/10.3390/math9192451
42. Hill, R.J.: Practice learnings from master planning Tasmania's gateway ports. In: Australasian Coasts and Ports 2019 Conference, pp. 576–595 (2019)
43. Dhanak, M., Parr, S., Kaisar, E.I., Goulianou, P., Russell, H., Kristiansson, F.: Resilience assessment tool for port planning. Environ. Plan. B Urban Anal. City Sci. **48**, 1126–1143 (2021). https://doi.org/10.1177/2399808321997824
44. Yang, Z., et al.: Risk and cost evaluation of port adaptation measures to climate change impacts. Transp. Res. Part D Transp. Environ. **61**, 444–458 (2018). https://doi.org/10.1016/j.trd.2017.03.004
45. Sirotić, M., Žuškin, S., Rudan, I., Stocchetti, A.: Methodology for the sustainable development of the Italy-Croatia cross-border area: sustainable and multimodal/cross-border passenger services. Sustainability **13** (2021). https://doi.org/10.3390/su132111895
46. Celik, M.S.: Determining maritime silk road strategies for ports along the route of belt and road initiative: a case of eastern Aegean ports. Res. Transp. Bus. Manag. **51** (2023). https://doi.org/10.1016/j.rtbm.2023.101056
47. Feng, H., Ye, C., Gao, R.: AHP-based evaluation of port logistics transportation efficiency. J. Coast. Res. **106**, 477–480 (2020). https://doi.org/10.2112/SI106-107.1
48. Mou, N., Wang, C., Yang, T., Zhang, L.: Evaluation of development potential of ports in the yangtze river delta using FAHP-entropy model. Sustainability **12**, (2020). https://doi.org/10.3390/su12020493
49. Pourebrahim, S., Mokhtar, M.: Bin: conservation priority assessment of the coastal area in the Kuala Lumpur mega-urban region using extent analysis and TOPSIS. Environ. Earth Sci. **75**, 1–12 (2016). https://doi.org/10.1007/s12665-016-5276-3
50. Madhumitha, R., Rajakumari, S., Deepika, B.: Investigation of port intervention to coastal and nearshore dynamics: a GIS and AHP approach. Environ. Monit. Assess. **196** (2024). https://doi.org/10.1007/s10661-024-12472-x

51. Amoura, R., Dahmani, K.: Visualization of the spatial extent of flooding expected in the coastal area of Algiers due to sea level rise. Horizon 2030/2100. Ocean Coast. Manag. **219** (2022). https://doi.org/10.1016/j.ocecoaman.2022.106041
52. Zanetti, V.B., Junior, W.C. de S., De Freitas, D.M.: A climate change vulnerability index and case study in a Brazilian Coastal City. Sustainability **8** (2016). https://doi.org/10.3390/su8080811
53. Messner, S., Moran, L., Reub, G., Campbell, J.: Climate change and sea level rise impacts at ports and a consistent methodology to evaluate vulnerability and risk. WIT Trans. Ecol. Environ. **169**, 141–153 (2013). https://doi.org/10.2495/CP130131
54. Hemmerling, S.A., et al.: A community-informed transdisciplinary approach to coastal restoration planning: maximizing the social and ecological co-benefits of wetland creation in Port Fourchon, Louisiana, USA. Front. Environ. Sci. **11** (2023). https://doi.org/10.3389/fenvs.2023.1105671
55. Fisher-Holloway, B., Mokhele, M.: Geographical patterns of warehousing facilities in the cape functional region. South Africa. Reg. Sci. Policy Pract. **15**, 1317–1338 (2023). https://doi.org/10.1111/rsp3.12630

Environmental, Social, Governance of Energy Planning (ESGEP 2024)

Assessing the Optimal Scenario of Collective Self-Consumption Groups in Net Zero Condominiums

Riccardo Trevisan[1](✉), Emilio Ghiani[1], Mara Ladu[2], and Ginevra Balletto[2]

[1] Department of Electrical and Electronic Engineering, University of Cagliari, Via Marengo 2, 09123 Cagliari, Italy
{riccardo.trevisan,emilio.ghiani}@unica.it

[2] Department of Civil and Environmental Engineering and Architecture, University of Cagliari, Via Marengo 2, 09123 Cagliari, Italy
{mara.ladu,balletto}@unica.it

Abstract. The energy transition promoted in European countries to improve the sustainability of the energy sector brings together citizens to generate, share and sell energy produced by locally owned renewable plants. Energy communities (ECs) and collective self-consumption (CSGs) groups are strategic power-sharing models to drive this democratization process in a flexible way. However, these two models face challenges such as regulatory barriers, technical limitations, and social acceptance. In addition, the optimal sizing in terms of consumption and production to locally balance the EC load and generate the best obtainable value for all stakeholders represents another pivotal challenge. In this framework, the study proposes a methodology to assess multiple sizing scenarios of the generation and storage system in condominiums, applying an innovative multi-criteria techno-economic approach that allows to analyze the potential benefits that can be obtained by simulating. A case study of a collective self-consumption group located in a smart building in Italy is developed to assess the relevance of the methodology. The investment case supported by a third-party investor is evaluated and analyzed.

Keywords: Energy Transition · Collective Self-Consumption Groups · Scenarios · Investment Evaluation

1 Introduction

Energy transition represents one of the main objectives of the European Green Deal strategy [1] to achieve climate neutrality by 2050. The Mid-century zero emissions strategy for the EU aims to create an energy sector based mainly on renewable sources (decarbonizing the European energy system) to guarantee a secure and affordable energy supply and to improve energy efficiency, especially of the existing building stock.

As a matter of fact, the building sector is responsible for approximately 40% of global Greenhouse Gases (GHG) emissions in the environment [2], More precisely, GHG

O. Gervasi et al. (Eds.): ICCSA 2024 Workshops, LNCS 14818, pp. 213–224, 2024.
https://doi.org/10.1007/978-3-031-65273-8_14

emissions from the building sector represented 35% of energy-related EU emissions in 2020 [3]. Similar percentages are recorded in Italy, where the poor energy performance of a significant asset [4] requires a deep renovation to reduce buildings energy consumption [5, 6], at different urban scales, in line with the circular economy perspective [7, 8].

However, international geoeconomical and geopolitical events, which are among the main causes of the increase in energy consumption from coal and oil in the most recent years, might threaten the "green" transition. This represents one of the reasons why the scientific debate on the energy transition [9–11] should consider the right to access to electricity and, therefore, to the primary services that energy makes possible [12, 13]. In this sense, the EU's legal energy policy framework and instruments such as EC may be able to face and alleviate energy poverty.

Energy communities (ECs), introduced by the RED II Directive [14], are emerging as an effective solution to mitigate energy poverty, to combat climate change and drive the transition to renewable energy sources. In line with the Integrated European Market Directive (IEM, 2019/944 [15]), ECs comprise individuals, households, businesses, or public institutions that collectively generate, consume, share, and manage locally produced energy, within a sustainable development perspective [16]. These collaborative groups or organizations aim to create a sustainable electricity system through active citizen participation [17], thus fostering democracy and resilience of local communities and places [18]. ECs developed at different locations and scales (from small-scale solutions to large-scale urban schemes) are supported by specific technologies, financing, and legal arrangements [19, 20].

However, ECs face several practical challenges, such as technical limitations, as well as regulatory barriers and social acceptance. In this study the authors propose a methodology aimed at obtaining optimal sizing of ECs, in terms of consumption and production, and the greatest benefits for stakeholders.

In Italy, where incentive policies (tax deductions) constitute the main strategic government policy to improve energy efficiency in the existing building stock, two models have been defined to promote energy communities: Renewable Energy Communities (RECs) and Collective Self-Consumption Groups (CSGs) [21, 22]. Given that both aim to increase the share of renewable energy in the Italian energy sector, there are significant differences between the two, in terms of scale of implementation (1), management model (2) and incentive framework (3):

1) RECs [23] usually involve multiple stakeholders to implement and manage large renewable energy projects; CSGs focus on single buildings like condominiums. It means that the distribution network also differ: while RECs are supported by a certain number of users dispersed over the territory, CSGs are instead based on a scheme that concentrates the consumption points within a building or a group of adjacent buildings. Figure 1, show a virtual model for a CSG, which allows the connection of the condominium network to the public distribution network using multiple fiscal energy meters, enabling separate electricity supply contracts for general services and individual housing units [24].
2) RECs require the establishment of a legal entity by virtue of the different spatial distribution and heterogeneous nature of participants; CSGs are considered a "management entity" associated with single buildings.

3) Both models are supported by incentives that encourage energy sharing among users, but the amount granted differs for CSGs and RECs.

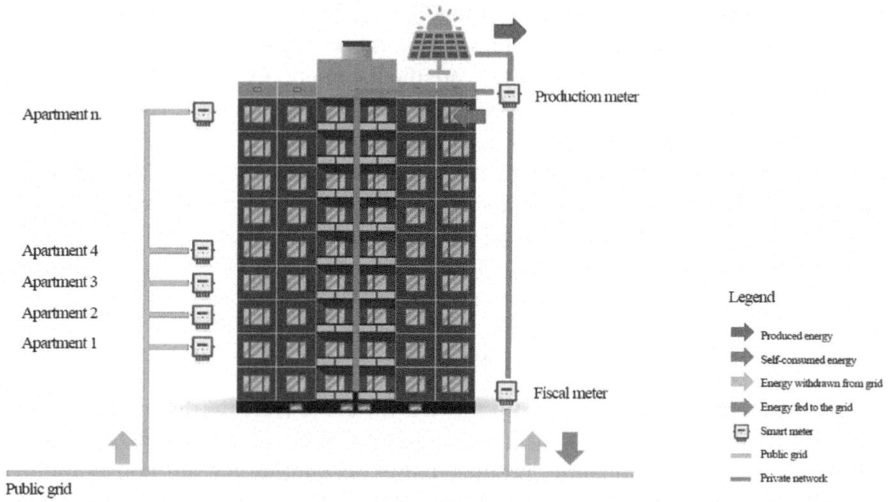

Fig. 1. A "virtual" configuration model chosen by Italian authorities for a CSG, where each user retains individual rights and obligations as an end customer.

The manuscript, that derives from previous studies conducted by the authors [25, 26], proposes a strategy for the optimal sizing and financial feasibility of collective self-consumption groups. This study is developed within the Ecosystem of Innovation for Next Generation Sardinia (e.INS) partnership, spoke 7 (Low emissions technologies for the efficient energy system), work package 4 with the objective of developing methodologies capable of bridging the complexity that can characterize the prototyping, implementation, and management of an energy community within a smart city district. Additional contributions to this study have been achieved as well within the Smart Communities for Resilient Energy Transition Project (SCoRET). An innovative multi-criteria techno-economic approach allows to explore the potential social and economic benefits for members at energy price variations. More precisely, Python is used to develop a simulation tool to assess different sizing alternatives in different scenario configuration. The possibility to easily simulate a significant number of scenarios that differ in the configuration parameters and the possibility to evaluate the generated scenarios under different perspectives that can consider social, economic and/or environmental factors represent a significant result that advance the current state of the art.

Within this framework, the manuscript is organized as follows: Sect. 2 describes the methodology employed; Sect. 3 focuses on the case study; Sect. 4 discusses the results of the methodology employed and their relevance in the disciplinary scientific debate.

2 Methodology

The study's methodology is described in this section. The authors developed a simulation tool for CSG groups leveraging Python language, particularly, NumPy and Pandas libraries have been used for data management. Using this tool, several techno-economic possibilities for combinations of solar power generation and storage system sizes can be evaluated. The main inputs of the software are the list of energy prices, demand profiles, lists of plant sizing under study, and the cost and incentive structure. The results of the simulation are conveniently stored as Excel files.

The Italian government has put in place a process for evaluating energy sold to the distribution network as well as an incentive scheme for energy shared by the CSG. The minimum between the energy injected and withdrawn from the distribution network within the same hourly period is known as the shared energy. It is mathematically expressed as:

$$SE_h = min(\Sigma w_{i,h}; \Sigma i_{i,h}) \tag{1}$$

$w_{i,h}$ and $i_{i,h}$ represent the energy withdrawn from and fed into the distribution network by the i-th user during the h-th hour. In the case of CSGs, the energy injected into the distribution network is the quantity produced after accounting for self-consumption by common loads. For each MWh of shared energy, the CSG is awarded with an incentive tariff worth 109 euros.

As previously mentioned, the energy injected into the grid is paid at market price as well. Namely, the price is the hourly zonal price (PZ). The software allows you to supply a file containing the hourly PZ or a specific year, or you may specify a constant value. The income generated by the selling of energy to the distribution network can be easily calculated like:

$$RID_h = PZ_h * \sum i_{i,h} \tag{2}$$

In the above formula, PZh represents the zonal price at time h.

The three stages of the devised methodology are depicted in Fig. 2.

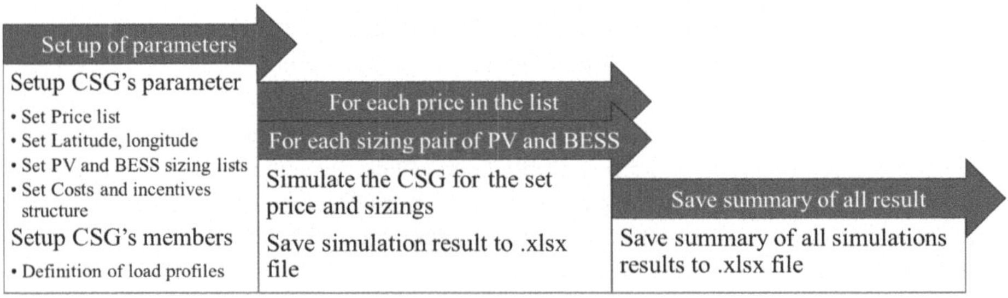

Fig. 2. Simulation workflow.

These are initialization of parameters, simulation, and results saving for further analysis. The PZ, and the geographic coordinates (latitude and longitude) of the CSG, must be specified during the initialization process to estimate the photovoltaic (PV) system output for which the PVGis API has been used. After that, a list of PV and BESS sizes for which simulations are to be run can be defined. The unit costs for the PV system (€/kWp), the electrochemical storage system (€/kWh), the infrastructure implementation (€/kW), and the labor (€/kW) must also be specified to perform the economic simulations. The costs of capital investments (CAPEX) are then estimated using the previously mentioned elements. The costs of management and insurance are then asked to be provided to correctly estimate operating expenses (OPEX). The type facilities can be chosen from pre-installed categories in the following phase. Since the load curve of common loads oversees the PV system's instantaneous self-consumption of energy, it is also crucial to characterize it. Next, a list is constructed that contains the cartesian product of the PV system powers and the designated storage system capacities. Numerous techno-economic data are calculated for every single option (that is, PZ, PV size, and BESS capacity combinations). The main techno-economic parameters that are computed are the energy injected into the distribution network at each period, the energy withdrawn from the distribution network at each period, the self-consumed energy during each period, the hourly shared energy, the total energy bill for the CSG and the revenues generated by the selling and sharing of energy.

The approach enables the possibility to analyze investment scenarios that involve a partnership with a third-party actor who contributes initial funding (such as an energy service company, or ESCo) in exchange for a share of the profits made. If a non-member stakeholder finances the development of the plant and facility, its earnings come from selling electricity to the distribution network and from a portion of the incentive gained from members sharing electricity. As a result, the allocation of incentives [27] is the subject of a private contract made between the CSG and the external investor. Various methods are put forth in the literature for a proportionate and meritocratic allocation, based on game theory [28] or key performance indicators [29]. The idea that the CSG always bears the operational costs (OPEX) is another crucial one. Various financial factors are calculated for each simulated scenario to conduct a single evaluation. The net present value (NPV), internal rate of return (IRR), profitability index (PI), and the payback time (PBT), which is the point at which the cash flow turns positive, have all been calculated using the discounted cash flow approach. An investor can customize the investment based on its financial viability and the anticipated return for the CSG's members by analyzing these scenarios.

The total revenue received in the first year of operation, the variation in CSG expenditure (calculated by comparing the situation in which the CSG is not created), and the shared to produced energy ratio are additional information that have been computed to evaluate the investment.

With the proposed methodology a vast number of scenarios can be simulated in a brief time. For this reason, it is often useful to skim the results of the simulation according to specific requirements. In the specific case, authors decided to filter results that do not ensure a shared energy to produced energy ratio greater than 65%. As a matter fo fact,this choice reduce the number of options that do not adequately utilize energy produced locally, thus providing only solutions that satisfy a certain level of adequacy. Only the instances where the cost reduction for CSG is higher than 5% were retained to further narrowing down the selection of findings. This way, a restriction is put in place with the intention of giving the participants in the collective self-consumption system an economic benefit baseline. Lastly, a second filter is applied to remove all simulations linked to a negative NPV value. Next, the new dataset is sorted by PI values that are getting lower. This makes it possible to quickly scan through profitable scenarios.

In the following section an analysis of the results generated for a case study is proposed.

3 Case Study

By examining the correlations between the variables and important system parameters like PV, BESS, and PZ, the analysis seeks to examine the landscape of created scenarios and the numerous aspects that define them. By examining the many KPIs of the system under consideration, the goal is to offer stakeholders in the energy industry relevant and useful information. To accomplish this goal, a CSG has been simulated using the approach presented in the preceding chapter. The simulations were conducted for various PV, BESS, and PZ values as well as under various investment realization scenarios. The case study foresees the economic assessment of an ESCo realizing the investment and providing advantages to the residents. It should be noted that operational expenses are in any case borne by the CSG. The sale of produced electricity to the distribution grid and a set portion of the incentives obtained from energy sharing among CSG members are the two main sources of income for the ESCo. This share has been set at 50% for the sake of this study, although it might be changed based on agreements reached between the parties.

Eight BESS capacities have been instantiated (0, 5, 7.5, 10, 12.5, 15, 17.5, and 20 kWh), while seven PV sizes have been assumed for the production and storage plants (5, 7.5, 10, 12.5, 15, 17.5, and 20 kWp). A total of 56 combinations are obtained by taking the cartesian product of the capacities. Various levels of energy prices in the wholesale market have been taken into consideration when evaluating the influence of energy prices: 50 €/MWh, 100 €/MWh, 150 €/MWh.

The other key simulation parameters used in the CSG simulation are displayed in Table 1. Thus, a CSG with twenty members has been established. Members were created using loads that reflect typical users. These typical users, which stand in for the CSG members, are a working couple between the ages of 30 and 64 (WC 30–64), a retired couple (RC), and a working couple with two kids where one of the parents works from home (CWK). Table 2 reports general information on both the annual usage and the total number of users instantiated in the setup.

The daily load profile for each typical user is displayed in Fig. 3. The disparity in users' attitudes toward energy utilization during the day is noticeable. Figure 4 shows an example of the same kind of graph as Fig. 3 for the load of common utilities.

Table 1. Simulation parameters

Parameter	Value
Latitude	39.205
Longitude	9.130
PV unitary cost (€/kWp)	1,100
BESS unitary cost (€/kWh)	1,000
Infrastructure unitary cost (€/kWx)	100
Labor unitary cost (€/kWx)	300
Management cost (€/kWx)	10
Insurance	0.5% of CAPEX
Discount rate	6%
Annual percentage rate	6%
Latitude	39.205

Table 2. Users' information

Name	Yearly consumption (kWh)	Number instantiated
WC 30–64	3,309	8
RC	2,351	6
CWK	5,841	6

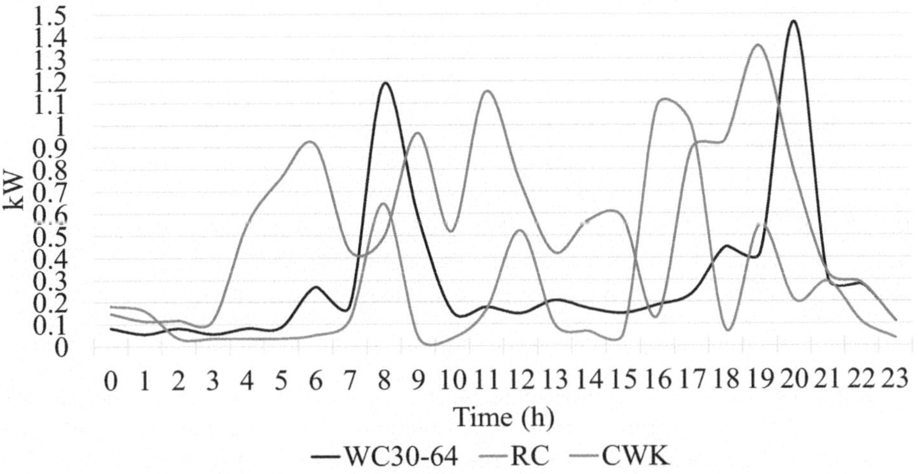

Fig. 3. Load profile of users' categories for a typical day.

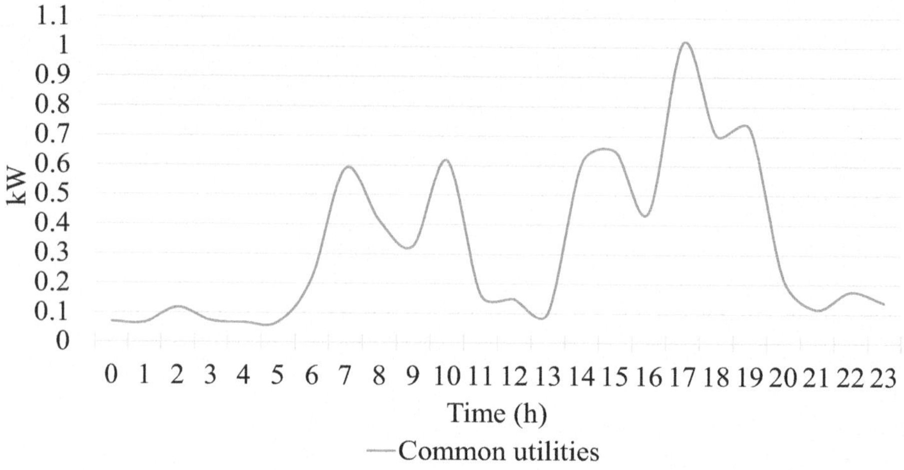

Fig. 4. Load profile of the common utilities for a typical day.

4 Results and Conclusions

After simulating and applying the procedure of results selection described in the methodology section, the total number of scenarios that foresees an ESCo assisting the investment for the CSG's facilities is reduced to 46. A statistical summary of results is reported in Table 3, while the first 5 results obtained with the sorting procedure (by descending value of PI) are reported in Table 4.

Table 3. Summary statistics: ESCo investing.

Statistic	CSGExpenditureVariation	NPV	TIR	PI	PBT	CAPEX	OPEX	SharedProducedRatio
mean	−9%	7,149.03	9%	126%	13.41	30,342.39	371.82	90%
std	2%	5,577.95	3%	22%	3.28	10,395.43	134.09	3%
min	−15%	12.38	6%	100%	7.80	7,500.00	87.50	82%
25%	−10%	2,643.03	7%	110%	11.21	23,500.00	292.50	88%
50%	−9%	5,910.30	9%	123%	13.13	30,375.00	376.25	91%
75%	−8%	10,780.98	10%	136%	15.73	38,000.00	465.00	92%
max	−5%	22,809.48	15%	176%	18.98	52,000.00	660.00	94%

Most interestingly, the average expenditure variation for the CSG stands at 9%. PBT for investment, with the current apportionment of the incentive, is slightly high and might not be appealing for an investor. Despite the not so inviting statistics, the first five results reported in Table 4 show some promising investment scenario for the ESCo and relevant returns for the CSG too. Interestingly, these are associated with a high price of energy which results in higher revenues generated by the sale of energy and with the absence of a storage system which, inevitably, increases the CAPEX and,

Table 4. First five results of the simulation ordered by descending PI, ESCo investing.

FirstYearRevenue	CSGExpenditureVariation	NPV	IRR	PI	PBT	CAPEX	OPEX	PV	BESS	PZ
6,615	−10%	22,809	15%	176%	7.80	30,000	350	20	0	0.15
5,058	−8%	17,097	15%	176%	7.80	22,500	262	15	0	0.15
3,406	−7%	10,917	15%	173%	7.99	15,000	175	10	0	0.15
5,692	−9%	18,687	14%	171%	8.09	26,250	306	17.5	0	0.15
4,079	−7%	12,695	14%	168%	8.32	18,750	218	12.5	0	0.15

consequently, the investment profitability. Results also show that the maximum capacity of PV is required by the investment scenario that provides the highest value of PI.

Another insightful analysis is to devise which variable most influences PI's value. To this end a feature analysis has been performed. The analysis's findings are presented in Fig. 5 for single terms, as well as for interacting terms.

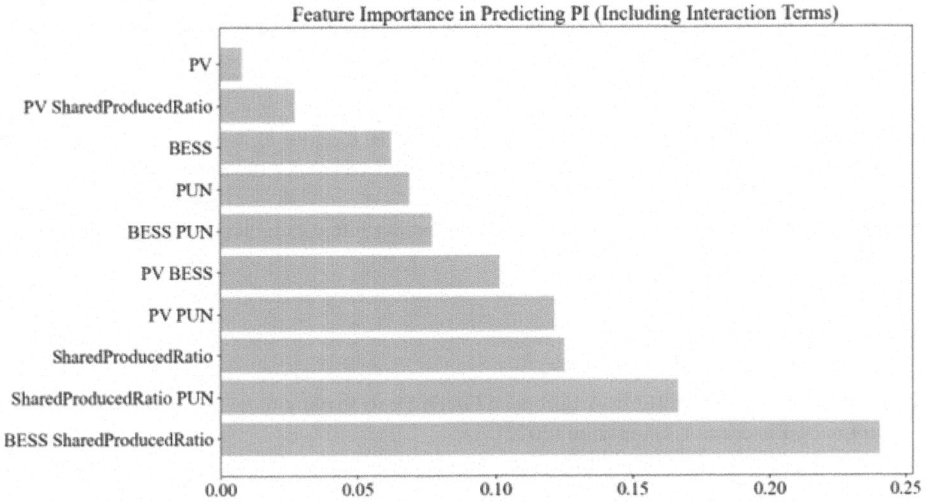

Fig. 5. Feature importance on PI

Interestingly, the most influential term on PI is given by the interaction of BESS and *SharedProducedRatio*: the installation of a storage system limits the sharing capability within the CSG, thus reducing the income generated by the incentive and increasing CAPEX due to the BESS presence, finally having a relevant negative impact on PI.

This research provides an in-depth analysis of energy communities, concentrating on Collective Self-Consumption Groups in Italy. The goal of the authors' multi-criteria techno-economic framework is to democratize local energy generation and consumption by assessing several condominium facilities sizing scenarios. The authors evaluated the social and financial benefits of various setups for a case study using simulation tools based on Python while taking financial indicators into account. In addition to adding to the body of knowledge on renewable energy communities, the study offers stakeholders—such as investors and policymakers—useful information that will help them make decisions as we move toward more decentralized and sustainable energy systems.

The findings of the simulation of a case study are described and summarized, and an analysis of the most relevant parameters is provided. Resulting scenarios show there are wide margins for investing that can satisfy various requirements concerning capital expenditure, saved costs, and return on investment.

Acknowledgments. This work has been carried out within the Project NATIONAL RECOVERY AND RESILIENCE PLAN (PNRR) - MISSION 4 COMPONENT 2, "From research to business" INVESTMENT 1.5, "Creation and strengthening of Ecosystems of innovation" and construction of "Territorial R&D Leaders. Project e.INS Ecosystem of Innovation for Next Generation Sardinia (e.INS) - MUR ECS00000038 - CUP F53C2200043000. The research was also supported within the Project Smart COmmunities for Resilient Energy Transition - SCORET NATIONAL RESEARCH AND RESILIENCE PLAN (PNRR) Mission 4 'Education and Research' - Component C2 Investment 1.1, 'Fund for the National Research Programme and Projects of Significant National Interest (PRIN)'. Project Code 2022YYMWTJ - CUP F53D23000760006 Directorial Decree no. 104 of 02.02.2022, Public Notice for the submission of Research Projects of Significant National Interest (PRIN) to be financed under the PNRR - PRIN 2022 - European Union Funding - NextGenerationEU.

Author Contributions:. G.B. and M.L. wrote Sect. 1; R.T. and E.G. wrote Sect. 2, Sect. 3 and Sect. 4. All authors have read and agreed to the published version of the manuscript.

Disclosure of Interests. The authors have no competing interests to declare that are relevant to the content of this article.

References

1. European Commission (EC): A European Green Deal: striving to be the first climate-neutral continent. European Commission (2022)
2. Circle Economy: The circularity gap report 2023 (pp. 1–64, Rep.). Circle Economy, Amsterdam (2023)
3. European Environment Agency Homepage. https://www.eea.europa.eu/ims/greenhouse-gas-emissions-from-energy. Accessed 30 Jan 2024
4. Palladino, D., Pagliaro, F., Del Fatto, V., Lavinia,C., Margiotta, F., Colasuonno, L. (eds.): Stato dell'arte del patrimonio edilizio nazionale e analisi degli Attestati di Prestazione Energetica (APE). Report RdS/PTR2019/037. Enea, Ministero dello Sviluppo economico (2019)
5. Manni, V., Valzano, L. S.: Comunità energetiche. Strumento per riqualificare l'edilizia della ricostruzione post-bellica. Techne **24**, 119–126 (2022)
6. Ladu, M.: Conservation and regeneration for a sustainable and circular city. In: Gervasi, O., et al. Computational Science and Its Applications – ICCSA 2023 Workshops. ICCSA 2023. Lecture Notes in Computer Science, vol. 14111, pp. 77–88. Springer, Cham. Springer Nature, Cham, Switzerland (2023). https://doi.org/10.1007/978-3-031-37126-4_6
7. Balletto, G., Ladu, M., Camerin, F., Ghiani, E., Torriti, J.: More circular city in the energy and ecological transition: a methodological approach to sustainable urban regeneration. Sustainability **14**(22), 14995 (2022)

8. Balletto, G., Ladu, M.: The role of spatial circular planning in urban governance. a set of indicators to evaluate performance in urban regeneration. In: International Conference on Computational Science and Its Applications, pp. 104–118. Springer Nature, Cham, Switzerland (2023). https://doi.org/10.1007/978-3-031-37126-4_8
9. Gatto, A.: The energy futures we want: a research and policy agenda for energy transitions. Energy Res. Soc. Sci. **89**, 102639 (2022)
10. Zakeri, B., et al.: Pandemic, war, and global energy transitions. Energies **15**(17), 6114 (2022)
11. Streimikiene, D., Kyriakopoulos, G.L.: Energy poverty and low carbon energy transition. Energies **16**(2), 610 (2023)
12. Walker, G., Cass, N.: Carbon reduction, 'the public' and renewable energy: engaging with socio-technical configurations. Area **39**(4), 458–469 (2007)
13. Fell, M.J.: Energy services: a conceptual review. Energy Res. Soc. Sci. **27**, 129–140 (2017)
14. Directive (EU) 2018/2001 of the European Parliament and of the Council - of 11 December 2018 - on the promotion of the use of energy from renewable sources (RED II Directive)
15. Directive (EU) 2019/944 of the European Parliament and of the Council of 5 June 2019 - on common rules for the internal market for electricity and amending Directive 2012/27/EU (IEM Directive)
16. Di Silvestre, M.L., Ippolito, M.G., Sanseverino, M.R., Sciumè, G., Vasile, A.: Energy self-consumers and renewable energy communities in Italy: new actors of the electric power systems. Renew. Sustain. Energy Rev. **151**, 111565 (2021)
17. Lode, M.L., te Boveldt, G., Coosemans, T., Ramirez Camargo, L.: A transition perspective on energy communities: a systematic literature review and research agenda. Renew. Sustain. Energy Rev. **163**, 112479 (2022)
18. Proto, D.: Renewable energy communities as an enabling framework to boost flexibility and promote energy transition. Energies **15**(23), 8874 (2022)
19. Lowitzsch, J., Hoicka, C.E., van Tulder, F.J.: Renewable energy communities under the 2019 European clean energy package – governance model for the energy clusters of the future? Renew. Sustain. Energy Rev. **122**, 109489 (2020)
20. Trevisan, R., Ghiani, E., Pilo, F.: Renewable energy communities in positive energy districts. a governance and realisation framework in compliance with the Italian regulation. Smart Cities **6** (1), 563–585 (2023)
21. Halkos, G.E., Gkampoura, E.C.: Evaluating the effect of economic crisis on energy poverty in Europe. Renew. Sustain. Energy Rev. **144**, 110981 (2021)
22. Sánchez, C.S.G., González, F.J.N., Aja, A.H.: Energy poverty methodology based on minimal thermal habitability conditions for low-income housing in Spain. Energy Buildings **169**, 127–140 (2018)
23. Autorità di regolazione per energia reti e ambiente (ARERA): Documento per la consultazione 112/2020/R/EEL. Ordinamenti per la regolazione delle partite economiche relative all'energia elettrica oggetto di autoconsumo collettivo o di condivisione nell'ambito di comunità energetica rinnovabile. ARERA (2020)
24. Hatziargyriou, N.: Differences and synergies between local energy communities and microgrids. Oxford Open Energy **2** (2023)
25. Trevisan, R., Ghiani, E., Ruggeri, S., Mocci, S., Pisano, G., Pilo, F.: Optimal sizing of PV and Storage for a Port Renewable Energy Community, In: 2022 2nd International Conference on Energy Transition in the Mediterranean Area (SyNERGY MED), Thessaloniki, Greece, pp. 1–5 (2022)
26. Trevisan, R., Ruffini, S., Ghiani, E., Caboni, O., Papa, R.: Optimal sizing of PV and storage in collective self-consumption groups in condominiums. In: 2023 International Conference on Future Energy Solutions (FES), Vaasa, Finland, pp. 1–6 (2023)

27. Trevisan, R., Ghiani, E., Pilo, F.: Economic benefits redistribution methodology for renewable energy communities. In: 13th Mediterranean Conference on Power Generation, Transmission, Distribution and Energy Conversion (MEDPOWER 2022), Hybrid Conference, Valletta, Malta, pp. 213–218 (2022)
28. Moncecchi, M., Meneghello, S., Merlo, M.: A game theoretic approach for energy sharing in the italian renewable energy communities. Appl. Sci. **10**, 8166 (2020)
29. Ghiani, E., Trevisan, R., Pilo, F.: Performance metrics for renewable energy communities. In: 2022 International Annual Conference (AEIT 2022), pp. 1–6, Rome, Italy (2022)

Urban Energy Redevelopment Projects: A Protocol for Economic and Financial Risk Assessment Under Public Incentive Policies

Luigi Dolores(✉) and Maria Macchiaroli

University of Salerno, Via Giovanni Paolo II, 132, Fisciano, SA, Italy
{ldolores,mmacchiaroli}@unisa.it

Abstract. Growing urbanisation and energy efficiency challenges pose the need for innovative strategies to reduce the environmental impact of cities. This study proposes to address this need through the definition of a protocol to assess the financial and economic risk of urban energy redevelopment projects implemented through public incentives (tax breaks or public funding), focusing on the Italian context but with an approach that can be replicated internationally. The methodology is based on a critical analysis of the existing regulatory framework and the synthesis of international best practices. The protocol envisages eleven phases, including the preliminary analysis of the urban context, the identification of interventions, the evaluation of incentive policies, the study of the main criticalities characterising these policies and the analysis of the economic and financial risk in probabilistic terms. In the next phase of risk assessment, objective thresholds of risk tolerability and acceptability are defined using the ALARP principle, originally employed in industrial engineering to estimate the probability of loss of life. The final considerations highlight the importance of a holistic assessment of the risks and economic benefits associated with urban energy efficiency projects to ensure sustainable investments and maximise the results.

Keywords: Urban Energy Projects · Energy Urban Policy · Economic Evaluation of Projects · Economic Evaluation Protocol · Risk Assessment · ALARP Principle

1 Introduction

The transition to renewable and sustainable energy sources has become a global priority in the context of the urgent need to tackle climate change and reduce greenhouse gas emissions [1]. Despite significant progress on this issue, there has been a slowdown in the implementation of energy efficiency policies on a global scale since 2015. This has resulted in significant negative impacts on citizens, the environment and the land, to the extent that efficient protocols for assessing the economic and financial risk of intervention strategies in the energy sector are necessary [2]. After a ten-year low in 2020,

L. Dolores and M. Macchiaroli—The two authors contributed equally to this work.

O. Gervasi et al. (Eds.): ICCSA 2024 Workshops, LNCS 14818, pp. 225–241, 2024.
https://doi.org/10.1007/978-3-031-65273-8_15

which was recorded following the COVID-19 pandemic, global energy efficiency trends in 2021 were close to the previous average. However, economic growth has led to an energy demand that far exceeds the efforts made to curb energy consumption, reducing the rate of improvement of energy intensity (i.e. primary energy consumption per unit of gross internal product) to 1.7 per cent in 2017. This rate is the lowest recorded in almost five years [3]. Fortunately, according to the International Energy Agency's (IEA) report "Energy Efficiency 2022", the situation has improved by the end of 2022, as global investments in energy efficiency (including building renovation, public transport and electric car infrastructure) reached USD 560 billion, up 16% on 2021. Arguably, the global energy crisis triggered by the Russian invasion of Ukraine has stimulated a global response through policies, investments and public campaigns to promote energy efficiency. Among the most ambitious initiatives implemented in recent years at a global level in the field of energy efficiency, the aforementioned report mentions: the Inflation Reduction Act of the United States, the REPowerEU plan of the European Union, the Green Transformation (GX) programme of Japan, the Superbonus of Italy [4]. In addition to the above-mentioned measures, the European Union's (EU) recently revised Energy Performance of Buildings Directive (EPBD) is expected to further propel energy transition policies. The directive stipulates that by 2028 all new public buildings must be 'zero-emission', while by 2030 all new buildings, including non-public buildings, must reach this standard. Member states must define minimum energy performance requirements by 2030, and from the same year, the installation of photovoltaic solar panels on new public and non-residential buildings is mandatory. The entire European building stock must be 'zero-emission' by 2050 [5].

It is clear from the framework of reference that the combination of architecture and environmental sustainability has now assumed a crucial role in the definition of current energy policies. In the context of global challenges such as climate change and the progressive depletion of natural resources, the demand for energy-sustainable buildings is inexorably growing. This has stimulated the search for concrete and effective solutions in the field of green building, offering design solutions that are functional, environmentally sustainable and optimise the use of energy resources [6]. Green building aims to minimise the negative impacts on the environment and society resulting from the construction process and, at the same time, strives to improve the well-being of building inhabitants through a higher quality of life [7]. Furthermore, green redevelopment projects that include energy efficiency measures can generate a positive impact on the local economy by reducing operational costs and financial risks related to energy price fluctuations. What is more, such interventions can foster the competitiveness of investors and attract new ones, as well as optimise the value of assets through greater operational efficiency and a lower carbon footprint [8]. To pursue these aims, it is necessary to integrate green building practices with an approach to energy planning at urban and territorial levels that is more sustainable in socio-environmental and economic terms, thus connoting a multidisciplinary paradigm in the fields of architecture, town planning and urban appraisal. This paradigm is embodied in ways of planning urban environments and building design capable of minimising both greenhouse gas emissions and energy expenditure and consumption. This approach, through the implementation of innovative design strategies and the adoption of advanced technologies, aspires to create healthy and sustainable living environments consistent with sustainable development goals [9]. Cities, regardless of their size, host a variety of facilities that support commercial, industrial, residential and community activities. Effectively addressing the energy needs of such facilities requires the implementation of targeted strategies and plans to balance energy supply and demand. Consequently, it is essential to develop energy policies that sustainably promote socio-economic development while ensuring the stability and prosperity of urban communities [10]. The convergence of sustainable building and urban energy planning issues with the economic evaluation of projects, especially with the economic and financial risk assessment of energy rehabilitation projects, is a crucial field of academic and scientific research. This integration requires an in-depth assessment of the economic and financial risks associated with energy rehabilitation measures, taking into account factors such as cost uncertainties, the instability of energy markets and the effectiveness of technologies. Appropriate models and methodologies need to be developed to manage these risks effectively, also considering the regulatory environment and collaboration between various stakeholders [11]. The approach to be adopted should therefore be integrated with the strategic legislative aspects of energy. Energy policies play a crucial role in supporting urban energy efficiency by providing financial incentives and tax breaks for energy upgrades of buildings and urban infrastructures [12]. These incentives stimulate the adoption of sustainable technologies and improve the economic sustainability of interventions by reducing initial costs and fostering the development of markets for innovative energy technologies. It is crucial to analyse the potential of these incentives to maximise both economic and socio-environmental benefits [13].

The goal of this work is to define a protocol for assessing the economic-financial risk relating to energy requalification interventions based on public energy policies applied in urban settings. The conceived protocol is divided into several key phases, including the preliminary analysis of the urban area subject to intervention, the identification and classification of the properties and infrastructures to be redeveloped, as well as the study of the tax breaks and financial incentives available. Subsequently, it involves the preparation of estimated metric calculations and project economic frameworks. Once the costs of the measures to mitigate the critical issues (of a technical, urban planning and administrative nature) potentially encountered have been defined, we proceed with the preparation of the forecast financial statement and the implementation of a cost–benefit analysis (CBA) for each type of intervention. This is followed by the risk analysis phase conducted via Monte Carlo simulation. The level of financial and/or economic risk of the interventions is then assessed through the introduction of objective thresholds of tolerability and acceptability of the risk borrowed from the theory on the ALARP (As Low As Reasonably Practicable) principle. The protocol, developed to evaluate the economic-financial risk of urban energy requalification operations, was conceived as an extension of another original protocol of the authors [14], which instead focuses on the economic-financial risk analysis to optimise the energy efficiency of individual buildings. Therefore, this study aims to broaden the evaluation scale from the building to the urban level. Although based on the Italian regulatory framework, the approach adopted is replicable and adaptable to other national contexts, thus providing a transferable methodological framework for the economic evaluation of urban energy redevelopment projects supported by public incentive policies. Another novelty consists in integrating for the first time the ALARP approach to the risk analysis of projects at an urban planning level.

The work is structured as follows: Sect. 2 contains a brief review of the literature on the most recent applied research in urban energy redevelopment, an analysis of the main energy policies promoted internationally and in Italy, and a description of the ALARP principle; in Sect. 3 the economic and financial risk assessment protocol is presented and defined; finally, Sect. 4 reports a synthetic discussion, the main conclusions as well as future research ideas.

2 Literature Review and Regulatory Framework

For the definition of the aforementioned protocol, the preliminary analysis included both the latest applied research in the field of energy efficiency concerning various dimensions of the urban context (residential buildings, infrastructure and transportation, productive buildings, etc.), and the main energy policies based on public incentives and tax benefits, related to recent years, undertaken internationally and within the Italian context. Furthermore, the ALARP principle is examined, here integrated into the protocol to introduce objective thresholds of tolerability and acceptability of economic-financial risk.

2.1 Recent Research on Energy Efficiency in Urban Areas

Globally, multiple research studies have focused on analysing the economic and environmental effectiveness of smart urban infrastructures, including energy infrastructures [15]. Below are some more recent examples. Xia et al. (2021) analyse the work carried out on energy in smart city models and planning, dividing the methodologies into active and passive approaches and evaluating their implications on economic performance and energy security requirements for efficient management of resources [16]. Berardi et al. (2014) highlight the role of green roofs in promoting more sustainable buildings and cities, underlining the importance of efficient integration that considers climatic conditions and building characteristics. The study includes economic considerations on life cycle costs and global policies to support green roofs, demonstrating the environmental benefits and economic feasibility of this approach [17]. Huovila et al. (2019) analyse and compare 7 indicator standards for smart and sustainable cities (413 indicators), also focusing on the sectoral domains of energy and economy [18]. Talandier (2018) conducts an empirical investigation on 20 European cities involved in the energy transition, examining the correlations between different energy management approaches and the socio-demographic, economic, geographical and urban characteristics of these cities [19]. Jiang et al. (2022) propose the implementation of a new semi-supervised deep learning method that uses both labelled and unlabelled data, also of an economic nature, to improve the prediction of the energy use intensity of urban buildings [20]. Chang et al. (2020), selecting Beijing as the study city, propose four urban solid waste management scenarios, evaluating the reduction of landfills, energy recovery, life cycle costs and benefits, and greenhouse gas emissions. [21]. In the context of urban transport, Letnik et al. (2018) develop a model for dynamic loading dock assignment for urban last-mile deliveries in energy-efficient and cost-effective ways [22].

From the literature analysis, a gap emerges regarding the analysis and evaluation of the economic and financial risk associated with tax incentive policies for urban energy efficiency.

2.2 International Energy Policies Based on Public Incentives and Tax Deductions

Tax incentive policies for urban energy efficiency have been widely adopted in several European countries such as Germany, France, Denmark, Greece, Belgium and the Netherlands [23]. These policies include the use of tax credits and tax breaks for the purchase of high-energy performance equipment and systems or for the implementation of energy retrofit interventions. However, there are significant differences in the approach adopted by each country, with some actively encouraging energy efficiency in homes through tax breaks, while others go so far as to implement measures that, paradoxically, discourage such interventions. For example, in Sweden, property tax is higher for buildings with excellent energy performance. Otherwise, in the Czech Republic, there is a tax break for those who replace the home heating system. Similarly, in Bulgaria, there is an exemption from property tax for energy-efficient buildings [24]. In other contexts, such as in Spain, grants and loans have been used to incentivize the renovation of existing buildings, but fiscal policy has not significantly contributed to the energy efficiency of the construction sector [25]. In the United States, tax credits for residential energy efficiency have been provided for several years, but have generated concerns regarding their equitable distribution and the possibility of excluding the most vulnerable segments of the population [26]. In Africa there is a lack of reliable energy data, hindering the implementation of sustainable energy policies. In Asia, several policies aimed at energy transition have recently been implemented in various sectors, including residential, but they often lack adequate regulation and are therefore fragmented [27, 28].

2.3 Italian Energy Policies Based on Public Incentives and Tax Deductions

In the Italian context, in recent decades, building bonuses have been implemented as an economic policy tool, with particular attention to energy consumption and seismic safety. These benefits are aimed not only at interventions on properties for residential use but also at works on properties intended for production purposes, owned by both citizens and legal entities. This expansion of tax breaks aims to promote interventions aimed at the modernization and renovation of the existing building stock, thus contributing to the sustainable development of the real estate sector [29]. The tax bonuses currently in force regarding energy efficiency interventions are as follows:

- Renovation Bonus (or Building Bonus): regulated by article 16-bis of Presidential Decree 917/86 (Consolidated Law on Income Taxes - TUIR), it offers a tax deduction of 36% of expenses incurred, without expiry, up to 48,000 euros per property. However, thanks to the special regime extended by the 2022 Budget Law, for expenses up to 31 December 2024, the benefit is 50% with a maximum limit of 96,000 euros. From 1 January 2025, unless extended, the deduction will return to 36% with a maximum ceiling of 48,000 euros. However, the benefit does not extend to properties considered as capital goods (land registry categories: B – properties with public functions; C – properties of a commercial nature or appurtenances; D – commercial properties of a special type or properties with industrial; E – properties of collective interest; A/10 – private studios and offices) or as merchandise goods (properties towards the construction or sale of which the company activity is directed). Among the various interventions envisaged, it also includes those carried out to achieve energy savings (even if the 65% Ecobonus is more convenient for this type of intervention);

- Ecobonus: offers a 50% or 65% tax deduction, as long as the building is pre-existing and not new. The 50% deduction is applicable for the purchase and installation of fixtures, solar shading, front doors and external doors, as well as for the replacement of winter air conditioning systems with condensing boilers or biomass heat generators. The other possible types of energy interventions benefit from a 65% deduction. For condominium works, the deduction can vary from 70 to 85%. In any case, the spending ceiling depends on the type of intervention. This relief extends to buildings of any cadastral category, including instrumental ones. Eligible entities include natural persons and various forms of companies, including partnerships and corporations. It is possible to intervene on merchandise goods, capital goods and residential property, and there is the possibility of transferring the credit or obtaining a discount on the invoice;
- Superbonus: the measure was introduced in 2020 to promote energy efficiency and the reduction of seismic risks in buildings. Born with a percentage of 110%, in 2023 the measure dropped to 90%. In 2024 the benefit was further reduced to 70% for most cases, and will drop to 65% in 2025 and then end at the end of the year. From February 2023, only the tax deduction is permitted and no longer the possibility of requesting the transfer of credit or the discount on the invoice (opportunities currently reserved only for a small number of subjects). From 2024, the Superbonus can no longer be used for single-family buildings. Instead, condominiums and natural persons who own properties with 2 to 4 distinctly registered property units benefit from the 70% tax relief. Third-sector bodies will continue to benefit from the 110% relief until the end of 2025 (ONLUS – Non-Profit Organizations of Social Utility; OdV – Volunteer Organisations; APS – Social Promotion Associations). In this case, the properties must fall into the following cadastral categories: B/1 – colleges, boarding schools, shelters, orphanages, hospices, convents, seminaries and barracks; B/2 – non-profit nursing homes and hospitals; D/4 – for-profit nursing homes and hospitals. To have access to tax relief, the buildings subject to the works must achieve an improvement of at least 2 energy classes;
- Facade Bonus: extended until 31 December 2024 with a 90% deduction of expenses incurred, it is one of the most interesting tax breaks for those who want to renovate the exterior of their property. Among the works admitted, many of them contribute to the energy efficiency of the external building envelope (renovation of the thermal insulation, replacement of fixtures and doors, installation of solar shading). Properties of any cadastral category, including capital ones, can benefit from the deduction. It can be requested by natural and legal persons, as well as by both public and private entities. To obtain the bonus the property must fall within zones A and B according to Ministerial Decree 1444/1968;

- Green Bonus: also extended until 31 December 2024, it consists of a deduction of 36% of the expenses incurred for interventions relating to: greenery of private uncovered areas, appurtenances or fences, construction of irrigation systems, construction of wells, construction of hanging gardens and green roofs. These interventions can indirectly impact, even if minimally, energy efficiency. The properties must not fall into the categories of capital or merchandise goods. It follows that the bonus concerns only residential property [14, 30].

Table 1. Types of tax relief and public financing provided in Italy differentiated by intended use (deduction percentages valid for the year 2024)

Type of subsidy or financing	Deduction	Residential property	Productive property	Commercial property	Property and infrastructure of collective interest
Building Bonus	50%	✓	-	-	-
Ecobonus	50% o 65%	✓	✓	✓	✓
Superbonus	70% o 110%[a]	✓[b]	-	-	✓[c]
Facade Bonus	90%	✓	✓	✓	✓[d]
Green Bonus	36%	✓	-	-	-
State or European funding	-	✓[e]	✓	✓	✓

[a] 110% deduction valid only for Third-Sector entities.
[b] Only for condominiums and natural persons who own properties with 2 to 4 real estate units.
[c] Only for cadastral categories B/1, B/2 and D/4 belonging to Third-Sector entities.
[d] The property must be in Zone A or Zone B (Ministerial Decree 1444/1968).
[e] Financing mainly for public housing.

Energy efficiency measures for buildings and public infrastructures can be financed and incentivized through various methods. For example, the European Union makes financing and incentives available for energy efficiency projects through various programs and initiatives, such as the European Regional Development Fund (ERDF), the European Investment Fund Strategic Funds (EFSI), the Cohesion Fund (CF), the European Agricultural Fund for Rural Development (EAFRD), the European Maritime and Fisheries Fund (EMFF), the URBACT program (to promote sustainable urban development), etc. [31]. Also, the National Recovery and Resilience Plan (PNRR) includes significant investments to improve energy efficiency. In particular, through Measure 1 of Mission 2/Component 3 (M2C3) the PNRR proposes the implementation of a program to improve the efficiency and safety of the public building heritage, with interventions concerning particular schools and judicial citadels. Instead, investment 1.2 (Promotion of renewables for energy communities and self-consumption) of Mission 2/Component

2 (M2C2), aims to provide support to energy communities, in particular to municipalities with fewer than 5,000 inhabitants, to allow the installation of at least 2,000 MW of additional capacity from renewable sources, for an indicative production of 2,500 GWh/year, coupled with energy storage systems [32].

Table 1 summarizes the main forms of tax relief and public financing differentiated by intended use (residential, commercial, productive and of collective interest).

2.4 The ALARP Principle

According to the ALARP principle, when making safety and health decisions, risks must be reduced to the lowest level reasonably practicable [33]. Further risk reductions would be prohibitively expensive. This principle, already outlined in the UK Health and Safety Executive (HSE) regulations, is of crucial importance in contexts where it is necessary to balance risks, costs and potential benefits [34]. The ALARP principle recognizes that zero risk is not practicable, and introduces risk tolerability and acceptability thresholds. The lack of specific numerical values suggests the need to calibrate the model: for instance, the HSE recommends a "broadly acceptable" threshold of 10–6 deaths per year. The ICAF represents the cost of an investment to avoid one death and is calculated as the ratio between the cost of the investment and the expected reduction in the number of deaths. The risk is considered tolerable if the additional costs to further reduce it would be excessive, based on the comparison between the estimated ICAF and the reference values. Risks above the tolerability threshold must be reduced as they are considered unacceptable, while those between the tolerability and acceptability levels fall within the ALARP area and require mitigation measures up to a reasonably practicable level. Risks below the acceptability threshold are considered "largely acceptable" and do not require further mitigation measures. These criteria, commonly applied in high-risk sectors such as industrial engineering, mean that managers must reduce risks to the lowest levels reasonably possible, recognizing that beyond a certain point, further reductions always become more expensive [35, 36]. La logica ALARP è sintetizzata in Fig. 1. The ALARP logic is summarized in Fig. 2.

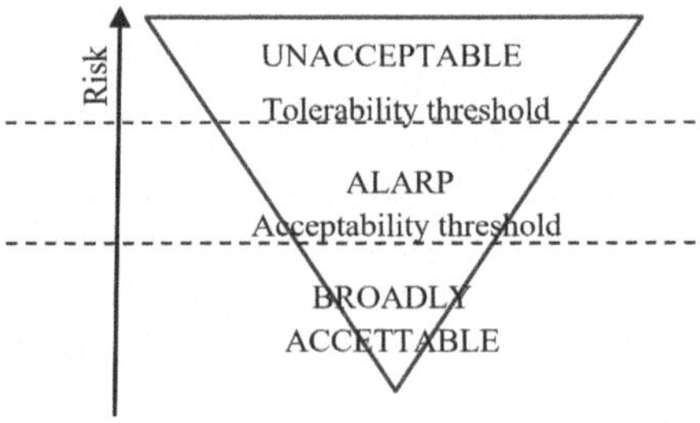

Fig. 1. The ALARP principle (own reworking from [14])

Recently, the ALARP approach has been applied in an original way, to evaluate the risk of financial failure of investment projects instead of the risk of loss of human life [37, 38]. In this work, the ALARP approach is integrated for the first time into an original risk assessment protocol to define tolerability and acceptability thresholds of the economic-financial risk of urban energy redevelopment projects.

3 The Protocol

This section presents the economic and financial risk assessment protocol relating to energy requalification interventions that can be implemented through public energy policies applied in urban areas. The logical-sequential steps of the protocol are summarized in Fig. 2.

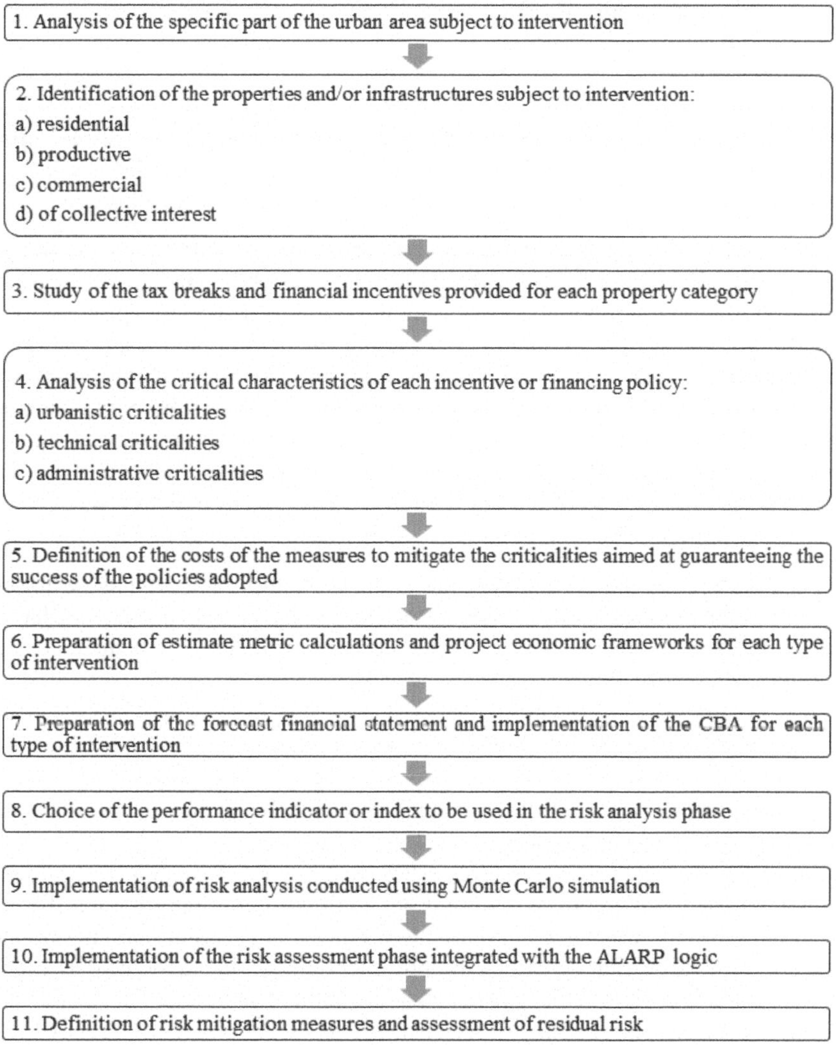

Fig. 2. Flowchart representative of the protocol steps

The following subsections describe the various logical steps in detail.

Phase 1. Analysis of the specific part of the urban area subject to intervention. In this phase, a detailed examination of the urban area designated for the building or urban planning intervention is carried out. This analysis includes the evaluation of the physical, environmental, socio-economic and urban characteristics of the area. Data relating to the structure of buildings, energy consumption, population density, existing infrastructure and any regulatory or environmental constraints are examined. The objective is to fully understand the specific context in which the intervention is inserted, identifying the opportunities and challenges that could influence the planning and implementation of energy efficiency measures.

Phase 2. Identification of the properties and/or infrastructures subject to intervention. This phase consists of specifically identifying the properties and/or infrastructures within the selected urban area subject to intervention. If the works involve multiple properties, it is essential to classify them based on their intended use (residential, commercial, productive, properties of collective interest). This categorization facilitates the management and planning of interventions, allowing a more detailed assessment of the needs and specific characteristics of each type of property or infrastructure.

Phase 3. Study of the benefits and financial incentives provided for each property category. This phase involves the study of the tax breaks and financial incentives available for each category of property identified in the previous phase. This process involves a detailed analysis of the tax regulations and financial incentive policies applicable to residential, productive, and commercial properties and infrastructures of collective interest. Tax credits, subsidized loans, grants and other incentives offered by public authorities or private institutions to promote energy requalification interventions are examined. The objective is to identify and evaluate the financing opportunities available to support energy efficiency interventions in each category of property, to maximize the economic benefits and incentivize the implementation of the interventions themselves.

Phase 4. Analysis of the critical issues characterizing each incentive/financing policy. This phase of the protocol concerns the analysis of the criticalities that characterize each incentive or financing policy, divided into three categories: urbanistic, technical and administrative criticalities.

- Urbanistic criticalities: to obtain funding or the benefits of building bonuses, buildings often must comply with state and local building and zoning regulations. Non-compliance leads to the loss of tax benefits and may be subject to severe sanctions and blocking of ongoing construction activities;
- Technical criticalities: these critical issues occur when the project requirements are not satisfied and the costs incurred are not considered appropriate. To access the Ecobonus, it is essential to present two crucial documents: the energy performance certificate, which describes the pre-intervention state and certifies the post-intervention energy improvement, issued by a qualified technician; the technical certification, transmitted exclusively electronically to ENEA (National Agency for New Technologies, Energy and Sustainable Economic Development). In the case of the

Superbonus, these two documents must demonstrate an improvement in at least two energy classes;

- Administrative criticalities: concern the bureaucratic procedures and regulatory compliance necessary to access the available incentives and financing. These issues can include extended wait times, complexity of the required documents and burdensome administrative requirements. It is necessary to issue a certificate of conformity of the data relating to the documentation which certifies the existence of the conditions to benefit from the relief. Only professionals authorized to electronically transmit the tax return, such as accountants, commercial experts or employment consultants, are authorized to issue this certificate [14].

Phase 5. Definition of the costs of criticality mitigation measures. This phase involves the analysis and estimate of the costs necessary to address the critical issues identified during the previous phases of the evaluation process. These costs, which represent with good approximation a quantitative measure of the three critical issues (i.e. the risks associated with the interventions), will be used as sensitive variables in the subsequent risk analysis phase. Corrective actions are therefore first identified to overcome the urbanistic, technical and administrative criticalities identified. We then proceed with the evaluation of the costs associated with the corrective actions. These generally concern the costs of consultancy by sector experts and, in the case of more complex projects, the costs for insurance policies, assistance from scientific institutions, and services offered by a general contractor. Finally, each type of cost must be associated with a probability distribution, established based on opinions formulated by experts or following empirical market research conducted to define the ranges of cost variability. For each mitigation measure, the minimum, maximum, average and modal market costs are therefore defined, as well as the type of probability function (triangular, rectangular, Gaussian, etc.). The main objective of this phase is to ensure that the policies adopted are supported by sufficient financial resources to address the challenges and ensure the success of urban energy redevelopment interventions [39].

Phase 6. Preparation of estimate metric calculations and project economic frameworks for each type of intervention. This phase involves the compilation of documents detailing the expected costs for the implementation of the various urban energy requalification interventions. During this phase, the estimate metric calculations are processed, which represent a quantitative estimate of the materials, labour and other resources necessary to carry out each intervention and their costs. Estimate metric calculations provide a detailed overview of the physical components and related costs associated with each type of intervention. Furthermore, project economic frameworks are developed, which represent a complete financial evaluation of the planned interventions. Economic frameworks include expected costs for materials, labour, consultancy, any additional costs, taxes and other related expenses.

Phase 7. Preparation of the forecast financial statement and implementation of the CBA. This phase consists of the drafting of a document that provides a detailed assessment of the costs and financial benefits expected for each type of urban energy requalification intervention. The forecast financial statements are prepared, which include the initial costs associated with the implementation of each intervention estimated in the previous phase of processing the estimate metric calculations and the project economic frameworks. Furthermore, the expected financial benefits deriving from the interventions are identified, such as energy savings, reduction of building operating costs, revenues deriving from tax incentives or government financing, and other related financial benefits. In addition, a cost–benefit analysis (CBA) is implemented to systematically and in detail evaluate the costs and benefits associated with each intervention. This analysis allows you to evaluate whether the planned investments are economically advantageous in the long term, comparing the expected benefits with the costs incurred. The CBA can lead to the estimation of a series of indicators and indices, such as the Internal Rate of Return (IRR), the capital recovery period (Payback Period), and the Net Present Value (NPV), useful for evaluating the economic and/or financial invalidity of the interventions [40]. A delicate phase concerns the estimate of the discount rate to be used to discount future cash flows at present. When the analysis is conducted from the point of view of the families benefiting from the interventions, the discount rate can be expressed in terms of the families' time preference rate. When the analysis is conducted from the point of view of a company, it is preferable to estimate the discount rate with the Weighted Average Cost of Capital (WACC) method. Finally, when the analysis is conducted from the point of view of a public body, it is preferable to express the discount rate in terms of the social rate of time preference of the entire community. Usually, the discount rates used in the energy sector vary in the following three ranges: high rates (5.5–7.5%) for owning families, intermediate (3.5–5.5%) for companies, and low (0–3.5%) when you want to incorporate the intergenerational effects of social and environmental sustainability [41].

Phase 8. Choice of indicator or performance index to be used in the risk analysis phase. In this phase, the objective is to select the right metric from a series of indicators and indices (Benefits, Revenues, Net Profit, NPV, IRR, Payback Period, Benefit–Cost Ratio, ROI, ROE etc.) to subsequently implement the analysis of the economic and/or financial risk of the planned interventions. The choice of the metric to be adopted is influenced by the type of analysis (whether economic or only financial), the scope of the intervention (residential, commercial, productive, of collective interest), the energy policy adopted (tax deduction, financing non-repayable fund, etc.) and the duration of the works. An often-used index is the IRR (Internal Rate of Return), as this metric is less influenced by the size of the project. The IRR is the value of the discount rate that makes the NPV zero [40]. When the duration of the intervention is rather limited (this is the case of an energy efficiency project for a single building), ROI is often preferred to IRR. ROI allows you to define the annual operating income of an investment compared to the total capital invested (both equity capital and debt capital) [42].

Phase 9. Implementation of risk analysis. In the implementation phase of probabilistic risk analysis via Monte Carlo simulation, computational techniques are used to evaluate the effects of uncertainty and variability in project results. This method involves the random generation of multiple possible scenarios, each with its inputs and parameters, to determine the probability of various outcomes and identify possible critical issues or opportunities associated with the project [43]. In particular, the costs associated with urbanistic, technical, and administrative criticalities are regarded as sensitive variables.

As these costs vary within their permissible range, according to the probability distributions defined during Phase 5, simulation enables us to obtain the cumulative probability distribution of the selected performance indicator or index.

Phase 10. Implementation of the risk assessment phase integrated with the ALARP logic. In this phase, by reading the graph representing the cumulative probability distribution we evaluate the probability that the chosen performance indicator/index is higher (or lower) than a certain threshold value. In this way, it is possible to evaluate the probability of financial and/or economic failure of urban energy redevelopment interventions. To introduce thresholds that are as objective as possible, in the protocol it was decided to integrate the traditional economic/financial risk assessment phase with the ALARP principle, in which threshold levels are established for risks considered "acceptable" and "tolerable" [44]. The tolerability threshold can be set concerning a probability of no less than 50% of having a performance indicator (or index) lower than or equal to its lowest useful value. For instance, if IRR is chosen as the metric, the risk is tolerable if there is a probability of no less than 50% of having an IRR lower than or equal to the discount rate used for the CBA. The acceptability threshold can be set concerning a probability of no less than 50% of having a performance indicator (or index) lower than or equal to the value of the same parameter obtainable in the absence of implementation of the project. For instance, if the IRR is chosen as the metric, the risk is acceptable if there is a probability of no less than 50% of having an IRR lower than or equal to the IRR obtained for scenario 0, i.e. when the efficiency interventions energy are not implemented [37, 38].

Phase 11. Definition of risk mitigation measures and assessment of residual risk. In the phase of defining the risk mitigation measures and evaluating the residual risk, actions aimed at reducing or managing the risk of economic/financial failure possibly recorded in the previous phase are identified and implemented. Subsequently, the residual risk is assessed, i.e. the level of risk remaining after the implementation of the mitigation measures, to evaluate the effectiveness of the strategies adopted. A new Monte Carlo simulation and a new risk assessment phase are therefore implemented, effectively making the protocol iterative.

4 Discussion and Conclusions

In the field of urban planning and energy management, efficiency and sustainability play a crucial role in the design and administration of modern cities [45]. In response to growing environmental awareness and the need to reduce greenhouse gas emissions, many countries are adopting policies and incentives to promote energy efficiency in buildings and urban infrastructure. However, the economic and financial evaluation of such interventions remains a complex challenge, requiring a strategic and systematic approach to balance long-term costs and benefits [46, 47]. The protocol conceived here represents an innovative attempt to address this challenge, offering a methodological guide for the economic evaluation of urban energy redevelopment projects supported by public incentive policies. The fundamental peculiarity of the proposed protocol lies in the adaptation of the ALARP principle to the financial context of urban redevelopment projects. Unlike the conventional approach, which focuses on assessing health risks and potential loss of life, our study applies ALARP logic to analyse the risk of financial failure associated with urban redevelopment projects. This innovation requires a redefinition of tolerable and acceptable risk levels, adapting them to the specific economic and financial context of the aforementioned projects. The comparison with other research that uses the ALARP principle [33–38], highlights the originality of its application in our field of investigation. The different phases of the protocol illustrate the complexity and exhaustive nature of the decision-making process, which goes beyond traditional cost–benefit and risk analyses, including a careful survey of the critical issues (urban planning, technical, administrative) that characterize each policy to incentivize or finance urban energy interventions.

Although the protocol has only been defined on a conceptual level and has not yet been applied in practice, its potential is promising. However, during its development, some critical issues and limitations emerged that require attention. For instance, the phase of defining financial incentive policies could be complex and susceptible to regulatory and political changes over time, making it necessary to constantly update the protocol. Furthermore, practical implementation may encounter operational and bureaucratic resistance, requiring adaptation to specific local circumstances. Despite these challenges, the potential of the protocol to drive informed and sustainable decisions in the area of urban energy efficiency is evident.

In conclusion, we underline the importance of a holistic and integrated approach to the economic evaluation of projects, considering not only the direct financial costs and benefits but also the social, environmental and long-term impacts [48, 49]. Future research areas should focus on the refinement and empirical validation of the protocol, as well as on the analysis of its effects on urban planning and governance.

Disclosure of Interests. The authors have no competing interests to declare that are relevant to the content of this article.

References

1. Vidadili, N., Suleymanov, E., Bulut, C., Mahmudlu, C.: Transition to renewable energy and sustainable energy development in Azerbaijan. Renew. Sustain. Energy Rev. **80**, 1153–1161 (2017). https://doi.org/10.1016/j.rser.2017.05.168
2. Chen, W., Alharthi, M., Zhang, J., Khan, I.: The need for energy efficiency and economic prosperity in a sustainable environment. Gondwana Res. **127**, 22–35 (2024). https://doi.org/10.1016/j.gr.2023.03.025
3. Zhang, C., Khan, I., Dagar, V., Saeed, A., Zafar, M.W.: Environmental impact of information and communication technology: unveiling the role of education in developing countries. Technol. Forecast. Soc. Chang. **178**, 121570 (2022). https://doi.org/10.1016/j.techfore.2022.121570.
4. ENEA Homepage. https://www.efficienzaenergetica.enea.it/vi-segnaliamo/rapporto-iea-2022-anno-di-svolta-per-l-efficienza-energetica-a-livello-globale-2.html. Accessed 20 Mar 2024
5. European Parliament Homepage. https://www.europarl.europa.eu/legislative-train/theme-a-european-green-deal/file-revision-of-the-energy-performance-of-buildings-directive. Accessed 20 Mar 2024

6. Umoh, A.A., Adefemi, A., Ibewe, K.I., Etukudoh, E.A., Ilojianya, V.I., Nwokediegwu, Z.Q.S.: Green architecture and energy efficiency: a review of innovative design and construction techniques. Eng. Sci. Technol. J. **5**(1), 185–200 (2024). https://doi.org/10.51594/estj.v5i1.743.
7. Cole, L.B.: Green building literacy: a framework for advancing green building education. Int. J. STEM Educ. **6**, 1–13 (2019). https://doi.org/10.1186/s40594-019-0171-6
8. Zakari, A., Khan, I., Tawiah, V., Alvarado, R.: Reviewing the ecological footprints of Africa top carbon consumer: a quantile on quantile analysis. Int. J. Environ. Sci. Technol. **19**(11), 11475–11486 (2022). https://doi.org/10.1007/s13762-021-03904-z
9. Moghadam, S.T., Delmastro, C., Corgnati, S.P., Lombardi, P.: Urban energy planning procedure for sustainable development in the built environment: a review of available spatial approaches. J. Clean. Prod. **165**, 811–827 (2022). https://doi.org/10.1016/j.jclepro.2017.07.142
10. Sampaio, H.C., Dias, R.A., Balestieri, J.A.P.: Sustainable urban energy planning: the case study of a tropical city. Appl. Energy **104**, 924–935. https://doi.org/10.1016/j.apenergy.2012.12.022.
11. Dolores, L., Macchiaroli, M., De Mare, G.: Financial impacts of the energy transition in housing. Sustainability **14**(9), 4876 (2022). https://doi.org/10.3390/su14094876
12. Liu, Z., et al.: Incentive initiatives on energy-efficient renovation of existing buildings towards carbon-neutral blueprints in China: advancements, challenges and perspectives. Energy Build., 113343 (2023). https://doi.org/10.1016/j.enbuild.2023.113343.
13. Chatterjee, S., Rafa, N., Nandy, A.: Welfare, development, and cost-efficiency: a global synthesis on incentivizing energy efficiency measures through co-benefits. Energy Res. Soc. Sci. **89**, 102666 (2022). https://doi.org/10.1016/j.erss.2022.102666
14. Macchiaroli, M., Dolores, L., De Mare, G., Nicodemo, L.: Tax policies for housing energy efficiency in Italy: a risk analysis model for energy service companies. Buildings **13**, 582 (2023). https://doi.org/10.3390/buildings13030582
15. Modrzyński, P., Karaszewski, R.: Urban energy management—a systematic literature review. Energies **15**, 7848 (2022). https://doi.org/10.3390/en15217848
16. Xia, X., Wu, X., BalaMurugan, S., Karuppiah, M.: Effect of environmental and social responsibility in energy-efficient management models for smart cities infrastructure. Sustain. Energy Technol. Assessments **47**, 101525 (2021). https://doi.org/10.1016/j.seta.2021.101525
17. Berardi, U., GhaffarianHoseini, A., GhaffarianHoseini, A.: State-of-the-art analysis of the environmental benefits of green roofs. Appl. Energy **115**, 411–428 (2014). https://doi.org/10.1016/j.apenergy.2013.10.047
18. Huovila, A., Bosch, P., Airaksinen, M.: Comparative analysis of standardized indicators for Smart sustainable cities: what indicators and standards to use and when? Cities **89**, 141–153 (2019). https://doi.org/10.1016/j.cities.2019.01.029
19. Talandier, M.: Are there urban contexts that are favourable to decentralised energy management? Cities **82**, 45–57 (2018). https://doi.org/10.1016/j.cities.2018.05.004
20. Jiang, F., Ma, J., Li, Z., Ding, Y.: Prediction of energy use intensity of urban buildings using the semi-supervised deep learning model. Energy **249**, 123631 (2022). https://doi.org/10.1016/j.energy.2022.123631
21. Chang, I.S., et al.: Evaluation on the integrated water resources management in China's major cities–Based on City Blueprint® Approach. J. Clean. Prod. **262**, 121410 (2020). https://doi.org/10.1016/j.energy.2021.120055
22. Letnik, T., Farina, A., Mencinger, M., Lupi, M., Božičnik, S.: Dynamic management of loading bays for energy efficient urban freight deliveries. Energy **159**, 916–928. https://doi.org/10.1016/j.energy.2018.06.125.

23. Ruijs, A., Vollebergh, H.: Lessons from 15 Years of Experience with the Dutch Tax Allowance for Energy Investments for Firms. Milan: Fondazione Eni Enrico Mattei (2013). https://www.econstor.eu/bitstream/10419/75430/1/750043776.pdf. Accessed 20 Mar 2024
24. Bertoldi, P., Economidou, M., Palermo, V., Boza-Kiss, B., Todeschi, V.: How to finance energy renovation of residential buildings: review of current and emerging financing instruments in the EU. Wiley Interdisc. Rev. Energy Environ. **10**, e384 (2021)
25. Villca-Pozo, M., Gonzales-Bustos, J.P.: Tax incentives to modernize the energy efficiency of the housing in Spain. Energy Policy **128**, 530–538 (2019)
26. Neveu, A.R., Sherlock, M.F.: An evaluation of tax credits for residential energy efficiency. East. Econ. J. **42**, 63–79 (2016)
27. Sebitosi, A.B., Pillay, P.: Renewable energy and the environment in South Africa: a way forward. Energy Policy **36**, 3312–3316 (2008)
28. Sarker, T., Taghizadeh-Hesary, F., Mortha, A.: The Role of Fiscal Incentives in Promoting Energy Efficiency in the Industrial Sector: Case Studies from Asia. ADBI Working Paper 1172. Tokyo: Asian Development Bank Institute (2020). https://www.adb.org/publications/role-fiscalincentives-promoting-energy-efficiency-industrial-sector-asia. Accessed 20 Mar 2024
29. ENEA Homepage. https://www2.enea.it/it/Ricerca_sviluppo/documenti/ricerca-di-sistema-elettrico/adp-mise-enea-2019-2021/tecnologie-efficienza-risparmio-edifici/report-rds_ptr_2021_132.pdf. Accessed 20 Mar 2024
30. Consiglio Nazionale del Notariato Homepage. https://notariato.it/it/news/online-ledizione-aggiornata-della-guida-immobili-e-bonus-fiscali-2024/. Accessed 20 Mar 2024
31. Commissione Europea Homepage. https://commission.europa.eu/eu-regional-and-urban-development/topics/cities-and-urban-development/funding-cities_it. Accessed 20 Mar 2024
32. Governo Italiano Homepage. https://www.governo.it/sites/governo.it/files/PNRR_0.pdf. Accessed 20 Mar 2024
33. Melchers, R.E.: On the ALARP approach to risk management. Reliab. Eng. Syst. Saf. **71**(2), 201–208 (2021). https://doi.org/10.1016/S0951-8320(00)00096-X
34. Ale, B.J.M.: Tolerable or acceptable, a comparison of risk regulation in the UK and in the Netherlands. Risk Anal. Int. J. **25**, 231–241 (2025). https://doi.org/10.1111/j.1539-6924.2005.00585.x
35. De Blaeij, A., Florax, R.J., Rietveld, P., Verhoef, E.: The value of statistical life in road safety: a meta-analysis. Accid. Anal.Prev. **35**(6), 973–986 (2003). https://doi.org/10.1016/S0001-4575(02)00105-7
36. Viscusi, W.K., Aldy, J.E.: The value of a statistical life: a critical review of market estimates throughout the world. J. Risk Uncertainty **27**, 5–76 (2003). https://doi.org/10.1023/A:1025598106257
37. Macchiaroli, M., Dolores, L., De Mare, G.: Risk analysis for the integrated water service manager in the selection of investments—the ALARP applied to the program of interventions. In: Naddeo, V., Choo, KH., Ksibi, M. (eds.) Water-Energy-Nexus in the Ecological Transition: Natural-Based Solutions, Advanced Technologies and Best Practices for Environmental Sustainability, pp. 47–49. Springer, Cham (2022). https://doi.org/10.1007/978-3-031-00808-5_12
38. Macchiaroli, M., Dolores, L.: Economic-financial sustainability and risk assessment in the water sector in Italy. In: Calabrò, F., Della Spina, L., Piñeira Mantiñán, M.J. (eds.) New Metropolitan Perspectives, NMP 2022. LNNS, vol. 482, pp. 1337–1346. Springer, Cham (2022). https://doi.org/10.1007/978-3-031-06825-6_130
39. Utica, G.: La Stima Sintetica del Costo di Costruzione. Il Computo Metrico e il Computo Metrico Estimativo per Classi di Elementi Tecnici. Maggioli Editore, Rimini, Italia (2011)
40. Berk, J., De Marzo, P., Venanzi, D.: Capital Budgeting, 1st edn. Pearson Paravia Bruno Mondadori, Milano, Italia (2009)

41. Copiello, S.: Economic viability of building energy efficiency measures: a review on the discount rate. AIMS Energy **9**(2), 257–285 (2021). https://doi.org/10.3934/energy.2021014
42. Ichsani, S., Suhardi, A.R.: The effect of return on equity (ROE) and return on investment (ROI) on trading volume. Procedia-Soc. Behav. Sci. **211**, 896–902 (2015). https://doi.org/10.1016/j.sbspro.2015.11.118
43. Sugiyama, S.: Monte Carlo simulation/risk analysis on a spreadsheet: review of three software packages. Foresight **9**, 8 (2008)
44. Pike, H., Khan, F., Amyotte, P.: Precautionary Principle (PP) versus As Low As Reasonably Practicable (ALARP): which one to use and when. Process. Saf. Environ. Prot. **137**, 158–168 (2020). https://doi.org/10.1016/j.psep.2020.02.026
45. Gerundo, R., Marra, A., Giacomaniello, O.: A methodology for analyzing the role of environmental vulnerability in urban and metropolitan-scale peripheralization processes. In: La Rosa, D., Privitera, R. (eds.) Innovation in Urban and Regional Planning, INPUT 2021. LNCE, vol. 146, pp. 459–468. Springer, Cham (2021). https://doi.org/10.1007/978-3-030-68824-0_49
46. Graziuso, G., Grimaldi, M., Mancini, S., Quartieri J., Guarnaccia, C.: Crowdsourcing data for the elaboration of noise maps: a methodological proposal. J. Phys. Conf. Ser. **1603**(1), 012030 (2020). https://doi.org/10.1088/1742-6596/1603/1/012030
47. Grimaldi, M., Sebillo, M., Vitiello, G., Pellecchia, V.: An ontology based approach for data model construction supporting the management and planning of the integrated water service. In: Misra, S., et al. (eds.) Computational Science and Its Applications – ICCSA 2019. ICCSA 2019. LNCS, vol. 11624, pp. 243–252. Springer, Cham (2019). https://doi.org/10.1007/978-3-030-24311-1_17.
48. Morano, P., Tajani, F., Guarini, M.R., Sica, F.: A systematic review of the existing literature for the evaluation of sustainable urban projects. Sustainability **13**(9), 4782 (2021). https://doi.org/10.3390/su13094782
49. Tajani, F., Guarini, M.R., Sica, F., Ranieri, R., Anelli, D.: Multi-criteria analysis and sustainable accounting. Defining indices of sustainability under Choquet's integral. Sustainability **14**(5), 2782 (2022). https://doi.org/10.3390/su14052782

A Conceptual Framework to Address Collective Energy Initiatives' Spatial Impacts at the City Level: An Ongoing Attempt

B. M. Frigione[1,2(✉)], M. Pezzagno[1,3], and C. Zuidema[4]

[1] Department of Civil, Environmental, Architectural Engineering and Mathematics—DICATAM, University of Brescia, 25123 Brescia, Italy
{barbara.frigione,michele.pezzagno}@unibs.it
[2] University School for Advanced Studies IUSS Pavia, 27100 Pavia, Italy
[3] The University Research and Documentation Center for the 2030 Sustainable Development Agenda (CRA 2030), Steering Committee, University of Brescia, 25123 Brescia, Italy
[4] Department of Spatial Planning and Environment, University of Groningen, Landleven 1, 9747 AD Groningen, The Netherlands
c.zuidema@rug.nl

Abstract. Energy is vital in sustaining human life and fostering development, yet the energy sector's significant contributions to greenhouse gas emissions and climate change necessitate transformative changes, commonly addressed as an energy transition.

As major energy consumers and greenhouse gas emitters, cities play a pivotal role in such transition. Urban plans and policy mechanisms offer a means to regulate urban processes and practices, influencing energy demand and emissions patterns; however, transformation processes also involve phenomena that extend beyond the confines of urban planning practices, for instance, collective actions.

Collective Energy Initiatives (CEIs) represent an emerging transformative social movement fostering the transition toward decentralized renewable energy production and consumption. The implementation of renewable energy technologies introduces spatial challenges, such as land competition and conflicting land uses, suggesting that CEIs practices may have spatial repercussions in urban areas. Yet, there is a lack of scholarly interest in the spatial impacts of CEIs.

This contribution presents early findings from an ongoing literature review aimed at identifying methodological approaches to study the spatial impacts of CEIs. Drawing upon existing publications on CEIs, a conceptual framework is formulated to stimulate new trajectories in understanding the complex interplay between energy transitions, collective actions, and their spatial implications. The resulting integrative framework highlights the multifaceted nature of CEI impacts and emphasizes the importance of a nuanced understanding of both internal dynamics and external contexts to effectively address spatial concerns at the city level.

Keywords: Urban Energy Transition · Urban Planning · Community Energy Initiatives

O. Gervasi et al. (Eds.): ICCSA 2024 Workshops, LNCS 14818, pp. 242–255, 2024.
https://doi.org/10.1007/978-3-031-65273-8_16

1 Introduction

Energy is vital for the life and development of humankind [1–3], yet the energy sector has a great impact on GHG emissions and climate change [4]. To mitigate the energy sector's impacts and, at the same time, adapt to climate change pressures, a substantial change in the energy systems is necessary [4].

Energy systems are historically addressed as techno-economic frameworks defined by the energy flows and the technical activities required to collect energy sources and supply consumers energy demand [5]. More recently, energy systems are also understood as complex socio-technical-economic configurations [6, 7, 10, 11] resulting from the decisions and actions of a variety of agents [12] deeply influenced by cultural values, social norms and identity, traditions, common visions, and imaginaries [6–9, 12, 13].

Changes in the energy system are commonly addressed as an energy transition [5, 9, 14, 15]. As energy systems are (re)shaped by individual actors and institutions' choices, which are rooted in a broad fan of cultural values and social norms, the current energy transition goes beyond swapping technologies and fuels. Energy transition also requires a shift in cultural paradigm among agents who finance, create, and use those technologies and fuels as well as a change in their role within the energy system itself.

Traditionally, cities were merely considered the end point of the energy supply chain [16]. Being the main energy consumers and GHG emitters [17], cities are now perceived as a crucial instrument to maintaining the global average temperature increase within 2 to 2.4 °C above pre-industrial levels [16, 18–20]. Through the exercise of urban policies and plans (for instance, in Europe there are the SECAP, the Climate City Contract, and the Carbon Neutral Cities Alliance) cities can govern urban processes and practices to reach climate neutrality, however urban environments are also historically profoundly influenced by phenomena out of urban planning control.

The different forms of social energy mobilization and cooperation that are currently emerging at the local level [21–24] represent an example of these phenomena. Citizens acting in their personal capacity (e.g., leaders, consumers, voters), professional capacity (e.g., town planners, builders, teachers, investors), or collective capacity (e.g., the youth climate movement) are actively participating in the current energy transition [25]. The activities span from people becoming active consumers and producers, i.e. prosumers [26], to the more complex forms of collective prosumerism such as collective self-consumption multi-tenancy buildings [26], energy neighborhoods and districts [27, 28], and energy communities [29]. In this contribution, the umbrella term collective energy initiatives (CEIs) is used to consider all collective initiatives of prosumers.

The movement of CEIs is already playing an important role in the evolving energy landscape [30], bringing real change between society and energy [31] through spurring collaborations, mobilizing resources, developing context-specific technological mixes and governance structures, and influencing public consensus [19, 32]. CEIs have also gained increasing prominence in scientific literature over the past two decades [33, 34], and various terms have emerged referring to these initiatives (e.g., civic renewable energy, grassroots energy initiatives, civic energy initiatives, cooperative energy initiatives, local energy initiatives, local energy communities, renewable energy communities, community energy initiatives, community energy) [19, 26, 35–41]. The European context itself is witnessing a clear proliferation of these initiative: Wierling et al. (2023) identified over 10,000 citizen-led energy initiatives across 29 European countries, while the European Federation of Renewable Energy Cooperatives (REScoop) counts 701 member cooperatives and about 1 500 000 citizens involved [42].

Their growth and number imply that CEIs can even be seen as a "transformative social movement" [43] that is upholding decentralized renewable energy production and consumption spurring the transition towards an electrified energy model. Nevertheless, the implementation of wind turbines, solar panels, sun boilers, thermal, and biogas systems needed to meet energy demands may require more land compared to fossil fuels-based technologies [44–46], exacerbating the global competition for land [47]. In the United States, for instance, the realization of renewable projects has already become one of the primary causes of change in land use [48]. Certainly, in urban areas physical space is both scarce and contested, urging for a proper understanding of where, how, and why it is changing.

In a recent investigation focusing on scenarios of solar penetration levels, Van de Ven et al. (2021) identified that the expansion of solar energy may primarily come to replace existing land use for commercial activities, including croplands or commercial forests utilized for timber products or biomass. According to the simulated solar penetration levels, for every 100 hectares of solar land in Europe, the extent of unmanaged forest clearance across the globe may range from 31 to 43 hectares. Whereas certain renewable energy sources may be considered inexhaustible, the land is limited [49]. Although the production of some renewable energies can be co-located with other land activities, such as installing solar panels on urban rooftops or integrating wind and solar infrastructure with agricultural land [45], the decarbonization process may lead to a series of conflicts [45, 46, 50]. The competition for land, arising from renewable energy projects, may result in conflicts with alternative land-use purposes such as agriculture, residential, commercial, and natural areas [46, 47, 50]. Additionally, conflicts may emerge among the pre-existing land uses in areas designated for renewable energy installations [45]. Concerns about property values also arise: the integration of renewable energy technologies on buildings may enhance their image and marketability [51] whereas, alternatively, the installation on land might lead to a depreciation of neighboring property values [45]. It is therefore reasonable to suppose that CEIs' practices may have spatial repercussions in our cities, although it's likely that the impacts vary depending on the specific type of initiative. Nevertheless, so far scholarly interest in the overall impacts of collective energy initiatives has been limited [52].

A question therefore naturally arises: spurring the upholding of renewable energies, how do CEIs impact at the city level from a spatial point of view?

This contribution presents early findings from an ongoing literature review aimed at identifying methodological approaches to study the spatial impacts of CEIs. Building on the existing work on the CEIs, the research develops an integrative conceptual framework, seeking to stimulate novel trajectories in understanding the complex interplay between energy transitions, collective actions, and their spatial implications.

2 Methodology

Because of the nature of the research question, the initial focus of the review regarded the strand of literature related to the impacts of collective energy initiatives (CEIs). The search process involved the selection and use of keywords (e.g., "energy initiative*" AND "impact*", "energy community*" AND "impact*", "energy project*" AND "impact*") within databases such as Scopus, Web of Science, and Google Scholar, and largely employed the snowballing technique[1] [53], encompassing both scientific and grey literature. The inclusion criteria did not have any time restriction. As the review progressed, it brought forth interconnected aspects that enriched the contextual understanding of the primary research subject. Consequently, supplementary thematic dimensions were incorporated.

The search has thus far yielded 39 publications, including 25 peer-reviewed articles, 5 review articles, 1 conference paper, 4 scientific reports, 1 policy brief, 1 Ph.D. thesis, and 2 master theses. The studies were subjected to thematic analysis and categorized based on the emergent themes.

To streamline the synthesis of ideas and promote innovative perspectives, an integrative review approach, as elucidated by Snyder (2019), is adopted [54]. The application of this approach to the emergent realm of CEIs does not seek to encompass the entirety of literature ever published on the subject. Rather, its objectives are (1) to draw a deductive conceptual framework combining different perspectives and (2) to provide a critical analysis of the literature [54, 55]. This contribution offers a preliminary version of the conceptual framework.

3 The Literature Review

The present contribution reviews 39 publications published between 2008 and 2023. The publications encompass a broad range of disciplines, spanning research areas from urban and social sciences to environmental and energy ones.

Five thematic dimensions were defined, and some publications cover more than one (Table 1). These dimensions are (1) CEIs' impacts; (2) CEIs' actors' participation; (3) CEIs' motives and ambitions; (4) CEIs' professionalization; and (5) CEIs' external context.

The analysis of CEIs' impacts included a total of 17 publications and revealed a widespread assumption among academia and policymakers that these initiatives have a vast array of impacts in advancing local energy transition [29, 32, 56–58]. Besides generating renewable energy [52], such impacts generally encompass social, environmental, and economic aspects. Literature reflects on many impacts, such as enhanced social cohesion, new job opportunities, acquisition of knowledge and skills [32, 57, 59, 60], reducing energy (fuel) poverty [61, 62], generating revenues from renewable energy systems [57], providing access to affordable energy [32, 57], increasing local support for renewable energy [32], enhancing communities' empowerment and involvement in the energy sector [19, 32, 63], reducing carbon emissions [64], impacts on social equity

[1] The "snowball sampling" method is a literature review approach that, starting from a set of articles, utilizes their citations to identify additional literature on the same topic [53].

Table 1. Thematic dimensions identified during the literature review process and included publications.

Dimension	Number of publications	Literature references
CEIs' impacts	17	19, 30, 33, 53, 57, 58, 59, 60, 61, 62, 63, 64, 65, 66, 67, 68, 73
CEIs' actors' participation	11	19, 59, 70, 71, 72, 73, 69, 74, 75, 77, 76
CEIs' motives and ambitions	8	34, 53, 71,73, 77, 78, 79, 80
CEIs' professionalization	9	22, 78, 79, 81, 82, 83, 84, 85, 87
CEIs' external context	16	22, 30, 34, 65, 70, 73, 76, 77, 78, 79, 80, 81, 87, 88, 89, 90

[62], visual impacts [57], and impacts on the existent electricity and heating system [65, 66]. However, one category goes overlooked: the spatial one.

Over the last decade, numerous studies have been highlighting strikingly insufficient empirical evidence of the impacts of CEIs and call for further research to develop formalized impact assessments, systemic and evaluative frameworks that consider the holistic nature of the energy transition, longitudinal studies, statistical analysis, and (quasi-) experimental designs [19, 32, 52, 57, 59–61, 63, 66]. These studies also offer alternative and interesting lenses. In their research, Germes et al. (2021) advocate an approach that considers the motives, objectives, and ambitions of the initiatives to comprehend and assess their impacts, thereby analyzing the specific impacts the initiatives themselves aspire to achieve [52]. This approach stems from the conceptualization put forth by Hasanov (2022), who delved into Dutch local energy initiatives (LECs) employing an "inside-out" perspective, encompassing an examination of both the internal dynamics of community initiatives and their external factors [67]. These studies have also contributed to the literature by illustrating that CEIs differ greatly from one another, both in their motives and outreach, and in organizational aspects such as size, internal structure, and professionalization.

The analysis of the actors' participation in CEIs covered a total of 11 publications. Besides discussing the capacity of CEIs to offer a vast array of social benefits and services, such as enhancing communities' empowerment and involvement in the energy sector [19; 30; 33; 64; 69] and helping mitigating energy poverty [30; 62; 63; 69], the publications reflect on the societal groups and actors involved in the initiatives [58, 68–72], the roles they can play in a CEI [72], and how they participate [73, 74]. However, it is worth mentioning that there is limited empirical proof supporting the efficacy of collective energy initiatives in engaging underrepresented and vulnerable communities and addressing energy poverty [68, 75].

For the analysis of CEIs' motives and ambitions 8 publications were examined. Some research works focus on the key reasons that spur the development of energy initiatives, namely the starting motives [52, 76, 77], whereas others on the ambitions and the objectives [34; 53, 79, 80]. In both cases categorizations are proposed. For instance, the study by Germes, Wiekens, Horlings (2021), the most recent of the five publications analyzed, firstly focused on the starting motives of Dutch local energy initiatives and subsequently noticed that in most cases motives were comparable with the objectives of the initiative [52]. Therefore, they delineated three main categories of local energy initiative's objectives: environmental (e.g., preserve environment, conserve, and produce energy, become independent of natural gas, etc.), economic (e.g., save money, strengthen local economy, generate income from local community), and social (e.g., increase social cohesion and awareness, take collective action).

The analysis of CEIs' professionalization included a total of 9 publications. Professionalization is recognized as one of the main success factors of collective energy initiatives [80]. Literature showcases various approaches in which energy initiatives are getting professionalized, together with some positive impacts and flipsides of the professionalization processes. For instance, Oteman et al. (2017) discuss the recruitment of professionals as permanent staff instead of relying on volunteers, along with collaboration with other initiatives [78]. Differently, Ahlemeyer, Griese, Wawer & Siebenhüner (2022) associate an initiative's professionalization with the level and difference of qualifications of its board members, linking it to the broader concepts of technical and economic professionalization [80]. Nevertheless, professionalization and its impacts are often poorly documented, making it challenging to develop a comprehensive framework [22, 77, 79–84].

For the analysis of CEIs' external context 16 publications were covered. Works explore different external contextual factors that may influence CEIs' development and some of them offer categorizations and definitions. A notable example is the study by Hicks & Ison (2018), which addresses four categories of the aforementioned factors [76].

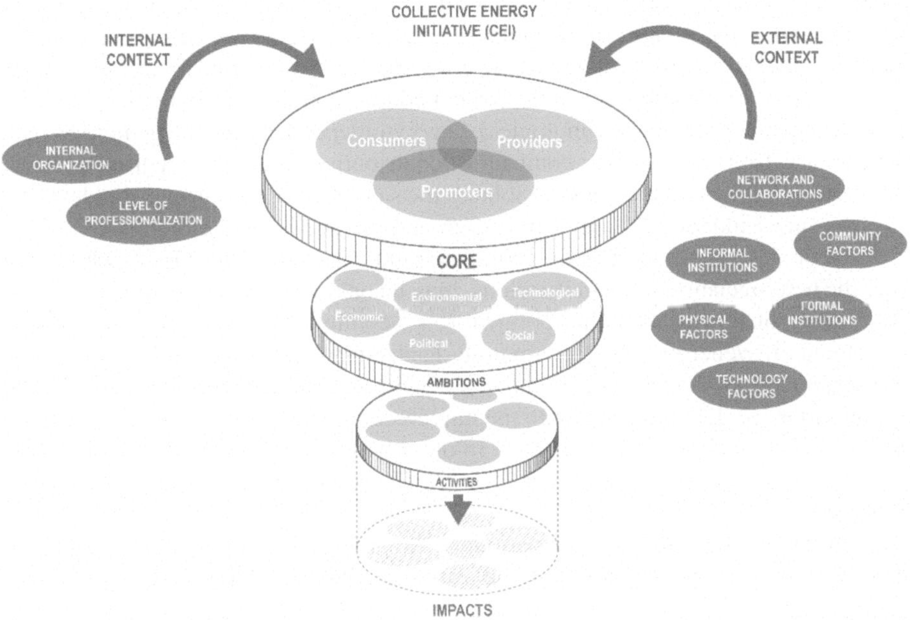

Fig. 1. The conceptual framework (Source: Authors own elaboration, adapted from [76]).

Institutional factors are a key factor, including both informal aspects (e.g. the culture within existing energy and other relevant institutions) and formal aspects (e.g. structure of the energy market, regulatory environment, laws governing legal structures, renewable energy policies). Community factors refer to aspects such as local history and culture, relationships or social capital, and skills and knowledge available, social perceptions of and appetite for certain technologies. Physical factors include topography, renewable resource availability, and existing energy infrastructure, whereas technology factors, are related the cost of different renewable energy technologies, the maturity and modularity of the technology, the energy needs and demand profile of the community/building). The authors conclude that whereas physical and technological contextual factors may be similar across different initiatives, the institutional and community factors are far more complex and nuanced.

The Integrative Conceptual Framework

Developing a framework for analyzing how the spatial impacts of collective energy initiatives (CEIs) take place first needs to acknowledge that CEIs themselves are complex systems[2] whose impacts are deeply influenced by multiple dimensions (Fig. 1).

CEIs have at their heart society [33]. Based on the collective action model, CEIs can involve a diverse array of actors, encompassing professionals, consumer cooperatives, academics, non-governmental organizations (NGOs), local authorities, national government, development firms, energy companies, financial institutions, and more [58, 68–72]. The actors involved (also called "members" or "project participants") assume different roles. According to Gjorgievski, Cundeva, and Georghiou (2021), three main roles can be identified: the initiators/promoters, the energy providers/practitioners, and the consumers [72].

Personal motives drive initiators to begin collective energy initiatives and people to join them and may span from financial (e.g., electricity cost reduction or investment opportunities) and environmental (e.g., lowering one's environmental footprint), to social (e.g., being part of a community, a movement addressing climate change), technological (e.g. engaging with new technologies), energy independence/security motives (e.g., independence from large power companies, contributing to energy security) [76, 77]. Studies have highlighted that, despite joining the same CEI, different stakeholders [70], as well as single individuals, may have different motivations and ambitions [52, 67].

Based on the motives, initiatives formulate joint ambitions (Fig. 1) [52] which, mirroring the heterogeneity of the motivations, may be only focused on energy, or embrace a wide range of topics, including political (e.g., community empowerment, influencing sustainability/energy policy, community leadership) and infrastructural ones (e.g., improving energy independence, community building refurbishment) [33, 52, 78, 79]. The ambitions are then translated into the implementation of an initiative's projects and activities [72], which generates impacts. In other words, impacts are the manifestation of different types of projects and activities, which are brought to existence by motives and objectives [52, 76].

Nevertheless, the path from motivations to the progression of activities is not linear; rather, it is shaped by both internal and external contexts. Being collective actions,

[2] A "complex system" can be defined as a "human-built world" that is beyond the comprehension of single disciplines and requires continuous interdisciplinary cooperation [85].

the effectiveness of CEIs heavily depends on civic participation and engagement [74]. However, as stated by Ison (2010), the level of internal participation and engagement is deeply shaped by the internal governance structure of the initiative, which may be hierarchical, a majority voting system, or based on a consensus approach [73].

The success, alias the impacts, of a CEI also deeply depends on its level of professionalization [80, 86]. Professionalization pathways encompass diverse trajectories, including the transition from volunteerism to remunerated employment, engagement with external consultants, and establishment of a well-organized internal structure [22, 77, 78, 80–84]. Higher levels of professionalization can help initiatives in better-attracting finances, customers, and members or in better supporting the members of the board [80]. Similarly, professionalization helps initiatives in managing increasing complexity following e.g., the augmentation of the projects, of the number of members, or of the services provided [80, 81, 83, 84]. Finally, professionalization may also enable the initiatives to play a greater role within the energy transition process [80]. It is through professionalization that collective energy initiatives can grow and take over utility services [22], which have been able to structure and optimize themselves over the past decades and now are not willing to see their investments deprecated. However, once they start professionalizing, initiatives can run the risk of "alienating their grassroots community" and losing the engagement of volunteers [77], they can "drift away" from initial ambitions, goals, and community's interests [77], or lose social credibility [80].

CEIs are deeply intertwined within their external context [67], which critically influences their development and their impacts [67, 72, 76, 87–89]. Firstly, the socio-cultural context may provide support in terms of knowledge, partnerships, or practical services. The professionalization of an initiative, for instance, deeply benefits from collaborations and networks. Local governments, organizations, businesses, and residents can assume significant roles as supporters, members, or also partners of an initiative [22]. Networks across various territorial levels also constitute a crucial element that supports the initiatives by facilitating valuable knowledge exchange [22; 34, 87]. An example at the European level is the REScoop organization, which promotes collaboration amongst European cooperatives, supports start-ups, and provides a vast array of services (e.g., community coaching, training sessions, financing services, advocacy services, etc.) [42]. The absence of support from the community or a constrained network can pose significant challenges to the cooperative's development [22], for instance in terms of limited community participation, limited expertise, and limited access to funds. It is reasonable to suppose that the higher the level of professionalization and size of an initiative, the more significant its impact becomes in the urban transformation process and the transition towards a more sustainable urban energy system.

The emergence and formation of community energy projects are also shaped by cooperative tradition and values, socially shared roles, and a strong sense of local responsibility, which have been proven to have influenced the willingness of citizens to start or join an initiative, and to have shaped cooperatives' decisions in terms of organizational form, and energy plant site and scale, helping better understand why certain regions display a higher density of community energy projects than others [87]. Alongside policies, laws, feed-in regulations, rules for grid connection, or tax incentives, energy market are factors conducive to the emergence of community energy projects [33, 76, 80, 87]. For instance, a policy framework that tends to focus on renewable technology and finance elements and lacks to support the value and quality of community engagement may constrain the development of some CEIs [76], as well as the delays (and instability) of the legal frameworks to recognize these new forms of energy production [77].

Moreover, as highlighted by Leone et al. (2023) in their work on Positive Energy Districts' projects, strong influences may also come from the existent energy infrastructure, but also by the urban fabric on which the project intervenes, the architectural-landscape constraints, the final function of the district (e.g., residential, commercial, mixed, etc.), the geographic and climatic characteristics, the availability of renewable energy resources and technologies, and the socio-economic fabric [89].

4 Conclusions

Recognizing the spatial impacts of Collective Energy Initiatives (CEIs) is a critical step towards effectively addressing spatial concerns related to the urban energy transition. Nevertheless, the literature review revealed a lack of comprehensive insights into the analysis of CEIs' impacts [52] and a notable absence of references to their spatial dimensions.

With the aim of better comprehending how to address the study of CEIs' spatial impacts, the present study reviewed published works on different types of CEIs and applied a holistic approach to analyze them.

Existing research on CEIs tends to focus on specific configurations and delves into sectoral analysis. An example is Oteman, Kooij, Wiering (2017) who studied the development of Dutch grassroot energy initiatives and their local and national network connections [78]. In contrast, Ison (2010) examined the governance of community projects within the UK Community Carbon Network [73]. Besides focusing on specific aspects, some studies may include brief references to other topics. The publications related to the professionalization dimension are a demonstration of this: most studies do not directly address it but offer some reflections.

However, this focus on individual sectors can hinder a broader understanding of CEIs as a whole. By analyzing studies on PEDs [89], energy cooperatives [79, 80, 82], the so-called local energy initiatives [52, 83, 84], energy communities [68, 81], community energy projects [73, 74] among others, and by distilling and categorizing the information contained in the publications, the present contribution "connects the dots" and captures the interrelations among the various aspects that characterize and affect a CEI, showing that CEIs' spatial impacts are intricately shaped by a multitude of both internal and external factors.

The resulting conceptual framework (Fig. 1) depicts the complex nature of CEIs' spatial impacts, emphasizing the importance of a nuanced understanding of CEIs to effectively address the impacts that they have at the city level.

The relevance of the study is therefore twofold. First, it considers sectorial studies on different types of CEIs. Second, by applying a holistic approach to the sectoral studies reviewed, it builds an integrated conceptual framework that facilitates the visualization of the multiple dimensions involved in CEIs' development and their interconnectedness.

Researchers, practitioners, and urban planners can already benefit from this preliminary version of the framework to gain a "bird's eye view" on a key topic in the global energy transition scenario. Moreover, the conceptualization lays the groundwork for further research to explore how and why CEIs generate spatial impacts.

Acknowledgments. This paper and related research have been conducted during and with the support of the Italian inter-university PhD course in Sustainable Development and Climate Change (link: www.phd-sdc.it, accessed on 5 January 2024).

The research activity is to be considered an integral part of the European H2020 Project "Institutionalized Integrated Sustainable Energy and Climate Action Plans" (2ISECAP). Grant Agreement (GA) No: 101033752.

Disclosure of Interests. The authors have no competing interests to declare that are relevant to the content of this article.

References

1. Hoffman, K.C., Wood, D.O.: Energy system modeling and forecasting. Ann. Rev. Energy **1**, 423–453 (1976)
2. Martínez, D.M., Ebenhack, B.W.: Understanding the role of energy consumption in human development through the use of saturation phenomena. Energy Policy **36**(4), 1430–1435 (2008)
3. Subramanian, A.S.R., Gundersen, T., Adams, T.A., II.: Modeling and simulation of energy systems: a review. Processes **6**, 238 (2018)
4. Bruckner, T., et al.: Energy systems. In: Edenhofer, O., et al. (eds.) Climate Change 2014: Mitigation of Climate Change; Contribution of Working Group III to the Fifth Assessment Report of the Intergovernmental Panel on Climate Change. Cambridge University Press, Cambridge, UK; New York, NY, USA (2014)
5. Sovacool, B.K.: How long will it take? Conceptualizing the temporal dynamics of energy transitions. Energy Res. Soc. Sci. **1**, 202–215 (2016)
6. Rohracher, H.: Energy systems in transition: contributions from social sciences. Int. J. Environ. Technol. Manage. (9), nos. 2/3 (2008). https://doi.org/10.1504/IJETM.2008.019026
7. Miller, C.A., et al.: The social dimensions of energy transitions. Sci. Cult. **22**(2), 135–148 (2013)
8. Miller, C.A., Richter, J.: Social planning for energy transitions. Curr. Sustain. Renew. Energy Rep. **1**, 77–84 (2014)
9. Laird, F.N.: Against transitions? Uncovering conflicts in changing. Energy Syst. Sci. Cult. **22**(2), 149–156 (2013). https://doi.org/10.1080/09505431.2013.786992
10. Bale, C.S.E., Varga, L., Foxton, T.J.: Energy and complexity: new ways forward. Appl. Energy **138**, 150–159 (2015)
11. Määttä, S.: Rethinking collaborative action and citizen empowerment: characterising a Whole-of-Society approach to the energy transition. Energy Res. Soc. Sci. **81**, 102277 (2021)
12. Balest, J., et al.: Local reflections on low-carbon energy systems: a systematic review of actors, processes, and networks of local societies. Energy Res. Soc. Sci. **42**, 170–181 (2018)
13. van de Poel, I., Taebi, B.: Value change in energy systems. Sci. Technol. Human Values **47**(3), 371–379 (2022)
14. Smil, V.: World history and energy. In: Encyclopedia of Energy, 6 (2010). https://vaclavsmil.com/wp-content/uploads/docs/smil-article-2004world-history-energy.pdf
15. Grubler, A., Wilson, C., Nemet, G.: Apples, oranges, and consistent comparisons of the temporal dynamics of energy transitions. Energy Res. Soc. Sci. **22**, 18–25 (2016)

16. Rutherford, J., Coutard, O.: Urban energy transitions: places, processes and politics of sociotechnical change. Urban Stud. **51**(7), 1353–1377 (2014)
17. Kamal-Chaoui, L., Robert, A. (eds.): Competitive Cities and Climate Change. OECD Regional Development Working Papers N° 2. OECD Publishing, © OECD (2009)
18. UN-Habitat. Cities and Climate Change Policy Directions. Global Report on Human Settlements 2011. Abridged Edition, UN-Habitat, Earthscan, London (2002)
19. Hoicka, C.E., MacArthur, J.L.: From tip to toes: mapping community energy models in Canada and New Zealand. Energy Policy **121**, 162–174 (2018)
20. IRENA. Rise of renewables in cities: Energy solutions for the urban future, International Renewable Energy Agency, Abu Dhabi (2020)
21. Hisschemöller, M., Sioziou, I.: Boundary organisations for resource mobilisation: enhancing citizens' involvement in the Dutch energy transition. Environ. Polit. **22** (5), 792-810 (2013)
22. Wagemans, D., Scholl, C., Vasseur, V.: Facilitating the energy transition—the governance role of local renewable energy cooperatives. Energies **12**(21), 4171 (2019)
23. Smith, I.D.: Energy transition and social movements: the rise of a community choice movement in California. In: Kurochkin, D., Shabliy, E., Shittu, E. (eds.) Renewable Energy. Palgrave Macmillan, Cham (2019)
24. Wahlund, M., Palm, J.: The role of energy democracy and energy citizenship for participatory energy transitions: a comprehensive review. Energy Res. Soc. Sci. **87**, 102482 (2022)
25. UNEP. Emissions Gap Report 2022: The Closing Window - Climate crisis calls for rapid-transformation of societies. Nairobi (2022). https://www.unep.org/emissions-gap-report-2022
26. Haf, S., Robison, R.: How Local Authorities can encourage citizen participation in energy transitions. UK Energy Research Centre, London (2020)
27. Ala-Juusela, M., Crosbie, T., Hukkalainen, M.: Defining and operationalising the concept of an energy positive neighbourhood. Energy Convers. Manage. **125**, 133–140 (2016)
28. Derkenbaeva, E., Vega, S.H., Hofstede, G.J., van Leeuwen, E.: Positive energy districts: mainstreaming energy transition in urban areas. Renew. Sustain. Energy Rev. **153**, 111782 (2022)
29. Caramizaru, A., Uihlein, A.: Energy Communities: An Overview of Energy and Social Innovation, EUR 30083. JRC Science for Policy Report, Publications Office of the European Union: Luxembourg (2020)
30. Leonhardt, R., Noble, B., Poelzer, G., Fitzpatrick, P., Belcher, K., Holdmann, G.: Advancing local energy transitions: a global review of government instruments supporting community energy. Energy Res. Soc. Sci. Res. Soc. Sci. **83**, 102350 (2022)
31. Hewitt, R.J., et al.: Social innovation in community energy in Europe: a review of the evidence. Front. Energy Res. **7**, 31 (2019)
32. Berka, A.L., Creamer, E.: Taking stock of the local impacts of community owned renewable energy: a review and research agenda. Renew. Sustain. Energy Rev. **82**, 3400-3419 (2018)
33. Seyfang, G., Park, J.J., Smith, A.: A thousand flowers blooming? An examination of community energy in the UK. Energy Policy **61**, 977–989 (2013)
34. Creamer, E., et al.: Community energy: entanglements of community, state, and private sector. Geogr. Comp. **12**(7), e12378 (2018)
35. European Economic and Social Committee (EESC). Changing the future of energy. Civil society as a main player in renewable energy generation. EESC study on the role of civil society in the implementation of the EU Renewable Energy Directive. Final Report, January 2015
36. McGovern, G., Klenke, T.: A process approach to mainstreaming civic energy. Energies **11**, 2914 (2018)

37. Hasanov, M., Zuidema, C.: Local collective action for sustainability transformations: emerging narratives from local energy initiatives in The Netherlands. Sustain. Sci. **17**, 2397–2410 (2022)
38. Otamendi-Irizar, I., Grijalba, O., Arias, A., Pennese, C., Hernández, R.: How can local energy communities promote sustainable development in European cities? Energy Res. Soc. Sci. **84**, 102363 (2022)
39. Gui, E.M., McGill, I.: Typology of future clean energy communities: an exploratory structure, opportunities, and challenges. Energy Res. Soc. Sci. **35**, 94-107 (2018)
40. Dóci, G., Vasileiadou, E., Petersen, A.C.: Exploring the transition potential of renewable energy communities. Futures **66**, 85–95 (2015). https://doi.org/10.1016/j.futures.2015.01.002
41. Wierling, A., Schwanitz, V.J., Zeiss, J.P., et al.: A Europe-wide inventory of citizen-led energy action with data from 29 countries and over 10000 initiatives. Sci. Data **10**(9) (2023)
42. REScoop Homepage. www.rescoop.eu. Accessed 05 Feb 2024
43. Campos, I., Marín-González, E.: People in transitions: energy citizenship, prosumerism and social movements in Europe. Energy Res. Soc. Sci. **69**, 101718 (2020)
44. Capellán-Pérez, I., de Castro, C., Arto, I.: Assessing vulnerabilities and limits in the transition to renewable energies: land requirements under 100% solar energy scenarios. Renew. Sustain. Energy Rev. **77**, 760–782 (2017)
45. Gross, S.: Renewables, land use, and local opposition in the United States, Brookings Institution. United States of America. Accessed https://policycommons.net/artifacts/4141267/renewables-land-use-and-local-opposition-in-the-united-states/4949427/
46. Jiwon, K., et al.: How to manage land use conflict between ecosystem and sustainable energy for low carbon transition?: Net present value analysis for ecosystem service and energy supply. Front. Environ. Sci. **10**, 1044928 (2022)
47. van de Ven, D.J., Capellan-Peréz, I., Arto, I., et al.: The potential land requirements and related land use change emissions of solar energy. Sci. Rep. **11**, 2907 (2021)
48. Gaur, V., Lang, C.: Property Value Impacts of Commercial-Scale Solar Energy in Massachusetts and Rhode Island. Submitted to University of Rhode Island Cooperative Extension on September 29, 2020. Accessed https://web.uri.edu/coopext/valuing-sitingoptions-for-commercial-scale-solar-energy-in-rhode-island/
49. Chen, Y., Kirkerud, J.G., Bolkesjø, T.F.: Balancing GHG mitigation and land-use conflicts: alternative Northern European energy system scenarios. Appl. Energy **310**, 118557 (2022)
50. Koelman, M., Hartmann, T., Spit, T.J.M.: Land use conflicts in the energy transition: Dutch dilemma's. TEMA J. Land Use Mobil. Environ. **11**(3), 273–284 (2018)
51. Leskinen, N., Vimpari, J., Junnila, S.: The impact of renewable on-site energy production on property values. J. Eur. Real Estate Res. **13**(3), 337–356 (2020)
52. Germes, L.A.M.H., Wiekens, C.J., Horlings, L.G.: Success, failure, and impact of local energy initiatives in The Netherlands. Sustainability **13**(22), 12482 (2021)
53. Johnson, T.P.: Snowball sampling: introduction†. In: Balakrishnan, N., Colton, T., Everitt, B., Piegorsch, W., Ruggeri, F., Teugels, J.L. (eds.) Wiley StatsRef: Statistics Reference Online (2014)
54. Snyder, H.: Literature review as a research methodology: an overview and guidelines. J. Bus. Res. **104**, 333–339 (2019)
55. Torraco, R.J.: Writing integrative literature reviews: guidelines and examples. Hum. Resour. Dev. Rev.Resour. Dev. Rev. **4**(3), 356–367 (2005)
56. IRENA Coalition for action. Community energy: Broadening the ownership of renewables (2018). https://coalition.irena.org/-/media/Files/IRENA/Coalition-for-Action/Publication/Coalition-for-Action_Community-Energy_2018.pdf
57. van der Waal, E.C.: Local impact of community renewable energy: a case study of an Orcadian community-led wind scheme. Energy Policy **138**, 111193 (2020)

58. Roberto, R., Ferruzzi, G., Negro, V., Noussan, M.: Mapping of energy community development in Europe: state of the art and research directions. Energies **16**(18), 6554 (2023)
59. Bere, J., Jones, C., Jones, S.: The economic and social impact of small and community hydro in Wales (2015). https://www.deg.wales/wp-content/uploads/2015/09/ImpactofSmalland-CommunityHydroinWales.pdf
60. Bielig, M., et al.: Evidence behind the narrative: critically reviewing the social impact of energy communities in Europe. Energy Res. Soc. Sci. **94**, 102859 (2022)
61. Walker, G., Devine-Wright, P.: Community renewable energy: what should it mean? Energy Policy **36**(2), 497–500 (2008)
62. Leicester, P., Goodier, C., Rowley, P.: Evaluating the impacts of community renewable energy initiatives. Loughborough University. Conference contribution (2011)
63. Schwanitz, V.J., et al.: Statistical evidence for the contribution of citizen-led initiatives and projects to the energy transition in Europe. Sci. Rep. **13**(1), 1342 (2023)
64. Interreg Europe. Empowering Citizens for Energy Communities. A Policy Brief from the Policy Learning Platform on Low-carbon economy (2022). Low-carbon economy https://www.interregeurope.eu/sites/default/files/good_practices/PolicyBrief_REContmu-nities_final.pdf
65. Weckesser, T., et al.: Renewable energy communities: optimal sizing and distribution grid impact of photovoltaics and battery storage. Appl. Energy **301**, 18 (2021)
66. Backe, S., et al.: Impact of energy communities on the European electricity and heating system decarbonization pathway: comparing local and global flexibility responses. Appl. Energy **323**(C), 119470 (2022). Elsevier
67. Hasanov, M.: All together now! community initiatives engaging in sustainability transformations in the Netherlands. [Thesis fully internal (DIV), University of Groningen]. University of Groningen (2021)
68. Hanke, F., Guyet, R.: The struggle of energy communities to enhance energy justice: insights from 113 German cases. Energ. Sustain. Soc. **13**, 16 (2023)
69. Allen, A.R., Hammond, G.P., Mc Manus, M.C.: Prospects for and barriers to domestic microgeneration: a United Kingdom perspective. Appl. Energy **85**(6), 528–544 (2008)
70. Li, L.W., Birmele, J., Schaich, H., Konold, W.: Transitioning to community-owned renewable energy: lessons from Germany. Procedia Environ. Sci. **17**, 719–772 (2013)
71. Verbong, G.P.J., Geels, F.W.: Exploring sustainability transitions in the electricity sector with socio-technical pathways. Technol. Forecast. Soc. Chang. **77**(8), 1214–1221 (2010)
72. Gjorgievski, V.Z., Cundeva, S., Georghiou, G.E.: Social arrangements, technical designs and impacts of energy communities: a review. Renew. Energy **169**(C), 1138–1156 (2021) . Elsevier
73. Ison, N.: Governance of Community Energy Projects. Independent Research Project, Lancaster Environment Centre (2010)
74. Radtke, J.: A closer look inside collaborative action: civic engagement and participation in community energy initiatives. People Place Policy **8**, 235–248 (2014)
75. Hanke, F., Guyet, R., Feenstra, M.: Do renewable energy communities deliver energy jus-tice? Exploring insights from 71 European cases. Energy Res. Soc. Sci. **80**, 102244 (2021)
76. Hicks, J., Ison, N.: An exploration of the boundaries of 'community' in community renewable energy projects: navigating between motivations and context. Energy Policy **113**(C), 523–534 (2018) . Elsevier
77. Horstink, L., et al.: Collective renewable energy prosumers and the promises of the energy union: taking stock. Energies **13**(2), 421 (2020)
78. Oteman, M., Kooij, H.-J., Wiering, M.A.: Pioneering renewable energy in an economic energy policy system: the history and development of Dutch grassroots initiatives. Sustainability **9**(4), 550 (2017)
79. Heras-Saizarbitoria, I., Sáez, L., Allur, E., Morandeira, J.: The emergence of renewable energy cooperatives in Spain: a review. Renew. Sustain. Energy Rev. **94**(C), 1036–1043 (2018) . Elsevier

80. Ahlemeyer, K., Griese, K.M., Wawer, T., et al.: Success factors of citizen energy cooperatives in north western Germany: a conceptual and empirical review. Energ. Sustain. Soc. **12**, 29 (2022)
81. Candelise, C., Ruggieri, G.: Status and evolution of the community energy sector in Italy. Energies **13**(8), 1888 (2020)
82. de Bakker, M., Lagendijk, A., Wiering, M.: Cooperatives, incumbency, or market hybridity: new alliances in the Dutch energy provision. Energy Res. Soc. Sci. **61**, 101345 (2020)
83. van der Weele, E.: Scaling of Local Frisian Energy Initiatives [Master Thesis, University of Groningen] (2022)
84. Dappe, K.: Master thesis: Expectations of Energy communities [Master Thesis, University of Utecht] (2022)
85. Munro, F.R., Cairney, P.: A systematic review of energy systems: the role of policymaking in sustainable transitions. Renew. Sustain. Energy Rev. **119**, 109598 (2020)
86. Petrovics, D., Huitema, D., Jordan, A.: Polycentric energy governance: under what conditions do energy communities scale? Environ. Policy Gov. **32**(5), 438–449 (2022)
87. Wirth, S.: Communities matter: institutional preconditions for community renewable energy. Energy Policy **70**(C), 236–246 (2014) . Elsevier
88. Koirala, B.P., et al.: Energetic communities for community energy: a review of key issues and trends shaping integrated community energy systems. Renew. Sustain. Energy Rev. **56**, 722–744 (2016)
89. Leone, F., et al.: Supporting cities towards carbon neutral transition through territorial acupuncture. Sustainability **15**(5), 4046 (2023)

Planning Renewable Energy Communities in Intermediate Areas. A Case of Study

Alessandra Marra(✉) and Domenica Luciano

Department of Civil Engineering, University of Salerno, 84084 Fisciano, Salerno, Italy
almarra@unisa.it

Abstract. Renewable Energy Communities (RECs) are a promising instrument for decarbonizing urban areas and reduce energy poverty. This contribution is part of a larger research program, where it was previously presented a methodology to identify priority areas in which to foster the RECs promotion through the Municipal Urban Plan. The specific objective of this work is to evaluate the transferability of this methodological path to the territorial context of intermediate areas, located halfway between urban poles and inland areas. So, the methodology is applied to the case study of Fisciano Municipality, in Campania Region (Italy), in the context of studies for the Municipal Urban Plan formation. From this application, it is possible to conclude that the transferability of the above methodology is also possible to intermediate contexts. However, the results obtained offer further insights, especially the need to diversify the incentives and reward mechanisms to be provided for the RECs development in the plan, in relation to the different territorial contexts in which the planning activity is carried out. In this way, it is possible to ensure a just energy transition in urban areas, avoiding the generation of further inequalities.

Keywords: Energy Poverty · Renewable Energy Communities · Urban Planning

1 Introduction

In the international scientific debate, as well as in recommendations and guidelines promulgated by public entities and global organizations, Renewable Energy Communities (RECs) are recognized as bringing innumerable social, environmental and economic benefits to the local areas in which they operate and to the populations living there. The main benefits concern the reduction of climate-changing emissions (environmental benefits) and energy poverty (social and economic benefits). These benefits derive indirectly from those that RECs directly bring: the increase in decentralised plants for the production of energy from renewable sources; the energy efficiency of existing buildings; the reduction of the bill cost for the community members [1–5].

According to the definition given by the European Commission with Directive EU/2001/2018, known as "RED II", RECs can be defined as spatially concentrated networks of users - citizens, small and medium-sized enterprises, municipal administrations - which, on a voluntary basis, aggregate to self-produce, self-consume and share

O. Gervasi et al. (Eds.): ICCSA 2024 Workshops, LNCS 14818, pp. 256–271, 2024.
https://doi.org/10.1007/978-3-031-65273-8_17

energy, produced locally by plants powered by renewable sources, with the main purpose of receiving the aforementioned benefits, rather than economic profits [6].

Since RECs respond to issues that afflict urban and metropolitan areas with increasing intensity, they are a promising tool for planners to put into practice the concept of Sustainable Cities, as advocated by the 2030 Agenda.

Promoting RECs in urban planning is therefore a way to achieve goals 1 "No poverty", 7 "Affordable and clean energy" and 11 "Sustainable cities and communities" [7], but also to reach the carbon neutrality goals set by the European Union [8].

Yet, the topic is approached by researchers with a prevalent interest in energy, technological or cost-effectiveness issues, rather than urban planning. Only recently, the RECs tool has sparked interest in academic research in the field of planning [9–11].

In Italy, this trend should be read in the light of a historically difficult integration between sectoral energy planning and general planning for development, at various levels, both at regional and urban scales [12]. The reversal of this trend is necessary so that planning itself does not become an obstacle to the spread of RECs, slowing down the approval processes, as stated both by scholars and in the Italian Energy and Climate Plan [13–15].

The work is part of a larger ongoing research project, aimed at promoting RECs through general urban planning. As part of this project, a methodology was proposed to spatially identify priority areas, or target areas, in which to direct this action. The methodological proposal is only briefly described in Sect. 2 for greater clarity, while for its deeper knowledge reference should be made to Gerundo & Marra [16], which is the study in which it was applied for the first time.

2 Methodological Path and Aim of the Work

The *priority areas*, mapped at an infra-urban scale, are intended to be located where a minimization of constraints and a maximization of the benefits deriving from the RECs establishment is expected, with particular reference to the reduction of energy poverty. Therefore, the proposed methodology is divided into three main Macro-phases (I, II, III), which consist, respectively, in the construction of the following maps:

- *Energy Poverty Map*, representative of the most critical urban areas from an energy and social point of view (I);
- *Constraints Map*, summarizing all the urban and regional constraints existing to the RECs establishment, with regard to the areas not suitable for the installation of RES plants, i.e. for the production of Energy from Renewable Sources (II);
- *Priority Areas map*, built from previously obtained maps (III).

The *Energy Poverty Map* is obtained by selecting urban areas with the worst energy consumption and low-income urban areas (I). The most energy-intensive areas are defined as the districts with a building-residential stock characterized by high consumption for heating and domestic hot water production. The identification of these areas is carried out by tracing the stock of residential buildings of each census section to a typical building, or archetype, representative of the entire section, on the basis of the prevailing building type and age class, calculated in a GIS environment. The energy consumption is then estimated, in accordance with current legislation, on the basis of the climatic zone and the stratigraphic, technological and plant characteristics of the identified archetypes. Existing Public Housing (ERP) and/or Social Housing (ERS) districts are instead intended to be representative of low-income urban areas.

The *Constraints Map* is constructed by selecting the constraint factors that characterize the so-called "no RES zones" according to the Italian reference legislation, Interministerial Decree (D.I.) of 10 September 2010, entitled "Guidelines for the authorization of plants powered by renewable sources": protected areas, forest areas, water bodies, areas with high agricultural potential, areas and buildings of landscape or cultural interest; areas of high hydrogeological hazard and/or risk (II).

In the last macro-phase, the *Priority Areas Map* is obtained through an overlay mapping between the Energy Poverty Map and the Constraints Map. This map makes it possible to identify different degrees of priority: urban areas characterized by critical energy and/or social performance and, at the same time, not affected by the areas where the installation of RES plants is not allowed, represent those areas where the greatest effectiveness is expected, thus presenting the highest degree of priority (III).

The source of the data for the application of the methodology described is represented by the graphic elaborations produced in the ordinary activity of municipal urban planning.

The proposed method has already been applied to the Municipality of Pagani, belonging to the populous conurbation of the Agro-Nocerino-Sarnese, in the Campania Region (Italy), and to the small town of Andretta, a typical inland area of the same region, as documented in two previous works, to which reference should be made for the respective in-depth studies [16, 17].

The specific objective of this paper is to evaluate the transferability of the proposed methodological path to a different territorial context, halfway among those examined above, in order to complete the testing phase of the model. To this end, the method is applied to the Municipality of Fisciano, a case study that can be defined as intermediate between urban poles and inland areas [18, 19] of the same region. The case study is presented in Sect. 3, while Sect. 4 describes the performed application and the main results, discussing them in relation to the aim of the work. Section 5 reports the conclusions and possible future developments of the research.

3 A Case of Study

The municipality of Fisciano, which covers 31.69 km^2, is located north of the city of Salerno, capital of the province of the same name, and on the border with the province of Avellino, in the Campania Region (Italy). Due to its strategic position, the municipal area was chosen as the location of the main campus of the University of Salerno, a reference point for higher education for the two provinces. In addition, there is an extensive Industrial Development Area (ASI), which extends for 2.4 km^2, including part of the territory of the neighboring municipality of Mercato San Severino.

These circumstances have led to a significant demographic evolution for the municipality, which has gone from 8,000 inhabitants in 1971 to 12,275in 2011, up to 14,064 current residents, according to data released annually by the Italian Institute of Statistics (ISTAT). The main inhabited centre of Fisciano, almost close to the university campus, is characterized as a watershed between two very different geographical and socio-economic contexts, both within the municipality in question and in the wider area including the neighboring municipalities: to the south-west the conurbation of the Irno River valley, developed in the municipalities around the coastal city of Salerno, with increasing demographic dynamics, albeit less rapidly over the last decade; to the north-east the Picentini Mountains, which lap the southern part of the province of Avellino, occupied by numerous inland municipalities, characterized by a severe demographic decline (Fig. 1). This barrier is reflected in the variation in the altitude of the territory, from a maximum of about 1,550 m above sea level to a minimum of 132 m above sea level, and in the articulation of the settlement system, denser in the south-eastern part of the municipality, which falls within the climatic zone D (1669 degree days).

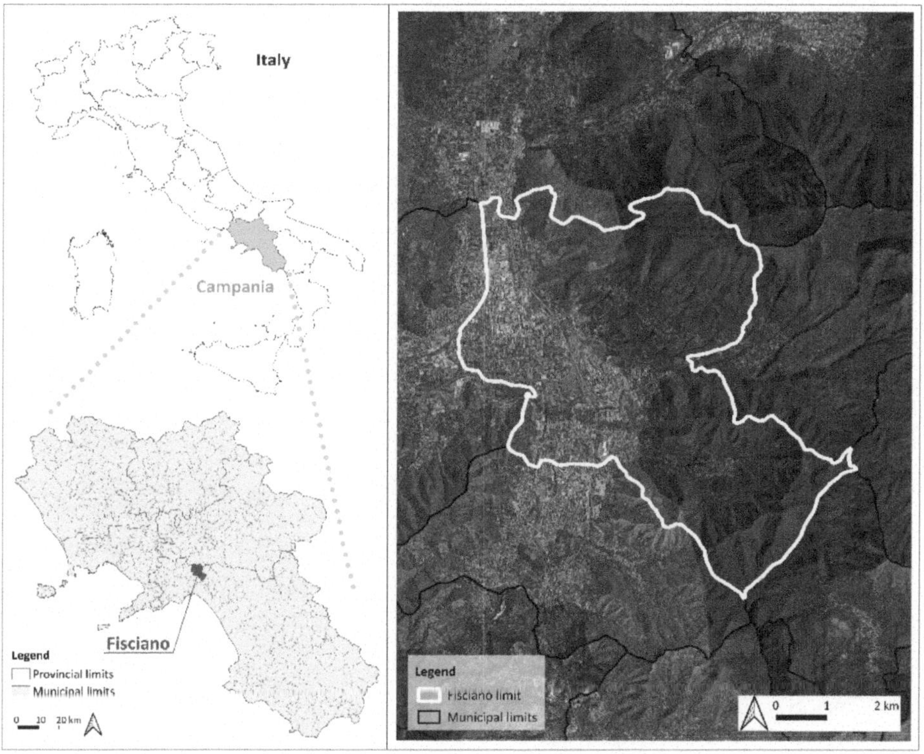

Fig. 1. The study case of Fisciano Municipality in Campania Region (Italy).

Fisciano has a General Regulatory Plan (PRG) approved in 2008. In 2018, the Municipality started the drafting of the new Municipal Urban Plan (PUC) to adapt to the changed regional legislation and to update the obsolete PRG, approving the preliminary proposal (Preliminary of PUC). In 2022, the Municipal Administration requested technical-scientific support from the Department of Civil Engineering of the University of Salerno (DiCiv), for the drafting of the PUC final design [20]. The application experimentation covered by this paper is part of the studies carried out during the above-mentioned support activity.

3.1 Input Data and Processing

As anticipated in the introduction of this paper, the input data necessary for the application of the proposed methodology can be extrapolated from the graphic elaborations of Fisciano Municipal Urban Plan. In particular, the following documents were used for the Energy Poverty mapping:

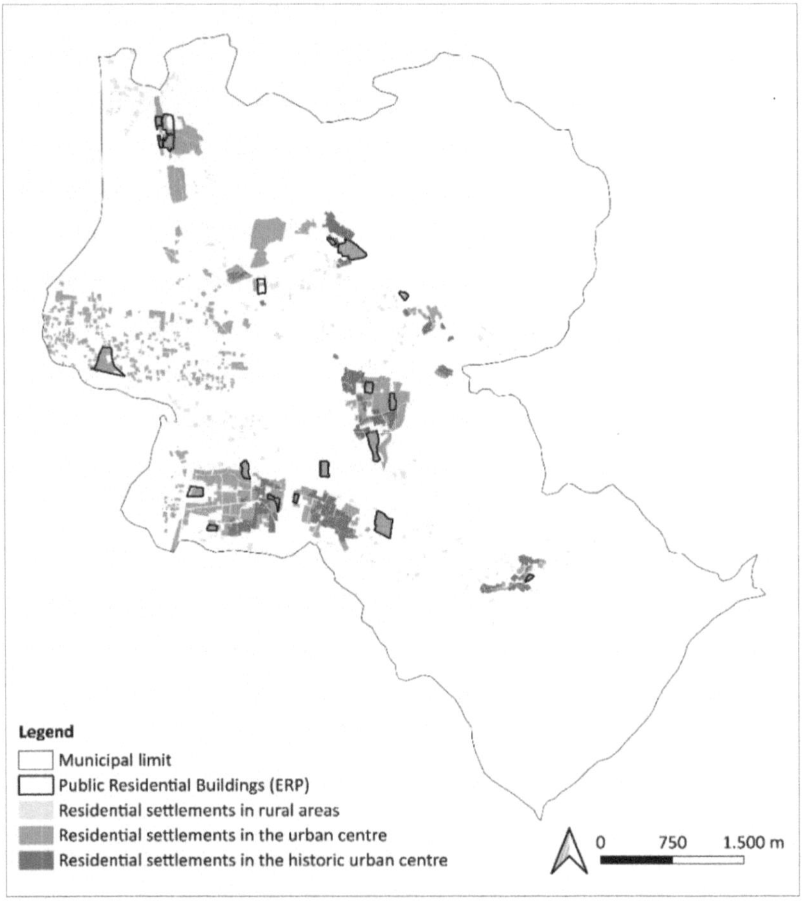

Fig. 2. Residential Urban Land Use Map. Source: Own elaboration on data from the "Urban Land Use Map" of Fisciano Municipal Urban Plan.

- Regional Technical Numerical Cartography, provided by the Campania Region to the Municipalities, from which to extract the information and geometric attributes of the buildings in the municipal area;
- Urban Land Use Map, from which to derive the prevailing use of the existing building, in order to isolate the mostly residential buildings from those for other uses. In addition, the 20 districts occupied by ERP housing, scattered throughout many of the municipality's urban centres, have been identified on that map (Fig. 2).

Subsequently, the data of some census variables were collected, selected from all census data provided by ISTAT [21], which divided the territory of Fisciano into 32 census tracts. However, it was possible to consider 27 of them in the analysis, as 3 lacked data and 2 had no residential buildings (Fig. 3).

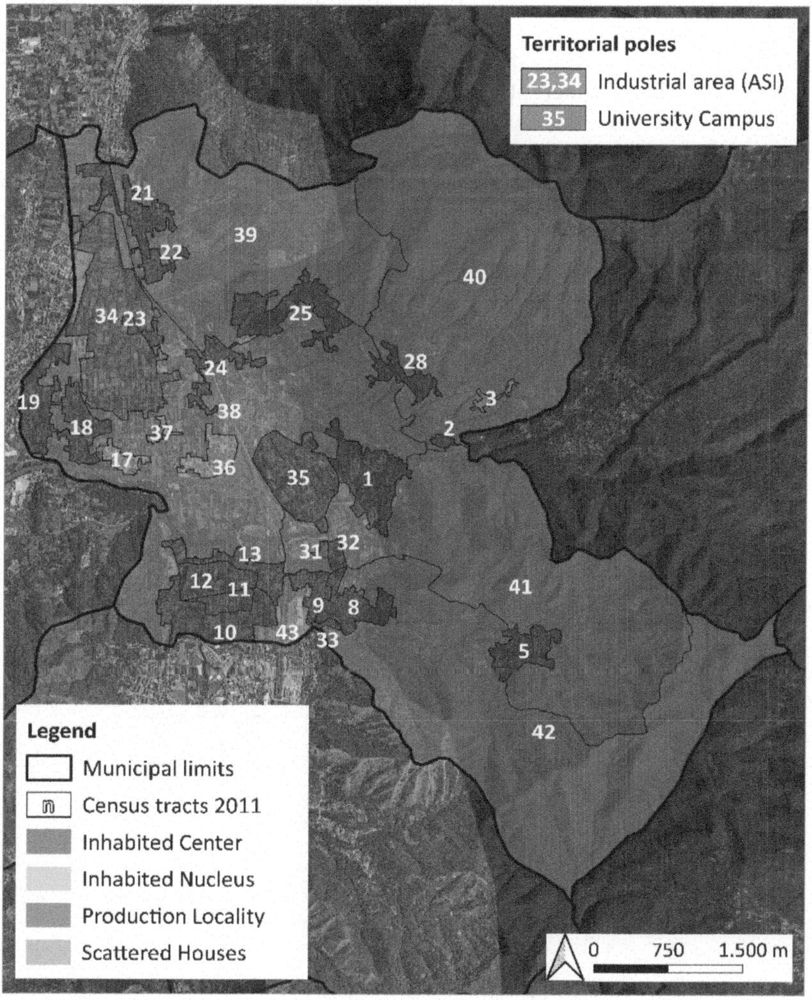

Fig. 3. Census tracts Map.

More precisely, the extrapolated data are as follows:

- number of residential buildings classified according to the number of floors;
- Number of residential buildings by age group.

Finally, for the identification of "No FER" areas and the mapping of Constraints, the data are extrapolated from the following graphic elaborations of the Fisciano Municipal Urban Plan:

- Hydrogeological hazard and risk maps, taken from the elaborations of the Hydrogeological Stability Plan, drawn up by the territorial competent Basin Authority, the one of the Sarno River;
- Map of agricultural land use, constructed on a preliminary basis by analyzing the Agricultural Land Use Map of the Campania Region;
- Protection system, a map made up of the constraints and buffer areas that insist on the municipal territory;
- Emergency system, i.e. resources of architectural and historic value;
- System of naturalness, inherent in the areas of environmental protection (Fig. 4).

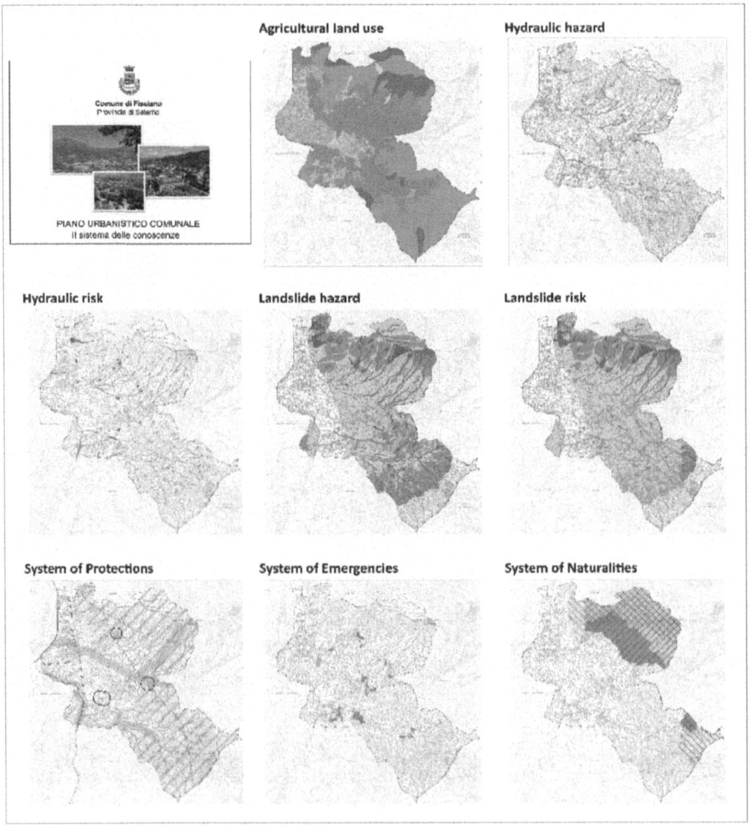

Fig. 4. Input data for Constraints Map. Source: Own elaboration on data from Fisciano Municipal Urban Plan.

The collected input data were processed in a GIS environment, using the commercial software ArcGis [22] and the open source software QGis [23].

4 Results and Discussion

The application of the proposed methodology to the Municipality of Fisciano involved the elaboration of the Energy Poverty Map (I), the Constraints Map (II) and the Priority Areas Map (III), as specified below.

4.1 Energy Poverty Map

From the analysis performed according to the first Macro-phase (I), it emerged how the prevailing building typology is that of multi-family buildings, more precisely buildings from 2 to 5 floors and up to 20 apartments. In addition, buildings for mainly residential use were built, on average, between the 1950s and 1990s, with a prevalence in the time interval 1961–1975. By intersecting the data relating to the building type and age class thus obtained, a building-type is associated with each census section, consisting of homogeneous stratigraphic, technological and plant characteristics.

For each building-type, the energy consumption for heating and domestic hot water production, expressed in kWh/m^2 per year, was assessed on the basis of the relevant technical literature [24].

The total consumption is calculated by multiplying the consumption value of the building-type representative of each section by the number of residential buildings in the same section. The resulted values were divided into five classes, from "Very Low" to "Very High", obtaining the map of the overall energy consumption. This classification was carried out in the QGIS environment using the Natural Breaks algorithm, which allows to identify natural groupings of the data, maximizing the variance between the classes [25].

The *Energy Poverty Map* is obtained by cross-referencing the Residential Land Use Map in Fig. 2 with the Map of the overall energy consumption. Of the latter, the 2 worst classes, "High" and "Very High", are considered.

From *the Energy Poverty Map* (Fig. 5) it is possible to deduce three types of priority areas, i.e. those most critical from an energy point of view, in the social dimension, or those where both profiles are problematic.

Compared to the case studies already analyzed, Pagani [16] and Andretta [17], the methodology proposed for Macrophase I was easily transferable to the case study of Fisciano. In fact, the input data necessary for the evaluation of energy consumption derive from official datasets, released and periodically updated by public bodies and entities, or from the Municipal Urban Plan ordinary elaborations.

It should also be noted that the complexity of the analysis is almost between those of the two cases previously analyzed. In fact, the Municipality of Fisciano has 32 census tracts, the spatial unit chosen for mapping, compared to the 125 into which the Municipality of Pagani is divided and the 15 of Andretta.

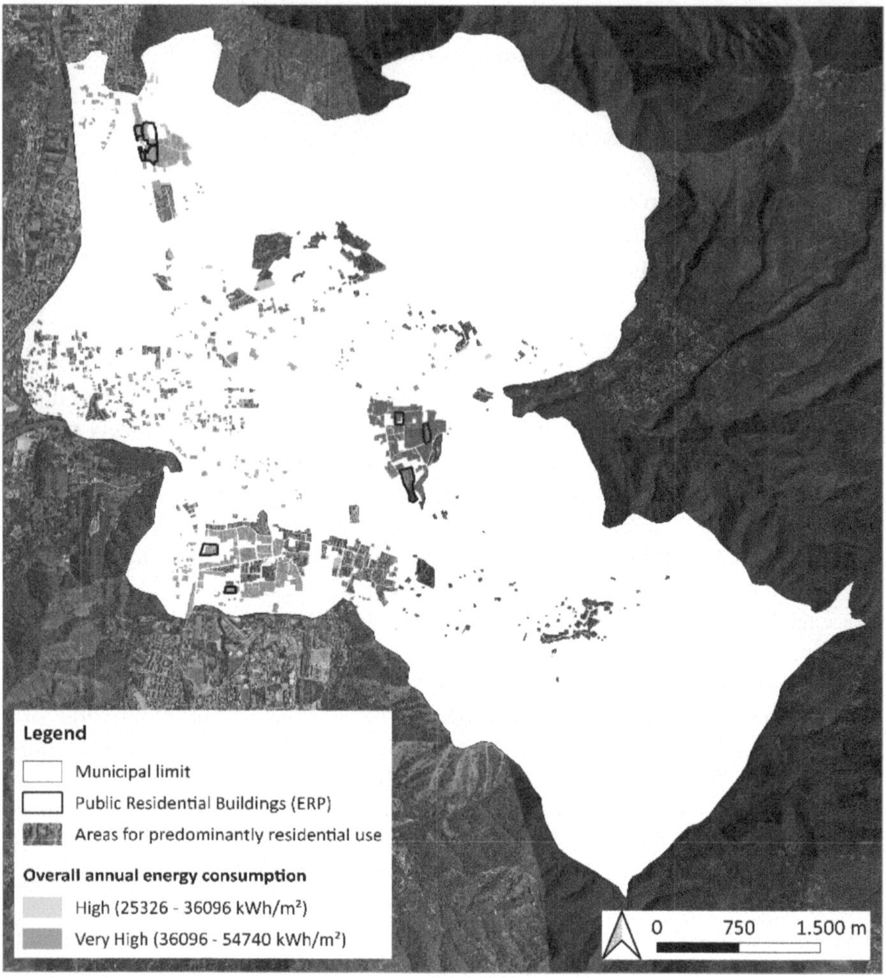

Fig. 5. Energy Poverty Map.

4.2 Constraints Map

The Constraints Map has been constructed considering the unsuitable areas for the localisation of RES plants, as mentioned in Annex 3 of the D.I. of 10 September 2010.

The map shows how the constraints are located mainly in the mountainous areas, home to the Monti Picentini Regional Park and the areas belonging to the Natura 2000 Network. In particular, the Municipality of Fisciano is crossed by the Site of Community Importance "Monte Mai and Monte Monna", and by the Special Protection Area "Picentini", which have a spatial relationship of partial overlap between them.

The areas at risk and hazard from landslides are also significantly extensive, even if only the worst classes have been considered (P3, P4; R3, R4).

In addition, most of the territory has areas affected by quality and/or particularly valuable agricultural-food production, especially in the types of orchards (Fig. 6).

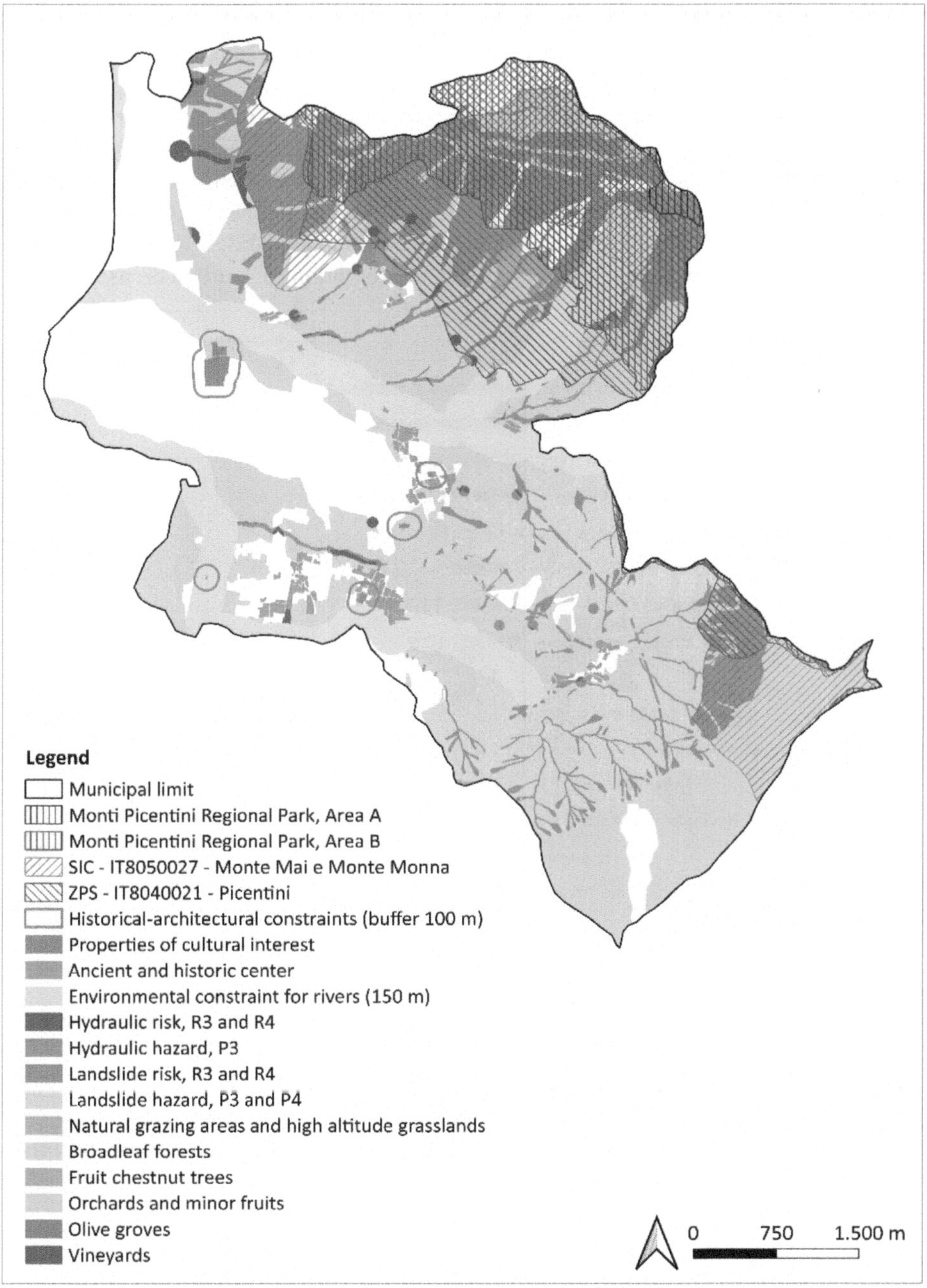

Fig. 6. Constraints Map.

On the contrary, historic centres and sites of cultural and architectural interest, as well as other types of cultivation, occupy a very small surface area, compared to the constraints mentioned above (Fig. 7).

From a constraint point of view, the case of Fisciano is more similar to that of Andretta, therefore to a typical context of the Italian inland areas, rich in natural and environmental resources. The difference is that the constraints are limited to the mountainous portion of the Fisciano territory, affecting the main inhabited centers to a limited extent. Exceptions are the areas at hazard and/or risk from landslides, which extensively involve the Fisciano municipal territory, also affecting numerous urban areas.

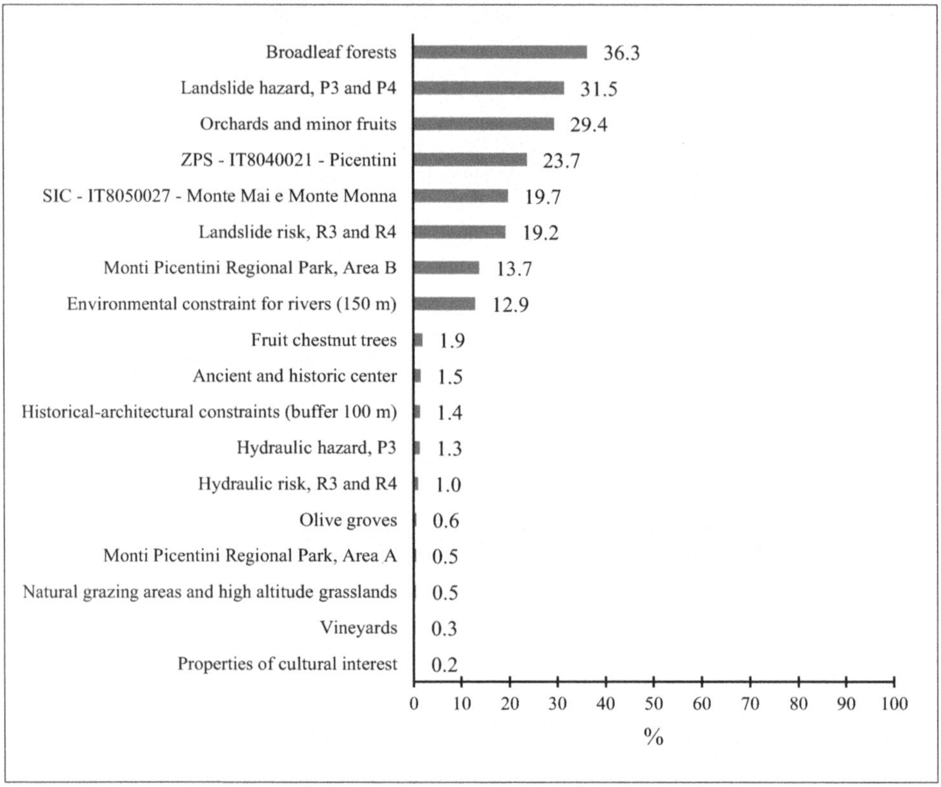

Fig. 7. Percentage of the area occupied by urban and regional constraints, compared to the surface of the entire municipal territory.

4.3 Priority Areas Map

The Priority Areas Map (Fig. 8) shows that the areas with the highest priority correspond to the low-income districts located in the urban centers of the localities of Fisciano, Soccorso and Bolano, included in census sections no. 1, 21 and 10, respectively. These are 7 districts out of a total of 20 ERP districts: 11 of them are located in areas characterized by modest energy consumption, while only 2 are burdened by constraint conditions. The inhabited centre of Fisciano, which was found to be the one with the highest energy consumption, is instead affected by large portions of the territory excluded from the possibility of locating RES plants.

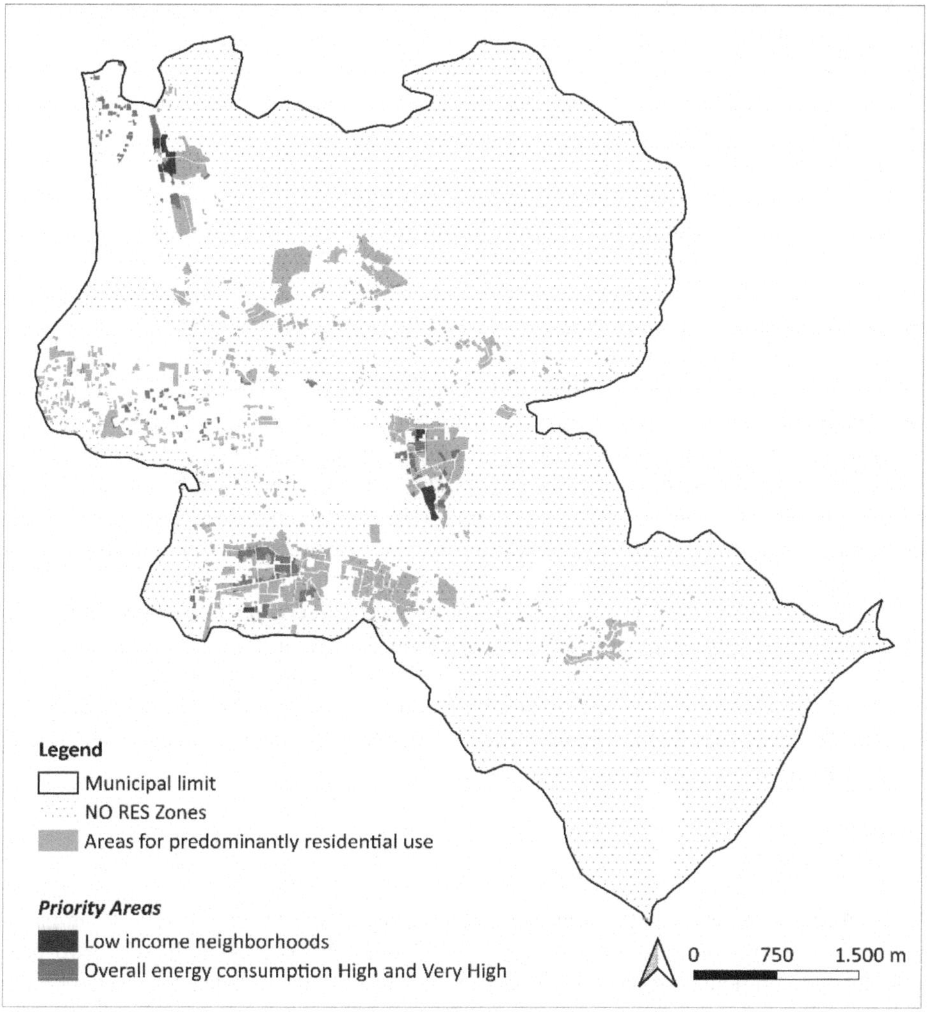

Fig. 8. Priority Areas Map.

In this scenario, it should be reiterated that this aspect does not represent a limit to the diffusion of RECs. In fact, the installations serving the RECs may be located in such a way as to admit to the community also members whose buildings are located in restricted areas.

Nevertheless, in order to receive state economic incentives, it is necessary to respect the spatial constraint that the energy produced is self-consumed within the area of the "primary" substation (high-voltage substation) of reference, in accordance with the most up-to-date Italian legislation. These are, more precisely, Legislative Decree 8 November 2021 n. 199, which transposed the European Directive "RED II", and Ministerial Decree 7 December 2023 n. 414, in implementation of the transposition decree and state aid provided for by Next Generation Italy [26, 27].

As can be seen from the map of the areas served by the primary substations [28], the municipality of Fisciano falls within a portion of the region in which these areas have an inter-municipal dimension. This characteristic is common to Italian inland areas, even if the number of municipalities included in the areas served by the same high-voltage substation is higher than in intermediate contexts.

Moreover, Fisciano is not included among the "critical" municipalities, i.e. saturated in terms of connection to the existing national grid [29].

However, the main inhabited centre of Fisciano, home to the municipal hall, straddles two different reference areas, shared with the neighboring municipalities, located respectively north-east and south-west of the demarcation border. This particular location reflects its position as a watershed between the populous Irno Valley and the municipalities of Picentini, close to the inland areas of the province of Avellino. Unless in the future the boundaries of the areas served by the primary substations are harmonized with the municipal administrative limits, this circumstance foreshadows a possible locational disadvantage of some districts compared to others, in addition to the social and energy problems considered here: take, for example, the case of a priority urban area for high energy consumption and/or inhabited by vulnerable households, which is both unsuitable for the installation of RES systems and is part of a primary substation falling almost entirely within a municipality other than the one to which it belongs.

These 'proximity' disadvantages should be taken into account when promoting RECs through municipal urban planning instrumentation, for example, by providing specific incentive mechanisms for the areas affected by them, in order to promote a fair energy transition, not a harbinger of further urban inequalities [30]. In this direction, the prior direct involvement of potential communities is desirable, also in innovative forms [31-33] and by activating inter-municipal synergies.

5 Conclusion

This contribution concerns the application to the Municipality of Fisciano (Campania Region, Italy) of a methodology to spatially identify priority areas in which to promote the development of Renewable Energy Communities through the Municipal Urban Plan, already applied to the case studies of Pagani [16] and Andretta [17], inserted in completely different geographical and socio-economic contexts.

The Priority Areas Map carried out integrates the cognitive framework of the aforementioned plan and allows to support the public decision-maker in the definition of incentive mechanisms, which differ according to the different priorities of intervention found. From the results obtained for the case study of Fisciano, it is possible to conclude that the transferability of the above methodology is also possible to intermediate contexts between urban poles and inland areas.

However, the results of the application performed in this work have offered further insights, especially with regard to the need to diversify the incentives and reward mechanisms to be provided for in the plan, in relation to the different territorial contexts in which the planning activity is carried out.

The definition of these rewards is a possible future development of the work [34, 35], in order to assure a just energy transition of urban areas, avoiding the generation of further inequalities and societal issues.

Acknowledgments. The input data necessary for the implementation of the proposed model derive from the elaborations carried out within the research project "Agreement between DiCiv and municipality of Fisciano for studies and research to support the formation of the Municipal Urban Plan". Roberto Gerundo is the project scientific responsible, while Alessandra Marra is principal investigator and coordinator of the research group. Domenica Luciano collaborated on these studies during the development of her Master Degree thesis, under the supervision of Alessandra Marra.

Authors Contributions. Conceptualization, A.M.; methodology, A.M.; software, A.M. and D.L.; validation, A.M.; formal analysis, A.M. and D.L.; investigation, A.M.; data curation, A.M.; writing—original draft preparation, A.M. (All Sections) and D.L. (Sects. 3, 4); writing—review and editing, A.M.; visualization, A.M. and D.L.; supervision, A.M.; project administration, A.M.; funding acquisition, A.M. All authors have read and agreed to the published version of the manuscript.

Disclosure of Interests. The authors have no competing interests to declare that are relevant to the content of this article.

References

1. Brummer, V.: Community energy - benefits and barriers: a comparative literature review of Community Energy in the UK, Germany and the USA, the benefits it provides for society and the barriers it faces. Renew. Sust. Energ. Rev. **94**, 187–196 (2018). https://doi.org/10.1016/j.rser.2018.06.013
2. Koltunov, M., Bisello, A.: Multiple impacts of energy communities: conceptualization taxonomy and assessment examples. In: Bevilacqua, C., Calabrò, F., Della Spina, L. (eds.) New Metropolitan Perspectives, NMP 2020. SIST, vol. 178, pp. 1081–1096. Springer, Cham (2021). https://doi.org/10.1007/978-3-030-48279-4_101
3. Barroco, F., Cappellaro, F., Palumbo, C. (eds.): Le comunità energetiche in Italia. Una guida per orientare i cittadini nel nuovo mercato dell'energia [Energy communities in Italy. A guide to orient citizens in the new energy market] (2020)
4. RSE - Research on the Energy System: Le Comunità Energetiche in Italia; Note per il Coinvolgimento dei Cittadini Nella Transizione Energetica [Energy Communities in Italy; Notes for Citizen Engagement in the Energy Transition]. Editrice Alkes, Milan (2021)
5. Ahmed, S., Wings, A., D'Angola, A.: A review of renewable energy communities: concepts, scope, progress, challenges, and recommendations. Sustainability **16**(5), 174 (2024). https://doi.org/10.3390/su16051749
6. EU- European Commission: Directive (EU) 2018/2001 of the European Parliament and of the Council of 11 December 2018 on the Promotion of the Use of Energy from Renewable Sources. EU, Brussels (2018)

7. UN – United Nations: Transforming our world: The 2030 Agenda for Sustainable Development. UN, New York, USA (2015)
8. EU, European Commission: Regulation 2021/1119 of the European Parliament and of the Council of 30 June 2021 establishing the framework for achieving climate neutrality and amending Regulations (EC) No 401/2009 and (EU) 2018/1999 («European Climate Law»). EU, Brussels (2021)
9. Curreli, S., Zoppi, C.: Carbone e pianificazione del territorio: retorica del declino e criticità della transizione energetica in Sardegna [Coal and spatial planning: Rhetoric of decline and critical issues within the energetic transition of Sardinia (Italy)]. Archivio di Studi Urbani e Regionali **52**(131), 166–185 (2021). https://doi.org/10.3280/ASUR2021-131-S1008
10. Balletto, G., Ladu, M., Camerin, F., Ghiani, E., Torriti, J.: More circular city in the energy and ecological transition: a methodological approach to sustainable urban regeneration. Sustainability **14**(22), 14995 (2022). https://doi.org/10.3390/su142214995
11. Gerundo, R., Marra, A.: Renewable energy communities: urban research and land use planning. BDC. Bollettino del Centro Calza Bini **22**(2), 160–311 (2022). http://www.serena.unina.it/index.php/bdc/issue/view/718. Accessed 27 Jan 2024
12. De Pascali, P., Bagaini, A.: Energy transition and urban planning for local development. A critical review of the evolution of integrated spatial and energy planning. Energies, **12**(1), 35 (2018). https://doi.org/10.3390/en12010035
13. De Lotto, R., Micciché, C., Venco, E.M., Bonaiti, A., De Napoli, R.: Energy communities: technical, legislative, organizational, and planning features. Energies **15**(5), 1731 (2022). https://doi.org/10.3390/en15051731
14. Moretti, E., Stamponi, E.: The renewable energy communities in Italy and the role of public administrations: the experience of the Municipality of Assisi between challenges and opportunities. Sustainability **15**, 11869 (2023). https://doi.org/10.3390/su151511869
15. Ministry of Economic Development, Ministry of the Environment and Protection of Land and Sea, Ministry of Infrastructure and Transport: Piano Nazionale Integrato per l'Energia e il Clima [Italian Integrated Energy and Climate Plan] (2019). https://www.mise.gov.it/images/stories/documenti/PNIEC_finale_17012020.pdf. Accessed 27 Jan 2024
16. Gerundo, R., Marra, A.: A decision support methodology to foster renewable energy communities in the municipal urban plan. Sustainability **14**(23), 16268 (2022). https://doi.org/10.3390/su142316268
17. Marra, A.: Pianificare le Comunità Energetiche per il rilancio delle aree interne [Planning Energy Communities for the revitalisation of inland areas]. In: Gerundo, R. (eds.) Città e Piani del Rischio Energetico e Alimentare. Collana Tecnica Urbanistica, vol. 9, pp. 217–236. Edizioni Scientifica Italiane, Naples, Italy (2023)
18. Lanzani, A., De Leo, D., Mattioli, C., Morello, E., Zanfi, F.: Nell'Italia di mezzo: rigenerazione e valorizzazione dei territori della produzione [In middle Italy: regeneration and enhancement of production territories]. In: Coppola, A., Del Fabbro, M., Lanzani, A., Pessina, G., Zanfi, F. (eds.) Ricomporre i divari. Politiche e progetti territoriali contro le disuguaglianze e per la transizione ecologica. Il Mulino, Bologna (2021). https://doi.org/10.12910/DOC2020-012
19. Istat - Italian Institute of Statistics: La Geografia delle aree interne nel 2020: Vasti territori tra potenzialità e debolezze [The Geography of Inland Areas in 2020: Vast Territories Between Potential and Weakness]. Istat, Rome (2022)
20. Fisciano Municipality: Dedicated website of Fisciano Municipal Urban Plan. Fisciano (2022). https://www.pucfisciano.it/index.php. Accessed 27 Jan 2024
21. Istat - Italian Institute of Statistics: Basi territoriali e variabili censuarie. Censimento della popolazione e delle abitazioni 2011 [Territorial bases and census variables. Population and Housing Census 2011]. Istat, Rome (2011). https://www.istat.it/it/archivio/104317. Accessed 27 Jan 2024

22. Esri: ArcGis Geographic Information System. Version 10.7.1. Esri, Redlands, California, USA (2019)
23. OSGeo - Open Source Geospatial Foundation: QGIS Geographic Information System. Version 3.28 (2023). http://qgis.osgeo.org. Accessed 27 Jan 2024
24. Ballarini, I., Corgnati, S.P., Corrado, V.: Use of reference buildings to assess the energy saving potentials of the residential building stock: the experience of TABULA project. Energ. Policy **68**, 273–284 (2014). https://doi.org/10.1016/j.enpol.2014.01.027
25. Jenks, G.F.: The data model concept in statistical mapping. In: Frenzel, K. (eds.) International Yearbook of Cartography, vol. 7. George Philip, London (1967)
26. Italian Government: Piano Nazionale di Ripresa e Resilienza (PNRR) [Next Generation Italy]. Italian Government, Rome (2021)
27. MASE - Italian Ministry of the Environment and Energy Security: Decree no. 414 of 07/12/2023 "Individuazione di una tariffa incentivante per impianti a fonti rinnovabili inseriti in comunità energetiche rinnovabili e nelle configurazioni di autoconsumo singolo a distanza e collettivo, in attuazione del decreto legislativo 8 novembre 2021, n.199 e in attuazione della misura appartenente alla Missione 2, Componente del 2, Investimento 1.2 del PNRR" [Identification of an incentive tariff for renewable energy plants included in renewable energy communities and in the configurations of individual remote and collective self-consumption, in implementation of Legislative Decree 8 November 2021, no. 199 and in implementation of the measure belonging to Mission 2, Component 2, Investment 1.2 of the PNRR]. MASE, Rome (2023). https://www.mase.gov.it/sites/default/files/Decreto%20CER.pdf. Accessed 27 Jan 2024
28. GSE - Manager of the Italian Energy Service: Mappa delle aree servite dalle cabine primarie [Map of the areas served by the primary substations]. GSE (2024). https://www.gse.it/servizi-per-te/autoconsumo/mappa-interattiva-delle-cabine-primarie. Accessed 15 Mar 2024
29. E-distribuzione: Mappa delle aree critiche [Map of critical areas]. E-distribuzione (2023). https://www.e-distribuzione.it/a-chi-ci-rivolgiamo/produttori/aree-critiche.html. Accessed 27 Jan 2024
30. Gerundo, R., Fasolino, I., Grimaldi, M.: ISUT model. A composite index to measure the sustainability of the urban transformation. In: Papa, R., Fistola, R. (eds.) Smart Energy in the Smart City. Green Energy and Technology, pp. 117–130. Springer, Cham (2016). https://doi.org/10.1007/978-3-319-31157-9_7
31. Dvarioniene, J., et al.: Stakeholders involvement for energy conscious communities: the Energy Labs experience in 10 European communities. Renew. Energ. **75**, 512–518 (2015). https://doi.org/10.1016/J.RENENE.2014.10.017
32. Graziuso, G., Grimaldi, M., Mancini, S., Quartieri, J., Guarnaccia, C.: Crowdsourcing data for the elaboration of noise maps: a methodological proposal. J. Phys. Conf. Ser. **1603**(1), art. no. 012030 (2020). https://doi.org/10.1088/1742-6596/1603/1/012030
33. Sebillo, M., Vitiello, G., Grimaldi, M., De Piano, A.: A citizen-centric approach for the improvement of territorial services management. ISPRS Int. J. Geo-Inf. **9**, 223 (2020). https://doi.org/10.3390/ijgi9040223
34. Dolores, L., Macchiaroli, M., De Mare, G.: Financial impacts of the energy transition in housing. Sustainability **14**(9), 4876 (2022). https://doi.org/10.3390/su14094876
35. Macchiaroli, M., Dolores, L., De Mare, G., Nicodemo, L.: Tax policies for housing energy efficiency in Italy: a risk analysis model for energy service companies. Buildings **13**, 582 (2023). https://doi.org/10.3390/buildings13030582

Ecosystem Services in Spatial Planning for Climate Neutral Urban and Rural Areas (ESSP 2024)

Strategic Environmental Assessment as a Tool to Embed Climate Change Adaptation in Urban Planning. The Case of Four Italian Cities

Federica Isola, Sabrina Lai, Federica Leone(✉), and Corrado Zoppi

Department of Civil and Environmental Engineering and Architecture, University of Cagliari, Cagliari, Italy
{federica.isola,sabrinalai,federicaleone,zoppi}@unica.it

Abstract. Adaptation to climate change and the need to deal with its impacts pose in evidence how important it is to identify and implement new planning practices that integrate these profiles into land-use policy-making. Within this conceptual framework, the identification of a system of plan objectives and actions that characterize the integration of climate change adaptation into planning policies, with particular reference to the local scale, is of particular importance. In this study, a methodology for implementing this integration is proposed through the establishment of a logical framework for the construction of municipal masterplans through strategic environmental assessment, as a pathway in which plans are formed and developed, as part of the assessment process, through the identification of a strategic system of objectives and an operational system of planning actions based on the integration of climate change adaptation into the plan formation process.

Keywords: Climate change adaptation · Strategic environmental assessment · Spatial policies

1 Introduction

The integration of climate change adaptation (CCA) into strategic environmental assessment (SEA) processes is a widely discussed and analyzed issue in the technical and scientific literature.

During the preparation of Directive no. 42/2001/EC, an analytical research report on the implementation of SEA in the decision-making processes of plans and programs, prepared by the Imperial College for the European Commission [1], emphasizes how the effectiveness of land-use policies at different scales, national, regional and local, is fundamentally linked to the integration of CCA issues, highlighting several profiles of these issues. Of particular relevance, among the many cases analyzed, are the SEA of the land-use plan of the city of Weiz (Austria), in which it is emphasized that "The targets, goals and objectives used in the SEA process are clearly defined, for example, carbon dioxide threshold according to the goals of the 'Climate Alliance'" [1, p. 9], the SEA of the National Environmental Policy Plan 3 of the Netherlands, which places CCA among

O. Gervasi et al. (Eds.): ICCSA 2024 Workshops, LNCS 14818, pp. 275–292, 2024.
https://doi.org/10.1007/978-3-031-65273-8_18

the reference themes for identifying the structure and framework of administrative and technical competencies of the decision-making process [1, p. 85], and the Regional Economic Strategy of the Yorkshire Regional Development Agency, which explicitly includes climate change among the components of the strategic framework [1, p. 153].

A very significant document is the Advisory Note on Environmental Assessment and CCA [2], which identifies four fundamental moments in the implementation of spatial planning processes that involve the integration of SEA and CCA: (i.) the precise and circumstantial identification of the universe of stakeholders and environmental components that reasonably will be affected, during the plan process, by the impacts of CCAs; (ii.) the implementation of the SEA; (iii) the process of informing and educating local communities, affected by the plan and climate change, in the proactive participation in the definition and implementation of planning policies; and, (iv) the continuous updating of the plan's strategic framework through the ongoing SEA and monitoring. A significant moment in the discussion proposed in the Advisory Note concerns the problematic nature of the conceptual and technical relationships between SEA and CCA. Indeed, it should be acknowledged and kept well in mind how SEA does not so much provide a conceptual and technical framework for the development of scientific research related to the impacts of climate change, but, rather, poses, in problematic terms, the need to fill knowledge gaps [2], which need to be kept in mind, if not resolved, as SEA is tasked with constructing frameworks and assessing impacts in reasonably plausible terms, which would not be possible if the gaps gave the assessment process a character of dramatic indeterminacy. SEA is effective in monitoring the implementation of plan processes, and, during this, the type and magnitude of climate change and, therefore, related adaptation measures, in relation to, for example, atmospheric precipitation, the genesis and likelihood of extreme weather events, the effects on water quality, and the generation of re-risk conditions, related to the hazard, vulnerability, and exposure associated with the impacts of climate change on natural resources and local societies.

An analytical discussion of this theoretical and technical conceptual vision is proposed by Wende et al. [3], who examine, in comparative terms, SEA processes which integrate CCA approaches with reference to regional land-use plans from Saxony and East England.

A more comprehensive and general view of the relationship between SEA and CCA is, on the other hand, proposed and discussed by Gonzáles Del Campo et al. [4] with reference to SEA processes that integrate, in the strategic device of the assessment, specific objectives that refer to Strategic Development Goal no. 13 of Agenda 2030 [5], "Take urgent action to combat climate change and its impacts."

The issue of integrating CCA into SEA processes is addressed, specifically, in the European Commission's document "Guidance on Integrating Climate Change and Biodiversity into Strategic Environmental Assessment" [6], which takes up and develops the contents of ENVIRONET's Advisory Note [2]. The issue is discussed, in systematic terms, in the fifth section of the paper, in which some basic conditions for the construction of SEA environmental reports (ERs) whose strategic device, i.e., whose hierarchical system of objectives, includes CCA-relatable goals, integrated into the plan strategy, are discussed.

Related to what is presented and discussed in this study, of particular relevance is what is indicated regarding the need to address, in a detailed and specific manner, in ERs, the issue of consistency between the strategic device of the plan and the systems of objectives regarding mitigation and CCA identified in national and local strategies and plans focused on these issues [6, p. 70]. In general, integrating the sustainability paradigm into public policy-making and implementation processes involves a careful assessment of economic and social equity issues in intra- and inter-generational terms [7]. With regard, in particular, to spatial planning, this integration is not made operational through measures identifiable in deterministic terms, but, rather, through practices that involve an open and continuous dialectic with local societies, based on mediation in relation to the instances and expectations they express, as well as on the contributions of spatial sciences, to be used not only as foundational references of spatial analysis, but, also, as sources of collective learning [8].

It is within this conceptual framework that the objective of this study is recognized and placed, which consists in the definition of a methodological approach for the integration of CCA in the ERs of the SEAs of the territorial plans of local governments that, in the Italian regulatory context, are identified as municipal masterplans (MMPs) [9]. It consists, therefore, in the implementation of a downscaling operation to the local level of strategies and plans concerning CCA, in force in the European and national contexts, that is, with reference to MMPs [10].

The starting point for the development of the downscaling process to municipal urban planning is the National Plan for Adaptation to Climate Change of Italy (NPCCA), whose strategic device declines the European Union Strategy for Adaptation to Climate Change (EUSCCA)[1] and the National Strategy for Adaptation to Climate Change of Italy (NSCCA)[2].

The EUSCCA has had two versions, the first dating back to 2013, the other to 2021 (see footnote 1). The second is a critical restatement of the 2013 EUSCCA following an in-depth evaluative review of its implementation through the national CCA-related strategies and plans adopted and implemented by many EU countries[3]. The EUSCCA promotes, in this regard, the implementation of the conceptual and technical operational

[1] The two reference documents of the European Commission concerning the EUSCCA are as follows: i) Communication from the Commission to the European Parliament, the Council, the European Economic and Social Committee and the Committee of the Regions "An EU Strategy on adaptation to climate change," COM(2013) 216 final; ii) Communication from the Commission to the European Parliament, the Council, the European Economic and Social Committee and the Committee of the Regions "Forging a climate-resilient Europe - the new EU Strategy on Adaptation to Climate Change," COM(2021) 82 final.

[2] The reference document for the NSCCA is Decree no. 86 of the Director of the Directorate General for Climate and Energy of the Ministry of the Environment and Land and Sea of June 16, 2015, adopting and approving the NSCCA (Art. 1). The NSCAA is contained in the Annex to the Decree and is an integral part of it.

[3] The evaluation document is the Commission Staff Working Document "Evaluation of the EU Strategy on adaptation to climate change," SWD(2018) 461 final.

framework of the EU Covenant of Mayors for Climate & Energy initiative[4], for whose local plans and actions the JRC has defined a specific approach in a three-part guideline manual, of which the first proposes a detailed itinerary for the definition of a municipal action plan aimed at CCA and mitigation, and energy savings, the second refers to emission census, and risk and vulnerability assessment, and the third to the identification of best practices and key actions concerning CCA and mitigation, as well as financial issues [11–13]. The NSCCA was defined, in 2015, on the basis of the 2013 EUSCCA, and, as far as it concerns the integration of CCA into local plans, it offers some very important pointers, in implementation of the technical device of the EUSCCA.

In particular, the approach of the NSCCA individuates the reference areas for defining the situation of climate dynamics and climate change and, in relation to these, identifies impacts and vulnerabilities, with respect to which the objectives and actions of the NPCCA will be defined. These will form the foundational strategic device for the downscaling operation that is the objective of this study.

The paper is structured into five sections. The second section provides a brief description of the methodology used. In the third section, the results of the downscaling operation are presented, in relation to the reference areas most relevant to this operation. The fourth section provides recommendations for spatial planning policies. The concluding section highlights the prospects for the development of the research. The complete description of the methodology and the discussion on the results are reported in the full paper from which this study was excerpted[5].

2 Methods and Materials

The methodology for the integration of the CCA into the ERs and, therefore, for the construction of the MMPs, consists of three phases, basically geared toward grafting the NPCCA's device of objectives into the systems of objectives of the MMPs. It should be noted that the application of the methodology proposed in this section, which refers to the adopted NPCCA, could be replicated with reference to any future drafts of the NPCCA, whose strategic framework, represented by the system of objectives of the updated version, would need to be incorporated. Among the objectives of the NPCCA, systematized in Annex IV of the NPCCA, called "Database of Actions"[6], the first phase aims to identify those that can be associated with the processes of defining MMPs, having significant implications on land governance. Based on the objectives of CCA, and

[4] An extensive and systematic information dossier on the initiative can be retrieved from the European Union website https://eu-mayors.ec.europa.eu/en/home. Accessed February 2024/03/11.

[5] See the related acknowledgment at the end of this study.

[6] Retrieved from: https://www.mase.gov.it/sites/default/files/archivio/allegati/clima/PNACC_AllegatoIV_database_azioni.ods. Accessed 2024/03/11.

also taking into consideration the 18 NPCCA Reference Sectors[7] to which they refer, a selection of significant objectives relevant to land-use and urban planning is made, followed by an identification of the actions/measures referring to them. The second stage is the construction of the system of specific objectives of the MMPs to which the ERs refer, that is, the construction of the logical frameworks (LFs) of the ERs of the MMPs, systems that are deduced directly from the ERs of the MMPs.

Finally, in the third phase, the objectives identified in the first phase are used in the drafting of the ERs for the construction of the MMPs as operational references to redefine the systems of the specific objectives and actions of the MMPs so that these systems integrate the CCA into the overall strategy of the MMPs. The methodological approach is implemented in relation to four case studies within the Sardinia Region, namely the municipalities of Capoterra, Nuoro, Sassari, and Selargius.

3 Results

The NPCCA reference sectors that stand out as the most significant within the MMPs of the four cities in Sardinia selected for the implementation of the methodology described in the second section of this study are hydrogeological instability, which is characterized by 31 specific objectives and 50 plan actions, urban settlements, with 29 specific objectives and 47 plan actions, and transportation, with 40 specific objectives and 62 plan actions. This section is divided into three parts and presents the results for each of these areas in relation to the definition of objectives and operational plan choices that integrate CCA into the LFs, i.e., strategic and implementation arrangements, of the MMPs.

3.1 Hydrogeological Instability

An extract from the full set of actions and measures contained in the LFs of the ERs of the four analyzed MMPs that integrate climate considerations and contribute to addressing hydrogeological instability is provided in Fig. 1 (third column), together with the objectives from which they descend within each LF (second column), and the NPPCA's goals that are pursued (directly or indirectly, to a larger or lesser extent) through the plans' objectives and action (first column). Three NPCCA's goals have been found to be pursued by the four the LFs of the ERs of the MMPs; two out of three aims at enhancing knowledge, either on areas that are prone to hydrogeological issues within the administrative boundaries, or on the conditions of buildings and infrastructure, while the third goal is action-oriented and paves the way for either revised planning choices or tangible actions.

7

Aquaculture; Agriculture and food production; Desertification, land degradation and drought; Geological, hydrological and hydraulic instability; Ecosystems and biodiversity in inland and transitional waters; Marine environments: Biodiversity, Functioning and Ecosystem Services; Energy; Terrestrial Ecosystems; Forests; Hazardous Industries and Infrastructure; Urban Settlements; Cultural Heritage; Marine Fisheries; Water Resources; Health; Transport; Tourism; Coastal Zones.

NPCCA goals	LF objectives	LF actions
To improve land management and maintenance.	NUORO – To regulate building expansion.	Analysis of the residential and service systems.
	NUORO – To restore areas currently hosting illegal buildings.	
To improve land management and maintenance.	SELARGIUS – To ensure soil conservation and protection.	Restrictive rules for Subareas C3.1 (residential) and G1.4 (services and facilities), classed as areas prone to high and very high geological hazard. Their implementation in terms of urban planning and construction is subject to the execution of hydraulic works of mitigation, regimentation and regularization of the current hydrogeological risk, so as to eliminate constraints arising from the current classification under the SPHF
	SELARGIUS – To mitigate and reduce current hydrogeological risks in the municipality.	
	SELARGIUS – To prevent new hydrogeological hazards.	
To improve knowledge on critical geological and hydraulic issues in the area and their associated risks, and to produce updated databases based on land monitoring.	SASSARI – To prevent hydrogeological risks through appropriate land use regulations.	Identification of areas characterized by hydrogeological hazard and risk.
	SELARGIUS – To ensure soil conservation and protection.	
	SELARGIUS – To mitigate and reduce current hydrogeological risks in the municipality.	
	SELARGIUS – To prevent new hydrogeological hazards.	

Fig. 1. Extract from the goals related to hydrogeological instability contained in the NPPCA and integrated within the four analyzed logical frameworks (LF) of the MMP's environmental reports, LF's objectives and actions and measures that contribute to pursuing the NPCCA objectives.

The first objective, concerning improved knowledge on critical geological and hydraulic issues, is integrated within Sassari's and Selargius' LFs, which both contain an action providing for the identification of areas characterized by hydrogeological hazard and risk. Such action stems from a single objective in Sassari's LF and is connected to three objectives in Selargius LF.

The second objective, concerning improved knowledge on the conditions of buildings and infrastructure with a view to increasing their resilience, is integrated within three LFs of the ERs of the MMPs (Capoterra, Sassari, and Selargius). As for Selargius, a single action, providing for mitigating hydrogeological risks, hence focusing on the resilience part, stems from two LF objectives, while in the two other LFs a one-to-one relation between action and objective can be observed.

In Capoterra, the action focuses on the coastal areas, where a critical infrastructure, the road connecting the regional capital with Capoterra and the southwestern part of Sardinia, lies over a narrow strip of land, constrained between the coastline and a large wetland.

In Sassari, the action focuses on former mining sites, which also include abandoned buildings forming the old mining hamlets, as knowledge on their status is a precondition for their recovery and reuse for tourism purposes.

Finally, the last objective, concerning improved land management and maintenance, is integrated within all of the four analyzed LFs of the ERs of the MMPs, by means of one action (connected to a single objective) in both Capoterra and Sassari, of three actions (connected to four objectives) in Selargius, and of seven actions (connected to seven objectives) in Nuoro.

The broad goal of enhancing land management and maintenance is variously pursued in the four LFs, whose actions range from studies and analyses, to the identification of rules to be included within the municipal planning implementation code, to tangible interventions aimed at addressing specific problems within the town, as in the case of the conversion of the former railroad and of landscaping actions in Nuoro, or of the recovery and reuse of former mining hamlets in Sassari, or of measures to mitigate hydrogeological hazards in Selargius and to improve the coastal area in Capoterra.

3.2 Urban Settlements

With reference to urban settlements, all four MMPs develop strategies aimed at defining objectives and actions aimed at mitigating climate change negative impacts, and at adapting urban environments (see Fig. 2).

NPCCA's goals	LF objectives	LF actions
To improve thermal comfort and living quality in periurban areas, suburbs, historic centers, and public spaces. To improve the efficiency of the water distribution system in periurban areas, suburbs, historic centers, and public spaces. To promote planning and design for hazard prevention and facilitating monitoring.	CAPOTERRA – To redevelop and re-organize the consolidated Urban Poles (urban center, coastal area, and the settlement of Poggio dei Pini).	Qualification of the main urban settlement: consolidation of the urban fabric, endowment, completion and rationalization of technological and road networks, integration of residential and public services in accordance with territorial safety and specific studies.
To improve thermal comfort and living quality in periurban areas, suburbs, historic centers, and public spaces. To improve the efficiency of the water distribution system in periurban areas, suburbs, historic centers, and public spaces. To promote planning and design for hazard prevention and facilitating monitoring.	CAPOTERRA – To redevelop and re-organize the consolidated urban areas (urban center, coastal area, and the settlement of Poggio dei Pini).	Restoration and completion of the existing urban settlement.

Fig. 2. Extract from the goals related to Urban settlements contained in the NPPCA and integrated within the four analyzed logical frameworks (LF) of the MMP's environmental reports, LF's objectives and actions and measures that contribute to pursuing the NPCCA objectives.

The NPCCA's objective concerning the improvement of thermal comfort and quality of living involves establishing measures for heat control, storage, and dissipation. The objective is fully implemented into the LFs of the ERs of the four analyzed MMPs through: i. two specific objectives in the LF of the ER of the Capoterra MMP; ii. eight specific objectives in the LF of the ER of the Nuoro MMP; iii. eight specific objectives in the LF of the ER of the Selargius MMP; and, iv. two specific objectives in the LF of the ER of the Sassari MMP. In relation to these objectives, two types of plan actions have been identified to achieve the NPCCA's objective: the first aims at redeveloping and recovering the built characteristics of the urban consolidated fabric, while the second concerns the redevelopment of peripheral, periurban and rural spaces.

This redevelopment pursues the strategic distribution of greenery to mitigate impacts due to solar radiation and the heat island effect [14].

As for Capoterra, four plan actions are of the first type, while three plan actions refer to the second type. Both clusters of actions are associated with the same LF objectives. In the case of Selargius, most of the plan actions are aimed at creating green areas and improving the conditions of existing ones.

The actions are associated with the objective "To pursue an environmental policy aimed at increasing the quantity and quality of green spaces present in the urban and suburban context and to encourage processes of reconfiguration and regeneration of the same through raising the building quality of public spaces and facilities." The actions of the LF of the ER of the Nuoro MMP are, in general, oriented toward the redevelopment of the landscape and built environment, with particular attention to the endowment of urban standards, the enhancement of the built urban fabric and the redevelopment of some of the most important sites of historical and cultural interest.

The LF of the ER of the Sassari MMP, on the other hand, defines a plan strategy that, with reference to the objective of the NPCCA, focuses on the partially unbuilt areas within the urban center, through the inclusion of a share of non-developable areas to make room for an urban network of green areas. These plan actions pursue the objective of reconnecting the most significant urban voids.

Regarding the NPCCA's objective concerning the improvement of the efficiency of the water supply system in periurban areas, suburbs, historic centers and public spaces, it is worth noting that the issue of water resource management is a nationwide problem. Regarding the LFs of the MMPs' ERs, the NPCCA goal is implemented into all of the four plans. In the case of Capoterra, six plan actions stem from a single LF objective alone related to the redevelopment and reorganization of consolidated urban hubs. As for Nuoro, ten plan actions referring to this NPCCA objective, and they derive from as many objectives of the LF of the MMP, largely aimed at enhancing the existing built and cultural heritage and ensuring the provision of new public services. The same aims are pursued in the LF of the Selargius MMP, in which thirteen plan actions implement the NPCCA objective and are associated with a set of specific objectives aimed at protecting the qualitative and quantitative state of water resources, preventing hydraulic and geological risk, and increasing the availability of quality green spaces in the urban, periurban and rural areas. The LF of the ER of the Sassari MMP targets two plan actions that relate to improving the efficiency of the water system; both refer to the need to define an urban ecological network and are linked to two different objectives of the LF.

The third and fourth objectives of the NPCCA aim to promote planning and design for risk prevention and to facilitate monitoring and increasing soil permeability and hydraulic system efficiency in periurban areas, suburbs, historic centers, and public open spaces. Both objectives aim to address these critical issues and are integrated into the LFs of the ERs of the analyzed MMPs.

The third objective is present in three of the LFs of the ERs of the analyzed MMPs, with the exception of Sassari, while the fourth is implemented into all four plans. As far as Capoterra is concerned, six plan actions contribute to the achievement of the NPCCA's third objective, all referring to the same specific objective of the LF of the ER of the MMP, namely the redevelopment and reorganization of the consolidated urban poles, i.e., the urban consolidated fabric, the Poggio dei Pini hamlet and the coastal strip. Also referring to Capoterra, six plan actions are associated with the fourth objective of the

NPCCA. Among them, some implement the LF objective focused on the protection and preservation of environmental, historical and cultural components, and the protection of areas of special landscape significance; others pursue the LF objective of spatial and environmental safety.

3.3 Transportation Network

The LF provides: i. in the first column, the NPPCA's goals directly or indirectly pursued by objectives and actions of the LFs of the ERs of the analyzed MMPs; ii. in the second column, the objectives of the four LFs of the ERs of the MMPs with which the objectives of the NPPCA are associated; and, iii. in the third column, the actions that implement the LFs' objectives, integrate climate considerations, and contribute to addressing transportation network (see Fig. 3).

NPCCA goals	LF objectives	LF actions
To experiment with materials, structures, facilities, and technologies that are more resilient to increasing temperatures and rainfall variability.	CAPOTERRA –To regenerate and reorganize the consolidated urban areas (city center, Poggio dei Pini, coastal urban settlement), to be integrated within a unitary, organically articulated functional design.	Urban, road and environmental requalification of the structure of the coastal urban settlement in line with the landscape characteristics of the context as well as with the new structure that will take on with the construction of new road "SS 195" in order to guarantee adequate coastal accessibility and a rebalancing and compensation of the urban load of coastal settlements; provision of networks, including public and private services, to compensate for the strong existing imbalance between services and housing; enhancement of the connection systems between the different suburban realities.
To integrate climate change risks into planning and design process in order to promote resilience and adaptation.		
To improve the effectiveness of monitoring, warning and emergency response systems for transport services.		
To experiment with materials, structures, facilities, and technologies that are more resilient to increasing temperatures and rainfall variability.	CAPOTERRA –To regenerate and reorganize the consolidated urban areas (city center, Poggio dei Pini, coastal urban settlement), to be integrated within a unitary, organically articulated functional design.	Development of coastal land-use plans allowing for greater qualification and diversification of services.
To integrate climate change risks into planning and design process in order to promote resilience and adaptation.		
To improve the effectiveness of monitoring, warning and emergency response systems for transport services.		

Fig. 3. Extract from the goals related to transportation network contained in the NPPCA and integrated within the four analyzed logical frameworks (LF) of the MMP's environmental reports, LF's objectives and actions and measures that contribute to pursuing the NPCCA objectives.

Four NPCCA's goals have been found to be pursued by the four LFs. Two objectives refer to prevention measures, such as promoting the securing of the territory against hydrogeological risk and the integrating climate change risks into planning and design processes. One objective refers to monitoring measures. Finally, the last objective refers to increasing knowledge in relation to materials, structures, plants and technologies that are more resilient to increasing temperatures and rainfall variability.

The first objective, which concerns the testing of materials, structures, plants and technologies more resilient to increasing temperatures and rainfall variability, is implemented into fifteen LF objectives: one from the Capoterra MMP's ER LF, eight from the Nuoro MMP's ER LF, five from the Selargius MMP's ER LF and one from the Sassari MMP's ER LF. With reference to the Capoterra LF, two actions, both connected to the LF objective, contribute to the achievement of the NPCCA objective in terms of transport network.

In reference to the Nuoro LF, twelve actions are connected to the eight LF objectives. With the exception of five LF objectives that are connected to a single action, the objectives "To restore areas currently hosting illegal buildings" and "To ensure the endowment of public services and facilities" are implemented through two actions each, while the objective "To enhance the historic center as a part of the city to be preserved and handed down to future generations in the most appropriate manner" is connected to five actions that contribute to the achievement of the NPCCA objective in relation to the transport network. With reference to the Selargius LF, the five objectives are implemented through a single action that contributes to the achievement of the NPCCA objective in terms of transport network. The action, concerning interventions aimed at mitigating and reducing the hydrogeological risk, is connected to three different LF objectives.

The only objective of the Sassari MMP's ER LF is linked to a single action that contributes to the achievement of the NPCCA objective in relation to the transport network.

The second objective concerning the integration of climate change risks into planning and design is implemented into nineteen objectives: one from the Capoterra MMP's ER LF, nine from the Nuoro MMP's ER LF, three from the Sassari MMP's ER LF, and six from the Selargius MMP's ER LF. With reference to the Capoterra LF, six actions, linked to a single LF objective, contribute to the achievement of the NPCCA objective in relation to the transport network. In reference to the Nuoro LF, thirteen actions contribute to the achievement of the NPCCA objective. In particular, with the exception of five LF objectives each connected to a single action, the LF objective "To enhance the historic center as a part of the city to be preserved and handed down to future generations in the most appropriate manner" is connected to five actions. The LF objectives "To ensure the endowment of public services and facilities," "To restore areas currently hosting illegal buildings," and "To enhance the area of the former powder mill in Prato Sardo," are each connected to two actions contributing to the achievement of the NPCCA objective.

In addition, three actions ("Preparation of a landscape-oriented redevelopment plan", "Conversion of the former railroad into a bicycle and pedestrian pathway", and "Conversion of existing areas into parking areas") refer each to two different LF objectives. With reference to the Sassari MMP's ER LF, three actions contribute to the achievement of the NPCCA objective in relation to the transport network.

With the exception of one LF objective, the remaining two LF objectives "To encourage sustainable nature-based tourism" and "Protection and Conservation of Sites of Community Importance" relate to two actions. Furthermore, two out of three actions relate to two different LF objectives. With reference to the Selargius MMP's ER LF, three actions contribute to the achievement of the NPCCA objective. With the exception of one action ("Redevelopment of the Is Corrias areas, the boundary areas with the municipality of Quartucciu, the boundary areas with the municipality of Monserrato"), which is connected to a single LF objective, the other two actions are connected, respectively, to three LF objectives in the case of the action "Interventions aimed at mitigating hydrogeological risks" and two LF objectives in the case of the action "Identification of the "Road of Parks" and concentration of areas handed over to the Municipality of Selargius through supplementary agreements pursuant to Law 241/90 within the areas "San Lussorio", "Paluna" and "Santa Lucia".

The third objective concerning the improvement of the effectiveness of monitoring, alerting and emergency intervention systems for transport services is implemented into a single LF objective related to the Capoterra MMP's ER LF. In addition, six actions, linked to the MMP objective, contribute to the achievement of the NPCCA objective.

The fourth objective concerning the securing of the territory in relation to hydrogeological risk is implemented into seven LF objectives: three of the LF of Nuoro, and four of the LF of Selargius. With reference to the Nuoro LF, three actions contribute to the achievement of the NPCCA objective.

Apart from one action ("Functional reconversion of the Artillery Barracks into a university campus (Campus in the Green)") which is connected to a single LF objective, the other two actions are both connected to two LF objectives. With reference to the Selargius MMP's ER LF, four actions contribute to the achievement of the NPCCA objective. Except for one LF objective ("To protect the qualitative and quantitative status of surface and groundwater resources") which is connected to a single action, the remaining three LF objectives are connected to all four LF actions.

4 Policy Implications

Implications concerning planning policies aimed at integrating CCA into local planning processes, based on the LFs of the ERs of the SEAs of the MMPs of the four cities of Sardinia, are related to the plan actions arising from the implementation of the specific objectives.

An effective classification of the plan actions, that is, the operational part of the SEAs of the MMPs, which integrate CCA into the LFs, can be adequately represented by three general themes, as follows. First, the establishment or expansion of green areas, parks, outdoor recreation areas, forests, wooded areas and trees. Secondly, the appropriate management of the water resources regime of the municipal spatial contexts. Finally, greening operations referred to existing buildings or new developments.

Parks, forests and urban wooded areas significantly reduce surface temperature [15, 16]. According to a study by Armson et al. [17], the shading and evapotranspiration of wooded areas results in a 5–7 °C drop in temperature, while an article by Bowler et al. [18] estimates the temperature difference between urban parks and built-up areas at just under one degree. Urban wooded areas, through runoff mitigation, are, in addition, very effective in limiting vulnerability in relation to extreme weather events and, therefore, flooding events, which are particularly severe in urban areas characterized by a significant human presence. For example, Pataki et al. [19] estimate how built-up areas, whose soils are made impermeable by urbanization, are characterized by runoff that falls within a range of 40–83% of rainfall, while in a wooded or forested urban area the phenomenon is about 13%.

With regard to planning policies referring to the construction or expansion of parks or wooded areas, the design profile is crucial for generating positive impacts in relation to heat waves, lowering temperatures, and for mitigating negative impacts of flood phenomena, including possible leakage of sewer lines [20]. There are, moreover, several studies available that estimate the economic impact of implementing these plan actions, although the outcomes are difficult to generalize (among many, [21, 22]).

The strengthening of ecosystems characterized by the presence of wooded and forested areas is based, in essence, on the protection of primary forests, the recovery of degraded forest systems, the sustainable management of wooded and forested areas, and tree plantings in contexts characterized by the presence of other ecosystems, such as, for example, urbanized areas or those dedicated to agricultural production: here, wooded areas operate effectively in relation to the mitigation of hydraulic and landslide risk, and the negative effects of heat waves.

Planning policies related to urban water resource management consist, fundamentally, of maintenance of riverbanks, and construction of retention basins, retention infrastructure, drainage, and release of water from the increasingly frequent extreme weather events, such as bioswales and rain gardens, and reservoirs [23]. Bioswales, rain gardens, and urban tree areas are also particularly effective vis-à-vis limiting pollution generated by sewage spills from pipes during extreme weather events [24]. Of great importance, as well, are plan actions aimed at re-permeabilizing soils, for example, by removing excess asphalt road covers, cleaning up riverbanks and riverbeds, and restoring tombed streams.

At the broad spatial scale, planning policies related to the relationships between CCA and water resource management generally have the function of regulating flood phenomena and consist of the re-functionalization and restoration of river courses and floodplains [7]. Floodplains and river belts provide the ecosystem service of mitigating the hardships that occur, in terms of water resource shortages, during periods of drought, as they promote, through drainage and water retention in the subsoil, the formation of water reservoirs, to be used, precisely, in the event of a shortage of the resource. Soil permeability allows, also, to decrease flood flows and the velocity of runoff, and to retain, at least partially, transported sediments: this function is also put in place by areas used for agricultural production, especially in valley bottoms [25]. Surface runoff, water retention, and natural drainage due to permeability, especially during flood events, are, also, facilitated by the effective management of forests and wooded areas, especially in or near river belts [26].

Regarding the management of transportation infrastructure, a relevant issue is the construction of roads, parking lots, pedestrian streets, and playgrounds whose permeable superstructures and pavements decrease runoff, especially during major storm events [24], and allow, as much as possible, its passage into underground aquifers, thereby promoting greater availability of the resource during drought periods, and, at the same time, decreasing the magnitude of flood flows, which often generate significantly negative impacts in urban areas [27]. Porous pavements also enact an important filtering action that improves the quality of water that, through percolation, is stored underground [28].

As for the interventions on individual buildings that can be traced back to the plan actions arising from the SEAs that integrate CCA-related policies, these are identified with the implementation of green roofs, walls and facades [26]. The greening approach to buildings enables the implementation of effective stormwater management. According to a study by Ruangpan et al. [29], green roofs increase water retention and decrease, and/or delay stormwater runoff during particularly significant events by up to 70% and 96%, respectively, during peak rainfall. Green roofs, facades and walls are, also, very effective in improving comfort inside residences and decreasing energy demand for building interior air conditioning [30]. In addition, green surfaces, whether they are outdoors or on the building envelope, have greater reflective power of sunlight than artificial surfaces in the built environment, with a difference that falls, according to an article by Perini and Rosasco [31], in the range of 15–25%, with a very important impact on the mitigation of the urban heat island phenomenon.

Finally, it should be emphasized that, when defining the plan actions that are identified in the LFs of the ERs of the SEAs of the MMPs, the spillover effects of the implementation of these planning policies, i.e., the positive impacts that go beyond the integration of CCA into the LFs of the SEAs of the MMPs, should also be taken into account. FAO, for example, defines sustainable forest management as an evolving conceptual and technical category that aims to conserve and enhance the economic, social and environmental values of all forest types for the benefit of current and future generations, a category that consists, fundamentally, of making sure that the available productions, with reference to both timber and food, come from production systems that ensure an equitable intergenerational distribution of supply [32].

Forest management is, also, of great importance with reference to the abatement of concentrations of pollutants in water from spills generated by any productive activities, agricultural or industrial, that occur upstream [33]. The increase in the availability of the resource in subsurface water reservoirs, as a result of the substantial in hydraulic and naturalistic engineering works, implemented according to the conceptual references recalled above, constitutes an important positive impact regarding the management of water resources at the large territorial scale.

Natural forest ecosystems are reservoirs of biodiversity and constitute a relevant defense against the negative impacts of climate change [34]. The restoration of degraded forest and wooded areas and the planting of new forests are important references for the definition and implementation of effective land policies, including in economic terms, in relation to the balance between costs and benefits, both in the medium and long term [35].

Plan actions related to the micro-area scale also include those aimed at phytodepuration, such as the establishment of adequate riparian vegetation in river belts, retention and filtration wells, and artificial wetlands: these can be extremely effective in treating effluent before it is discharged into rivers [24]. In conclusion, it should be emphasized how a fundamental issue is represented by the inclusion, in prescriptive terms, of the plan actions concerning CCA, generated within the SEA processes, in the implementation codes of the MMPs.

One feasible approach to incorporate a set of rules explicitly linking CCA-related planning policies with the implementation of land-use regulations into the MMPs' statutory code is to establish project worksheets as prescriptive guidelines integrated within the planning code. These worksheets should provide detailed instructions on how to implement each identified plan action to address the specific objectives outlined in the MMPs' LFs. The use of such project worksheets is a well-established method, widely adopted in the planning regulations of various Italian cities, and extensively discussed in scientific and technical literature.

A notable example demonstrating the application of these project worksheets can be found in the Trieste Master Plan [36]. In this Master Plan, the project worksheet approach is used on a large spatial scale, particularly in areas earmarked for significant spatial transformations, urban regeneration zones, areas slated for urban renewal, and the creation of a new garden city. According to the Trieste Master Plan, project worksheets pertain to areas and subareas identified through the delineation of the relevant urban zones. Unless specified otherwise (with such specifications included in the worksheets), graphic elements are not prescriptive. However, the key project elements, quantitative parameters, implementation procedures, and land-use regulations are legally binding [36, p. 3].

These "key project elements" encompass quantitative parameters, urbanization efforts, and ecological criteria related to urban planning. These criteria are identified by factors such as permeability rates, tree and shrub densities, and should be integrated with the public service framework across different urban sectors, considering their endowments and planned interventions on existing settlements and expansion. The mandatory nature of these project worksheets concerning plan actions, as dictated by the Trieste Master Plan, necessitates a high level of technical detail to ensure the binding nature of the project rules, thereby ensuring the effective execution of these plans.

Nevertheless, the planning approach centered on project worksheets, which can be seamlessly integrated into standard spatial planning practices without the need for additional legislation, requires expertise not only in CCA-related technologies but also, perhaps most importantly, in financial aspects, project timelines, and the trained personnel essential for managing the operational phases of project implementation.

5 Conclusions

In the construction of MMPs which implement CCA into their LFs, SEA is connoted as a process in which planning and evaluation are progressively integrated, leading to the identification of a system of objectives and operations aimed at their pursuit, which constitute a strategy.

Within this strategy, the evaluation of the impacts of choices, i.e., of plan actions, is aimed at incrementally refining this system in such a way as to arrive at an overall result that is identified as the best strategy with reference to the implementation of local development processes that are configured as expressions of the best compromise, i.e., the most effective integration, between the instances aimed at nature conservation and the protection of archaeological, historical and landscape resources, social equity, economic development and, certainly, the CCA [16].

Information and participation, on the part of public administrations vis-a-vis local communities, are, likewise, fundamental characteristics for the effectiveness of the evaluation and planning process, as factors that favor, in a relevant way, the recognition of their demands [37].

The methodological approach proposed and applied in this study could serve as a valuable tool to support decision-making processes related to spatial planning, and it has the potential to be applied in various European Union member states.

This adaptability is attributed to its alignment with national planning regulations based on legislation stemming from Directive 2001/42/CE. Within these regulatory frameworks, there is ample room for the incorporation of SEAs into the decision-making processes of MMPs, where they can seamlessly integrate questions and policies related to CCA.

The employed methodology may effectively help civil servants, practitioners, and local authorities in addressing the repercussions of changes in land cover and land use. From this perspective, the inclusion of CCA-related measures within MMPs' policies can serve as a foundation for steering local decision-making towards prevention or, at the very least, mitigation of damages caused by climate change impacts [15].

Looking ahead, promising avenues for future research become apparent. One crucial focus should be on the implementation of SEAs for MMPs that entail legal provisions encompassing CCA-related activities within MMP implementation codes. Furthermore, an important area to explore is the role of local communities in shaping and executing environmental hazard management policies aimed at mitigating climate change impacts [38]. These initiatives should be built upon the gradual enhancement of scientific, technical, and cultural knowledge within local societies regarding climate change, climate change impacts and CCA [39].

Within this conceptual framework, the evolving awareness within communities can be identified as a key driver of the qualitative enhancement of local spatial, environmental, and landscape assets. From this standpoint, policies related to CCA can be seamlessly integrated into planning practices undertaken by local governments, reflecting societies that fully grasp the significance of nature and natural resources in terms of enhancing quality of life [40].

Acknowledgments. F.I., S.L, F.L., and C.Z. collaboratively designed this study. F.I. wrote Sect. 3.2. S.L. wrote Sect. 3.1. F.L. wrote Sect. 3.3. C.Z. wrote Sects. 1, 2, 4, and 5. This study was carried out within the RETURN Extended Partnership and received funding from the European Union Next-Generation EU (National Recovery and Resilience Plan –NRRP, Mission 4, Component 2, Investment 1.3 –D.D. 1243 2/8/2022, PE0000005), with the financial support under the National Recovery and Resilience Plan (NRRP), Mission 4, Component 2, Investment 1.1, Call for tender No. 1409 published on 14.9.2022 by the Italian Ministry of University and Research (MUR), funded by the European Union – NextGenerationEU– Project Title "Definition of a guidelines handbook to implement climate neutrality by improving ecosystem service effectiveness in rural and urban areas" – CUP F53D23010760001 - Grant Assignment Decree No. 1378 adopted on September 1, 2023, by the Italian Ministry of Ministry of University and Research (MUR). And within the framework of the research project "Implementation and revision of the regional strategy for climate change adaptation (SRACC). Macro-Action 2, Mainstreaming of the SRACC spatial framework: governance and environmental assessment" (2021–2023), funded by the Autonomous Region of Sardinia. This article was excerpted from Isola F., Lai. S., Leone F., Zoppi C. (2024).

Integrating climate change adaptation into municipal masterplans through Strategic Environmental Assessment (SEA). A case study concerning Sardinia. TeMA – Journal of Land Use, Mobility and Environment, Special Issue 1.2024, 61–78. https://doi.org/10.6093/1970-9870/10438.

Disclosure of Interests. The authors have no competing interests to declare that are relevant to the content of this article.

References

1. Sheate, W., Dagg, S., Richardson, Aschemann, R., Palerm, J., Steen, U.: SEA and Integration of the Environment into Strategic Decision-Making, Volume 3 (Case Studies). Final Report to the European Commission Contract no. B4–3040/99/136634/MAR/B4. ICON, London, United Kingdom (2001). https://discomap.eea.europa.eu/map/Data/Milieu/OURCOAST_047_DK/OURCOAST_047_DK_Doc1_SEAintegrationVol1.pdf. Accessed 11 Mar 2024
2. ENVIRONET (DAC) (Development Assistant Committee) (Network on Environment and Development Cooperation): Strategic Environmental Assessment and Adaptation to Climate Change. OECD, Paris, France (2010). http://content-ext.undp.org/aplaws_publications/1769217/SEA%20and%20Adaptation%20to%20CC%20full%20version.pdf. Accessed 11 Mar 2024
3. Wende, W., Bond, A., Bobylev, M., Stratmann, L.: Climate change mitigation and adaptation in strategic environmental assessment. Environ. Impact Assess. Rev. **32**, 88–93 (2012). https://doi.org/10.1016/j.eiar.2011.04.003
4. Gonzáles Del Campo, A., Gazzola, P., Onyango, V.: The mutualism of strategic environmental assessment and sustainable development goals. Environ. Impact Assess. Rev. **82**, 106383 (2020). https://doi.org/10.1016/j.eiar.2020.106383
5. Partidário, M., Verheem, R.: Impact Assessment and the Sustainable Development Goals (SDGs). IAIA FasTips n. 19. (2019). https://www.iaia.org/uploads/pdf/Fastips_19%20SDGs.pdf. Accessed 11 Mar 2024
6. McGuinn, J., Hernandez, G., Eales, R., Sheate, W., Baker, J., Dusik, J.: Guidance on integrating climate change and biodiversity into Strategic environmental assessment. Publications Office of the European Union, Luxembourg (2013). https://doi.org/10.2779/11869
7. Francini, M., Chieffallo, L., Gaudio, S.: Climate change as stressor in rural areas. vulnerability assessment on the agricultural sector. TeMA – J. Land Use Mob. Environ. Special Issue **1.2021**, 53–71 (2021). https://doi.org/10.6093/1970-9870/7422
8. Gambino, R.: Prefazione. In: Minucci, F. (Ed.): L'evoluzione del governo del territorio e dell'ambiente. IX–XII. UTET Libreria: Turin, Italy (2005)
9. Isola, F., Lai, S., Leone, F., Zoppi, C.: Land take and landslide hazard: spatial assessment and policy implications from a study concerning Sardinia. Land **12**(2), 359 (2023). https://doi.org/10.3390/land12020359
10. Frigione, B.M., Pezzagno, M.: The strategic environmental assessment as a "Frontline" tool to mediate regional sustainable development strategies into spatial planning: a practice-based analysis. Sustainability **15**(3), 2378 (2023). https://doi.org/10.3390/su15032378
11. JRC (Joint Research Center, European Commission): Guidebook How to develop a Sustainable Energy and Climate Action Plan (SECAP) Part 1 – The SECAP process, step-by-step towards low carbon and climate resilient cities by 2030. Publications Office of the European Union, Luxembourg (2018). https://doi.org/10.2760/223399

12. JRC (Joint Research Center, European Commission): Guidebook How to develop a Sustainable Energy and Climate Action Plan (SECAP) PART 2 – Baseline Emission Inventory (BEI) and Risk and Vulnerability Assessment (RVA). Publications Office of the European Union, Luxembourg (2018). https://doi.org/10.2760/118857
13. JRC (Joint Research Center, European Commission): Guidebook How to develop a Sustainable Energy and Climate Action Plan (SECAP) PART 3 – Policies, key actions, good practices for mitigation and adaptation to climate change and Financing SECAP(s). Publications Office of the European Union, Luxembourg (2018). https://doi.org/10.2760/58898
14. Isola, F., Leone, F., Pittau, R.: Evaluating the urban heat island phenomenon from a spatial planning viewpoint. a systematic review. TeMA – J. Land Use Mob. Environ. Special Issue **2.2023**, 75–93 (2023). https://doi.org/10.6093/1970-9870/10306
15. Lai, S., Leone, F., Zoppi, C.: Land surface temperature and land cover dynamics. a study related to Sardinia, Italy. TeMA – J. Mob. Land Use Environ. **13**(3), 329–351 (2020). https://doi.org/10.6092/1970-9870/7143
16. Lai, S., Zoppi, C.: Factors affecting the supply of urban regulating ecosystem services. empirical estimates from Cagliari, Italy. TeMA – J. Mob. Land Use Environ. **Special Issue 2**, 7–32 (2023). https://doi.org/10.6093/1970-9870/10194
17. Armson, D., Stringer, P., Ennos, A.R.: The effect of tree shade and grass on surface and globe temperatures in an urban area. Urban Forestry Urban Greening **11**(3), 245–255 (2012). https://doi.org/10.1016/j.ufug.2012.05.002
18. Bowler, D.E., Buyung-Ali, L.M., Knight, T.M., Pullin, A.S.: A systematic review of evidence for the added benefits to health of exposure to natural environments. BMC Public Health **10**(1), 456 (2010). https://doi.org/10.1186/1471-2458-10-456
19. Pataki, D.E., et al.: Coupling biogeochemical cycles in urban environments: ecosystem services, green solutions, and misconceptions. Front. Ecol. Environ. **9**(1), 27–36 (2011). https://doi.org/10.1890/090220
20. Berland, A., Hopton, M.E.: Comparing street tree assemblages and associated stormwater benefits among communities in metropolitan Cincinnati, Ohio, USA. Urban Forestry Urban Greening **13**(4), 734–741 (2014). https://doi.org/10.1016/j.ufug.2014.06.004
21. Calfapietra, C.: Nature-Based Solutions for Microclimate Regulation and Air Quality: Analysis of EU-Funded Projects. Publications Office of the European Union, Luxembourg (2020). https://doi.org/10.2777/383904
22. Roy, S., Byrne, J., Pickering, C.: A systematic quantitative review of urban tree benefits, costs, and assessment methods across cities in different climatic zones. Urban Forestry Urban Greening **11**(4), 351–363 (2012). https://doi.org/10.1016/j.ufug.2012.06.006
23. UNaLab: Nature based solutions – Technical handbook Part II (2019). https://unalab.eu/system/files/2020-02/unalab-technical-handbook-nature-based-solutions2020-02-17.pdf. Accessed 11 Mar 2024
24. Wild, T.: Research and innovation pritorities in horizon Europe and beyond. In: Wild, T., Freitas, T., Vandewoestijne, S. (eds.): Nature-based Solutions: State of the Art in EU-funded Projects, pp. 223–233. Publications Office of the European Union, Luxembourg (2020). https://doi.org/10.2777/236007
25. Reberski, J.L., Terzić, J., Boljat, I., Patekar, M., Baniček, I., Čupić, D.: Proline-CE Work Package T1, Activity T1.2 – D.T1.2.2-Transnational best management practice report (2017). https://programme2014-20.interreg-central.eu/Content.Node/CE110-PROLINE-CE-D.T1.2.2.-Transnational-best-management-pra.pdf. Accessed 11 Mar 2024
26. EEA (European Environment Agency): Nature-based solutions in Europe: Policy, knowledge and practice for climate change adaptation and disaster risk reduction. EEA Report no. 1/2021. Publications Office of the European Union, Luxembourg (2021). https://doi.org/10.2800/919315

27. Du, S., et al.: Mapping the capacity of concave green land in mitigating urban pluvial floods and its beneficiaries. Sustain. Cities Soc. **44**, 774–782 (2019). https://doi.org/10.1016/j.scs.2018.11.003
28. Depietri, Y., McPhearson, T.: Integrating the grey, green, and blue in cities: nature-based solutions for climate change adaptation and risk reduction. In: Kabisch, N., Korn, H., Stadler, J., Bonn, A. (eds.): Nature-based Solutions to Climate Change Adaptation in Urban Areas, Theory and Practice of Urban Sustainability Transitions, pp. 91–109. Springer International Publishing, Switzerland (2017). https://doi.org/10.1007/978-3-319-56091-5_6
29. Ruangpan, L., et al.: Nature-based solutions for hydrometeorological risk reduction: a state-of-the-art review of the research area. Nat. Hazard. **20**(1), 243–270 (2020). https://doi.org/10.5194/nhess-20-243-2020
30. Francis, L.F.M., Jensen, M.B.: Benefits of greenroofs: a systematic review of the evidence for three ecosystem services. Urban Forestry Urban Greening **28**, 167–176 (2017). https://doi.org/10.1016/j.ufug.2017.10.015
31. Perini, K., Rosasco, P.: Cost-benefit analysis for green façades and living wall systems. Building Environ. **70**, 110–121 (2013). https://doi.org/10.1016/j.buildenv.2013.08.012
32. FAO (Food and Agriculture Organization of the United Nations): Global Forest Resources Assessment 2020. Main Report. FAO, Rome, Italy (2020). https://doi.org/10.4060/ca9825en
33. Mysiak, J., Torresan, S., Perez Blanco, D.: Cambiamenti climatici e agricoltura nel Nordest. Fondazione Nordest, Venice, Italy (2019). https://www.fnordest.it. Accessed 11 Mar 2024
34. Forest Europe: State of Europe's Forests 2015 (2015). https://foresteurope.org/wp-content/uploads/2022/02/soef_21_12_2015.pdf. Accessed 11 Mar 2024
35. Mansourian, S., Doncheva, N., Valchev, K., Vallauri, D.: Experiences in Forest Landscape Restoration. Lessons Learnt from 20 Years of Floodplain Forest Restoration: The Lower Danube Landscape. WWF Field Series (2019). https://www.wwf.fr/sites/default/files/doc-2020-01/202001_Report_20_years_of_floodplain_forest_restoration_the_lower_danube.pdf. Accessed 11 Mar 2024
36. Regione Autonoma Friuli-Venezia Giulia-Comune di Trieste: Piano Regolatore Generale Comunale – Schede Progetto – P03 (2018). http://documenti.comune.trieste.it. Accessed 11 Mar 2024
37. Zoppi, C., Lai, S.: Assessment of the regional landscape plan of Sardinia (Italy): a participatory-action-research case study type. Land Use Policy **27**(3), 690–705 (2010). https://doi.org/10.1016/j.landusepol.2009.09.004
38. Lai, S., Isola, F., Leone, F., Zoppi, C.: Assessing the potential of green infrastructure to mitigate hydrogeological hazard. TeMA – J. Land Use Mob. Environ. **2021 s.i.**, 109–133 (2021). https://doi.org/10.6093/1970-9870/7411
39. Magnaghi, A. La bioregione urbana nell'approccio territorialista. Contesti. Città, Territori, Progetti **1**, 26–51 (2019). https://doi.org/10.13128/contest-10629
40. Magnaghi, A.: I caratteri innovativi dei Contratti di fiume: il Contratto di fiume del torrente Pesa (Toscana). In: Angrilli, M. (Ed.): BikeFlu. Atlante dei Contratti di fiume in Abruzzo. 30–39, Gangemi Editore international, Rome, Italy (2020)

From Global to Local Scenarios of Climate Change Impact on Natural Resources. The Role of Regional Planning in an Inner Area of South of Italy

Giuseppe Mazzeo(✉)

Department of Engineering, PEGASO Telematic University, Centro Direzionale, Isola 2, 80143 Napoli, Italy
giuseppe.mazzeo@unipegaso.it

Abstract. The global scenarios associated with climate change predict a rise in temperatures over the coming decades, with serious consequences in terms of economics and quality of life. The scenarios outlined by the IPCC and the European Environment Agency foresee the Mediterranean basin as one of the areas potentially most affected by these phenomena. The paper, starting from the global scenarios, hypothesizes a local scenario in a region of significant relevance in Southern Italy, the Province of Avellino. Due to its implications, the paper proposes to use the scenarios as a primary tool in the building of regional planning for this area.

Keywords: Climate change · Scenario analysis · Mediterranean basin · South of Italy · Regional planning

1 Introduction and Structure of the Paper

Climate change, given the high probability of negative impacts stemming from it, has become a central issue in shaping the majority of national and international medium to long-term policies, especially concerning natural resources. This leads to an increasing role of scenarios envisioning future transformations, starting from the fact that they already influence current local events and situations, as the growing frequency of extreme weather events and the damages they cause to territories.

The policies pursued by states, based on scientific considerations from supranational bodies – in particular, on the various reports of Intergovernamental Panel on Climate Change (IPCC) –, are directed towards mitigation actions (addressing the causes of climate change) and adaptation actions (addressing the current response to risks). However, these policies cannot swiftly reverse the trends of rising temperatures, hence negative impacts must still be anticipated in the near future.

However, it is to emphasize the tension surrounding the concept of climate change, mitigation, and adaptation. The international debate reflects the complexity and urgency

O. Gervasi et al. (Eds.): ICCSA 2024 Workshops, LNCS 14818, pp. 293–309, 2024.
https://doi.org/10.1007/978-3-031-65273-8_19

of addressing this global challenge. Starting from the condition that there's often disagreement on the severity and causes of climate change, which can hinder consensus on appropriate actions, the disputes on mitigation arise over how to reduce greenhouse gas emissions effectively and fairly. This involves balancing the responsibilities of developed and developing nations, as well as considering economic impacts and technological feasibility. At the same time the dispute about adaptation revolves around actions to adapt to the impacts of climate change, such as rising sea levels, extreme weather events, and shifts in agricultural patterns. Other questions arise about resource allocation, equity, and the role of governments versus local communities.

The interest in adaptation has increased significantly because anthropogenic climate change appears unavoidable and mitigation agreements are difficult to achieve. But the discourse over climate change adaptation can be critical. Taking it for granted as a strategy for coping with climate change simplifies the political and economic contexts in which this strategy could be applied, and it might also unintentionally subvert the vulnerable communities it wants to benefit. This because adaptation strategy is favoured by neoliberal policies, which shift responsibilities for social and environmental problems to the individual [1].

Many cities and territories around the world are increasingly developing plans and policies for either mitigation and adaptation to climate change. This approach can be problematic and result in trade-offs because efforts aimed at enhancing mitigation may hinder achieving adaptation aims, and viceversa. Mitigation measures may have negative impacts on adaptation by increasing exposure to risks. In contrast, adaptation measures may increase greenhouse gas emissions by, among other things, reducing efficiency and increasing energy demand [2].

Other emphasize the lack of clarity and multiplicity among and between these concepts allows them to continue to be widely adopted in different sectors of society [3]. In addition, limited information and insufficient resources are inherent challenges for climate policy, and policy makers must cope with actions to design and implement adaptation policies under conditions of scarcity. Also, if adaptation policy will be designed under these conditions, the perspective calls for greater attention to the social justice implications in the policy design process [4].

Overall, these tensions underscore the need for collaboration, innovation, and compromise on a global scale to address climate change effectively. It requires navigating diverse interests and perspectives while prioritizing the long-term health of the planet and its inhabitants.

Attention to natural resources assumes particular relevance in the mitigation and adaptation perspective. They provide a range of fundamental services for life on the planet and are studied from various viewpoints, focusing on one or more of them at a time.

The aim of the paper is to apply a scenario analysis, albeit in preliminary form, on the possible impacts of climate change on a territory (the case study) characterized by natural resources that need to be preserved and protected for their strategic importance, and on the responses in terms of territorial governance.

The paper starts from an analysis of the general scenarios proposed by the latest IPCC report and their relevance for the Mediterranean area (Sect. 2). Section 3 highlights the case study (the Province of Avellino, Italy) analysing some key indicators of the territory. These indicators are strategic for the territory, but they are also the most exposed to adverse scenarios related to climate change. Section 4 presents a local scenario concerning the case study. Finally, recommendations for the future regional planning follow (Sect. 5).

2 IPCC Scenarios

The IPCC Synthesis Report, presented in 2023, highlights a series of evidence and scenarios characterizing the current state and foreseeable evolution of climate change [5].

Evidence and forecasts are grouped into three temporal situations:

- Actual state.
- Long-term scenario.
- Short-term responses.

Then, the IPCC develops a specific scenario for the Mediterranean area, which is reported in the final part of the chapter.

2.1 Actual State

Human activities, primarily through greenhouse gas emissions, are the cause of global warming, with the global surface temperature reaching 1.1 °C above 1850–1900 levels during the period 2011–2020. Global greenhouse gas emissions continued to rise during the period 2010–2019, with contributions stemming from unsustainable energy production, land use, lifestyle and consumption patterns, and production models among regions, countries, and within countries and individuals. Human-induced climate change is already influencing a range of extreme weather and climate events in every region of the world. This has led to widespread negative impacts on food and water security, human health, economies, and society, with associated losses and damages to nature and people (Fig. 1).

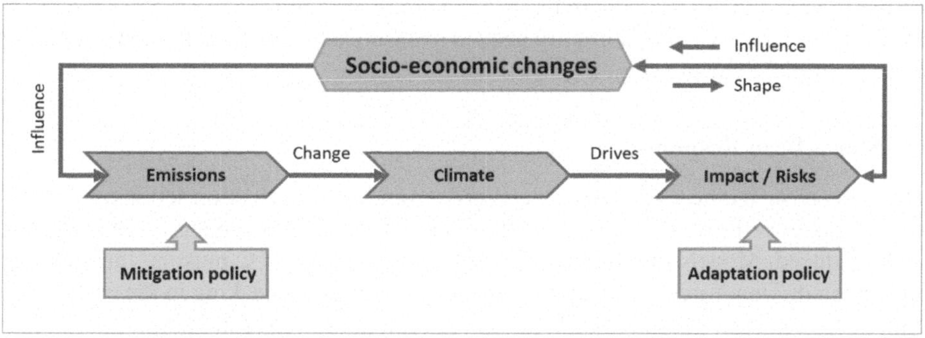

Fig. 1. The cause-effect chain joining the emissions and the impacts for climate change [5]

The responses of the international community have seen the signing of climate agreements and the of growing national ambitions for climate action. They, along with increasing public awareness, are accelerating efforts to address climate change across multiple levels of governance. Mitigation policies have contributed to a decrease in global energy and carbon intensity. Low-emission technologies are becoming increasingly accessible, with many low or zero-emission options now available for energy, buildings, transportation, and industry. Adaptation planning and implementation progress have yielded multiple benefits, with effective adaptation options having the potential to reduce climate risks and contribute to sustainable development.

However, current mitigation and adaptation actions and policies are not sufficient as there are gaps between global ambitions and the sum of declared national ambitions. In terms of mitigation, projected global greenhouse gas emissions for 2030 would lead to warming exceeding 1.5 °C over the course of the 21st century. This would make it more challenging to limit warming to below 2 °C. Gaps persist in adaptation, with many initiatives prioritizing short-term risk reduction but falling short of substantial adaptation.

2.2 Long-Term Scenario

The future warming will be driven by future emissions. It will affect all major components of the climate system, as each region will undergo multiple and concurrent changes. Many of the climate-related risks will increase compared to assessments made in past years, and the long-term impacts are hypothesized to be many times greater than those currently observed. The multiple climate and non-climate risks will interact, resulting in overall and cascading damages across all sectors and regions.

With increasing warming, adaptation actions will become more limited and ineffective. At higher levels of warming, losses and damages will increase, and additional areas will reach adaptation limits. This will increase vulnerability, exposure, and risks, although long-term planning may have beneficial effects if it includes flexible, multisectoral, and inclusive adaptation actions.

Regarding mitigation processes, the report emphasizes that limiting global warming caused by humans requires net anthropogenic CO_2 emissions to reach zero. On the other hand, pathways even with temperature increases of 1.5 °C and 2.0 °C entail rapid, profound, and immediate reductions in greenhouse gas emissions across all sectors. However, the windows of opportunity needed to ensure a liveable and sustainable future for all are rapidly closing (Fig. 2).

2.3 Short-Term Responses

In the short term, it is necessary to implement deep and rapid mitigation actions and accelerate adaptation efforts. This way, the risks of climate change to humans and ecosystems can be reduced. Models that limit warming to 1.5 °C and 2.0 °C must be immediately implemented. They foresee global greenhouse gas emissions peaking in the early years of the second decade of the 2000s followed by rapid and deep reductions.

Accelerated implementation of adaptation actions will enhance well-being by reducing losses and damages. Rapid and sustained mitigation actions entail higher initial

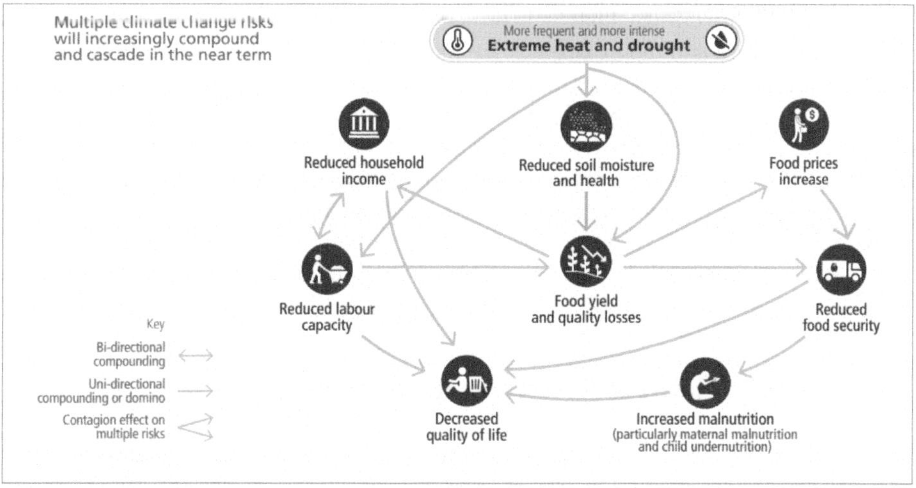

Fig. 2. Example of complex risk, where impacts from climate extreme events have cascading effects on food, nutrition, livelihoods, and well-being of smallholder farmers [5]

investments but will reduce future adaptation costs and improve co-benefits associated with sustainable development. Nevertheless, even in the short term, extreme events will increase, interacting with each other. Continuing with current development patterns means increasing the exposure and vulnerability of ecosystems and people to climate risks.

2.4 Scenarios for Mediterranean Region

Climate change strongly interacts with other environmental problems in the Mediterranean Basin, resulting from urbanisation, land use change, overfishing, pollution, biodiversity loss and degradation of land and marine ecosystems [6]. Main phenomena are:

- surface temperature in the Mediterranean region is 1.5 °C above the pre-industrial level;
- increasing of high-temperature extreme events;
- variable trends in precipitation across the basin;
- more frequent and intense droughts, especially in the north Mediterranean;
- warming of the sea surface by 0.29 °C–0.44 °C per decade since the early 1980s with stronger trends in the eastern basin;
- acceleration of the rising of sea level;
- increasing of sea acidity.

Climate change is the main cause of a growing number of impacts across the entire basin. These impacts include multiple consequences as longer and/or more intensive heat waves, droughts, floods, sea acidification and sea level rise, such as cascading impacts on marine and terrestrial ecosystems, as well as on land and sea use (agriculture, forestry, fisheries, tourism, recreation, etc.) and human health.

In the next future climate change is projected to intensify throughout the region. Air and sea temperature and their extremes are likely to continue to increase more than the global average.

The report underlines some of the expected main changes: the growing warming on land, the decrease of precipitation, the growing rainfall extremes in the northern part of the region, the growing of droughts. "Due to its particular combination of multiple strong climate hazards and high vulnerability, the Mediterranean region is a hotspot for highly interconnected climate risks. The main economic sectors in the region (agriculture, fisheries, forestry, tourism) are highly vulnerable to climatic hazards, while socioeconomic vulnerability is also considerable" [6, 2235].

3 Case Study. Natural Resources in the Province of Avellino

This paper focuses on the use of scenarios to build relationship between natural resources and land use planning. Natural resources are in a fragile state in the face of climate change and the scientific community pays considerable attention to the topic, which is being developed through several different lines of study.

The following analysis use three types of resources: food, water and energy. They are fundamental natural resources, and also the topic of the FEW (Food-Energy-Water) Nexus model [7–9], a theoretical framework used to analyse the interconnections and interdependencies among these resources. It recognizes that these three essential resources are strictly linked and that changes in one system can have significant impacts on the others. Furthermore, this model deepens the methodological and analytical tools to quantify the intensity of the relationships between them both locally and globally.

The province of Avellino is an inland province of the region of Campania in southern Italy (Fig. 3). The area is rich in natural resources: it contains regional nature reserves and protected areas, its soil favours high quality agricultural production [10], and weather conditions are conducive to a significant generation of electricity from renewable sources. Furthermore, the main springs feeding the Apulian Aqueduct [11] and supplying water to the Metropolitan City of Naples originate in southern Irpinia.

Part of the province lies within the Alta Irpinia Pilot Area, one of the four inland areas selected by the Campania Regional Authority as part of the National Strategy for Inland Areas (SNAI). In 2017, the relative Strategy Document was approved by the Regional Authority [12].

3.1 Soil, Agriculture, Food

The province of Avellino is 2,806.07 km^2. Soil consumption for Campania and Irpinia are 10.49 and 7.36% of surface area, respectively (for Italy the value amounts to 7.13%). The total amount of soil consumed to 20,536 hectares [13]. Soil not consumed comprises basically natural and agricultural soils.

Irpinia has extensive forest cover while its farmland is noted for its high-quality products [10]. For example, in the central part of the province there are internationally acclaimed winemakers.

Fig. 3. Province of Avellino in the Italian peninsula

3.2 Water

In terms of the organization of water resources, the provinces of Avellino and Benevento are part of the optimal territorial area ATO-1 of Campania, known as the Calore Irpino. ATO is the acronym of Ambito Territoriale Omogeneo, a territory on which integrated water services are organized. The Plan of (ATO-1) Campania [15] shows that the province of Avellino has significant water resources deriving from the presence of numerous natural springs. Most of Irpinia's water resources are used to supply drinking water to areas outside the province: the water of the Caposele and Cassano springs are channelled towards East (Puglia and Basilicata), while the water of the Serino springs are channelled towards the metropolitan area of Naples. Overall, 91% of water resources come from springs and 9% from wells. Water production in 2012 was estimated at approximately 9,524 litres per second.

The water balance of the ATO-1 in 2012 (Fig. 4) shows that 33% of the spring waters are used to supply drinking water for the ATO-1 itself, while the remaining part (67%) is used outside the province.

Overall, the average flow rate available in the ATO-1 area should be 3,244 L per second. In practice, given that water infrastructure censuses reveal average losses of

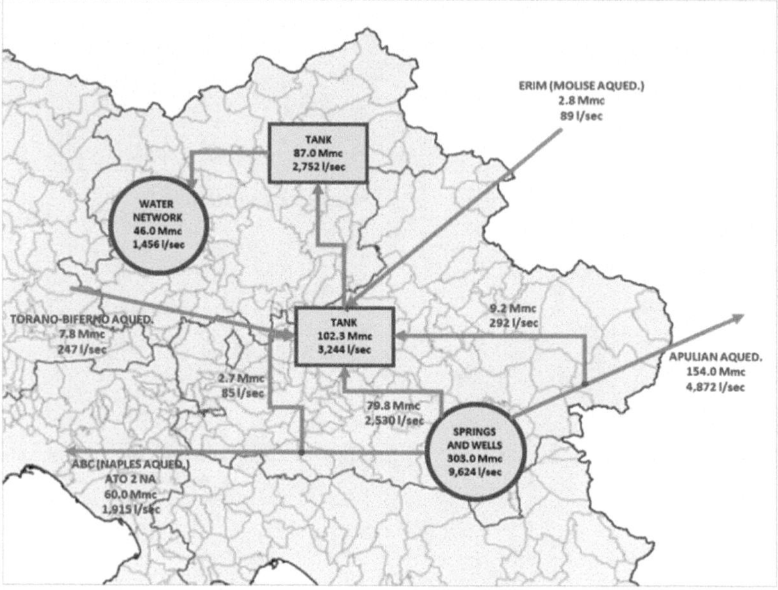

Fig. 4. Water balance of ATO-1 of Campania (2012) [15]

Table 1. Full data of the resources available in ATO-1 [15].

Flow rate	Springs	Wells	Surface water	Total
Supplied water (l/s)	3,965	1,955	0.0	5,920
Mean derived (l/s)	6,642	1,153	0.0	7,795
Maximum derived (l/s)	9,245	1,121	0.0	10,367
Minimum derived (l/s)	3,652	483	0.0	4,135
Annual mean derived volume (mc x 1,000)	204,664	33,855	0.0	238,519

15% for the adduction network and 47% for the distribution network, the effectively distributed flow to users is approximately 1,458 L per second. Table 1 shows the total size of the flow rates within the area of the ATO-1.

3.3 Energy

Production of electricity from sustainable sources represents a sector with significant potential impact [16]. The inland areas of Campania and stretches of Molise and Puglia extending as far as the Adriatic coast account for some of the main production areas of renewables, namely solar and wind power. The plants built so far in Irpinia produce approximately 14% more energy from renewable sources than that actually consumed

(1,513.7 GWh vs. 1,329.3 GWh) [17]. Wind power generation represents the dominant share of renewable energy production.

4 Local Scenario

The analysed natural resources can be the main topic for the development of a scenario analysis, with the aim to aid stakeholders and decision-makers in exploring potential future outcomes and evaluating the potential consequences of different actions (plans) and policies, as suggested by [18, 19], and [20].

4.1 Elements Influencing the Scenario

The construction of the scenario begins with the identification of the elements that can influence future trajectories in some way. These impactful conditions can be internal and external to the territory. Below, they are related to the three resource systems.

The local system of agricultural production must be framed within a broader horizon that sees agriculture being transformed into a sustainable sector able to reverse the current state in which it is viewed as a non-secondary source of emissions of climate-changing gases. In this regard, Community policies are very pressing and the European Green Deal and its Farm to Fork Strategy consider that it will contribute to sustainability goals such as social well-being, ecosystem health, and food and nutrition security [21]. Climate change, however, may negatively impact the expansion of areas designated for agriculture and the yield per hectare of crops.

Regarding water, the plan of ATO-1 highlights several critical elements. Despite the presence of numerous underground water bodies, the province's hydrological balance is negative, and it cannot be modified by using springs not yet connected to the water network. According to the Campania Regional Authority [15], in this context it is therefore essential to adopt measures to rationalize the management of water resources, aimed at partial recovery of the resource currently piped out of the region and focus on its regional distribution, as well as modernisation of existing works, to ensure the reduction of losses from aqueducts.

An aspect of great importance is annual rainfall (Fig. 5). Irpinian springs owe their water flows to winter and spring precipitation affecting the southern and western mountain areas. In a paper [22] showed that in the period 1981–1999 rainfall in Campania decreased by 15% compared to the previous 20 years. This decrease has a greater impact in mountain and hill areas. Over the same period, the number of rainy days has decreased by a similar percentage.

The authors attribute this negative variation in the rainfall to global climate change. In the same period, temperatures in Campania reveals an overall average increase of 0.3 °C, resulting from +0.2 °C in the flat coastal areas and +0.5 °C in inland and mountainous areas. Furthermore, in Campania the water resources have been decreased in flows from sources and lowering of the piezometric surface. If this trend continues, they conclude that "a 70% decrease in groundwater resources could occur in the next 50 years, and groundwater management in Campania should be overhauled. In 2050, according to the scenario set out above, a population of several million people (population of Campania

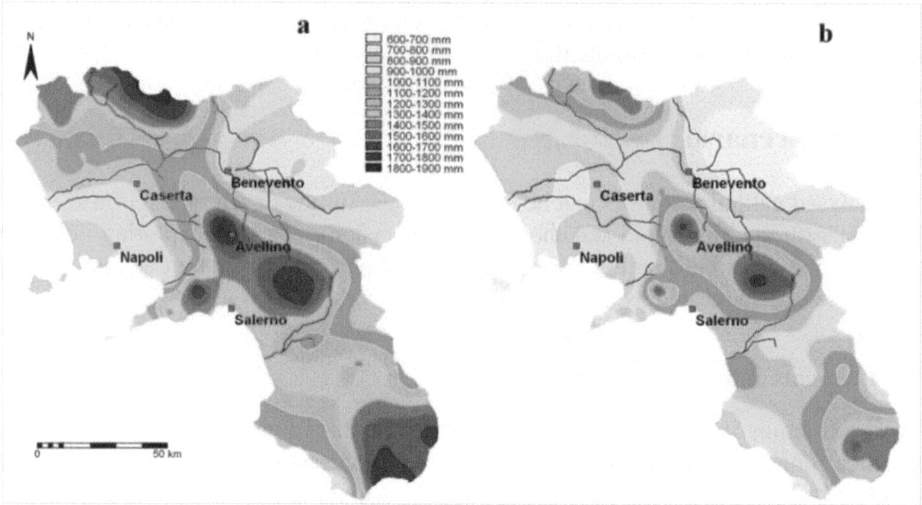

Fig. 5. Average annual rainfall (mm/year) for the period 1951–1980 (a) and 1981–1999 (b) [22]

and Puglia, currently supplied with groundwater from Campania) may be in a critical situation" [11, 22].

Regarding the energy, the development of this production chain has also triggered a rise in local protests against the construction of new wind farms accused of generating a strong environmental impact, of being a vector of landscape fragmentation and, ultimately, of sometimes being based on poorly transparent business initiatives.

The current European directive on the promotion of renewable energy (Directive 2018/2001) establishes that Member States must achieve by 2030 a share of energy from renewable sources equal to 32% of gross final consumption, and that the share of the transport sector must reach 14%. The Integrated National Plan for Energy and Climate [23] sets the Italian targets at 30 and 22%, respectively.

At the Community level, there is discussion on an acceleration of this process ("Fit for 55") that will raise the general target from 32 to 40% in 2030. This means that the focus on renewable energy can only grow further and will make it necessary to build new production plants, as well as boost research activities in the sectors of energy production, conservation and distribution.

4.2 The Scenario

Taking what has been said into consideration, the scenario is split two sections: the first starts from the current state of resources, while the second assumes possible evolutionary trajectories and it refers to a possible future scenario. Both the first and second use the same four types of qualitative indicators: (1) Flows of goods; (2) Quantity of goods; (3) Local impact; and (4) Outlook. For each indicator, one opinion was expressed for the current state and the future state. Some judgments are followed by trend indications placed in square brackets. Figure FF shows the results of the assessment both verbally and graphically.

The building of the scenario is based on the data presented in the sections above, and on the hypothesis that in the near future there will be a strong step change in the decision-making system, such as to modify current area trends.

Current state - At present the province produces food resources exported in significant quantities. Yet imports are also significant. The local impact of this resource can be considered significant, even if it does not yet express all its potential in terms of impact on the province. The outlook is positive.

Since the energy produced is higher than that consumed in the province, the latter may be said to be a net exporter even if the quantities are not significant. Although the local impact of this resource is limited in economic terms, it is critical from the social point of view due to opposition from some local communities. The outlook is positive.

The province is a net water exporter, even if in this case the exported quantities are decidedly greater and more significant than those consumed locally. Of the resources analysed, water is the most important and strategic, especially in consideration of its impact on some communities outside the province. Prospects for the future are stable.

Potential forecast - Turning to the outlook, the considerations that can be hypothesized give rise to substantially different scenarios from those relating to the current state. As regards food, import/export flows between the province and outside areas will continue. Quantities could change significantly: they could grow if agriculture gains weight within land use development policies. This could have a high local impact both in economic and social terms. The positive outlook could consider the negative impacts of climate change or other events that could create critical situations within one or more agricultural production sectors.

The energy sector confirms the judgment on flows, with a province that can strengthen its position as a net exporter. Due to its characteristics, it should significantly increase the amount of energy produced. The local impact of the sector is more nuanced, which should be taken into due account within future investment policies. Local impact means the possibility that energy will be used locally to develop specific economic sectors, but also the possibility that local communities can self-produce and market increasing amounts of energy. The prospects, however, appear positive (Fig. 6).

Finally, the water sector confirms the status of Avellino as an exporting province, even if there has been a reduction in spring flows in recent decades as a result of changing climatic conditions and lower rainfall. The resource will continue to have a significant local impact, which will grow if sustainable water use becomes the basis of economic development. In the case of this resource, the outlook can be considered negative and potentially critical because significant reductions in water supplies from springs due to climate change cannot be excluded. Hence the urgent need to draw up policies that are much more attentive to the use of the resource.

The scenario presented earlier is qualitative and should be considered as the initial step of a future research, aimed at deepening the data and at developing the scenario structure that is useful for applying territorial policies and plans.

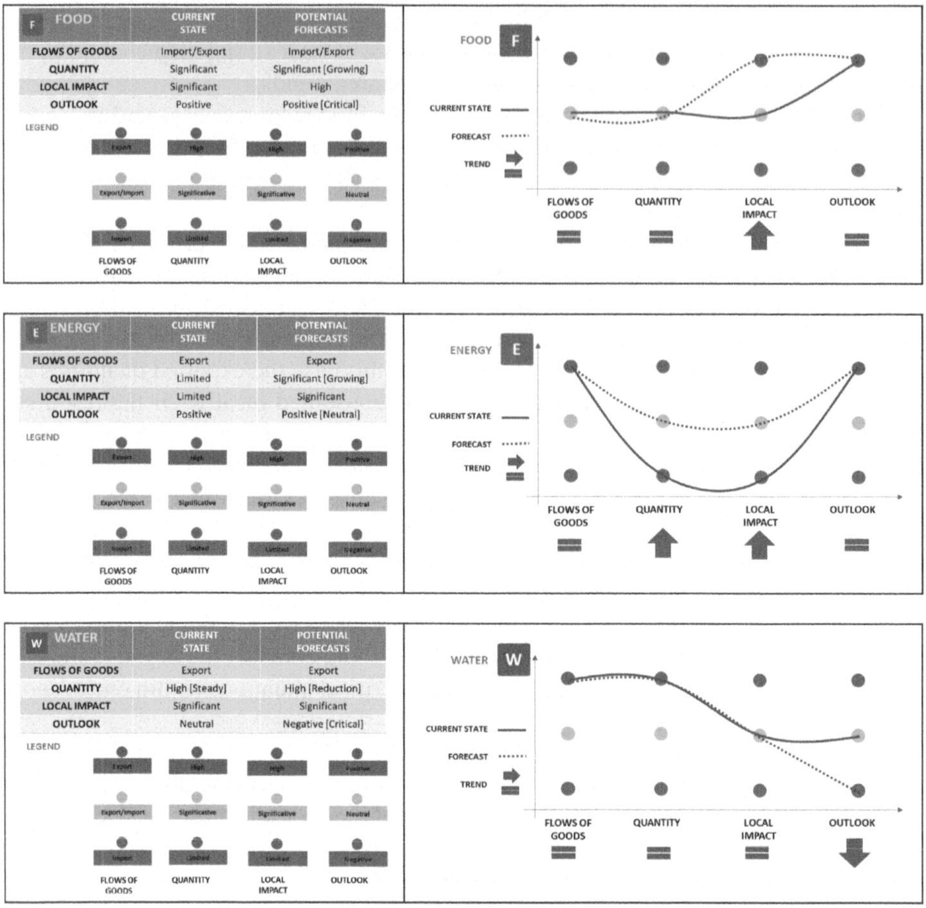

Fig. 6. State and scenario qualitative indicators as applied to the case study of the FEW Nexus model [24]

5 Recommendation for the Future Regional Planning

Planning addresses the necessity of incorporating sustainability and resilience elements into the tools [25]. These elements must have an environmental focus, guiding urban and land-use systems toward active management of natural capital and ecosystem services [26, 27]. They should also encompass social aspects. In this context, a particular challenge affecting inland areas, such as the province of Avellino, is their livability. This extends beyond overall environmental quality (which is crucial) and includes the quality of services provided to the resident population, enabling them to lead their best possible lives. These considerations are contained in the Territorial Coordination Plan (PTCP) of the Avellino Provincial Authority, approved in 2014. According to the Provincial Authority [28], ensuring that people live better means offering them appropriate services, thereby avoiding depopulation of marginal lands and concentrations of population only in the largest urban areas.

The purpose of the plan is to organise the province so that the system of services and production activities operating within it can support the existing population and, if possible, lead it to increase in the future. To combat the demographic crisis pre-existing territorial strengths need to be enhanced and improved.

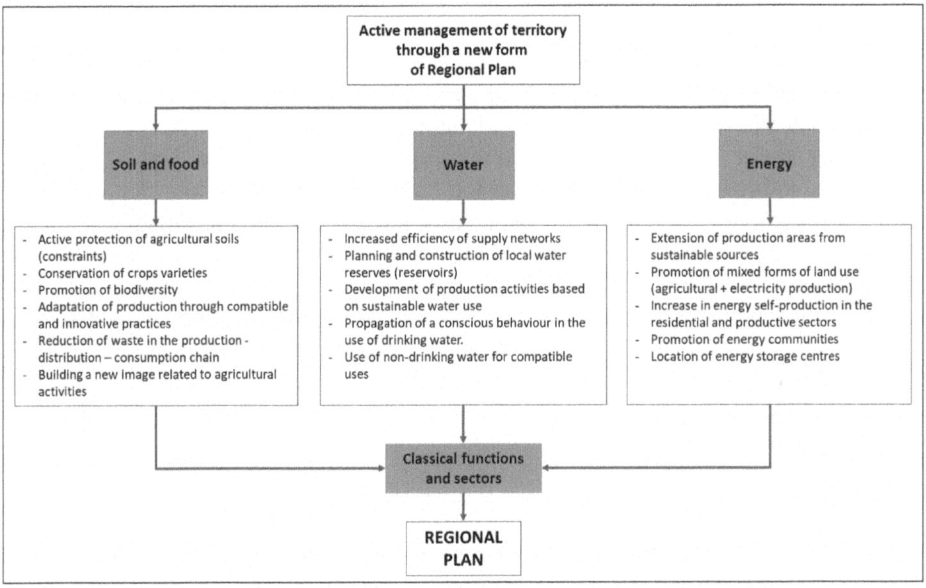

Fig. 7. New strategic elements for a regional plan of the province

Classical planning defines land use methods, urban loads, functions, and interrelationships between spaces and functions, with little attention to the impacts that such forecasts may have on primary area resources, such as water, food and soil. Such plans fail in these basic aspects: they lack a vision of the systemic impacts of sectoral policies, while a rigid system of actions and procedures persists, and they unfold along separate lines that do not intersect [29].

The need for a substantial change in pace is evident from the analysis of Avellino provincial planning (but the reasoning can probably be extended to many other plans).

In an area such as Irpinia, the basic conditions require targeted actions able to achieve at least two important results. The first is to better preserve the system of existing resources, to be considered strategic because they are used to obtain the necessary goods. To ensure that such resources continue to be used in the future, they must be taken into consideration in all land management policies and in all plans, from the regional level to the project level involving major environmental changes [30]. The second result would be to reverse the trend in depopulation and the reduction in existing services and in life quality [31].

To this end, the plan to use must respond to specific characteristics: 1) systemic attention to the strategic resources; 2) the highlighting of relationships between such resources; 3) the ability to link resources and typical functions connected with regional planning, such as domestic and production sectors; 4) finally, the ability to reveal how such resources may be connected with areas outside the province.

Redefining the connections between land use planning and natural resources entails giving new meanings to the plan (Fig. 7). The latest generations of plans are distinguished by some specific characteristics such as landscape protection, soil consumption reduction or elimination, rewilding, urban regeneration, renewable energy production, and buildings with low carbon footprints. All the above aspects are important to be able to confer greater sustainability. However, it is necessary to take a step forward and ascertain the possibility of the plan, with its general systemic vision, becoming the tool that sets the balance between consumption and natural resource use.

Another aspect to consider is the need for the plan to use scenario analysis [32, 33]. A plan devised today unfolds in the next decade; therefore, it needs to consider alternative development scenarios. Construction of such scenarios, based on the strategic resources present in the area, creates a new type of land use plan in which visions of the future are identified and worked towards through specific actions.

Active management of a region therefore means giving the right weight to new elements. They must be the backbone of land-use planning which will have to expand its range of action from the simple regulation of land use to the ways in which this use is brought about.

6 Conclusions

The paper has underlined the role of scenarios in constructing interrelationships among different temporal state of a region. Scenarios, despite their inherent uncertainty, build knowledge that contributes to refining models for predicting changes. This knowledge can be directed towards the identification and test of a model of the relationship between resources that have so far been considered autonomous and analysed separately; the application of a systemic analysis, which involves exploring the relationships and intensity of flows generated due to these interconnections; and the investigation of the impacts of climate change on resources by determining the share of impact resulting from these interrelationships.

Starting from in-depth knowledge of the natural resources and from the relationships existing between them and the area in which they act, new guidelines may be drawn up for regional planning, which must change to keep abreast of new attention to land use and to natural resources. Such planning for the near future must increasingly become strategic, changing from a static to a dynamic structure that accompanies evolving land use, employing tools for the various scenarios envisaged.

The new form of planning should be capable of achieving three objectives [24]:

- increasing awareness of how the protection of natural capital can translate into an economic advantage and how it can affect settlement trends. Each of the components (water, soil, energy and others) has characteristics that can have a significant impact on the region and can lead to the construction of ecological districts that, in turn, can generate significant economic and settlement impacts;
- clarifying the meaning of natural resources, and modifying the perspective with which they are normally considered, especially highlighting the benefits resulting from usage

patterns that contribute to resource enhancement and that have a positive impact on environmental sustainability, on the quality of the area in question and on liveability;
– drawing up guidelines to update land-use planning tools in terms of effective application of the principles of sustainability and neutrality of the local carbon footprint and as a contribution to the parallel process of decarbonisation [34–36].

References

1. Jennings, T.L.: Transcending the adaptation/mitigation climate change science policy debate: unmasking assumptions about adaptation and resilience. Weather Climate Soc. **3**(4), 238–248 (2011). https://doi.org/10.1175/WCAS-D-11-00056.1
2. Sharifi, A.: Trade-offs and conflicts between urban climate change mitigation and adaptation measures: a literature review. J. Clean. Prod. **276**, 122813 (2020). https://doi.org/10.1016/j.jclepro.2020.122813
3. Zanotti, L., Ma, Z., Johnson, J.L., Johnson, D.R., Yu, D.J.B., M., Carothers C.: Sustainability, resilience, adaptation, and transformation: tensions and plural approaches. Ecol. Soc. **25**(3), 4 (2020). https://doi.org/10.5751/ES-11642-250304
4. Kuhl, L.: Policy making under scarcity: reflections for designing socially just climate adaptation policy. One Earth **4** (2021). https://doi.org/10.1016/j.oneear.2021.01.008
5. Lee, H., Romero, J. (eds.): Climate Change 2023. Synthesis Report - Contribution of Working Groups I, II and III to the Sixth Assessment Report of the Intergovernmental Panel on Climate Change. IPCC, Geneva, Switzerland (2023). https://doi.org/10.59327/IPCC/AR6-9789291691647
6. Ali, E., et al.: Cross-Chapter Paper 4 - Mediterranean Region. In: Climate Change 2022-Impacts, Adaptation and Vulnerability. Contribution of Working Group II to the Sixth Assessment Report of the Intergovernmental Panel on Climate Change. Cambridge University Press, Cambridge (2022). https://doi.org/10.1017/9781009325844.021
7. Hoff, H.: Understanding the Nexus. Background Paper for the Bonn 2011 Conference - The Water, Energy and Food Security Nexus. Stockholm Environment Institute, Stockholm (2011)
8. Zhang, P., et al.: Food-energy-water (FEW) nexus for urban sustainability - a comprehensive review. Resour. Conserv. Recycl. **142**, 215–224 (2019). https://doi.org/10.1016/j.resconrec.2018.11.018
9. Abdi, H., Shahbazitabar, M., Mohammadi-Ivatloo, B.: Food, Energy and Water Nexus - a Brief Review of Definitions, Research, and Challenges. Inventions **5**(56), 14 (2020). https://doi.org/10.3390/inventions5040056
10. Provincia di Avellino. Studi propedeutici al Preliminare del Piano Territoriale di Coordinamento Provinciale. Assessorato alla Pianificazione Territoriale, De Angelis Editore, Avellino (2004)
11. Balacco, G., Carbonara, A., Gioia, A., Iacobellis, V., Piccinni, A.F.: Evaluation of peak water demand factors in Puglia (Southern Italy). Water **9**(2), 96 (2017). https://doi.org/10.3390/w9020096
12. Regione Campania: Delibera della Giunta Regionale no. 305 del 31/5/2017, Strategia Nazionale Aree Interne - Approvazione del documento di strategia d'area dell'Alta Irpinia. BURC n. 50, 22/6/2017. (2017)
13. Munafò, M. (Ed.): Consumo di suolo, dinamiche territoriali e servizi ecosistemici. Edizione 2022. Report SNPA 32/22. ISPRA, Roma (2022)
14. Mazzeo, G.: Lo sviluppo dell'agricoltura di qualità in Irpinia. Urbanistica Informazioni **201**, 83–84 (2005)

15. Regione Campania, ATO-1: Piano d'Ambito - Rapporto ambientale - Aggiornamento 2012 (2012). http://www.atocaloreirpino.it/area-tecnica/piano-d-ambito/documenti-pda-2012.html
16. Mazzeo, G.: City and Energy Infrastructures between Economic Processes and Urban Planning. TeMA – J. Land Use Mob. Environ. **6**(3), 311–324 (2013). https://doi.org/10.6092/1970-9870/1929
17. Terna: Pubblicazioni statistiche. (2022). https://www.terna.it/it/sistema-elettrico/statistiche/pubblicazioni-statistiche
18. Kosow, H., Gaßner, R.: Methods of Future and Scenario Analysis - Overview, Assessment, and Selection Criteria. DIE Studies, 39. Deutsches Institut für Entwicklungspolitik gGmbH, Bonn (2008)
19. Alcamo, J. (ed.): Environmental Futures - The Practice of Environmental Scenario Analysis. Elsevier, Amsterdam (2009)
20. Paltsev, S.: Energy scenarios - the value and limits of scenario analysis. Wires Energ. Environ. **6**(4), 242 (2017). https://doi.org/10.1002/wene.242
21. European Environment Agency: Rethinking agriculture. Briefing no. 25/2021. European Environment Agency, Copenhagen (2021). https://doi.org/10.2800/090009
22. Ducci, D., Tranfaglia, G.: L'impatto dei cambiamenti climatici sulle risorse idriche sotterranee in Campania. Geologi. Bollettino dell'Ordine dei Geologi della Campania, **1**(4), 13–21 (2005)
23. Italian Government: PNIEC - Piano Nazionale Integrato per l'Energia e il Clima. Ministero dello Sviluppo Economico, Ministero dell'Ambiente e della Tutela del Territorio e del Mare, Ministero delle Infrastrutture e dei Trasporti, Roma (2019)
24. Mazzeo, G.: Analysis of strategic natural resources - the FEW Nexus model applied to Irpinia (Italy) and implications for regional planning. Tema. J. Land Use Mob. Environ. **2**, 123–142 (2023). https://doi.org/10.6093/1970-9870/10257
25. Colucci, A.: The Potential of Periurban Areas for the Resilience of Metropolitan Region. TeMA – J. Land Use Mob. Environ. 103–122 (2015). https://doi.org/10.6092/1970-9870/3655
26. Griggs, D., et al.: Sustainable development goals for people and planet. Nature **495**, 305–307 (2013). https://doi.org/10.1038/495305a
27. Lai S., Zoppi C.: Factors affecting the supply of urban regulating ecosystem services. Empirical estimates from Cagliari, Italy. TeMA – J. Land Use Mob. Environ. **2**, 7–32 (2023). https://doi.org/10.6093/1970-9870/10194
28. Provincia di Avellino: PTCP Avellino, Relazione generale (2014). http://old.provincia.avellino.it:8090/web/guest/p.t.c.p.
29. Pahl-Wostl, C.: Governance of the water-energy-food security nexus - a multi-level coordination challenge. Environ. Sci. Policy **92**, 356–367 (2019). https://doi.org/10.1016/j.envsci.2017.07.017
30. Mazzeo, G., Polverino, S.: Nature-based solution for climate change adaptation and mitigation in urban areas with high natural risk. TeMA – J. Land Use Mob. Environ. **16**(1), 47–65 (2023). https://doi.org/10.6093/1970-9870/9736
31. Friedman, E., Solecki, W., Troxler, T.G., Paganini, Z.: Linking quality of life and climate change adaptation through the use of the macro-adaptation resilience toolkit. Clim. Risk Manag. **39**, 100485 (2023). https://doi.org/10.1016/j.crm.2023.100485
32. Önkal, D., Sayım, K.Z., Gönül, M.S.: Scenarios as channels of forecast advice. Technol. Forecast. Soc. Chang. **80**, 772–788 (2013). https://doi.org/10.1016/j.techfore.2012.08.015
33. Meissner, P., Wulf, T.: Cognitive benefits of scenario planning - Its impact on biases and decision quality. Technol. Forecast. Soc. Chang. **80**, 801–814 (2013). https://doi.org/10.1016/j.techfore.2012.09.011

34. European Environment Agency: Sustainable Development Goals and the environment in Europe: a cross-country analysis and 39 country profiles. European Environment Agency, Copenhagen (2000)
35. United Nations: Transforming our World - The 2030 Agenda for Sustainable Development. A/RES/70/1. United Nations, New York (2015). https://sdgs.un.org/2030agenda
36. Ulpiani, G., Vetters, N., Melica, G., Bertoldi, P.: Towards the first cohort of climate-neutral cities - expected impact, current gaps, and next steps to take to establish evidence-based zero-emission urban futures. Sustain. Cities Soc. **95**, 104572 (2023). https://doi.org/10.1016/j.scs.2023.104572

Exploring Urban Metabolism Through Ecosystem Services Mapping: The Case Study of Turin (Italy)

Matteo Giacomelli(✉), Silvia Ronchi, and Stefano Salata

Department of Architecture and Urban Studies, Politecnico di Milano, Via Bonardi 3, 20133 Milano, Italy
matteo.giacomelli@polimi.it

Abstract. In a scenario of global urbanization and increasing impact of cities on climate change and loss of biodiversity, urban planning has a critical role in the path towards sustainability. The pathway towards climate neutrality requires novel tools and approaches for accounting for different components of urban metabolism relating cities to the surrounding territories. This paper explores the integration of a climate neutrality framework to address climate and biodiversity challenges in urban planning. Through the mapping of crucial ecosystem services (ES), the study presents a methodology to account for the (Eco)system capacity of the urban system, stimulating considerations in respect of urban impact towards inland territories in the case study of Turin, Italy. The study reveals the critical role of ES in enhancing urban sustainability and resilience, offering valuable insights for policymakers toward achieving climate neutrality.

Keywords: Urban Metabolism · Ecosystem Services · Climate neutrality

1 Introduction

Global policy is in a critical race to find strategies for addressing the escalating climate and biodiversity emergencies. Although urban areas are smaller in size compared to forests and agricultural lands, they account for over 70% of global emissions and cities are in urgent need to adopt low-carbon and resilient urban development strategies to mitigate climate change emissions and adapt to its impacts (Rezvani et al. 2023). This challenge is addressed in European policies through the concept of climate neutrality, which aims for net-zero greenhouse gas emissions by balancing emissions with their removal from the atmosphere (IPCC 2014). Within the EU Mission "Climate-Neutral and Smart Cities", the commission aims to deliver 100 climate-neutral and smart cities by 2030 (European Commission 2021), aligning with the broader objectives of the European Green Deal to achieve climate neutrality by 2050 (Salvia et al. 2023).

To address this challenge, urban planning is encouraged to utilize Nature-Based Solutions (NBS) for mitigating the urban heat island effect or using a less intrusive engineering approach to drainage management (Albert et al. 2019; Pauleit et al. 2019;

O. Gervasi et al. (Eds.): ICCSA 2024 Workshops, LNCS 14818, pp. 310–322, 2024.
https://doi.org/10.1007/978-3-031-65273-8_20

Ronchi et al. 2020). Among regulatory and planning solutions, Urban Green Infrastructure represents a network of green spaces and ecological systems planned with the scope of supporting the delivery of multiple Ecosystem Services (ES) such as water purification, air quality improvement, and climate mitigation and adaptation, vital for achieving climate neutrality (Ronchi et al. 2020).

Analyzing and managing the flow of energy and materials within cities concerns the concept of Urban Metabolism (Rosado et al. 2016; Perrotti and Stremke 2020) that transform these inputs into growth, energy, and waste. This approach can be applied to the development of strategies that address energy efficiency, waste reduction, and resource conservation in urban areas, thereby contributing to the achievement of climate neutrality. In the context of urban planning, this study considers cities as dynamic systems that both contribute to mitigation and adaptation to climate change. It advocates for actions that not only mitigate climate change by reducing emissions but also enhance the city's resilience and adaptability to climate-related challenges and extreme events. This approach ensures that efforts towards climate neutrality also contribute to the broader goal of sustainable urban development (Perrotti and Stremke 2020). A metabolic perspective can help decision-makers understand the socio-technical processes associated with the harvesting and using resources (especially natural) in cities. Enhanced awareness can facilitate a more resource-efficient urban planning.

Adaptation to foster resilience requires measurable targets and means of improving the efficiency and effectiveness of using resources, including energy. As mentioned above, reduction remains a pivotal aspect, encompassing minimizing waste from environmental pollution and reducing greenhouse gas emissions, among others (Lampinen et al. 2022). Additionally, optimizing water consumption and promoting sustainable transportation practices are integral facets of fostering resilience in resource management. The adaptive approach in urban planning requires the adoption of a multidisciplinary perspective accounting the relation between urban poles and the surrounding territory (Giacomelli et al. 2023) Specifically, a climate-neutral approach considers urban metabolism assessment in cities to define spatial policies and adaptive solutions should take into account the input-output flow of natural resources involving urban poles. Moreover, specific actions/urban design solutions should be socially inclusive and integrate the needs of urban stakeholders as well as vulnerable groups for urban planning solutions put in place (Lampinen et al. 2022; Giacomelli et al. 2024). However, increasing the adaptation of the system means developing an in-depth knowledge of the multiple dimensions that involve urban metabolism.

At the core of this framework is the notion that urban planning must balance between enhancing the system's capacity for reducing emissions and augmenting the ecosystem's capacity for offering benefits to people (among others, carbon sequestration and pollutant reduction). In this sense, through the concept of ES, it highlights the critical role of natural processes in improving air quality, regulating water, and supporting biodiversity within urban environments.

Considering this theoretical framework, the research aims to test and practically apply a methodology for urban metabolism assessment in the case study of Turin, Italy. Five regulatory ES were mapped within urban environments to inform urban planning and promote climate neutrality and draw consequences with respect to the relationship between the city of Turin and neighboring natural ecosystems. In this sense, the research aims to bridge the gap between theoretical ES assessment and practical urban planning needs, providing actionable insights applicable to other cities.

The study is organized as a research paper, the methodology introduces the study area and the models used for ES mapping and assessment. The results illustrate the distribution of the mapped ES (Carbon sequestration; Habitat Quality; Cooling Capacity; Water Regulation) and the total values for Air Purification. Within the discussion, a series of considerations on urban metabolism are developed in order to offer insights to urban planning. The last paragraph aims to offer shortcomings and draw directions for future researches.

2 Methodology

2.1 Study Area

The city of Turin (Piedmont, North-west of Italy) extends for 130.2 km^2, with 878,074 inhabitants (ISTAT, 2021). It is located in the Po Valley and develops over a flat urban area (239 m a.s.l.) and in a hilly area that reaches an altitude of 715 m. The choice as a case study is related to its unique environmental challenges, as the city has high levels of air pollution, with an average PM10 of 2018 measured in the center of Turin of 33.0 $\mu g\ m^{-3}$, and the maximum daily value, set at 50 $\mu g\ m^{-3}$, exceeded 55 times. Urban planning faces the challenges of managing urban sprawl, addressing the urban heat island effect, towards the improvement of an Urban Green Infrastructure supporting biodiversity and ES provision (Vitulano 2024).

Turin's historical and contemporary urban development presents an ideal context for applying and evaluating the developed methodology as the city is now working on the new land use plan (Piano Regolatore Generale). Furthermore, the city recently approved the Climate Resilience Plan (2020) and the strategic plan of Green Infrastructure (2021), showing a strong commitment to the integration of environmental and ES subjects into urban planning. Among the aspects taken into account by urban policies, it is well-known that imperviousness, among compactness, salinization, desertification, or the loss of organic matter, is a cause of degradation that the soil can receive from a land-use alteration reducing ecological functions (Salata et al. 2021). The so-called "land take" is defined as an unsustainable kind of land use alteration that compromise the possibility of the soil returning to its original natural properties (European Environmental Agency, 2019). Therefore, urbanization and the related sealing process is considered an irreversible dynamic since it does not allow to return to natural conditions while maintaining the initial stock of the (limited) soil resource (Fig. 1).

Turin is a deep "compact" urban catchment since the soil sealing rate for each land use demonstrates that the city has been subject to a concentrated anthropic transformation while creating a densely built-up core urban area with a low rate of permeability. Despite the eastern hilly area, the "Parco del Valentino" and the confluence between the Po and Stura rivers, the remaining areas have fewer natural unsealed areas. This distinctive characteristic is confirmed by the Natural Difference Vegetation Index (NDVI), which assesses the "health" of the vegetation even inside the compact built-up area. Recent research based on this indicator demonstrates that a few urban districts are "porous"

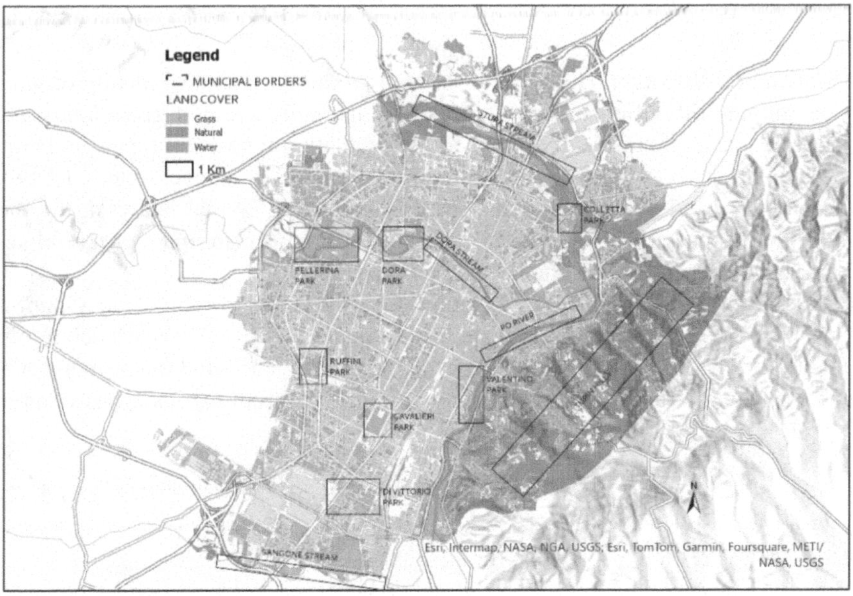

Fig. 1. City of Turin and location of the major streams and green areas. Authors own elaboration.

with healthy vegetation in the private space such as the Vanchiglia, the Vanchiglietta, Crocetta, and Borgo San Secondo.

Urban adaptation can be realistically pursued if the capacity to measure and evaluate the urban system is precise and reliable while reaching an accurate comprehension of the biophysical performance of the urban environment.

2.2 Ecosystem Services Mapping

The analysis of ES capacity for the city of Turin was developed through the mapping of four regulative ES: Carbon sequestration; Habitat Quality; Cooling Capacity; Water Regulation. The spatial assessment was developed through available data using the software Integrated Evaluation of ES and Tradeoffs (InVEST) (Natural capital project, 2024). InVEST is a suite of free, open-source software used to map and assess the goods and services from nature that sustain and fulfill human life. If properly managed, ecosystems yield a flow of services that are vital to humanity, including the production of goods (e.g., food), life-support processes (e.g., water purification), and life-fulfilling conditions (e.g., beauty, opportunities for recreation), and the conservation of options (e.g., genetic diversity for future use). In this sense InVEST aims to support assessments towards scarcely monitored natural capitals and, in many cases, undergoing degradation and depletion.

The methodology integrates InVEST's models with a novel approach to derive LULC data. The model employs land cover data derived from the digital topographic dataset (BDTRE 2023) of the city freely available at Geoportale Piemonte. To reach an updated and more grounded evaluation of this model, we self-produced an NDVI index of the city from an original acquisition of Copernicus L2A image processed on the 04 of

January 2024. We used the NDVI value to integrate LULC classes with the degree of soil permeability important for the assessment of ES provision.

ES1 Carbon Storage: The Carbon Storage model focuses specifically on quantifying the amount of carbon stored in vegetation and soils within a given landscape. It utilizes various data inputs such as land cover, vegetation types, soil types, and climate data to estimate carbon storage in different pools, including aboveground and belowground biomass, and soil organic carbon. By conducting spatial analysis, the model generates maps and tabular outputs that support efforts to prioritize conservation and restoration actions for climate change mitigation.

ES2 Habitat quality: InVEST's Habitat Quality model combines information on LULCs and biodiversity threats to create habitat quality maps. This approach produces two main sets of information that are useful for making an initial assessment of conservation needs: the relative size and degradation of different habitat types in an area,

Table 1. List of mapped Ecosystem Services (ES) and input data

ES	Ecosystem Service	Input	Source
ES1	Carbon Sequestration	-Amount of carbon stored in 4 different carbon pools	-Italian National Inventory of Forests and Forest Carbon Sinks
ES2	Habitat Quality	-Land Use/Land Cover	-BDTRE 2023
		-Anthropic threats (LULC)	-BDTRE 2023
		-Sensitivity of each land use to anthropic threats	-Based on Arcidiacono et al. 2015; Salata et al. 2017
ES3	Cooling Capacity	-Land Use/Land Cover Shade (%)	-BDTRE 2023
		-Evapotranspiration (mm)	-Ministry of Agriculture, Food and Forestry database
		-Albedo (K)	-Copernicus Global Land Services
		-Air temperature (°C)	-ARPA Piemonte
		-Imperviousness (K)	-High resolution layer – ISPRA
ES4	Water regulation	-Land Use/Land Cover	-BDTRE 2023
		-Map of soil hydrologic groups (USGS soil groups)	- Regional geological map
		-Map of average annual precipitation (10 years)	-ARPA Piemonte Regione Piemonte
		-Stormwater runoff and percolation coefficients for each Soil group (CN K)	-Map of soil protection capacity

and changes over time. This approach also allows rapid assessment of habitat status and trends as a proxy for more detailed measurements of biodiversity status.

ES3 Heat mitigation: Based on the InVEST Urban Cooling model, the indicator identifies the contribution of vegetation in reducing urban temperatures, which is crucial for enhancing urban resilience against climate-induced temperature increases. The model calculates an index of heat mitigation based on shade, evapotranspiration, and albedo, as well as distance from cooling islands (e.g. parks). The index is used to estimate a temperature reduction by vegetation. Finally, the model estimates the value of the heat mitigation service using two (optional) valuation methods: energy consumption and work productivity.

ES4 Water regulation: The urban flood risk mitigation InVEST model addresses cloudburst impacts, particularly in densely populated hillside cities. It calculates the runoff reduction, i.e. the amount of runoff retained per pixel compared to the storm volume, based on land cover, soil type, and rainfall data. Key inputs include watershed delineation, rainfall depth, land use/cover maps, soil hydrologic groups, and biophysical values for land classes. The model integrates various datasets, including imperviousness levels and tree cover density, to classify urban areas and assess their vulnerability to flooding, aiming to inform urban planning and mitigation strategies (Table 1).

2.3 Air Purification Assessment

The assessment of the **ES5 Air Purification** is based on the I-Tree Eco model, which allows to estimate the urban forest structure and its function on air pollutant removal. (i-Tree, n.d.). The software allows to estimate the grass effects on air pollutant removal and specifically CO, NO2, SO2, O3 and PM10. For running the Air Purification assessment, the Turin tree cataster has been collected and codified according to the I-tree model. The GPS coordinates of the individual trees were also entered to spatialize the outcome subsequently. Furthermore, some limited simplifications and adjustments were done for records missing any land registry data. The reference year for the meteorological and pollutants data is 2015, with data up to 2021 used for the models, excluding the

Table 2. Input data for Air Purification assessment

ES	Ecosystem Service	Input	Source
ES5	Air Purification	-Species (Identify and record the species and genus names of each tree assessed) -Diameter at breast height - DBH (Measure and record the tree stem diameter at breast height for each tree) -Total tree height (Height from the ground to the top (alive or dead) of the tree)	-Tree Cataster, Città di Torino

years impacted by the COVID-19 pandemic (2020, 2021) due to significant pollution reductions (Table 2).

3 Results

In this section, we present the findings of the assessment exploring the distribution and significance of selected regulating ES, identified as crucial for urban sustainability and climate resilience. The analysis encompasses four spatial mappings (Carbon Storage, Habitat Quality, Heat Mitigation, and Water Regulation) (Fig. 2) and one quantitative evaluation (Air Purification) (Table 3). By examining the results in the perspective of Urban Metabolism, we aim to provide valuable insights into the environmental dynamics of Turin and inform evidence-based urban planning strategies aimed at promoting climate neutrality and enhancing the city's ecological resilience.

3.1 The Distribution of Ecosystem Services

ES 1 Carbon Storage: The Carbon Sequestration model calculated an average pooling capacity of Turin urban soils to store 47 tons per hectare, including both soil and vegetation functions. The maximum pooling capacity, as expected, is recorded in the hills (see Fig. 2) with a maximum pooling capacity of 218 tons per hectare. This first result demonstrates the crucial role of seminatural areas surrounding cities, accounting the main share of CO2 absorption and oxygen generation.

ES 2 Habitat Quality: The Habitat Quality index in Torino presents a minimum value of 24.29 and a maximum value of 92.17. Well-maintained habitats are mainly distributed in the hills and along the main tributaries of the Po Rives that create an ecological connection between the western and eastern borders of the city (Dora and Stura). Although the hill displays higher values, it is noticeable how they are clusterized and not diffused, meaning that the area is a highly anthropized system with threats for the environment. The Dora Stream displays good values of Habitat Quality around the Pellerina and Parco Dora parks, but except from here, there is no trace of higher values, possibly meaning that the Riverbed has been highly channelized and heavily anthropized, leaving no space for natural habitats. Differently, the Stura stream is more natural, and along its course, it can be well-recognized as a higher Habitat Quality.

ES 3 Heat mitigation: As for heat mitigation, Torino presents a minimum value of 27.92 and a maximum value of 92.87. Four main core mitigation areas can be detected as hotspots to cool down temperatures during hot days. The hills are first, including the northeastern part where the Stura tributary reaches the Po River. Parco della Colletta also plays a vital role in connecting the natural areas of the hill with the peri-urban agricultural system. Also, the Basse di Stura cluster can be recognized as a strong mitigation point. In the south, Sangone Park seems to have a dimension that provides enough cooling capacity, thus representing the fourth hotspot of the city. It should be mentioned that, for this specific ES, the Po River creates a channel that crosses the city from south to north and provides air circulation with a freeze even during hot, stagnant days. Unfortunately, here the model does not account for wind interaction because it has been designed only

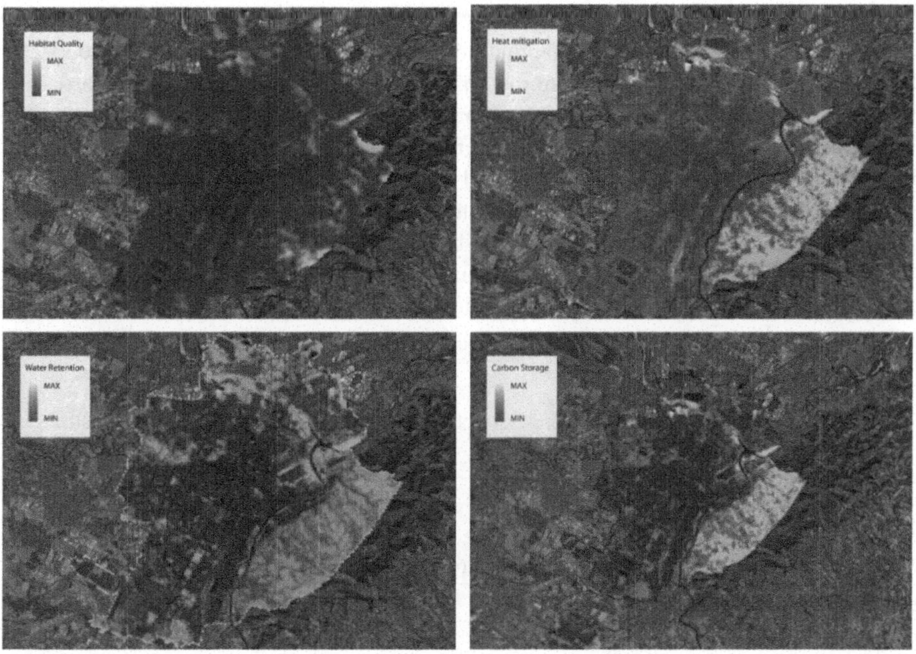

Fig. 2. Mapping of four Ecosystem Services for the City of Turin. Carbon Storage, Habitat Quality, Heat Mitigation, and Water Regulation. Authors elaboration

to assess the capacity of natural soil features (such as vegetation) to provide cooling capacity.

ES 4 Water regulation: As for heat mitigation, the Urban Flood Risk Mitigation integrates the information on land uses with soil hydraulic conductivity. The retention index indicates, for each pixel, the amount of total yearly water volume that can be retained by soil. Of this amount a small fraction goes to recharge the aquifer (percolation) while the remaining part evapotranspirates through roots absorption. In Turin, the model indicates that the 54% of the annual rain volume infiltrates on soil, while the 46% goes on runoff. Only a fraction of 1,34% of the infiltrated soil percolates and recharge the deep acquifer of the city (this amount correspond to 1.636.020 cube meters which is a valuable measure that can be used to see if the water demand is satisfied by the infiltration). Other studies on the city of Turin (Salata et al. 2020, 2021) show how the average infiltration capacity in the dense urban areas is lover of the average, often falling below the 10%. This might suggest that the large majority of the rainwater goes on runoff in the urban catchment. As expected, the hill has in general a good retention capacity, even though the soils are not performing well in terms of water conductivity.

3.2 The Assessment of Air Purification

The **ES Air Purification** Model estimated the pollution removal by the urban forest of Turin composed by 116.300 trees with a cover of 594.48 ha (Table 3). The most common species of Trees are Oriental planetree, Basswood, European hackberry. The estimation of pollution removal is 55 tons/year including Ozone (O3), Carbon monoxide (CO), Nitrogen dioxide (NO2), Particulate matter less than 2.5 microns (PM2.5), Particulate matter less than 10 microns and greater than 2.5 microns (PM10), and sulfur dioxide (SO2). In Turin, the carbon storage (i.e. the amount of carbon bound up in woody material above and below ground) is 40.59 thousand tons, the trees are estimated to store 40,600 tons of carbon. Oriental planetree stores and sequesters the most carbon, accounting approximately 16.1% of the total carbon stored and 20.5% of all sequestered carbon. Finally, according to I-Tree the removal of carbon (in the form of carbon dioxide) from the atmosphere through the process of photosynthesis is 1.566 thousand tons of carbon per year.

Table 3. Summary of Urban Forest effects and values from i-Tree Eco assessment.

Number of trees: 116.372
Tree Cover: 1471 acres
Most common species of trees: Oriental planetree, Basswood spp, European hackberry
Pollution Removal: 55.82 tons/year
Carbon Storage: 40.61 thousand tons
Carbon Sequestration: 1.568 thousand tons
Oxygen Production: 4.18 thousand tons/year
Avoided Runoff: 12.33 million gallon/year

4 Discussion

4.1 Urban Metabolism and Ecosystem Services Relationships

The application of the Urban Metabolism approach to the city of Turin can provide insights into the values of the urban environment and its implications for climate-neutral urban planning. Specifically, the framework can support the integration of system dynamics in addressing urban challenges and can help moving towards a knowledge base tool to support urban planners (Elliot et al. 2019). Knowledge on Ecosystem capacity can support the definition of spatial actions towards NBS designed to address site-specific situations (Salata and Grillenzoni 2021).

In this discussion, we address the implications of the ES assessment findings by comparing them with data on actual urban consumption. The analysis revealed significant patterns in the distribution of ES across Turin as the city benefits from substantial carbon storage capacity, particularly in semi-natural surrounding areas, which play a role in environmental regulation and habitat quality.

As concerns water balance, statistics indicate that the daily water consumption per inhabitant in the City of Turin stands at 197 L/day (Comune di Torino, 2016). Comparing this with the infiltration capacity of the urban ecosystem reveals a significant disparity, as only approximately 2.6% of the annual water usage by the urban population is actually infiltrated into the city's ecosystem.

In terms of CO_2 emissions, our assessment indicates that the Carbon Sequestration capacity of the urban ecosystem accounts for roughly half of the total greenhouse gas (GHG) emissions (Comune di Torino 2021). This contrast between urban ecosystem capacity and actual consumption underscores an unbalanced metabolism, where the demand for ES far exceeds what urban ecosystems can currently provide.

These disparities underscore the need for strategic interventions to rebalance the urban metabolism and enhance the resilience of Turin's urban ecosystem. Fostering greater integration of nature-based solutions within urban planning frameworks can help bridge the gap between demand and supply of ES, ultimately promoting a more sustainable and resilient urban environment.

4.2 Urban Planning Towards Climate Neutrality

The mapping exercise highlights environmental challenges in the context of Turin, including habitat degradation and flood risk. To address these issues, information on ES presents a possible base for developing targeted interventions that enhance the city's ecological resilience (Ronchi et al. 2020).

In response to the challenge of stormwater mitigation in the flat areas of Turin, proactive measures in the upslope regions are crucial. Establishing NBS, such as water reservoirs and seminatural basins with Bioretention systems, before streams reach densely populated settlements can significantly mitigate flood risks (Rezvani et al. 2023). Accompanying these measures with a comprehensive afforestation program in the upslope areas aligns with broader objectives of creating afforestation initiatives in peri-urban zones, acting as a natural buffer against urban sprawl (Salata and Grillenzoni 2021).

Moreover, re-naturalizing the riverbeds of the Po, Dora, Stura, and Sangone rivers offers an opportunity to create a natural pedestrian and slow-mobility connection between the urban fringe and central areas of the city. This approach seamlessly integrates peri-urban spaces with the urban context, fostering connectivity and enhancing biodiversity corridors (Salata et al. 2020). In densely urbanized zones where establishing seminatural water reservoirs may pose challenges, repurposing areas along streams for rainwater harvesting systems, filter strips, and green roofs can be explored as alternative solutions (Manes et al. 2016).

Integration of spatial urban information in the habitat quality allows defining ecological hotspots while supporting a new perspective for an ecosystem-oriented planning (Salata, 2020). Furthermore, confirming findings for other urban contexts (Manes et al, 2016), the urban forest of Turin contributes significantly to air purification, with specific tree species playing key roles in pollution removal. Manes et al (2016) further demonstrates how PM10 deposition on forests mainly depends on Leaf Area Index (LAI) while O3 absorption is mainly influenced by forest functional diversity. This information may be useful to manage urban green spaces more efficiently.

Initiating a sponge district necessitates a meticulous assessment of the natural drainage pattern using basic GIS tools like Digital Elevation Models for catchment areas (Salata et al. 2020). Visualizing streams with flow direction and accumulation tools aids in understanding topography and ground conditions, informing the identification of suitable areas for infiltration. Defining surface water sub-catchments involves considerations of flow routes and development clusters, ensuring strategic placement of parks and open spaces downstream to accommodate surface water attenuation and controls. These measures aim to integrate surface water conveyance paths harmoniously with the landscape, prioritizing integrated storage and treatment to maximize benefits while minimizing disruptive infrastructure.

4.3 Shortcomings and Further Research

The analysis relies on available data at the city level and the impact of some physical factors are not included. It's important to acknowledge that the analysis overlooks the influence of construction materials as the reflectivity of sealed vertical surfaces in tall buildings, which can significantly affect urban microclimates and contribute to the urban heat island. Future research should explore how these factors impact the urban climate.

Furthermore, while we assessed overall pollution removal by trees in Turin, extra research could spatialize the Air Purification model to identify areas within the city benefiting most from urban forests. Mapping pollutant removal rates across neighborhoods can inform targeted tree planting initiatives for improving air quality. In this context, investigating the relationship between urban parameters (e.g., land use, vegetation cover) and ES could provide insights into effective heat mitigation strategies. Understanding how socioeconomic factors and land use policies influence cooling capacity can guide interventions to enhance urban resilience to heat stress.

Finally, the urban metabolism perspective could be further explored by mapping the demand for ES, identifying matches and mismatches with areas of supply. The widening of the study area to the metropolitan region would allow to map ES budget in a larger territory, allowing to draw implications in terms of environmental and spatial justice.

5 Conclusions

In conclusion, this study bridges the gap between theoretical ES assessment and practical urban planning needs, providing actionable insights applicable to other cities. By embracing an urban metabolism approach and integrating ES considerations, cities like Turin can pave the way towards sustainable and resilient urban development, aligning with broader global objectives of climate neutrality and biodiversity conservation.

The disparities between the ecosystem capacity and the emission budget highlight the need for strategic interventions to rebalance urban metabolism and support the resilience of its ecosystem. Integrating nature-based solutions more extensively into urban planning frameworks can serve as a crucial step in bridging this gap, ultimately fostering a more sustainable and resilient urban environment. By integrating system dynamics and ecosystem capacity assessments, urban planners can develop knowledge-based tools to address urban challenges. Integrating NBS within urban planning frameworks can bridge the gap between demand and supply of ecosystem services, addressing pressing environmental challenges in Turin such as habitat degradation and flood risk. Proactive measures, including establishing NBS in upslope regions for stormwater mitigation and afforestation initiatives in peri-urban zones, can mitigate flood risks and act as natural buffers against urban sprawl. Renaturalizing riverbeds offers opportunities for creating connectivity between peri-urban and central areas, enhancing biodiversity corridors. In densely urbanized zones, repurposing areas along streams for rainwater harvesting systems and green infrastructure can provide alternative solutions. Integrating spatial urban information in habitat quality assessment supports ecosystem-oriented planning, while efficient management of urban green spaces contributes to air purification.

Although the analysis provides valuable insights, certain factors such as construction materials and vertical surface reflectivity's influence on urban microclimates require further exploration. Future research should spatialize air purification models to identify areas benefiting most from urban forests and inform targeted tree planting initiatives. Investigating the relationship between urban parameters and ES can provide insights into effective heat mitigation strategies, considering socioeconomic factors and land use policies.

Acknowledgements. This work was developed within Climaborough Horizon Europe Project, funded by the European Union. Views and opinions expressed are however those of the author(s) only and do not necessarily reflect those of the European Union or CINEA. Neither the European Union nor the granting authority can be held responsible for them.

References

Albert, C., Schröter, B., Haase, D., et al.: Addressing societal challenges through nature-based solutions: How can landscape planning and governance research contribute? Landsc. Urban Plan. **182**, 12–21 (2019). https://doi.org/10.1016/j.landurbplan.2018.10.003

Arcidiacono, A., Ronchi, S., Salata, S.: Ecosystem Services Assessment Using InVEST as a Tool to Support Decision Making Process: Critical Issues and Opportunities. In: Gervasi, O., Murgante, B., Misra, S., et al. (eds.) Computational Science and Its Applications – ICCSA 2015, pp. 35–49. Springer International Publishing, Cham (2015)

Comune di Torino: Consumi giornalieri di acqua per abitante 2000–2016 - aperTO. (2016) http://aperto.comune.torino.it/es/dataset/consumi-giornalieri-di-acqua-per-abitante-2000-2016. Accessed 27 Mar 2024

Comune di Torino: Turin Action Plan for Energy - Terzo rapporto di monitoraggio. (2021) http://www.comune.torino.it/ambiente/bm~doc/terzo-report-monitoraggio-tape-2019.pdf. Accessed 27 Mar 2024

Elliot, T., Babí Almenar, J., Niza, S., et al.: Pathways to Modelling Ecosystem Services within an Urban Metabolism Framework. Sustainability **11**, 2766 (2019). https://doi.org/10.3390/su11102766

European Commission. Directorate General for Research and Innovation: Communication from the Commission to the European Parliament, the Council, the European Economic and Social Committee and the Committee of the Regions on European missions. Publications Office, LU (2021)

Giacomelli, M., Pierantoni, I., Perna, P.: Un'infrastruttura verde regionale basata sulla domanda e offerta di Bundles di Servizi Ecosistemici. Una sperimentazione per le Marche. Contesti Città, territori, progetti, pp. 108–127. (2023) https://doi.org/10.13128/contest-14823

Giacomelli, M., Sargolini, M., Felipe-Lucia, M.R.: Including the perspective of stakeholders in landscape planning through the Ecosystem Services co-production framework: an empirical exploration in Le Marche. Italy. Regional Environmental Change **24**, 24 (2024). https://doi.org/10.1007/s10113-024-02184-w

Intergovernmental Panel On Climate Change: (ed) Summary for Policymakers. In: Climate Change 2013 – The Physical Science Basis, 1st edn. Cambridge University Press, pp 1–30 (2014)

Lampinen, J., García-Antúnez, O., Olafsson, A.S., et al.: Envisioning carbon-smart and just urban green infrastructure. Urban Forestry & Urban Greening **75**, 127682 (2022). https://doi.org/10.1016/j.ufug.2022.127682

Manes, F., Marando, F., Capotorti, G., et al.: Regulating Ecosystem Services of forests in ten Italian Metropolitan Cities: Air quality improvement by PM 10 and O 3 removal. Ecol. Ind. **67**, 425–440 (2016). https://doi.org/10.1016/j.ecolind.2016.03.009

Pauleit, S., Andersson, E., Anton, B., et al.: Urban green infrastructure – connecting people and nature for sustainable cities. Urban Forestry & Urban Greening **40**, 1–3 (2019). https://doi.org/10.1016/j.ufug.2019.04.007

Perrotti, D., Stremke, S.: Can urban metabolism models advance green infrastructure planning? Insights from ecosystem services research. Environment and Planning B: Urban Analytics and City Science **47**, 678–694 (2020). https://doi.org/10.1177/2399808318797131

Rezvani, S., De Almeida, N., Falcão, M.: Climate Adaptation Measures for Enhancing Urban Resilience. Buildings **13**, 2163 (2023). https://doi.org/10.3390/buildings13092163

Ronchi, S., Arcidiacono, A., Pogliani, L.: Integrating green infrastructure into spatial planning regulations to improve the performance of urban ecosystems. Insights from an Italian case study. Sustainable Cities and Society **53**, 101907. (2020) https://doi.org/10.1016/j.scs.2019.101907

Rosado, L., Kalmykova, Y., Patrício, J.: Urban metabolism profiles. An empirical analysis of the material flow characteristics of three metropolitan areas in Sweden. J. Clean. Prod. **126**, 206–217 (2016). https://doi.org/10.1016/j.jclepro.2016.02.139

Salata, S., Giaimo, C., Barbieri, C.A., et al.: The Utilization of Normalized Difference Vegetation Index to Map Habitat Quality in Turin (Italy). Sustainability **12**, 7751 (2020). https://doi.org/10.3390/su12187751

Salata, S., Grillenzoni, C.: A spatial evaluation of multifunctional Ecosystem Service networks using Principal Component Analysis: A case of study in Turin. Italy. Ecological Indicators **127**, 107758 (2021). https://doi.org/10.1016/j.ecolind.2021.107758

Salata, S., Ronchi, S., Arcidiacono, A., Ghirardelli, F.: Mapping Habitat Quality in the Lombardy Region. Italy. OE **2**, e11402 (2017). https://doi.org/10.3897/oneeco.2.e11402

Salata, S., Ronchi, S., Giaimo, C., et al.: Performance-Based Planning to Reduce Flooding Vulnerability Insights from the Case of Turin (North-West Italy). Sustainability **13**, 5697 (2021). https://doi.org/10.3390/su13105697

Salvia, M., Pietrapertosa, F., D'Alonzo, V., et al.: Key dimensions of cities' engagement in the transition to climate neutrality. J. Environ. Manage. **344**, 118519 (2023). https://doi.org/10.1016/j.jenvman.2023.118519

Vitulano, V.: ntegrating green infrastructure in Italian urban plans. Lessons from Turin and Bologna. Proceedings of the Institution of Civil Engineers - Urban Design and Planning, pp. 1–12. (2024) https://doi.org/10.1680/jurdp.22.00049

ISTAT: Il Censimento permanente della popolazione in Piemonte. (2021) https://www.istat.it/it/archivio/253517. Accessed 27 Mar 2024

Methodology to Planning Green Infrastructure to Mitigate Landslide Risk

Federica Cicalese(✉), Isidoro Fasolino, and Michele Grimaldi

University of Salerno, via Giovanni Paolo II 132, 84084 Fisciano, SA, Italy
fcicalese@unisa.it

Abstract. *Green Infrastructures* (GIs) can mitigate the effects of climate change and extreme events that they pose, by managing, for example, the devastating power of floods or landslides, re-establishing spaces and functions. Many tools are available and many more are emerging. The challenge is ensuring that well-planned GIs provide functions which will meet numerous planning objectives.

We propose a methodology for the effective planning of a GIs network that can help public policies. The methodology's application highlights how it is possible to identify the contact points between the protection of ecological integrity and the mitigation of landslide risk, in line with *European Union* (EU) strategies.

Keywords: green infrastructures · ecosystem services · landslide risk

1 Introduction

Green Infrastructure (GI) consists of a network of natural and semi-natural areas strategically planned with other environmental features, designed and administered to provide a wide range of *Ecosystem Services* (ESs) [1, 2], delivering a set of environmental benefits while maintaining and improving the ecological functions [3].

GIs can mitigate the effects of climate change and the extreme events that they pose, managing, for example, the devastating power of floods or landslides by re-establishing spaces and functions [4–6].

In general, the main elements of a GI include parks, private gardens, agricultural fields, hedges, trees, woodland, green roofs, green walls, rivers and ponds [7]. The theme of GIs is closely related to ESs, which are a set of functions naturally provided by ecosystems [8–10] and are fundamental to maintaining the resilience of a territory. ESs are classified by Millennium Ecosystem Assessment (MEA) [11] into four main categories: *Ecological Integrity* (EI), *Regulating* (SR), *Provisioning* (SA) and *Cultural* (SC).

Human-induced climate change, combined with other man-made pressures such as the conversion of soil uses, has altered the functions of ecological systems and has consequently altered the flow of ESs in terms of scale, timing and position.

For decades, European Union (EU) policies have aimed to provide strategies to make Europe more resilient to climate change by ensuring ecosystem-based approaches.

O. Gervasi et al. (Eds.): ICCSA 2024 Workshops, LNCS 14818, pp. 323–333, 2024.
https://doi.org/10.1007/978-3-031-65273-8_21

For instance, the *Seventh Environment Action Programme – 7EAP* (Decision No 1386/2013/EU) was published, which aimed to increase ecological and climatic resilience, so that the restoration of ecosystems and GIs could have significant socio-economic benefits, including public health. Then, in 2011, the *EU Biodiversity Strategy* (COM/2011/244 final) envisaged the restoration of at least 15% of the degraded ecosystems in Europe and aims to expand the use of GIs; this strategy was confirmed by the United Nations General Assembly (UNGA) [12]. In 2013, the *European Commission Strategy on Green Infrastructure* (COM/2013/0249 final) emphasized the important contribution of GIs to the effective implementation of all policies where some or all of the desired objectives could be achieved, at least in part through natural solutions.

To date, the cornerstone for sustainable development is the 2030 Agenda. Since 2016, with its 17 Sustainable Development Goals (SDGs), the Agenda has constituted the new global and universal reference framework for national and international efforts to find common solutions to the planet's major challenges.

In this framework, an important contribution can be carried out by green infrastructures (GIs) that, at all scales, emphasize nature-based solutions (NBS) and the use of local resources and materials to build community self-reliance and resilience. NBS "involve the innovative application of knowledge about nature, inspired and supported by nature" [13].

The document Exploring Nature-based Solutions of the European Environment Agency (EEA) – EU [14], shows the need for GIs to mitigate vulnerability to atmospheric agents, climate change and, in particular, landslide risk.

Soil erosion can cause significant losses in agricultural and forest productivity as it depends mainly on precipitation as well as more surface properties of the soils [15].

According to a study prepared by the EEA, many tests determine how much the increase in landslides is due to climate change. To provide a specific example, forests, along with other vegetation, can reduce the presence of surface landslides.

However, their global increase is indeed due to the excessive use of natural resources and deforestation, as well as increasing urbanization and uncontrolled land use.

2 Objective

Plant cover plays an important role in soil conservation and landslide prevention [16].

Although a large part of European literature has recognized the link between plant cover and slope stability, the ability of GIs to mitigate the risk of landslides is poorly evaluated locally.

Studies show the ability of ESs to mitigate the risk of landslides based on the presence of protective forests and the potential hazard of landslides. Integrating both maps (hazard of landslides and protective forests) involves delineating a potential GIs network to mitigate vulnerability from landslides.

The application of GIs as a protection function can be evaluated.

The aim of this study is to develop a methodology that makes it possible to provide an in-depth description of the procedure outlined in the European document *Exploring Nature-based Solutions* by EEA and translate this into more detailed actions in territorial and urban planning.

3 Materials and Methods

3.1 Methodology

The methodology proposed to achieve this aim is based on a spatial analysis implemented in the GIS environment and consists of three macro-phases:

1) Evaluation of ESs offered by the different types of land use;
2) Identification of land uses that maximize the *ecosystem erosion control* (SRec) service in areas of high and very high hazard;
3) Designing a network that maximizes the multi-functionality of the GIs.

For *macro-phase 1*, a matrix was created to evaluate the capacities of the different types of land cover to provide the ESs.

Several tables in the current literature [17, 18] containing ESs, as defined by MEA, were compared to derive the values for each class of soil according to the EU project *CORINE Land Cover* (CLC).

The different values found in the literature for each ES, which are related to the four macro-categories (EI, SR, SA, SC) according to the land cover type, have been homogenized by performing an arithmetic average operation.

Table 1. Classification of the capacity of different land cover types to provide an individual service.

Level	Capacity
0	No
1	Low
2	Relevant
3	Average
4	High
5	Very high

The values assigned are based on expert evaluations, both conceptual and derived from different case studies and were normalized on a scale of 0–5 (Table 1) [8, 19, 20]. These values can be considered as research hypotheses that need to be tested in further applications.

Concerning the *macro-phase 2*, the methodology provides:

2a) the classification of land uses providing the ecosystem service SRec;
2b) the identification of high (P3) or very high (P4) hazard areas mapped by *Hydrogeological Basin Authorities*;
2c) the intersection between the patches that maximize the SRec service in combination with the high-hazard areas (P3) and very high-hazard areas (P4).

The *macro-phase 3* provides for the design of networks that can connect the patches that maximize the combination of IE and SRec.

3.2 Techniques and Instruments

The analysis implemented in the GIS environment, upon which the methodology is based, refers in detail to the following techniques of spatial analysis: cost-distance, network analysis and map algebra.

The cost-distance analysis is based on the fundamental geographic principle of friction of distance. Two species were analysed, representative of an ecologically homogeneous group, linked to specific habitats: *Sitta europaea* and *Triturus Carnifex* [21].

Two different algorithms were used for the analysis: for *Sitta europaea*, connectivity was evaluated using an isotropic model [22]; in the case of *Triturus carnifex* the use of a simple isotropic model was deemed insufficient. The species' limited ability to move around necessitated a modelling approach that could factor in any geographical obstructions, so an anisotropic model [23] was applied.

The result of the cost-distance models is a raster matrix in which each cell assumes a specific equivalent distance from a target cell class. In the case of isotropic models, the equivalent distance matrix is produced by: a raster image that identifies the target cells; and an impedance matrix that expresses the ratio of impedance of each cell to base cost. Impedance, therefore, acts with the same intensity regardless of the direction of movement. In an anisotropic model, on the other hand, a single impedance matrix cannot represent the forces at play, forces acting differently in different directions.

The impedance value associated with each cell is calculated as:

$$I_e = i^{cos^k \alpha} \tag{1}$$

where:

I_e = resultant value of impedance;

i = modulus of impedance;

α = angle equal to the difference between the direction of motion and the direction of impedance;

k = coefficient; as the coefficient k increases, the function becomes increasingly direction-specific.

The cost-distance analysis was carried out using the Graphab software [24], which is dedicated to modelling ecological networks based on graph theory. It allows for the creation of connections between the same patch types based on the cost-distance by taking into account the type of patches to be crossed or based on the minimum distance between patches. It is then compatible with GIS, through which maps can be modelled to create ecological networks.

The GIS overlay procedure generates a new layer (output level) depending on two or more input layers. The attribute value assigned to each position (e.g. raster or polygon) on the output layer is a function of the independent values associated with that position on the input levels. Two commonly used classes for combining overlays in GIS are *Boolean overlay* (such as AND and OR) and *Linear Weighted Combination* (WLC).

4 Case Study

4.1 Hydrogeological Risk in the Campania Region

The importance of green infrastructures already emerged in Campania in the Rural Development Program (PSR) 2014–2020. Indeed, it is worth mentioning the measure foreseen within the programme "*Creation and/or restoration and/or expansion of green infrastructure and agri-rural landscape elements*" and sub-measure "*Support for non-productive investments related to the fulfilment of agro-environmental objectives*".

The objective of the measure is the sustainable development of agri-food businesses in Campania, providing, where necessary, for the improvement of the environmental conditions of the territory from the agro-climatic point of view.

The Campania Region (Fig. 1), in Southern Italy, has high lithological and geological-structural variability that makes the territory susceptible to different types of landslides. An analysis of the historical events and investments that have affected the Italian territory, and Southern Italy in particular, reveals a lack of knowledge of the places where intervention is called upon [25].

It is ranked second among the Italian regions for the number of victims due to the hydrogeological phenomena. Almost all of these were due to the rapid castings of loose pyroclastic covered by the carbonate masses of the Campanian Apennine and the stone sequences present in the volcanic areas of Mount Vesuvius and the Phlegrean Fields.

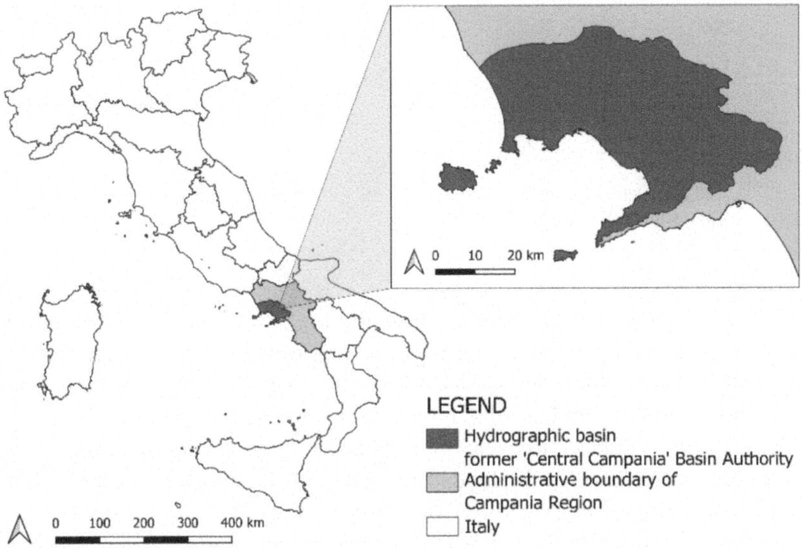

Fig. 1. Geographical framework: Italy with the Campania Region (in light grey) and the hydrographic basin of the former 'Central Campania' basin authority (in dark grey).

The distribution of the hazard from landslides, identified by the *Hydrogeological Basin Authorities*, is highlighted according to the territorialization and use of the soil expressed as the second level of the CLC legend. In the Campania region, there are 135,800 hectares of land in high hazard (P3) and 131,500 hectares of land in very high hazard (P4). Regarding the P3 and P4 hazardous areas, the wooded areas constitute the most significant land use type (126,279 hectares), equal to 20,5% of the total landslide areas.

However, it should be noted that the application section, refers to the territory included in the former basin authority "Central Campania" because the landslide hazard map was only available for this portion.

5 Application

Macro-phase 1 led to the elaboration of a matrix translated, then graphically, into a series of maps to assess the capacity of different land cover types to provide ESs in the specific case of the Campania Region.

As explained in the previous section, scores, normalized on a scale of 0–5, were associated with each land use according to the CLC classification [8, 19, 20].

Table 2. Percentage of attributing score for each soil cover

Level	Capacity	EI [%]	SR [%]	SRec [%]	SA [%]	SC [%]
0	No	8	8	11	10	36
1	Low	2	33	24	37	3
2	Relevant	2	17	–	53	26
3	Average	49	14	23	–	4
4	High	11	–	11	–	2
5	Very high	28	28	31	–	29

The results (see Fig. 2 and Table 2) show the high capacity of different soil types (forests, swamps, bogs and heaths) to provide a wide range of ESs. Whereas land cover types that have been profoundly modified by man, such as urban fabric, industrial or commercial areas, mines and landfills, have a very low capacity to provide ESs.

For the next macro-phase, the selection of land uses providing the *Regulating Erosion* (SRec) ecosystem service for landslide hazard mitigation was carried out using a geodatabase consisting of information from the Soil Map for Agricultural Use (CUAS) of the Campania Region (2009); while the identification of high (P3) or very high (P4) landslide hazard areas took place after consultation with the Hydrogeological Plan of the Central Campania Hydrogeological Basin Authority (2015) (Fig. 3).

Then, the intersection of the patches maximizing SRec service in combination with the high (P3) and very high (P4) hazard areas was carried out (Fig. 4).

The *macro-phase 3* involves mapping the networks by using cost-distance analysis.

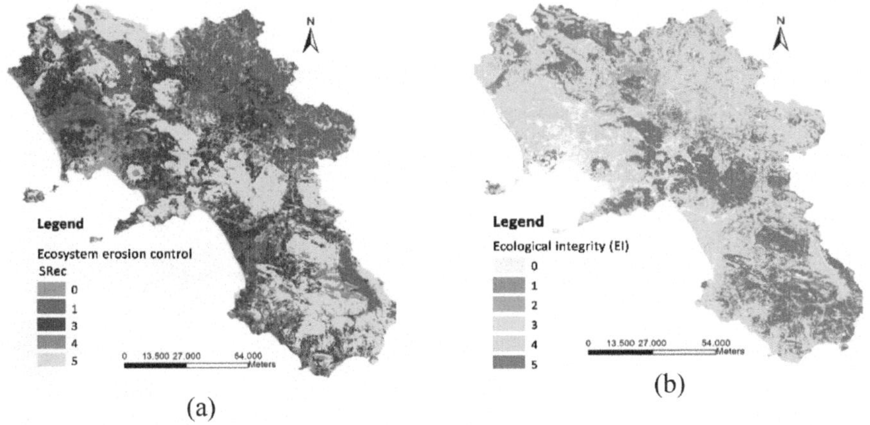

Fig. 2. Scoring based on soil cover: (a) Ecosystem erosion control services map (SRec); (b) Ecological integrity (EI) map

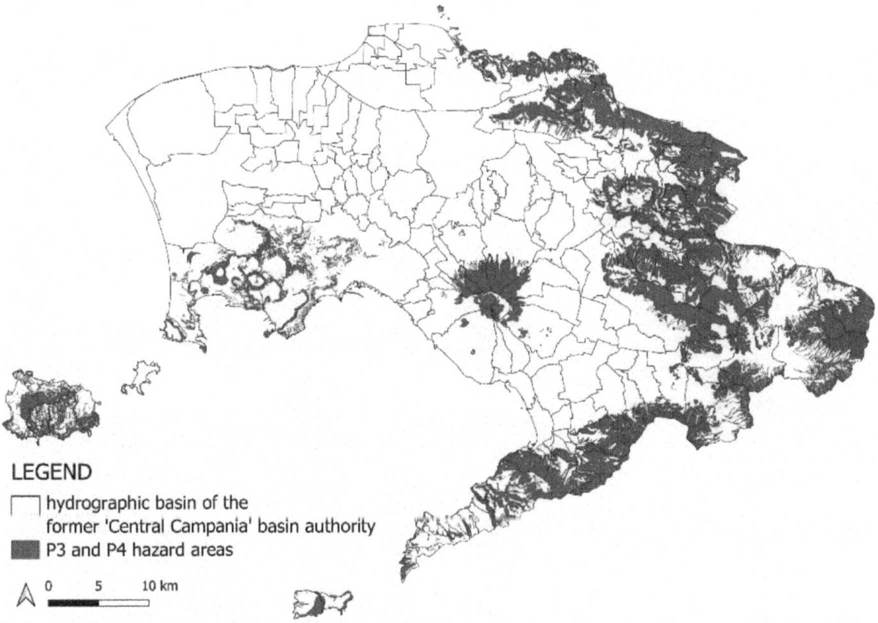

Fig. 3. Identification of high (P3) or very high (P4) landslide hazard areas

To design the networks, the Graphab software was used, using as the resistance values the corresponding scores for soil utilization related to *Ecological Integrity* (IE) and the combination of IE with *Regulating Erosion* (SRec).

Finally, as the last step of macro-phase 3, the overlap map (Fig. 5) of these networks was elaborated, identifying the network which maximizes IE and SRec.

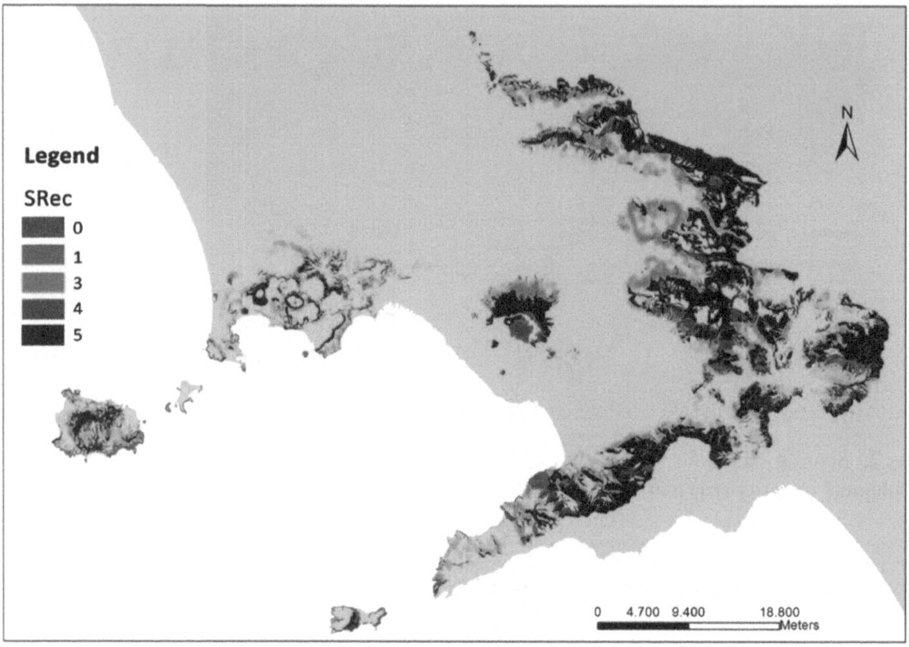

Fig. 4. Intersection of the patches that maximize the SRec and the P3 and P4 areas

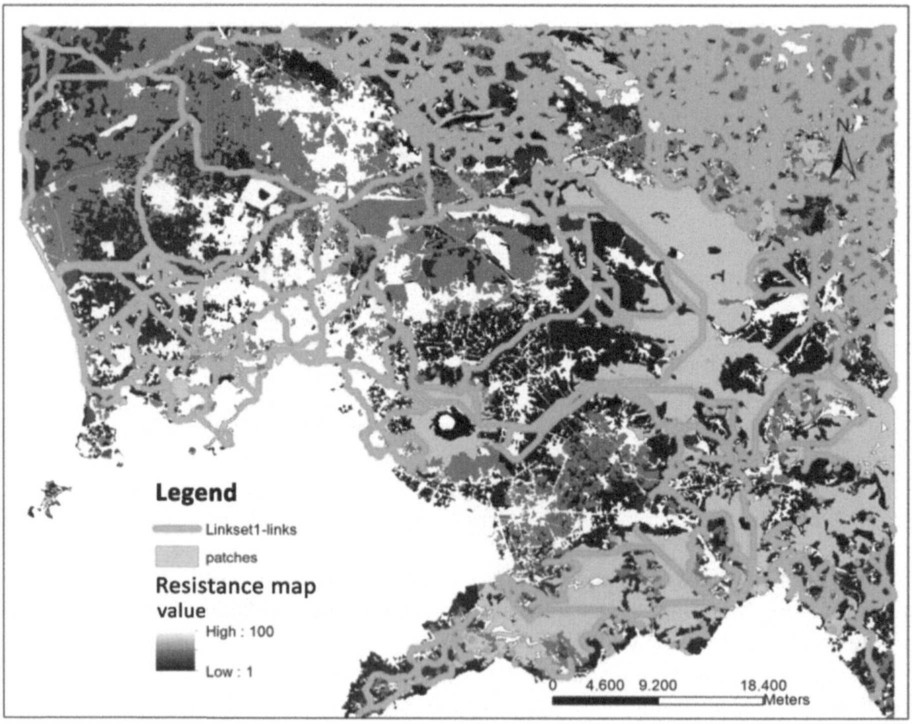

Fig. 5. Overlap map: network that maximises the ESs IE – SRec

6 Discussion and Perspectives

In conclusion, to achieve a territory's resilience, risk assessments must be widely promoted in the development of urban land-use policies.

Integrating the concept of functions and ESs into land management and planning decisions is essential so that local administrators can control the pressures that could threaten the ecosystem.

The potential assessment of soils to establish their ability to provide ESs is considered an important tool to address the current difficulty of systematically considering ESs in spatial planning. However, since ESs have never been quantified in monetary terms, they have always had little or no weight in political decisions.

The proposed methodology addresses the effective planning of GIs network that can help achieve numerous benefits, including the reduction of risks to people and property.

This methodology, therefore, provides a new way of mapping GIs, which is generally based on the specific ability to favour biotic and abiotic flows. In this case, the aim was to optimize this pattern by maintaining and increasing other ESs, starting from the regulation of soil disruption.

The challenge is to ensure that well-planned GIs provide functions that meet numerous planning objectives and become an integral part of public policy; however, this requires thorough planning, design and management.

The case study analyzed shows how fundamental it is for the Campania Region to provide assessment mechanisms in areas with a high risk of landslides and erosion, given the considerable resources allocated to this environmental condition by the PSR 2014–2020.

Most input data were taken from published and reclassified European sources for analysis purposes.

The choices made during the data processing and analysis are based on expert opinions and are open to public control.

On analysis of the overlay map of the SRec distribution and the map of the landslide hazard distribution, the concomitant use of high-value SRec soil with P3 and P4 hazard levels emerges. This information indicates that, on the one hand, the use of the soil performs a protective action, while, on the other hand, that alone does not contribute to the resolution of the problem, mainly due to geological and geotechnical conditions.

From a future perspective, the study should consider identifying strategic priorities [26]. Therefore, using network analysis, based on the theory of graphs, this additional step could provide a comparison between the different networks to define their maximization and mitigation strategies.

Acknowledgements. This study was funded by MIUR, with the PRIN 2022 "Nature for sustainable cities: planning cost-effective and just solutions for urban issues" (NatSolis) (grant number 2022ZLE8HC).

Disclosure of Interests. The authors have no competing interests to declare that are relevant to the content of this article.

References

1. European Commission (EC): Green Infrastructure (GI) – Enhancing Europe's Natural Capital. Communication from the Commission to the European Parliament, the Council, the European Economic and Social Committee and the Committee of the Regions, Brussels (2013)
2. Mapping and Assessment of Ecosystems and their Services – MAES: Mapping and assessment of ecosystems and their services – Indicators for ecosystem assessments under Action 5 of the EU Biodiversity Strategy to 2020. Technical report, European Commission, Brussels (2014)
3. Lennon, M., Scott, M.: Delivering ecosystems services via spatial planning: reviewing the possibilities and implications of a green infrastructure approach. Town Plan. Rev. **85**(5), 563–587 (2014). http://www.jstor.org/stable/24579269
4. Austin, G.: Green Infrastructure for Landscape Planning: Integrating Human and Natural Systems. Routledge, New York (2014)
5. Benedict, M.A., McMahon, E.T.: Green Infrastructure: Linking Landscapes and Communities. Island Press, Washington, DC (2006)
6. Lafortezza, R., Davies, C., Sanesi, G., Konijnendijk, C.: Green Infrastructure as a tool to support spatial planning in European urban regions. iForest **6**(3), 102–108 (2013). https://doi.org/10.3832/ifor0723-006
7. Naumann, S., Davis, M., Kaphengst, T., Pieterse, M., Rayment, M.: Design, Implementation and Cost Elements of Green Infrastructure Projects. Final Report. European Commission, Brussels (2011)
8. Costanza, R., et al.: The value of world's ecosystem services and natural capital. Nature **387**, 253–260 (1997). https://doi.org/10.1038/387253a0
9. Burkhard, B., Petrosillo, I., Costanza, R.: Ecosystem services – bridging ecology, economy and social sciences. Ecol. Complex. **7**, 257–259 (2010). https://doi.org/10.1016/j.ecocom.2010.07.001
10. Burkhard, B., Kroll, F., Müller, F., Windhorst, W.: Landscapes capacities to provide ecosystem services – a concept for land-cover based assessments. Landsc. Online **15**, 1–22 (2009). https://doi.org/10.3097/LO.200915
11. Millennium Ecosystem Assessment (MEA): Ecosystems and Human Well-Being: A Framework for Assessment. Island Press, Washington DC (2003)
12. United Nations General Assembly (UNGA) Resolution 73/284. United Nations Decade on Ecosystem Restoration (2021–2030), A/RES/73/284 (2019)
13. European Commission (EC): Towards an EU Research and Innovation policy agenda for Final Report of the Horizon 2020. Expert Group on "Nature-Based Solutions and Re-Naturing Cities", Brussels (2015)
14. European Environment Agency (EEA): Exploring nature-based solutions. In: The Role of Green Infrastructure in Mitigating the Impacts of Weather- and Climate Change-Related Natural Hazard. Publications Office of the European Union, Luxembourg (2015)
15. Crozier, M.J.: Deciphering the effect of climate change on landslide activity: a review. Geomorphology **124**(3–4), 260–267 (2010). https://doi.org/10.1016/j.geomorph.2010.04.009
16. Stokes, A., Raymond, P., Polster, D., Mitchell, S.J.: Engineering the ecological mitigation of hillslope stability research into the scientific literature. Ecol. Eng. **61**, 615–620 (2013). https://doi.org/10.1016/j.ecoleng.2013.07.065
17. Müller, F., de Groot, R., Willemen, L.: Ecosystem services at the landscape scale: the need for integrative approaches. Landsc. Online **23**, 1–11 (2011). https://doi.org/10.3097/LO.201023
18. Fisher, B., Turner, K.R.: Ecosystem services: classification for valuation. Biol. Conserv. **141**(5), 1167–1169 (2008). https://doi.org/10.1016/j.biocon.2008.02.019
19. de Groot, R., et al.: Global estimates of the value of ecosystems and their services in monetary units. Ecosyst. Serv. **1**(1), 50–61 (2012). https://doi.org/10.1016/j.ecoser.2012.07.005

20. Koschke, L., Fürst, C., Frank, S., Makeschin, F.: A multi-criteria approach for an integrated land-cover-based assessment of ecosystem services provision to support landscape planning. Ecol. Ind. **21**, 54–66 (2012). https://doi.org/10.1016/j.ecolind.2011.12.010
21. Vannuccini, M., Geri, F.: Valutazione di connettività ambientale attraverso analisi di cost distance su specie target. In: XVI Congresso Società Italiana di Ecologia, SItE (2007)
22. Eastman, J.R.: Pushbroom algorithms for calculating distances in raster grids. In: Autocarto 9, pp. 288–297. Clark University, Worcester (1989)
23. Eastman, J.R.: IDRISI 32 Guide to GIS and Image Processing. Clark Labs, Clark University, Worcester (1999)
24. Foltête, J.C., et al.: Graphab: an application for modeling and managing ecological habitat networks. Softw. Impacts **8**, 100065 (2021). https://doi.org/10.1016/j.simpa.2021.100065
25. Russo, F., Marra, A., Gerundo, R., Nesticò, A.: On the phenomenon of depopulation of inland areas. In: Gervasi, O., Murgante, B., Misra, S., Rocha, A.M.A.C., Garau, C. (eds.) Computational Science and Its Applications – ICCSA 2022 Workshops. ICCSA 2022. Lecture Notes in Computer Science, vol. 13382, pp. 381–391. Springer, Cham. (2022). https://doi.org/10.1007/978-3-031-10592-0_28
26. Gerundo, R., Marra, A., Grimaldi, M.: A preliminary model for promoting energy communities in urban planning. In: Calabrò, F., Della Spina, L., Piñeira Mantiñán, M.J. (eds.) New Metropolitan Perspectives. NMP 2022. Lecture Notes in Networks and Systems, vol. 482, pp. 2833–28 40. Springer, Cham. (2022). https://doi.org/10.1007/978-3-031-06825-6_270

Regulating Urban Ecosystem Services Unveiled: Empirical Insights on Driving Factors in Cagliari, Italy

Sabrina Lai(✉) and Corrado Zoppi

Department of Civil and Environmental Engineering and Architecture, University of Cagliari, Cagliari, Italy
{sabrinalai,zoppi}@unica.it

Abstract. This study aims at analyzing the relationships between supply of ecosystem services, features of green areas and characteristics of settlements in urban contexts, by taking the Italian city of Cagliari as study area. The services offered by the urban ecosystems that are identified as the most relevant in association with the spatial framework of green areas in urban environments are heat mitigation, carbon capture and storage, and runoff control, with particular reference to flood-related events. The features of green areas are identified with reference to the height of vegetation, by distinguishing between grasslands, shrubby cover, and trees and woodland cover. Finally, we characterize the urban settlement through the building and population densities, and through the education level, as a proxy for the residents' social statuses. The assessment of performances of the urban ecosystem services shows negative correlations with the intensity of urbanization, whereas the size of the enhancement in the supply of ecosystem services can be associated with different types of green areas. In terms of policy implications, the outcomes of the study show that there is plenty of room for improvement in the ecosystem services performance based on fine-tuning measures which involve building and population densities and vegetation cover.

Keywords: Ecosystem services · Carbon capture and storage · Urban runoff control · Urban heat mitigation

1 Introduction

According to figures available from the World Bank, around 56% of the world's population is presently concentrated in urban areas, whereas the population living in cities is forecast to grow up to nearly 70% by 2050. The increasing urban expansion trend generates radical transformations in city landscapes, which entail, for example, decrease in urban biodiversity that causes destruction of habitats and widespread development of alien invasive species, loss of habitats and addition of new habitats, increase in vegetated land cover fragmentation [1], decrease in agriculture production, increase in air temperature [2], loss of carbon storage and capture potential [3], and increase in flood risk due to the growing trend of soil sealing [4]. A well-grounded approach to mitigation

O. Gervasi et al. (Eds.): ICCSA 2024 Workshops, LNCS 14818, pp. 334–351, 2024.
https://doi.org/10.1007/978-3-031-65273-8_22

of negative impacts of intensive urbanization on natural ecosystems is represented by the paradigm of the "compact city" [5], which aims at minimizing land-taking processes by concentrating city services in the central locations of urban fabrics. By doing so, decrease in urban sprawl and important ecological and social positive impacts are implemented in the long run, so as to boost living quality in urban, periurban and rural environments [6]. A less investigated profile is represented by the effects of intensive urbanization processes on the areas where such processes take place, with reference to ecosystem services endowment and provision [7]. For this purpose, three ecosystem services are identified in this study, which are particularly important as regards city environments, such as carbon capture and storage (CCS), runoff control (ROC) and heat mitigation (HEM).

CCS through biotic processes can be identified with photosynthesis, which takes place in ecosystems such as forests, woodlands, shrubs, garrigues, natural grasslands, and heathlands. Photosynthesis eliminates carbon dioxide from the air, and stores carbon in soil and biomass [8] while releasing oxygen in the atmosphere. Soil and air composition, and their interactions, significantly contribute to global climate regulation, and are heavily related to land covers and their transitions [9]. Thus, green areas and soil play key roles in carbon cycle processes at the global level by supplying the CCS ecosystem service. Soil is the largest carbon terrestrial pool, since the carbon quantity stored in the soil is quite larger than in the biomass located overground [10]. That is why even low changes in carbon concentration in the soil may generate relevant effects on the aerial concentration of carbon dioxide and on the temperature at the global level. Climate and soil show a mutual connection. On the one hand, climate changes influence geophysical and geomorphological conditions and environmental processes related to soil; on the other hand, soil transitions generate important effects on the terrestrial climate. In this perspective, urban planning policies should play a relevant role as regards conservation and enhancement of the CCS ecosystem service provided by soils, which entails the implementation of specific and effective measures aimed at improving its capacity of storing carbon.

ROC is related to regulation of flood events, which can be operationalized through shaping and maintenance of river courses and floodplains so as to implement effective drainage and adequate runoff, underground aquifers recharge and generation of groundwater tanks to be used in case of drought as well. Soil permeability is fundamental as regards regulation of flood flow rates and related runoff speed, and, that being so, it represents a key factor in the provision of the ROC ecosystem service, which entails a reduction in sediment downstream flows as well. Permeability makes agricultural areas perform quite well as ROC ecosystem service providers [11]. Effective management of forests, as well as of other wooded and tree areas, provides an adequate ROC in case of flood events, especially through the implementation of appropriate drainage and runoff into floodplains [12]. Forest management coupled with effective ROC is also very important with reference to reduction of the share of polluted wastewater from industrial and agricultural productive activities which flows downstream through waterways [13]. Nature-based solutions aimed at conveying surface waters into underground aquifers are very effective in maximizing the positive impacts of the ROC ecosystem service [12]. ROC can be appropriately implemented in urban contexts through rainwater collection and storage, like for instance raingardens, swales and bioswales, green roofs and walls, and collection tanks [14].

In this study, land surface temperature (LST) is targeted as a key point of reference to address the issue of HEM. In other words, decrease in LST is taken as a proxy for the provision of the HEM ecosystem service. As per Hulley et al. [15], LST is a fundamental climate-related and biological factor which influences the ecosystems viability at different geographical scales. In particular, LST can be assumed as a measure of the thermal radiance generated by the interaction between the land surface, or the canopy in case of tree areas, and the sun power [16]. Furthermore, LST can be considered as the temperature felt to the contact and can be identified by the temperature of what you can see from a satellite by looking through the atmosphere, be it the soil surface or the canopy of an urban woodland; thus LST is different from air temperature. Under this perspective, it is evident that the spatial taxonomies of land cover and LST are interrelated, and that land cover transitions influence climate change trends [17].

The three ecosystem services described so far are generated through peculiar characteristics of highly urbanized contexts and work as relevant factors related to the quality of life in cities. The provision of CCS, ROC and HEM is associated with urban land cover, and it basically depends on two intertwined aspects. First, as discussed above, the three ecosystem services are supplied by unsealed soils, therefore the building density structure, which identifies the urban taxonomy of sealed and unsealed soils, is a focal point to detect if, and to what extent, urban environments can be providers of such ecosystem services. Second, the intrinsic characteristics of urban unsealed soils play a fundamental role in the effectiveness of the provided ecosystem services. In other words, to detect the correlations between provision of ecosystem services and urban land cover characteristics is highly important. Such correlations can lead to relevant implications in terms of planning policies aimed at increasing the provision of urban ecosystem services and at improving the living quality of urban communities.

The research question that this study aims at addressing is, therefore, identified as follows: how and to what extent is the endowment of sealed and unsealed areas related to the supply of ecosystem services in intensively urbanized urban areas?

The study develops as follows. In the second section, the spatial data used to identify urban land cover characteristics and building structure are presented, and the approach to derive them is explained. Moreover, the methodology used to identify the correlations between ecosystem services supply and urban characteristics is described. The third section assesses the outcomes of the correlation analysis and of the estimates of the regression models. In the fourth section, such outcomes are discussed. The concluding section remarks the value added of the study, with particular reference to planning policy implications.

2 Materials and Methods

2.1 Study Area

The area chosen for this study is Cagliari (Fig. 1). The historic districts host only a few tiny green areas, and the parts of the city that surround the old districts are also characterized by a dense urban fabric; however, some large green areas and parks can be found therein, and larger roads are often lined with rows of trees. The outer districts are characterized by a larger share of green areas. This reflects on the uneven spatial pattern of the residential density: as of December 2022, Cagliari's resident population was around 150,000 people and the population density was 1,751, with peaks of nearly 19,000 residents/km^2 in the most central districts. Therefore, the complex urban morphology and uneven endowment of green areas make the city of Cagliari a good case study to investigate the relationship between sealed/unsealed areas and the supply of ecosystem services in an urban context.

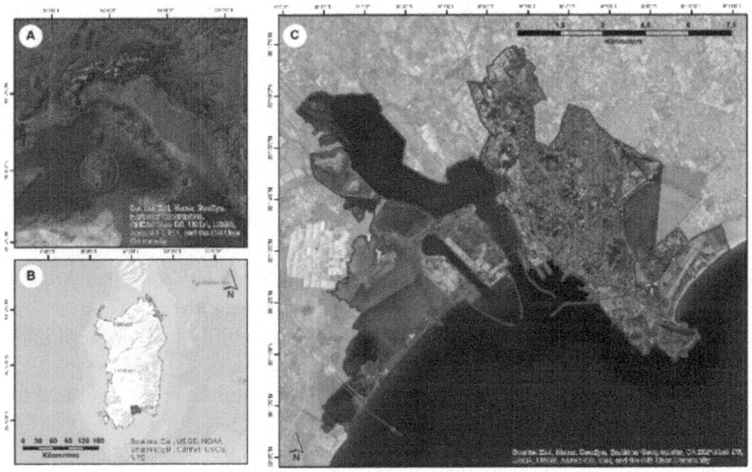

Fig. 1. Study area: Location of Sardinia (A), location of Cagliari (B), city boundaries (C)

2.2 Spatial Data

As per the introduction, factors concerning the urban settlement that might influence the provision of the selected urban ecosystem services can be grouped into three main classes: land cover types; aspects related to the built environment; socio-economic characteristics of the resident population. Spatial references for each factor are the 1,285 census tracts in which the municipality of Cagliari is divided for statistical purposes. While all the factors were mapped across the whole municipality of Cagliari, the correlation analysis and the regression model next described in Sect. 2.3 were implemented by looking only at the 1,114 census tracts inhabited at least by ten residents. The indicators for the three selected urban ecosystem services are listed in Table 1, together with their definition and data sources.

Concerning CCS, the InVEST "Carbon Storage and Sequestration" model, part of the InVEST suite, was used. This model focuses on carbon stored in four terrestrial pools: above-ground biomass, below-ground biomass, dead organic matter, and soil, i.e., the top 30-cm layer. INVEST estimates the amount of carbon stored in the area of interest through a land cover map and a lookup table that provides carbon density values for each land cover type and for each carbon pool. The result is a carbon density map, measured in megagrams (Mg) of carbon per pixel. No information is available regarding below-ground biomass in the study area; therefore, the model was here applied solely focusing on the three remaining carbon pools.

Table 1. Selected urban ecosystem services: list, definition, and data sources.

Variable	Definition and unit of measurement	Data sources
C_Stor	Average density of carbon capture and storage in a census tract [Mg/hectare]	Regional land cover map 2005 National Inventory of Italian Forests Regional pilot project on land units and soil capacity in Sardinia
LST_Cmax	Maximum land surface temperature in a census tract [°C]	Landsat Collection 2 Surface Temperature
Dens_Ret	Density of runoff retention in a census tract [m^3/m^2]	City of Cagliari's 2021 land cover dataset Sardinian Hydrological Annals as for rainfall depth Permeability map of Sardinian substrates Curve number table, tailored to Sardinia

As for LST, raster maps are currently available off-the-shelf from the United States Geological Survey, which makes it possible to retrieve Landsat Collection 2 Level-2 products, including 30-m land surface temperature raster maps. A four-month interval, from June to September 2022, was selected. Among the twenty LST maps retrieved from Earth Explorer, the one having the highest LST mean value across the study area (once the wetlands were removed) was selected. As regards ROC, this was mapped using the InVEST "Urban Flood Risk Mitigation" model, which enables to estimate both runoff levels and the water volume retained by permeable soils and green areas, hence providing an assessment of the flood regulation ecosystem service. The model requires the following input data: the area of interest; the rainfall depth; a raster map of the soil hydrologic group; a land use/land cover raster map; a biophysical table.

Finally, for each of the three variables representing as many ecosystem services, their spatially lagged variables (C_Stor_lag, LST_Cmax_lag, and Dens_Ret_lag) were assessed to control for their spatial autocorrelation. Anselin's approach was followed, as implemented in the freeware GeoDA [18], by taking census tracts as the minimum spatial units and by considering the first-order queen contiguity to determine weights. The spatial distribution of the three selected proxies for CCS, HEM, and ROC, together with their indicators within each census tract, is shown in Fig. 2. Factors that can affect the provision of the three selected urban ecosystem services are listed in Table 2, together with their definition and data sources.

As well as the boundaries of census tracts, spreadsheets are available from the National Census website providing several data related to the built environment and to resident population at the tract level; this made it possible to retrieve the variables Dwel_Dens, Res_Dens and Perc_Degr within each census tract. As for the built volume, a dataset available from the regional geoportal and termed "Geotopographic dataset of Sardinian urban centers and inhabited areas" was used.

Table 2. Factors that can affect the supply of the selected urban ecosystem services.

Variable	Definition and unit of measurement	Data sources
Perc_Tree	Share of a census tract occupied by tall trees [%]	Public green areas and tree lined roads from the City of Cagliari's geoportal City of Cagliari's 2021 land cover dataset Sentinel-2 Copernicus satellite imagery
Perc_Bush	Share of a census tract occupied by medium-height shrubs [%]	
Perc_Lowveg	Share of a census tract occupied by low-height vegetation [%]	
Vol_Dens	Built volume density (i.e., built volume per unit of land) within a census tract [m^3/m^2]	Geotopographic dataset of Sardinian urban centers and inhabited areas
Sup_Nvp	Share of sealed area within a census tract [%]	Public green areas and tree lined roads from the City of Cagliari's geoportal City of Cagliari's 2021 land cover dataset Sentinel-2 Copernicus satellite imagery Geotopographic dataset of Sardinian urban centers and inhabited areas
Dwel_Dens	Density of housing units (i.e., number of dwellings per unit of land) within a census tract [no./m^2]	2011 National Census dataset
Res_Dens	Population density within a census tract [residents/hectare]	
Perc_Degr	Share of residents holding a college degree within a census tract [%]	

Concerning the three types of vegetation (tall trees, medium-height shrubs, low vegetation), the vegetation map for the study area was developed by using the city's land cover map and the city's map of green areas as primary sources of information, by reclassing the vegetation types into three types: trees, shrubs, and low-height vegetation, of which the latter was complemented with green spots within built-up areas, identified through an NDVI (Normalized difference vegetation index) raster map. Once the vegetation map was built, the variables Perc_Tree, Perc_Bush, and Perc_Lowveg could be calculated, as well as the variable Sup_Nvp, which represents the share of the census tract that is sealed, i.e., non-vegetated.

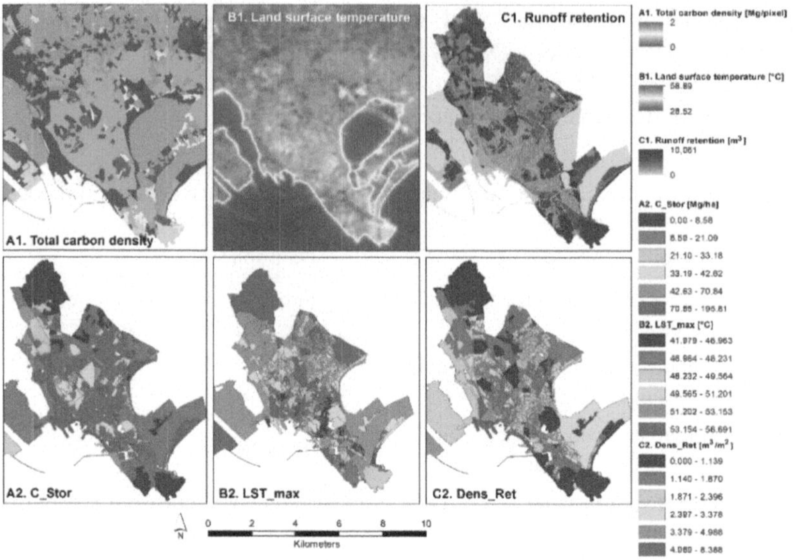

Fig. 2. Spatial layout of the selected ecosystem services (A1-B1-C1) and corresponding indicators within census tracts (A2-B2-C2)

2.3 Methods

The relationships between the supply of ecosystem services in intensively urbanized environments, and the presence and quantity of sealed and unsealed areas, are analyzed and assessed through correlation analysis and regression models.

Correlation Analysis. The associations of the distributions of the supply of the three selected ecosystem services (CCS, ROC, and HEM) with the distributions of the variables which represent land covers concerning sealed and unsealed areas can be identified through correlation analysis, which makes it possible to associate the distributions of two variables with each other so as to detect if, and to what extent, the two phenomena show similar trends. The Spearman's rank correlation coefficients (SRCCs) are estimated to detect correlations between the distributions of the variables, related to census tracts, associated with CCS, ROC and HEM, that is, density of CCS (C_Stor), density of runoff retention (Dens_Ret) and maximum land surface temperature (LST_Cmax), and the distributions of those representing sealed and unsealed areas. Unsealed areas are associated with the following: percentage share of a census tract covered with: i. tall plants (Perc_Tree); ii. medium-height shrubs (Perc_Bush); or, low-height vegetation (Perc_Lowveg). Sealed zones area associated with the following: population density (Res_Dens), built volume density (Vol_Dens), density of housing units (Dwel_Dens), and percentage share of sealed area (Sup_Nvp). Since the forms of the distributions are unknown, SRCC can be identified as the most reliable measure to detect the presence and size of correlations between the variables at stake. The assessment of the results of the SRCCs estimates as regards the expectations on such estimates is based on three items, that is, the sign, the size, and the related p-values.

Regression Model. Three linear regression models are implemented to identify the correlations between the variables associated with the provisions of CCS (C_Stor), ROC (Dens_Ret) and HEM (LST_Cmax) ecosystem services, and the covariates associated with vegetated (Perc_Tree, Perc_Bush) and sealed (Sup_Nvp) land covers. Among the variables related to land covers Perc_Lowv in not included in the sets of the explanatory variables of the regression models since it is the complement to 1 of the sum of Perc_Tree, Perc_Bush and Sup_Nvp, and, that being so, it has to be omitted to avoid multicollinearity. As it is shown by the SRCCs results reported in the next section, the variables representing sealed land cover (Res_Dens, Vol_Dens, Dwel_Dens, Sup_Nvp) are strictly correlated with each other and, therefore, just one of them can be included in the set of the explanatory variables. Sup_Nvp is selected among the variables associated with sealed land cover since it is the one most correlated with the three dependent variables (C_Stor, Dens_Ret and LST_Cmax), as shown in the next section. Moreover, Sup_Nvp identifies sealed soils in the most complete way, since it takes account not only of buildings, dwellings and inhabitants, but also of paved areas such as roads, squares, sidewalks, parking lots, pedestrian paths, playgrounds, and the like. The three regression models operationalize as follows:

$$\begin{aligned}\{\text{C\_Stor}|\text{Dens\_Ret}|\text{LST\_Cmax}\} &= \beta_0 + \beta_1\text{Perc\_Tree} + \beta_2\text{Perc\_Brush} + \beta_3\text{Sup\_Nvp}\\ &+ \beta 4\text{Perc\_Degr} + \beta 5\{\text{C\_Stor\_lag}|\text{Dens\_Ret\_lag}|\text{LST\_Cmax\_lag}\}\end{aligned} \tag{1}$$

where labels are associated with dependent and explanatory variables, as follows:

- C_Stor is for density of CCS in a census tract;
- Dens_Ret is for density of runoff retention in a census tract;
- LST_Cmax is for maximum land surface temperature in a census tract;
- Perc_Tree is for percentage share of area covered with tall plants in a census tract;
- Perc_Bush is for percentage share of area covered with medium-height shrubs in a census tract;
- Sup_Nvp is for percentage share of sealed area in a census tract;
- Perc_Degr is for percentage share of residents holding a college degree in a census tract;
- C_Stor_lag, Dens_Ret_lag and LST_Cmax_lag are the spatially lagged dependent variables that control for spatial autocorrelation of C_Stor, Dens_Ret and LST_Cmax.

The estimates of the coefficients of the regression models reveal the correlations between the supply of the CCS, ROC and HEM ecosystem services and the presence of vegetated and sealed parcels in census tracts. Multiple regression models are used since there are not any prior hypotheses associated with the relation between dependent and explanatory variables, which is in line with the current literature concerning this issue. In this perspective, a spatial variable related to *n* factors can be associated with a surface, which develops in a space with n dimensions, whose equation is not identified. The functional form of such surface can be approximated, in a very small neighborhood of each of its points, by the hyperplane which is tangent at that point [19]. That is why the linear equations estimated through the multiple regression models (1) are assumed as adequate approximations, in six-dimensional spaces, of the relationships between C_Stor, Dens_Ret and LST_Cmax, and the covariates which identify models (1).

Variable Perc_Degr, which represents the share of graduates within a census tract, is used as a proxy for the residents' economic welfare, since no data on household or personal income are available at the census tract level, which is likely to provide information on such item. This covariate controls for a possible income effect related to the provision of the three ecosystem services at stake. An increase in the income level is expected to be correlated with an increase in the supply of ecosystem services since public goods such as CCS, ROC and HEM ecosystem services should be seen as luxury goods in urban contexts and, in so doing, as drivers of gentrification phenomena.

The covariates C_Stor_lag, Dens_Ret_lag and LST_Cmax_lag are associated with the spatially lagged values of C_Stor, Dens_Ret and LST in models (1), and control for their spatial autocorrelation on the basis of the methodological approach by Zoppi and Lai [20]. Finally, the statistical significance of the estimated coefficients of the regression models are tested through their p-values.

3 Results

3.1 Spearman's Rank Correlation Coefficients

The SRCCs estimated with reference to the correlations between the variables associated with the supply of the three targeted ecosystem services and the covariates representing the characteristics of sealed and unsealed soils, generally show the expected sign and significant p-values, as reported in Table 3.

As for C_Stor, variable Perc_Tree shows a very low SRCC associated with a high p-value, which indicates no correlation between wooded and tree covers and CCS. This is consistent with the features of wooded and tree covers in the highly urbanized urban area of Cagliari, where wooded and tree areas are characterized by tree rows planted in small garden beds with limited soil size. The other SRCCs are in a range between 19% and 28%, with very low and significant p-values; moreover, they show the expected correlation signs. Perc_Bush and Perc_Lowv are positively correlated to the supply of CCS, which is consistent with expectations, since unsealed shrubby soils covered with medium-height plants or low-height ground covers generally perform as important sinks for CCS. On the other hand, Vol_Dens, Res_Dens, Dwel_Dens and Sup_Nvp are negatively correlated with C_Stor, which is also consistent with expectations, since all of them are measures of the size of the soil-sealing phenomenon. Moreover, such four variables are highly correlated with each other (Table 4). These findings are consistent with the SRCCs estimates associated with the ROC and HEM ecosystem services. Therefore, this indicates that the densities of population, built-up volume and housing units, and the size of the unsealed area, can be considered four measures, consistent with each other, of the size of land subtracted from the potential provision of important ecosystem services in the densely urbanized urban context of Cagliari.

With reference to Dens_Ret, the three variables associated with unsealed soil are positively correlated with the supply of the ROC ecosystem service, and the estimated SRCCs are highly significant in terms of p-values, which indicates that the larger the size of the unsealed soil, the higher the potential runoff retention, as expected. However, the SRCCs of the four variables related to sealed soil are all negatively correlated with Dens_Ret and highly significant in terms of p-values, as expected as well.

Finally, the variable LST_Cmax, associated with the supply of the HEM ecosystem service, shows a negative and significant correlation with Perc_Tree and a positive and significant correlation with Perc_Lowv, whereas the estimate of the SRCC with respect to Perc_Bush is very low and not significantly different from zero in terms of p-value. Such findings entail the following: i., lower LSTs are associated with larger woods and tree canopies, whose shading is generally recognized as an outstanding factor as regards heat waves mitigation; ii., low-height vegetated and sparsely vegetated or bare soils characterize areas with LST comparatively higher than zones featured by shading canopies, which explains a positive LST gradient from such zones towards low-height vegetated or bare soils; iii., since the SRCC concerning LST_Cmax and Perc_Bush is very close to zero, there is no evidence of correlation between the supply of the HEM

Table 3. Spearman's correlation coefficients of targeted ecosystem services.

Variable	Coefficient	t-Statistic	p-Value
Carbon capture and storage (C_Stor)			
Perc_Tree	−0.00603	−0.20110	0.84066
Perc_Bush	0.20989	7.15852	0.00000
Perc_Lowv	0.25416	8.76325	0.00000
Vol_Dens	−0.25830	−8.91600	0.00000
Res_Dens	−0.19024	−6.46204	0.00000
Dwel_Dens	−0.21294	−7.26748	0.00000
Sup_Nvp	−0.27868	−9.67642	0.00000
Water retention (Dens_Ret)			
Perc_Tree	0.04398	1.46801	0.14238
Perc_Bush	0.22418	7.67089	0.00000
Perc_Lowv	0.48682	18.58474	0.00000
Vol_Dens	−0.52649	−20.65031	0.00000
Res_Dens	−0.41839	−15.36086	0.00000
Dwel_Dens	−0.46388	−17.46138	0.00000
Sup_Nvp	−0.45443	−17.01156	0.00000
Mitigation of land surface temperature (ref.: Land surface temperature, LST_Cmax)			
Perc_Tree	−0.19263	−6.54632	0.00000
Perc_Bush	−0.02725	−0.90892	0.36359
Perc_Lowv	0.17642	5.97667	0.00000
Vol_Dens	−0.07881	−2.63617	0.00850
Res_Dens	0.02683	0.89487	0.00000
Dwel_Dens	−0.05421	−1.81036	0.00000
Sup_Nvp	0.22581	7.72974	0.00000

ecosystem service and land cover characterized by medium-height plants, which implies that such areas show neither a shading effect due to more-or-less extended canopy, nor a positive LST gradient associated with the vegetation height.

Table 4. Spearman's correlation coefficients of variables associated to sealed soil.

Variable	Coefficient	t-Statistic	p-Value
Built volume density (Vol_Dens)			
Res_Dens	0.74078	36.77303	0.00000
Dwel_Dens	0.81686	47.22204	0.00000
Sup_Nvp	0.68789	31.60424	0.00000
Population density (Res_Dens)			
Dwel_Dens	0.94272	94.23846	0.00000
Sup_Nvp	0.58721	24.19157	0.00000
Density of housing units (Dwel_Dens)			
Sup_Nvp	0.66033	29.32140	0.00000

3.2 Regressions

The estimates of the coefficients of the Perc_Degr control variable are not significant in terms of p-values in two out of three cases. As indicated above, the share of the graduates is identified as a proxy for the income level, and public goods such as the CCS, ROC and HEM ecosystem services are likely to work as luxury goods in urban contexts and eventually as drivers of gentrification phenomena. According to the estimates of the regression models, this only happens in the case of the HEM ecosystem service, represented by the LST_Cmax dependent variable, since the estimated coefficient of Perc_Degr shows a very low and significant p-value and a negative sign. This implies that an increase in land surface temperature, which identifies a decrease in the supply of the HEM ecosystem service, is negatively correlated with the share of graduates (a proxy for the income level), consistently with the expectations. On the other hand, changes in Perc_Degr do not show any influence on the supply of the CCS and ROC ecosystem services, which entails that there is no evidence of an income effect as regards the supply of these ecosystem services. Moreover, the spatially lagged covariates C_Stor_Lag, Dens_Ret_Lag and LST_Cmax_Lag reveal positive and significant estimated coefficients as regards p-values, which indicates an adequate control for the spatial autocorrelation associated with the dependent variables. Since the estimated coefficients of the spatially lagged variables provide evidence of an effective control for the autocorrelation phenomena related to the dependent variables, and the existence of income effects is adequately investigated, the impacts of the explanatory variables associated with permeable and sealed soils can be considered reliable and well-grounded on the basis of the estimates reported in Table 5.

As for Perc_Tree, the results show there is no evidence of a significant impact of the covariate associated with the share of wooded and tree areas on the supply of the CCS ecosystem service, which is consistent with the estimate of the SRCC concerning C_Stor and Perc_Tree. This is consistent with the fact that throughout the city of Cagliari wooded and tree areas are featured by the limited size of the small garden beds and planting pits where trees are usually planted, hence allowing for a limited capacity of CCS. Moreover, the limited size of the small garden beds associated with wooded and tree land covers can explain why the ROC ecosystem service, represented by the Dens_Ret dependent variable, is negatively and significantly correlated with the covariate Perc_Tree, since such limited size allows a large part of the runoff waters to flow over the ground and minimizes underground runoff retention. Finally, Perc_Tree shows a negative and significant influence on LST_Cmax, which implies that the impact of the

Table 5. Estimates of regression models.

Variable	Coefficient	t-Statistic	p-Value
Dependent variable: Carbon capture and storage (C_Stor)			
Perc_Tree	−0.04090	−0.86124	0.38929
Perc_Bush	0.21349	2.86994	0.00418
Sup_Nvp	−0.04283	−2.84009	0.00459
Perc_Degr	0.01913	0.87229	0.38324
C_stor_Lag	10.52142	16.87334	0.00000
Mean and Standard deviation of dependent variable: 45.11756, 10.90465 – Adjusted R-squared: 0.21705			
Dependent variable: Runoff retention (Dens_Ret)			
Perc_Tree	−0.02728	−9.93164	0.00000
Perc_Bush	0.04036	9.52346	0.00000
Sup_Nvp	−0.01721	−17.07541	0.00000
Perc_Degr	−0.00049	−0.39350	0.69403
Dens_Ret__Lag	0.77435	26.87518	0.00000
Mean and Standard deviation of dependent variable: 2.36299, 0.99759 – Adjusted R-squared: 0.69620			
Dependent variable: Mitigation of land surface temperature (ref.: Land surface temperature, LST_Cmax)			
Perc_Tree	−0.03426	−6.87215	0.00000
Perc_Bush	0.04804	43.68462	0.00000
Sup_Nvp	0.00001	7.80715	0.00000
Perc_Degr	−0.03800	−15.12484	0.00000
LST_Cmax_Lag	1.07469	206.51714	0.00000

size of the shading tree canopies play a decisive role in mitigating heating waves, which is particularly important in the hottest days of the year.

With reference to Perc_Bush, the estimated coefficients are positive and significant in terms of p-values, which implies that an increase in medium-height shrubby land cover areas is associated with an increase in the supply of the CCS and ROC ecosystem services, which is consistent with expectations, since such land cover entails large room for soil and subsoil carbon storage, and for drainage capacity, especially in case of relevant meteorological events. Moreover, the results of the regression concerning the HEM ecosystem service show evidence of a positive impact of Perc_Bush on land surface temperature, which entails that an increase in the share of shrubby medium-height land cover is associated with an increase in LST_Cmax. This indicates that shrubby land cover areas, which do not grant an adequate canopy such as tree and wooded areas, have a negative impact in terms of mitigation of urban heating waves.

Finally, the impacts of Sup_Nvp on the supply of the CCS, ROC and HEM ecosystem services are always negative, meaning that the higher the share of sealed soil, the lower the provision of CCS, ROC and HEM. This is consistent with the results related to the SRCCs, which show significant negative values as for C_Stor and Dens_Ret, and a positive value with reference to LST_Cmax.

4 Discussion

This section focuses on the discussion of main findings, therefore looking at the relationships between, on the one hand, CCS, HEM, and ROC, and, on the other hand, trees, bushes, and low-height vegetation and their significance in affecting the provision of the three selected ecosystem services.

Firstly, as for the relationship between the share of a census tract covered by trees and carbon storage, both coefficients in Tables 3 and 5 are negative and, as far as the regression is concerned, not significant. The importance of urban trees in removing carbon from the atmosphere and storing it in their biomass is acknowledged by many authors. Urban tree's effectiveness in providing CCS is affected by several factors, first and foremost the size of land covered by trees [21], but also other factors such as tree density, limitation to tree growth due to management such as pruning, species type, age, structure, and overall condition, Most studies focus on urban forests and urban parks, while the role of street trees is under-researched; such role is highly important in our study area because most records concerning trees in the city dataset used in this study refer to street trees. In Cagliari, they usually grow in small planting pits in sidewalks, which can hamper carbon sequestration by affecting the soil substrate and its quality [22], as well as water dynamics and water retention, hence impacting on trees' growth and overall condition [23]. Soares et al. [24] report carbon sequestration by street trees to be fairly low in Lisbon; this might explain the low value of the coefficient linking Perc_Tree to C_stor in both Tables 3 and 5. Moreover, a possibly relevant observation for Cagliari is that by Havu et al. [25], according to whom street trees can act as carbon emitters or sinkers depending on their age and on seasonal variations, which might also help explain the lack of significance of the regression coefficient in Table 5, since no information concerning age is available in the study area.

Contrary to Perc_Tree, both Perc_Bush and Perc_Lowv show positive and significant correlation with C_Stor (Table 3) and the impact of Perc_Bush on C_Stor is positive and significant as far as the regression model is concerned (Table 5). As with trees, factors such as species type, size, and structure are important in CCS provided by urban shrubs [26]. Although it is well known that shrubs store less carbon than trees, a study by Edmondson et al. [27] shows that carbon storage in the topsoil is higher under trees, followed by topsoil beneath urban shrubs and herbaceous vegetation, which both perform better than non-urban agricultural fields, hence highlighting the importance of mid- and low-height vegetation in fostering carbon storage in soils. Furthermore, a relevant study on CCS by shrubs is that by Baraldi et al. [28], who, despite estimating their CCS capacity lower than that by trees because of their smaller structure and leaf coverage, report that evergreen shrubs can perform better than deciduous trees because their removal activity through photosynthesis spans across the year. This is of importance in Cagliari, where large areas are covered by evergreen shrubby vegetation.

Secondly, as far as ROC is concerned, Table 3 highlights the positive significant correlation between Dens_Ret and Perc_Bush and Perc_Lowv, as well as the positive, but far less significant, correlation between Dens_Ret and Perc_Tree. This signals that, in general, ROC is positively correlated with whichever type of vegetated land cover, not only because of the soil's higher porosity, but also because vegetation, even when herbaceous, channels water fluxes through the roots [29]. Moreover, Table 5 singles out Perc_Bush as the most important factor in positively driving Dens_Ret, while, and counterintuitively, the coefficient of Perc_Tree is negative, although quantitatively modest. Trees recorded in the city dataset are mostly street trees in planting pits surrounded by paved surfaces; this highlights the significance of permeable, vegetated areas in retaining rainwater, therefore also shedding light on the quantitatively relevant, negative, and significant Pearson coefficient reported in Table 3 as for the relationship between Dens_Ret and Sup_Nvp. The negative impact of impervious areas is fully acknowledged in several other studies, some of which propose mitigating solutions that result in increased vegetated areas, such as planting bushes and trees in paved areas to improve infiltration [30] or unsealing paved areas and planting native shrubs to improve water retention [31].

Finally, in regard to HEM, Tables 3 and 5 highlight the positive impact of Perc_Tree on HEM, although the magnitude of the impact is small, as signaled by the low regression coefficient. Zardo et al. [32] assess the effects of tree canopy coverage, soil cover, and unsealed land size on the cooling capacity of three climatic regions; one of the most important, and straightforward, finding, is that size matters: in the Mediterranean area, unsealed soils smaller than two hectares have limited cooling capacity even when having a good canopy coverage. This helps explain the counterintuitive positive coefficient that relates LST_max and Perc_Lowv in Table 3, which, in the built-up and dense part of the city, include all of the small green areas within building blocks. Marando et al. [33] confirm that the HEM capacity of urban trees in the Mediterranean context is affected by the size of the area they stand on: by taking Rome as a case study, they show that the mitigating capacity of tree canopies in road tree lines is approximately half that of urban and periurban forests because street trees' roots are mostly constrained within planting pits surrounded by sealed soils which hamper evapotranspiration, and ultimately the cooling effect. The limited cooling capacity of tree lines can be improved through appropriate design, which includes not only choosing the most appropriate species to increase both evapotranspiration and the size of the shadow, but also selecting the appropriate interval between individual trees depending on their height and width [34]. What is more important, notwithstanding their lower cooling capacity compared to urban forests, tree lines or even isolated trees within built-up areas are significant in reducing heat-related illnesses in the elderly, thus improving human well-being. Finally, with reference to the low-height vegetation land-cover type, which includes grassland, lawns and gardens, contrary to the findings from this study its HEM service has been found to be observable, although much lower than that of trees, in previous literature (for instance, [35]).

5 Conclusions

Policy implications can be straightforwardly identified, which entail that increases in the supply of the CCS, ROC, and HEM ecosystem services target the expansion of shrubby and low-height ground covers, and of wooded and tree areas grounded on large garden beds, or, even better, on continuous vegetated areas. The increased size of vegetated areas, be they characterized by forests or woodlands, or medium- or low-height vegetation, would be associated with increases in CCS and in the effectiveness of ROC, whereas the presence of trees and wooded areas would be correlated with heating mitigation. That being so, the integrated increases in the size of vegetated areas and of tree-covered areas would generate a comprehensive significant improvement in the city quality of life [36, 37], in terms of the three targeted ecosystem services. Policies aimed at pursuing such integrated increases should be associated with virtuous behaviors on behalf of the local communities, organized citizen groups, building companies, and public administrations [38].

An important issue concerns the narrow relationship between the value of land and the maximum permitted volume, in terms of either residential or service buildings. In this perspective, the plantation of new vegetated areas, be they covered by trees, shrubs, or low vegetation, or the expansion of areas characterized by such land covers, would imply the implementation of appropriate planning measures. Said measures should carefully integrate the landowners' demands and the public interest which would entail a plea for a robust increase in the size of urban green areas, in face of a significant decrease in the value of a share of the urban land. In view of this, rigorous rules should be operationalized to control over building permits related to new and existing urban settlements, which should be associated with an adequate provision of wooded and tree areas or vegetated areas, such as green roofs, walls and facades, rain gardens, swales, and related collection tanks [14], as much as green and blue urban grids, as it occurred in East London [39]. Moreover, a structured system of public financial incentives should be planned to make the implementation of urban greening policies attractive for landowners and investors [40]. Such operational framework can be based on the integration of several instruments, such as grants made available to building firms on condition that new developments are endowed with an agreed-upon amount of wooded, tree or green areas, impact fee reductions, and VAT and property tax allowances [41]. Lastly, the commitment of local governments towards increasing the supply of the CCS, ROC, and

HEM ecosystem services should be made visible and recognizable to the local societies through flagship financial operations such as public purchasing of areas where to implement urban greening policies [42].

The methodological approach defined in this study, and the outcomes provided by its implementation with reference to the city of Cagliari, effectively address the research question highlighted in the Introduction, namely, how and to what extent is the endowment of sealed and unsealed areas related to ecosystem supply in intensively urbanized urban areas? The findings indicate that an adequate endowment of vegetated areas and of tree-covered areas would be associated with a comprehensive and relevant urban living quality, which builds on suitable levels of the CCS, ROC, and HEM ecosystem services. The value added of the methodology here implemented is identified by the fact that it is readily exportable to other urban contexts of different demographic sizes, and physical and social conditions, since it develops through databases whose structure is easy to build and whose data are straightforward to retrieve and collect, across cities located in different regions and countries. Moreover, not only is the methodological approach easy to export, but also the results of its implementation as regards cities located in different national and international urban contexts can be readily and effectively compared. Finally, this study leaves plenty of room for future research concerning other ecosystem services supplied in and by urban environments, such as food production, recreational services, natural and cultural heritage, and so on, whose analysis can build on the identification and assessment of their relations with urban land covers and uses, whose specification can be different and possibly more detailed than the characteristics considered in this article.

Acknowledgments. This study was carried out: i. within the RETURN Extended Partnership funded by the European Union – NextGenerationEU (National Recovery and Resilience Plan – NRRP, M4, C2, Investment 1.3 – D.D. 1243 2/8/2022, PE0000005); ii. with the financial support under the NRRP, M4, C2, Investment 1.1, Call for tender no. 1409 published on 14.9.2022 by the Italian Ministry of University and Research (MUR), funded by the European Union – NextGenerationEU – Project Title "Definition of a guidelines handbook to implement climate neutrality by improving ecosystem service effectiveness in rural and urban areas" – CUP F53D23010760001 – Grant Assignment Decree no. 1378 adopted on September 1, 2023, by MUR. This article was excerpted from Lai, S., Zoppi, C. (2023) Factors affecting the supply of urban regulating ecosystem services. Empirical estimates from Cagliari, Italy. TeMA Journal of Land Use, Mobility and Environment, S.I. 2/2023, pp. 7–32.

Authors' Contribution. Sabrina Lai (S.L) and Corrado Zoppi (C.Z.) collaboratively designed this study. Individual contributions are as follows: C.Z. wrote Sections 1; 2.3, 3, and 5. S.L. wrote Sections 2.1, 2.2, and 4.

Disclosure of Interests. The authors declare no competing interests.

References

1. Tratalos, J., Fuller, R.A., Warren, P.H., Davies, R.G., Gaston, K.J.: Urban form, biodiversity potential and ecosystem services. Landsc. Urban Plan. **83**, 308–317 (2007)

2. Donovan, R.G., Stewart, H.E., Owen, S.M., MacKenzie, A.R., Hewitt, C.N.: Development and application of an urban tree air quality score for photochemical pollution episodes using the Birmingham, United Kingdom, area as a case study. Environ. Sci. Technol. **39**(17), 6730–6738 (2005)
3. Floris, M., Zoppi, C.: Ecosystem services and spatial planning: A study on the relationship between carbon sequestration and land-taking processes. Archivio di Studi Urbani e Regionali **51**(Suppl. 127), 11–33 (2020)
4. Isola, F., Lai, S., Leone, F., Zoppi, C.: Land take and landslide hazard: spatial assessment and policy implications from a study concerning Sardinia. Land **12**(2), 1–23, 359 (2023)
5. Bibri, S.E., Krogstie, J., Kärrholm, M.: Compact city planning and development: emerging practices and strategies for achieving the goals of sustainability. Dev. Built Environ. **4**(100021), 1–20 (2020)
6. Hofstad, H.: Compact city development: high ideals and emerging practices. Eur. J. Spatial Dev. **10**(5), 1–23 (2012)
7. Geneletti, D., Cortinovis, C., Zardo, L., Adem Esmail, B.: Planning for Ecosystem Services in Cities. Springer, Dordrecht (2019)
8. Lal, R.: Carbon sequestration. Philos. Trans. R. Soc. B **363**(1492), 815–830 (2008)
9. Jobbagy, E.G., Jackson, R.B.: The vertical distribution of soil organic carbon and its relation to climate and vegetation. Ecol. Appl. **10**(2), 423–436 (2000)
10. European Environment Agency: Climate Change, Impacts and Vulnerability in Europe. An Indicator-Based Report. Publications Office of the European Union, Luxembourg (2012)
11. Patekar, M., et al.: Assessing climate change and land-use impacts on drinking water resources in karstic catchments (Southern Croatia). Sustainability **13**(9), 1–20, 5239 (2021)
12. European Environment Agency: Nature-Based Solutions in Europe: Policy, Knowledge and Practice for Climate Change Adaptation and Disaster Risk Reduction. Publications Office of the European Union, Luxembourg (2021)
13. Mysiak, J., Torresan, S., Perez Blanco, D.: Cambiamenti climatici e agricoltura nel Nordest. Fondazione Nordest, Venice (2019)
14. Berland, A., et al.: The role of trees in urban stormwater management. Landsc. Urban Plan. **162**, 167–177 (2017)
15. Hulley, G.C., Ghent, D., Göttsche, F.M., Guillevic, P.C., Mildrexler, D.J., Coll, C.: Surface temperature. In: Hulley, G.C., Ghent, D. (eds.) Taking the Temperature of the Earth Steps Towards Integrated Understanding of Variability and Change, pp. 57–127. Elsevier, Amsterdam (2019)
16. Salata, K.D., Yiannakou, A.: Green Infrastructure and climate change adaptation. TEMA J. Land Use Mobil. Environ. **9**(1), 7–24 (2016)
17. Alfraihat, R., Mulugeta, G., Gala, T.: Ecological evaluation of urban heat island in Chicago City, USA. J. Atmos. Pollut. **4**(1), 23–29 (2016)
18. Anselin, L., Syabri, I., Kho, Y.: GeoDa: an introduction to spatial data analysis. Geogr. Anal. **38**(1), 5–22 (2006)
19. Byron, R.P., Bera, A.K.: Linearised estimation of nonlinear single equation functions. Int. Econ. Rev. **24**(1), 237–248 (1983)
20. Zoppi, C., Lai, S.: Land-taking processes: an interpretive study concerning an Italian region. Land Use Policy **36**, 369–380 (2014)
21. Nowak, D.J., Greenfield, E.J., Hoehn, R.E., Lapoint, E.: Carbon storage and sequestration by trees in urban and community areas of the United States. Environ. Pollut. **178**, 229–236 (2013)
22. Schütt, A., Becker, J.N., Reisdorff, C., Eschenbach, A.: Growth response of nine tree species to water supply in planting soils representative for urban street tree sites. Forests **13**(936), 1–21 (2022)

23. Nielsen, C.N., Bühler, O., Kristoffersen, P.: Soil water dynamics and growth of street and park trees. Arboricult. Urban Forest. **33**(4), 231–245 (2007)
24. Soares, A.L., Rego, F.C., McPherson, E.G., Simpson, J.R., Peper, P.J., Xiao, Q.: Benefits and costs of street trees in Lisbon, Portugal. Urban Forest. Urban Green. **10**(2), 69–78 (2011)
25. Havu, M., Kulmala, L., Kolari, P., Vesala, T., Riikonen, A., Järvi, L.: Carbon sequestration potential of street tree plantings in Helsinki. Biogeosciences **19**, 2121–2143 (2022)
26. Khan, N., Jhariya, M.K., Yadav, D.K., Banerjee, A.: Structure, diversity and ecological function of shrub species in an urban setup of Sarguja, Chhattisgarh, India. Environ. Sci. Pollut. Res. **27**, 5418–5432 (2020)
27. Edmondson, J.L., Davies, Z.G., McCormack, S.A., Gaston, K.J., Leake, J.R.: Land-cover effects on soil organic carbon stocks in a European city. Sci. Total. Environ. **472**, 444–453 (2014)
28. Baraldi, R., et al.: An integrated study on air mitigation potential of urban vegetation: from a multi-trait approach to modeling. Urban Forest. Urban Green. **41**, 127–138 (2019)
29. Técher, D., Berthier, E.: Supporting evidences for vegetation-enhanced stormwater infiltration in bioretention systems: a comprehensive review. Environ. Sci. Pollut. Res. **30**, 19705–19724 (2023)
30. Salata, S., Ronchi, S., Giaimo, C., Arcidiacono, A., Pantaloni, G.G.: Performance-based planning to reduce flooding vulnerability insights from the case of Turin (North-West Italy). Sustainability **13**(5697), 1–25 (2021)
31. Salata, S., Velibeyoğlu, K., Baba, A., Saygın, N., Couch, V.T., Uzelli, T.: Adapting cities to pluvial flooding: the case of Izmir (Türkiye). Sustainability **14**(16418), 1–19 (2022)
32. Zardo, L., Geneletti, D., Pérez-Soba, M., Van Eupen, M.: Estimating the cooling capacity of green infrastructures to support urban planning. Ecosyst. Serv. **26**, 225–235 (2017)
33. Marando, F., Salvatori, E., Sebastiani, A., Fusaro, L., Manes, F.: Regulating ecosystem services and green infrastructure: assessment of Urban Heat Island effect mitigation in the municipality of Rome, Italy. Ecol. Model. **392**, 92–102 (2019)
34. Park, C.J., et al.: Variations in pedestrian mean radiant temperature based on the spacing and size of street trees. Sustain. Cities Soc. **48**(101521), 1–9 (2019)
35. Park, C.Y., Park, Y.S., Kim, H.G., Yun, S.H., Kim, C.-K.: Quantifying and mapping cooling services of multiple ecosystems. Sustain. Cities Soc. **73**(103123), 1–12 (2021)
36. Lai, S., Leone, F., Zoppi, C.: Policies to decrease land surface temperature based on land cover change: an assessment related to Sardinia, Italy. TeMA J. Mobil. Land Use Environ. **13**(3), 329–351 (2020)
37. Gómez-Baggethun, E., Barton, D.N.: Classifying and valuing ecosystem services for urban planning. Ecol. Econ. **86**, 235–245 (2013)
38. Zoppi, C., Lai, S.: Differentials in the regional operational program expenditure for public services and infrastructure in the coastal cities of Sardinia (Italy) analyzed in the ruling context of the Regional Landscape Plan. Land Use Policy **30**, 286–304 (2013)
39. Jenning, V., Larson, L., Yun, J.: Advancing sustainability through urban green space: cultural ecosystem services, equity, and social determinants of health. Int. J. Environ. Res. Public Health **13**(2), 1–15, 196 (2016)
40. Webster, C.: The new institutional economics and the evolution of modern urban planning: insights, issues and lessons. Town Plan. Rev. **76**(4), 455–484 (2005)
41. Slätmo, E., Nilsson, K., Turunen, E.: Implementing green infrastructure in spatial planning in Europe. Land **8**(4), 1–21, 62 (2019)
42. Fors, H., Frøik Molin, J., Murphy, M.A., van den Boschab, C.K.: User participation in urban green spaces – for the people or the parks. Urban Forest. Urban Green. **14**(3), 722–734 (2015)

Urban Heat Island Phenomenon and Ecosystem Services. A Systematic Review

Federica Isola, Federica Leone(✉), and Rossana Pittau

Department of Civil and Environmental Engineering and Architecture, University of Cagliari, Cagliari, Italy

{federica.isola,federicaleone,rossana.pittau}@unica.it

Abstract. The increasing rate of urbanization and continuous population growth in urban areas leads to several problems, including the emergence of urban heat islands (UHI), defined as urban areas where temperatures are higher than in the surrounding rural areas. UHIs have negative impacts on the health of populations and lead to increased energy consumption for cooling. One of the main causes of higher temperatures in urban areas and, therefore, the creation of UHIs is impervious surfaces, which in turn lead to poor thermal comfort in cities. Ecosystem services and, in particular, the ecosystem service of local climate regulation are a valuable tools to mitigate the effects of UHI. The contribution reviews the existing literature concerning the mitigation of heat island effects through ecosystem services, in order to understand how they are studied and analyzed in the international scenario. The proposed methodological approach is based on a framework of analysis of the scientific contributions published in the last fifteen years on the subject of UHI, investigating the phenomenon through an interpretation key based on the issues addressed, the methods used and the spatial scales to which these methods have been applied.

Keywords: Urban Heat Island · Local Climate Regulation · Spatial Planning

1 Introduction

According to the United Nations Report [1] in 2018 more than 55% of the world's population lived in urban areas and it is expected that by 2050, the percentage will grow to 64–84%. The increasing rate of urbanization and the continuous increase of population in urban areas have entailed several problems which impact negatively on health and well-being of populations and the environment in general [2, 3]. Furthermore, urbanization often leads to negative impacts that are linked to other threats in areas with heightened natural risk, like hydrogeological issues [4].

Urban heat island (UHI) is defined as "...an urban area that is significantly warmer than its rural surroundings due to artificial infrastructure and human activities" [5]. UHIs have negative impacts on the health of populations [6] and lead to increased energy consumption required for cooling [7]. One of the main causes of rising temperatures in urban areas and, thus, the occurrence of UHIs is impervious surfaces [8], which in turn

O. Gervasi et al. (Eds.): ICCSA 2024 Workshops, LNCS 14818, pp. 352–369, 2024.
https://doi.org/10.1007/978-3-031-65273-8_23

lead to poor thermal comfort in cities [9]. The thermal behavior of urban structures can be significantly improved through the implementation of measures that affect only the surface cover of the buildings [10]. In literature, several methodological approaches are used to study and assess the UHI phenomenon and climate vulnerability and risks, such as methods based on field observations [11, 12], empirical methods [13, 14], remote sensing techniques [15–17], modelling [18, 19], and simulation [20, 21].

The determination of thermal comfort in cities represents a fundamental element in the assessment of the urban microclimate [22]. For this reason, the implementation of appropriate land management policies and modelling techniques aimed at analyzing these phenomena is of paramount importance [23]. The current scientific debate focuses on the need to formulate effective policies for climate change adaptation and mitigation [24]. In fact, several studies focus on the mitigation of impacts due to heat islands. Some studies are based on physical-microclimatic models that allow to compare two or more different scenarios, such as in the case of the ENVI-met simulation tool [25]. Other studies propose the use of materials and design methods based on nature-based solutions (NBSs) and Green Infrastructure (GI) in new urban settlements and in restoration sites. Green transition of urban areas makes cities more adaptable and more resilient to the effects of UHI. With regard to the risks associated with UHIs, appropriate short- and long-term planning and management measures are required [26]. GIs represent a strategically planned network of high quality natural and semi-natural areas [27], and according to the European Commission, GIs "…are designed and managed to deliver a wide range of ecosystem services" (ESs) [28, p. 3]. Subsequently, EU 2030 Biodiversity Strategy encourages Member States to map and assess ecosystems and their services [29].

Ecosystem services and, in particular, regulating ecosystem service such as local climate regulation, are a valuable tool for mitigating the effects of UHI. Numerous studies have highlighted the relationship between land cover types and its impact on surface temperature [30]. The increase of vegetation in urban areas allows the regulation of the macro- and microclimate and the reduction of surface temperatures [31]. In general, the increase of permeable areas is implemented through the realization of new public green spaces, the requalification of existing green areas and the tree planting in the most sealed areas. Ecosystem-based adaptation approach to urban spatial planning can be a key element of establishing urban policies focused on natural resource management [32]. As a result, this approach incorporates the advantages provided by ESs also in terms of socio-economic development impacts [33]. On the other hand, the integration of ES within spatial planning is still limited. Longato et al. [34] evaluate the use of ES in spatial planning by means of a literature review in order to understand the integration level, and planning strategies addressed. According to Longato et al. [34], although the number of studies that focus on the integration of ES in spatial planning have increased, the systematic review of the literature highlights how the main benefits of integrating ES into spatial planning concern only practical issues, such as the promotion of stakeholder participation.

Starting from this theoretical framework, the study presents an in-depth analysis of the existing literature on the mitigation of heat island effects through ecosystem services, in order to understand how these phenomena are studied and analyzed in the international scenario. The proposed methodological approach is based on a systematic review of scientific studies published in the last fifteen years on the subject of UHI, investigating the phenomenon through an interpretation key based on four aspects: general context, reasons, methods and data, and implications for spatial and urban planning. Moreover, the proposed method represents a preliminary analysis aimed at defining key elements on which to base spatial planning policies at the municipal scale.

The contribution is structured in four sections. The first, introductory section focuses on the theoretical background. The second describes the proposed methodological approach, structured in two specific evaluation phases. The third section presents the results which are discussed in the fourth section, where final considerations are provided.

2 Materials and Methods

The paper proposes a systematic analysis of a part of the existing scientific literature on the mitigation of heat island effects through ecosystem services, with the aim to understand how this phenomenon is internationally studied and analyzed. The sample of papers to be analyzed was selected from the Scopus database, (2023/06/01) using the combination of words "Urban AND heat AND island AND ecosystem AND services" within the field "Article Title; Abstract; Keywords". The preliminary paper selection resulted in the identification of 331 papers spanning a period from 2007 to 2023.

The paper selection phase was structured into two evaluation steps. In the first step, all abstracts were analyzed in order to exclude papers not related to spatial and urban planning. As a result of the first step, 122 papers were selected. The second evaluation step involved a content analysis in which all 122 selected papers were analyzed. As a result of the second step, 57 papers were selected.

The remaining 65 papers were excluded because: only the abstract was freely available or only the abstract was written in English (24 papers); they were theoretical and did not deal with application cases (four papers); they focused on the building scale (29 papers); and they focus on the influence of specific typologies of plant species on the mitigation of urban heat island effects (8 papers). In relation to the temporal distribution of the 57 analyzed papers, most of the analyzed papers were published from 2017 to 2019, excluding the year 2023 because the results are only partial as they refer to papers published until 1 June.

In relation to the type of source and the number of papers published in each journal or series. With the exception of two papers, all the analyzed papers were published in 34 different scientific journals. The journals where were published the highest number of the analyzed papers were published are: Landscape and Urban Planning, Science of the Total Environment, Sustainability, and Sustainable Cities and Society.

With regard to the data analysis, the main research question was broken down into four sub-questions that guided the content analysis of the papers. Table 1 shows the sub-questions, specified through a description of the interpretation keys.

Table 1. Sub-question and description of interpretation keys for the analysis of selected papers.

Sub-question	Interpretation key
A – General context	A1 – The scale to which the analysis refers (region, city, part of the city)
	A2 – Type and number of case studies analysed
B – Reasons	B1 – Description of the reasons why the urban heat island phenomenon is studied
	B2 – It is described whether the phenomenon was studied individually or within a larger study
C – Methods and data	C1 – Description of the methods used (in situ observations, remote sensing techniques, modelling and simulation)
	C2 – Description of data retrieval with reference to three parameters concerning the season in which the data was retrieved, the duration of data collection and the period of the day to which the data refers
D – Implications for spatial and urban planning	D1 – Does the study define strategies and/or suggest recommendations for planners and policy maker
	D2 – Do you dedicate a special section to such strategies?
	D3 – What kind of strategies are defined?

3 Results

Below are the results for each sub-question shown in Table 1.

3.1 General Context

In relation to the general context, 49 papers refer to the municipal scale, two concern the metropolitan scale [35, 36], two focus on an inter-municipal scale, between municipal and regional territory [37, 38], three refer to the regional scale [39–41], and one concerns the national scale [42]. For example, Bassett et al. [42] study the thermal contribution of land cover change to climate regulation between 1975 and 2014 in Great Britain.

Furthermore, with reference to the 49 cases that refer to the urban scale, 29 cases study the entire municipal territory while 20 focus on parts of the territory. With reference to the 29 papers that consider the entire municipal territory, in six cases a comparison between several cities is made [19, 43–47].

For example, Marando et al. [19] analyze the mitigation of the local microclimate offered by urban green infrastructure through a model that combines land surface temperatures (LST) with air temperature, taking 601 European cities with less than 50,000 inhabitants as case studies. With reference to studies analyzing parts of the city, twelve papers consider permeable areas such as urban parks [12, 48, 49], cemeteries [50], the urban green system [51–56], forest areas [57] and urban gardens [11]. In the remaining eight cases [58–65], urban areas characterized by a mixed fabric comprising both sealed and permeable areas are considered. For example, Murtinová et al. [58] study the effect of the Surface Urban Heat Island (SUHI) in the city of Zvolen (Slovakia), analyzing the LST within three urban areas during the summer months in the period 2010–2021.

Figure 1 shows the location of the case studies analyzed in those papers that focus on the municipal scale with the exception of the 601 European cities analyzed in the study of Marando et al. [19]. The most studied is Beijing, followed by Shenzen (China), Manila (Philippines), and Milan and Rome (Italy).

Fig. 1. Localization of case studies in the case of papers focusing on the municipal scale.

3.2 Reasons

With reference to the second sub-question 'motivations', 47 papers focus their study on the phenomenon of urban heat islands. In the other ten cases [26, 31, 36, 46, 51, 66–70], the study of the urban heat island phenomenon is embedded within a broader study. For example, van Oorschot et al. [31] study the best location of green infrastructure based on its ability to provide three benefits, that is three ecosystem services, one of which is the reduction of the urban heat island effect.

In reference to the papers that focus on the urban heat island phenomenon, 24 papers focus on the influence of permeable areas on the temperature reduction and, thus, urban heat island effects. Different types of permeable areas are studied such as urban green areas [43, 58], parks [52], forests [57] and urban green infrastructure [19, 71]. In addition, nine papers [12, 21, 39–42, 48, 72, 73] focus on the influence of land use and/or land cover and their changes on temperatures and, thus, on reducing urban heat island effects.

For example, Elliot et al. [21] propose a study in which a simulation model of land cover changes was used to assess the effects of urban heat islands over time in the city of Lisbon. Five papers [35, 38, 47, 59, 74] study how landscape conformation influences temperatures and, thus, the reduction of urban heat islands.

Two papers [60, 61] study the influence of water bodies on the urban heat island phenomenon. The remaining six papers focus on different factors [37, 45, 64, 75–77]. For example, Ramaiah et al. [45] study how LST is influenced by three factors: built-up areas, vegetated areas and water bodies.

3.3 Methods and Data

With reference to the "methods and data" sub-question (Fig. 2), 37 papers [50, 66] use remote sensing techniques to assess LST. In six papers [11, 12, 36, 41, 56, 78] air temperatures are obtained through direct measurements. Two cases [19, 43] assess both ground land surface temperatures through remote sensing techniques and air temperatures through direct measurements. In the remaining twelve papers, the urban heat island phenomenon and local climate regulation are studied through other methods such as INVEST [26, 76] or the i-Tree cool air model [79].

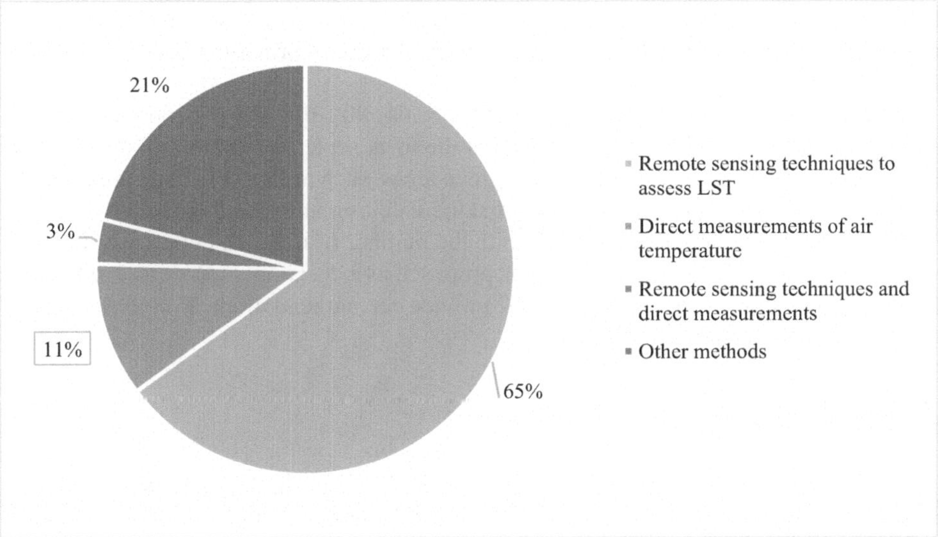

Fig. 2. Methods used to study the urban heat island phenomenon.

In relation to the 37 papers using remote sensing techniques, in twelve cases the images analyzed refer to a single date. In relation to the twelve papers referring to a single date, in five cases [31, 50, 53, 61, 80] the image refers to daytime, in one case to nighttime [47], while in the remaining six cases the time of day to which the assessment refers is not made explicit [48, 49, 51, 59, 67, 69].

The remaining 25 papers that use remote sensing techniques analyze images that refer to several, often very distant years, of which nine [37, 45, 52, 57, 58, 60, 61, 74, 77] have acquired images that refer to daylight hours, two [54, 72] to nocturnal hours, two [75, 81] to both nocturnal and diurnal hours, and in the remaining twelve cases [35, 38–40, 44, 55, 66, 69, 71, 73, 81, 82] is not made explicit.

3.4 Implications for Spatial and Urban Planning

With reference to the fourth sub-question "Implications for spatial and urban planning", 32 papers suggest strategies and/or recommendations for policy makers and planners. In thirteen cases [12, 35, 36, 43, 50, 51, 53, 55, 59, 64, 71, 77, 79] the authors decided to dedicate a special section to them. The 32 papers suggesting strategies and/or recommendations can be grouped into four types: strategies at the territorial scale, guidelines on individual interventions at the local scale, the methodological approach as decision support, and strategies and guidelines that address both the territorial and local scales.

In nine instances [26, 35, 36, 57, 59, 68, 71, 77, 82], these recommendations and/or strategies refer to a spatial scale. For example, Shih and Mabon [77] suggest favoring natural ventilation and protecting areas that promote temperature reduction.

In order to improve natural ventilation, the study suggests preserving mountain valleys from intensive urbanization by preserving green spaces along pathways where wind is channeled, especially those located along river corridors. With regard to areas that favor temperature reduction, such as agricultural land, wetlands and ponds, their continuity should be ensured within development plans also by providing forms of transfer of building rights.

In ten instances [11, 12, 40, 47, 51, 53, 55, 64, 80], the recommendations and/or strategies refer to individual interventions at the local scale. For example, Ronchi et al. [76] suggest design parameters for urban green areas, such as tree cover and area size. In relation to tree cover, the study suggests that the species planted, and the relative canopy cover are most significant elements despite the number of trees. In eight cases [19, 39, 41, 43, 58, 65, 66, 72] the methodology is proposed as a decision-support tool, while the remaining five cases [50, 54, 64, 76, 79] provide recommendations at both the spatial and individual intervention scales.

4 Discussion

The results of the content analysis conducted on the 57 papers highlight several important aspects. Firstly, the study of the urban heat island phenomenon in relation to ecosystem services has been a widely discussed topic in the literature for several years. The oldest analyzed paper dates back to 2011. Howard was the first scholar to describe the urban heat island phenomenon and its causes within his work "The Climate of London" where he described the study conducted from 1806 to 1830 during which he recorded the temperature every day for a year at three sites outside London (Plaistow, Tottenham and Stratford) and one within the city (Royal Society) [83]. Howard defines the urban heat island as the difference between the air temperature within the urban settlement and the air temperature measured in the rural areas and assumes that this value increases as one moves closer from the city limits towards the center [83]. Howard also defines possible causes, arriving at the conclusion that "the temperature of the city is not to be considered as that of the climate; it partakes too much of an artificial warmth, induced by its structure, by a crowded population, and the consumption of great quantities of fuel in fire" [84, p. 2].

The second aspect concerns the factors influencing urban heat islands. According to the study conducted by Voogt [85], they can be traced to six aspects: i. geographic location; ii. time; iii. synoptic weather; iv. city form; v. city function; and vi. city size. Geographic location takes into account climate, topography and rural surrounds. Time considers time of day (day and night) and season (winter, spring, summer, and autumn). Synoptic weather considers wind and the presence of clouds. City form concerns materials, geometry, and green spaces. City function concerns the activities that take place in terms of energy use, water use and pollution and, finally, city size is connected with the shape of the city and the functions that take place in it. However, according to Voogt [85], mitigation measures relate to city form and city function are the only parameters on which human being can operate.

For example, Okumus and Terzi [50] study the cooling effect of 309 cemeteries in the city of Istanbul. The study showed that the optimal distance for cooling, beyond which this regulation service decreases, is 200 m for very small cemeteries, 250 m for medium-small cemeteries and 400 m for larger cemeteries. In terms of the time factor, the intensity of urban heat islands is influenced by the seasons and day and night variations. The influence of seasons concerns two aspects: amount of solar radiation reaching the earth and metabolic activity of vegetation [86]. In addition, the influence of the seasons also depends on geographical location. In the case of cities with arid and semi-arid climates, the intensity of temperatures is lower in urban areas than in surrounding areas characterized, for example, by the presence of desert areas [87]. This phenomenon is called an urban cold island [88, 89]. With regard to night and day variations, the cooling process in cities after sunset is slower than in rural areas. This behavior is due to the high thermal capacity of certain materials such as concrete and asphalt that characterize urban areas. These materials absorb high amounts of heat during the day, which is then slowly released during the night, resulting in very high UHI values during the night hours.

The third aspect concerns the techniques used. The vast majority of papers analyzes and surface temperatures through remote sensing techniques. The analysis of urban heat islands requires a reliable and consistent dataset. Prior to the use of remote sensing techniques, the process to obtain such data was particularly laborious and costly as it required the installation of meteorological stations [90]. The process has been simplified by the use of remote sensing satellites, which have made it possible to study large urban areas due to frequent wall-to-wall coverage [86]. Furthermore, thermal information is provided both as a scene/image and as individual products making the process of deriving urban heat island intensity simpler [91].

Several satellites are used to acquire the images needed to obtain urban heat island intensities, including Landsat TM, Landsat ETM, MODIS, Landsat OLI (8), etc. According to a study conducted by Deilami et al. [86], Landsat imagery is the most widely used as it is freely available, it has worldwide coverage with an acceptable resolution of 30 ×

30 m, it has temporal coverage dating back to 1972 and it provides both thermal and thematic spectral bands. However, Landsat images also have disadvantages. Firstly, it takes sixteen days for the satellite to re-image the same area. As a consequence, assessing the effects of urban heat islands during a day or a week is not possible. The second disadvantage relates to processing time, which could be significant if mosaicking is required in order to process analyses on a national or regional scale, given the size of the images [92, 93]. The use of MODIS imagery solves some of these critical issues, such as the time between two images of the same location (1–2 days for the Terra and Aqua MODIS satellite) and the processing time, as it provides some immediately available products that significantly reduce the time. For example, LSTs are produced daily and weekly. However, MODIS images have a low spatial resolution.

Despite the widespread use of remote sensing techniques at the expense of direct measurements, these techniques have limitations. Firstly, obtaining cloud-free images is difficult in some geographical areas [86]. In addition, the thermal and thematic bands of remote sensing have a spatial resolution of 30 m to 1 km that do not make it possible to process very detailed analyses, which would require costly downscaling and fusion approaches [94]. Finally, the accuracy of the data is affected by several factors, including atmospheric effect, sensor noise, etc. For example, the study conducted by Jiménez-Muñoz and Sobrino [95] showed that the main source of error for the determination of LST from satellite images is atmospheric effects. Krehbiel and Henebry [96] conducted a study in which they compared urban heat island intensities obtained from direct measurements and those obtained from MODIS imagery. The urban heat island intensity during the day is much higher in the case of data derived from MODIS images than that determined from direct measurements, while the values for the night are quite similar.

The third aspect concerns the implications for spatial and urban planning. At the international level, the 2030 Agenda for Sustainable Development and the 17 Sustainable Development Goals (SDGs) provide an important framework for the identification of climate change adaptation measures. In particular, Goal 11 "Sustainable cities and communities" places resilience as a priority aspect in the future development of cities that adopt and implement climate change adaptation policies [97]. Furthermore, Goal 13 "Climate action" in Target 13.2 calls for "Integrate climate change measures into national policies, strategies and planning" [98]. However, strategies defined at the international level have an indirect impact on adaptation at the local scale as these agreements have been negotiated and signed by national governments [99]. In fact, the adoption of these international frameworks does not lead to the implementation of concrete adaptation actions by cities. For example, the study proposed by Aguiar et al. [100] examined 147 Local Adaptation Strategies/Plans in Europe, highlighting that the United Convention on Climate Change process promoted the elaboration of only 21 plans. Furthermore, the study shows that the issue of extreme temperatures is addressed by 35 plans.

At the national level, Athens and Stockholm represent two cities that have addressed heat islands in their climate change adaptation plans [101]. Athens adopted the Athens Resilience Strategy for 2030 in 2017 and the latest Climate Action Plan was published in 2022. In particular, one of the main objectives of the Athens Strategy is to increase the area of green areas in order to decrease vulnerability to heat islands. To this end, Athens has quadrupled the budget for green areas, also thanks to the Greek state and European funds [102]. In 2019, Athens obtained funding on the Natural Capital Financing Facility, a new financing instrument of the European Investment Bank that finances cities for blue and green infrastructure projects. In addition, in 2021, the city of Athens signed the Urban Nature Declaration, which envisages de-impermeabilizing between 30% and 40% of impermeable surfaces and ensuring that at least 70% of the population has access to green or blue spaces within a 15-min radius.

According to the EEA [97], green infrastructure is one of the most effective adaptation measures to combat high temperatures. Several projects have been aimed at increasing green areas. For example, the LIFE Green Heart Project envisaged a significant increase in green areas in Toulouse, France, in order to reduce the temperature by approximately 3 °C during urban heat islands [103]. Furthermore, the LIFE@Urban Roof Project aims to increase green roofs in the city of Rotterdam, Netherlands.

The RESIN (Climate resilient cities and infrastructures) project funded by the Horizon 2020 Programme for the years 2015–2018 aims to improve climate resilience in European cities. The RESIN Adaptation Options Library is one of the tools developed by the Project. The library is a database on the effectiveness of adaptation measures, divided into three types (structural, social and institutional) based on an analysis of scientific studies published up to 2017 [97]. The database is structured in four parts: general information, organization, effectiveness and cost-efficiency. The general information section provides the main information on each adaptation measure (type, scale, climate region, sector, etc.). Organization defines who should implement the adaptation measure. Effectiveness measures in terms of reducing the vulnerability and risk components. Finally, Efficiency considers the costs and benefits of implementing the adaptation measure [104].

5 Conclusions

In conclusion, the study is a preliminary research that attempts to provide an overview of how urban heat islands are studied in literatures by analyzing some key issues related to the general context, rationale behind the study, materials and data, and implications for land-use and urban planning.

With reference to the general context, almost all papers focus on the municipal scale. In relation to reasons, only a small percentage of the papers analyzed study the heat island phenomenon within a broader approach that investigates other aspects. In relation to methods and data, the vast majority analyze land surface temperatures through remote sensing techniques. Finally, not all papers use their findings to propose strategies and recommendations for policy makers and planners.

The originality of the proposal consists in the construction of a framework that connects the issues due to UHI with the ES and GI topic at the municipal scale. In fact, if the urban setting represents the place where the effects of climate change are most perceived, the construction of an analytical system on which to base a climate-proof urban planning policy turns out to be a useful and important tool. It must be emphasized that adaptation will not be cross-cutting and ubiquitous [105], so specific analysis will be needed for each urban context.

Future research developments could concern the definition of climate-proof methodological approaches that integrate ES within spatial planning with particular reference to the local scale. The principle of climate proof-planning at the local level can be a guiding factor in territorial governance processes but, above all in the definition of a regulatory framework aimed at improving the quality of space and life in urban contexts.

Acknowledgments. Federica Isola (F.I.), Federica Leone (F.L.), and Rossana Pittau (R.P.) collaboratively designed this study. Individual contributions are as follows: F.I. wrote Section 1; F.L. wrote Sections 2, 3, 3.1, 3.2, 3.3, and 3.4. F.I. and F.L. jointly wrote Sections 4 and 5. F.L and R.P. jointly analyzed data.

This article was excerpted from Isola F., Leone F., Pittau R. (2023). Evaluating the urban heat island phenomenon from a spatial planning viewpoint. A systematic review. TeMA – Journal of Land Use, Mobility and Environment, Special Issue 2.2023, 75–93. 10.6093/1970–9870/10306.

This study was carried out: i. within the RETURN Extended Partnership and received funding from the European Union Next-GenerationEU (National Recovery and Resilience Plan – NRRP, Mission 4, Component 2, Investment 1.3 – D.D. 1243 2/8/2022, PE0000005); ii. with the financial support under the National Recovery and Resilience Plan (NRRP), Mission 4, Component 2, Investment 1.1, Call for tender No. 1409 published on 14.9.2022 by the Italian Ministry of University and Research (MUR), funded by the European Union – NextGenerationEU– Project Title "Definition of a guidelines handbook to implement climate neutrality by improving ecosystem service effectiveness in rural and urban areas" – CUP F53D23010760001 – Grant Assignment Decree No. 1378 adopted on September 1, 2023, by the Italian Ministry of Ministry of University and Research (MUR).

For Rossana Pittau. This pubblication was producted while attending the PhD programme in Civil Engineering and Architecture at the University of Cagliari, Cycle XXXVIII, with the support of a scholarship financed by the Ministerial Decree no. 351 of 9th April 2022, based on the NRRP – funded by the European Union – NextGenerationEU – Mission 4, Component 1, Investment 4.1.

Disclosure of Interests. The authors have no competing interests to declare that are relevant to the content of this article.

References

1. United Nations, Department of Economic and Social Affairs, P.D.: World Urbanization Prospects: The 2018 Revision. United Nations, New York (2019)
2. EEA (European Environmental Agency): Climate Change, Impacts and Vulnerability in Europe 2016. An Indicator-Based Report. EEA Report No. 1/2017. European Environmental Agency, Copenhagen: (2017). https://doi.org/10.2800/534806
3. Rosenzweig, C., Solecki, W.D., Romero-Lankao, P., Mehrotra, S., Dhakal, S., Ibrahim, S.A.: Climate Change and Cities Second Assessment Report of the Urban Climate Change Research Network. Cambridge University Press, Cambridge (2018)
4. Mazzeo, G., Polverino, S.: Nature-based solution for climate change adaptation and mitigation in urban areas with high natural risk. TeMA J. Land Use Mobil. Environ. **16**(1), 47–65 (2023). https://doi.org/10.6093/1970-9870/9736
5. Copernicus: Demonstrating heat stress in European cities (2021). https://climate.copernicus.eu/demonstrating-heat-stress-european-cities. Accessed 05 March 2024

6. Heaviside, C., Macintyre, H., Vardoulakis, S.: The urban heat island: implications for health in a changing environment. Curr. Environ. Health Rep. **4**, 296–305 (2017). https://doi.org/10.1007/s40572-017-0150-3
7. Santamouris, M., Cartalis, C., Synnefa, A., Kolokotsa, D.: On the impact of urban heat island and global warming on the power demand and electricity consumption of buildings – a review. Energy Build. **98**, 119–124 (2015). https://doi.org/10.1016/j.enbuild.2014.09.052
8. Irmak, M.A., Yilmaz, S., Dursun, D.: Effect of different pavements on human thermal comfort conditions. Atmósfera **30**, 355–366 (2017). https://doi.org/10.20937/atm.2017.30.04.06
9. Morris, K.I., et al.: Impact of urbanization level on the interactions of urban area, the urban climate, and human thermal comfort. Appl. Geogr. **79**, 50–72 (2017). https://doi.org/10.1016/j.apgeog.2016.12.007
10. Icaza, L.E., van der Hoeven, F., van den Dobbelsteen, A.: Surface thermal analysis of North Brabant cities and neighbourhoods during heat waves. TeMA J. Land Use Mobil. Environ. **9**(1), 63–87 (2016). https://doi.org/10.6092/1970-9870/3741
11. Nolte, A.C., Buchholz, S., Pernat, N., Egerer, M.: Temporal temperature variation in urban gardens is mediated by local and landscape land cover and is linked to environmental justice. Front. Sustainable Food Syst. **6**, 826437 (2022). https://doi.org/10.3389/fsufs.2022.826437
12. Li, Y., Fan, S., Li, K., Zhang, Y., Dong, L.: Microclimate in an urban park and its influencing factors: a case study of Tiantan Park in Beijing, China. Urban Ecosyst. **24**, 767–778 (2021). https://doi.org/10.1007/s11252-020-01073-4
13. Oliveira, A., Lopes, A., Correia, E., Niza, S., Soares, A.: An urban climate-based empirical model to predict present and future patterns of the Urban Thermal Signal. Sci. Total Environ. **790**, 147710 (2021). https://doi.org/10.1016/j.scitotenv.2021.147710
14. Su, M.A., Ngarambe, J., Santamouris, M., Yun, G.Y.: Empirical evidence on the impact of urban overheating on building cooling and heating energy consumption. iScience **24**(5), 102495 (2021). https://doi.org/10.1016/j.isci.2021.102495
15. Almeida, C.R.d., Teodoro, A.C., Gonçalves, A.: Study of the urban heat island (UHI) using remote sensing data/techniques: a systematic review. Environments **8**, 105 (2021). https://doi.org/10.3390/environments8100105
16. Chen, X.L., Zhao, H.-M., Li, P.-X., Zhi-Yong Yin, Z.-Y.: Remote sensing image-based analysis of the relationship between urban heat island and land use/cover changes. Remote Sens. Environ. **104**(2), 133–146 (2006). https://doi.org/10.1016/j.rse.2005.11.016
17. Despini, F., Ferrari, C., Santunione, G., Tommasone, S., Muscio, A., Teggi, S.: Urban surfaces analysis with remote sensing data for the evaluation of UHI mitigation scenarios. Urban Clim. **35**, 100761 (2021). https://doi.org/10.1016/j.uclim.2020.100761
18. Mahdavi, A., Kiesel, K., Vuckovic, M.: Methodologies for UHI analysis. In: Musco, F. (ed.) Counteracting Urban Heat Island Effects in a Global Climate Change Scenario. Springer, Cham (2016). https://doi.org/10.1007/978-3-319-10425-6_3
19. Marando, F., et al.: Urban heat island mitigation by green infrastructure in European Functional Urban Areas. Sustain. Cities Soc. **77**, 103564 (2022). https://doi.org/10.1016/j.scs.2021.103564
20. Sobocká, J., et al.: Mapping of urban environmentally sensitive areas in Bratislava city. J. Soils Sediments **21**, 2059–2070 (2020). https://doi.org/10.1007/s11368-020-02682-4
21. Elliot, T., Babí Almenar, J., Rugani, B.: Modelling the relationships between urban land cover change and local climate regulation to estimate urban heat island effect. Urban Forest. Urban Green. **50**, 126650 (2020). https://doi.org/10.1016/j.ufug.2020.126650
22. Kalogeropoulos, G., Dimoudi, A., Toumboulidis, P., Zoras, S.: Urban heat island and thermal comfort assessment in a medium-sized Mediterranean city. Atmosphere **13**, 1102 (2022). https://doi.org/10.3390/atmos13071102

23. Evola, G., et al.: UHI effects and strategies to improve outdoor thermal comfort in dense and old neighbourhoods. Energy Procedia **134**, 692–701 (2017). https://doi.org/10.1016/j.egypro.2017.09.589
24. Pinto, F.: Urban planning and climate change: adaptation and mitigation strategies. TeMA – Rivista di uso del territorio, mobilità e ambiente, 829–840 (2014). https://doi.org/10.6092/1970-9870/2547
25. Magliocco, A., Perin, K.: La vegetazione in ambiente urbano: comfort e riduzione del fenomeno isola di calore (2014). https://www.rivistadistoriadelleducazione.it/index.php/techne/article/download/4397/4397. Accessed 05 March 2024
26. Biasin, A., Masiero, M., Amato, G., Pettenella, D.: Nature-based solutions modeling and cost-benefit analysis to face climate change risks in an urban area: the case of Turin (Italy). Land **12**(2), 280 (2023). https://doi.org/10.3390/land12020280
27. Fauk, T., Schneider, P.: Does urban green infrastructure increase the property value? The example of Magdeburg, Germany. Land **12**, 1725 (2023). https://doi.org/10.3390/land12091725
28. European Commission: Communication from the Commission to the European Parliament, the Council, the European Economic and Social Committee and the Committee of the Regions. SWD (2013) 155 final (2013). https://eur-lex.europa.eu/resource.html?uri=cellar:d41348f2-01d5-4abe-b8174c73e6f1b2df.0014.03/DOC_1&format=PDF. Accessed 05 March 2024
29. Córdoba, H.R., Camerin, F.: Assessment of ecological capacity for urban planning and improving resilience in the European framework: an approach based on the Spanish case. Cuadernos de Investigación Geográfica Geogr. Res. Lett. **49**(2), 119–142 (2023). https://doi.org/10.18172/cig.5638
30. Semenzato, P., Bortolini, L.: Urban heat island mitigation and urban green spaces: testing a model in the city of Padova (Italy). Land **12**, 476 (2023). https://doi.org/10.3390/land12020476
31. van Oorschot, J., Sprecher, B., van't Zelfde, M., van Bodegom, P.M., van Oudenhoven, A.P.E.: Assessing urban ecosystem services in support of spatial planning in the Hague, the Netherlands. Landsc. Urban Plan. **214**, 104195 (2021). https://doi.org/10.1016/j.landurbplan.2021.104195
32. Bush, J., Doyon, A.: Building urban resilience with nature-based solutions: how can urban planning contribute? Cities **95**, 102483 (2019). https://doi.org/10.1016/j.cities.2019.102483
33. Chen, S., Wang, Y., Ni, Z., Zhang, X., Xia, B.: Benefits of the ecosystem services provided by urban green infrastructures: differences between perception and measurements. Urban Forest. Urban Green. **54**, 126774 (2020). https://doi.org/10.1016/j.ufug.2020.126774
34. Longato, D., Cortinovis, C., Albert, C., Geneletti, D.: Practical applications of ecosystem services in spatial planning: lessons learned from a systematic literature review. Environ. Sci. Policy **119**, 72–84 (2021). https://doi.org/10.1016/j.envsci.2021.02.001
35. Kowe, P., Mutanga, O., Odindi, J., Dube, T.: Effect of landscape pattern and spatial configuration of vegetation patches on urban warming and cooling in Harare metropolitan city, Zimbabwe. Geosci. Remote Sens. **58**(2), 261–280 (2021). https://doi.org/10.1080/15481603.2021.1877008
36. Lonsdorf, E.V., Nootenboom, C., Janke, B., Horgan, B.P.: Assessing urban ecosystem services provided by green infrastructure: golf courses in the Minneapolis-St. Paul metro area. Landsc. Urban Plan. **208**, 104022 (2021). https://doi.org/10.1016/j.landurbplan.2020.104022
37. Gohr, C., Blumröder, J.S., Sheil, D., Ibisch, P.L.: Quantifying the mitigation of temperature extremes by forests and wetlands in a temperate landscape. Eco. Inform. **66**, 101442 (2012). https://doi.org/10.1016/j.ecoinf.2021.101442

38. Chaudhuri, S., Kumar, A.: Evaluating the contribution of urban ecosystem services in regulating thermal comfort. Spat. Inf. Res. **29**, 71–82 (2021). https://doi.org/10.1007/s41324-020-00336-8
39. Ashwini, K., Sil, B.S.: Impacts of land use and land cover changes on land surface temperature over Cachar Region, Northeast India – a case study. Sustainability **14**(21), 14087 (2022). https://doi.org/10.3390/su142114087
40. Bindajam, A.A., Mallick, J., Talukdar, S., Shahfahad Shohan, A.A.A., Rahman, A.: Modeling the spatiotemporal heterogeneity of land surface temperature and its relationship with land use land cover using geo-statistical techniques and machine learning algorithms. Environ. Sci. Pollut. Res. **30**, 106917–106935 (2022). https://doi.org/10.1007/s11356-022-23211-5
41. Schwarz, N., Bauer, A., Haase, D.: Assessing climate impacts of planning policies – an estimation for the urban region of Leipzig (Germany). Environ. Impact Assess. Rev. **31**(2), 97–111 (2011). https://doi.org/10.1016/j.eiar.2010.02.002
42. Bassett, R., Young, P.J., Blair, G.S., Cai, X.-M., Chapman, L.: Urbanisation's contribution to climate warming in Great Britain. Environ. Res. Lett. **15**(11), 114014 (2020). https://doi.org/10.1088/1748-9326/abbb51
43. Han, B., Wu, T., Cai, Z., Meng, N., Wang, H., Ouyang, Z.: Evaluating the benefits of ecosystem-based urban cooling using a dynamic "on-site" method. Sci. Total Environ. **880**(1), 162908 (2023). https://doi.org/10.1016/j.scitotenv.2023.162908
44. Degefu, M.A., Argaw, M., Feyisa, G.L., Degefa, S.: Effects of urbanization on the relationship between greenspace patterns and evolution of regional heat island in cities of Ethiopia. Chin. J. Popul. Resour. Environ. **19**(4), 330–343 (2021). https://doi.org/10.1016/j.cjpre.2022.01.006
45. Ramaiah, M., Avtar, R., Rahman, M.M.: Land cover influences on LST in two proposed smart cities of India: comparative analysis using spectral indices. Land **9**(9), 292 (2020). https://doi.org/10.3390/land9090292
46. Meerow, S.: A green infrastructure spatial planning model for evaluating ecosystem service tradeoffs and synergies across three coastal megacities. Environ. Res. Lett. **14**(12), 125011 (2019). https://doi.org/10.1088/1748-9326/ab502c
47. Estoque, R.C., Murayama, Y., Myint, S.W.: Effects of landscape composition and pattern on land surface temperature: an urban heat island study in the megacities of Southeast Asia. Sci. Total Environ. **577**, 349–359 (2017). https://doi.org/10.1016/j.scitotenv.2016.10.195
48. Permatasari, P.A., Trissanti, V.N., Amalo, L.F., Effendi, H.: Understanding the impact of riparian corridors on microclimate (case study: Bogor City, Indonesia). In: 42nd Asian Conference on Remote Sensing, ACRS 2021 (2021)
49. Cheng, X., Wei, B., Chen, G., Li, J., Song, C.: Influence of park size and its surrounding urban landscape patterns on the park cooling effect. J. Urban Plan. Dev. **141**(3), A4014002 (2015). https://doi.org/10.1061/(ASCE)UP.1943-5444.0000256
50. Okumus, D.E., Terzi, F.: Ice floes in urban furnace: cooling services of cemeteries in regulating the thermal environment of Istanbul's urban landscape. Urban Clim. **49**, 101549 (2023). https://doi.org/10.1016/j.uclim.2023.101549
51. Feng, L., Mi, X., Yuan, D.: Optimal planning of urban greening system in response to urban microenvironments in a high-density city using genetic algorithm: a case study of Tianjin. Sustain. Cities Soc. **87**, 104244 (2022). https://doi.org/10.1016/j.scs.2022.104244
52. Wo, R., Dong, T., Pan, Q., Liu, Z., Li, Z., Xie, M.: Ecological performance evaluation of urban agriculture in Beijing based on temperature and fractional vegetation cover. Urban Ecosyst. **25**, 341–353 (2022). https://doi.org/10.1007/s11252-021-01157-9
53. Yan, L., Jia, W., Zhao, S.: The cooling effect of urban green spaces in metacities: a case study of Beijing, China's capital. Remote Sens. **13**(22), 4601 (2021). https://doi.org/10.3390/rs13224601

54. Wang, X., Dallimer, M., Scott, C.E., Shi, W., Gao, J.: Tree species richness and diversity predicts the magnitude of urban heat island mitigation effects of greenspaces. Sci. Total Environ. **770**, 145211 (2021). https://doi.org/10.1016/j.scitotenv.2021.145211
55. Yu, Z., Guo, X., Jørgensen, G., Vejre, H.: How can urban green spaces be planned for climate adaptation in subtropical cities? Ecol. Ind. **82**, 152–162 (2017). https://doi.org/10.1016/j.ecolind.2017.07.002
56. Vaz Monteiro, M., Doick, K.J., Handley, P., Peace, A.: The impact of greenspace size on the extent of local nocturnal air temperature cooling in London. Urban Forest. Urban Green. **16**, 160–169 (2016). https://doi.org/10.1016/j.ufug.2016.02.008
57. Yosef, R., Rakholia, S., Mehta, A., Bhatt, A., Kumbhojkar, S.: Land surface temperature regulation ecosystem service: a case study of Jaipur, India, and the urban island of Jhalana Reserve Forest. Forests **13**(7), 1101 (2022). https://doi.org/10.3390/f13071101
58. Murtinová, V., Gallay, I., Olah, B.: Mitigating effect of urban green spaces on surface urban heat island during summer period on an example of a medium size town of Zvolen, Slovakia. Remote Sensing **14**(18), 4492 (2022). https://doi.org/10.3390/rs14184492
59. Chen, X., Wang, Z., Bao, Y.: Cool island effects of urban remnant natural mountains for cooling communities: a case study of Guiyang, China. Sustain. Cities Soc. **71**, 102983 (2021). https://doi.org/10.1016/j.scs.2021.102983
60. Ruiz-Aviles, V., Brazel, A., Davis, J.M., Pijawka, D.: Mitigation of urban heat island effects through "Green Infrastructure": integrated design of constructed wetlands and neighborhood development. Urban Sci. **4**, 78 (2020). https://doi.org/10.3390/urbansci4040078
61. Wu, Z., Zhang, Y.: Water bodies' cooling effects on urban land daytime surface temperature: ecosystem service reducing heat island effect. Sustainability **11**(3), 787 (2019). https://doi.org/10.3390/su11030787
62. Petri, A.C., Wilson, B., Koeser, A.: Planning the urban forest: adding microclimate simulation to the planner's toolkit. Land Use Policy **88**, 104117 (2019). https://doi.org/10.1016/j.landusepol.2019.104117
63. Bartesaghi-Koc, C., Osmond, P., Peters, A.: Spatio-temporal patterns in green infrastructure as driver of land surface temperature variability: the case of Sydney. Int. J. Appl. Earth Obs. Geoinf. **83**, 101903 (2019). https://doi.org/10.1016/j.jag.2019.101903
64. Zhang, X., Estoque, R.C., Murayama, Y.: An urban heat island study in Nanchang City, China based on land surface temperature and social-ecological variables. Sustain. Cities Soc. **32**, 557–568 (2017). https://doi.org/10.1016/j.scs.2017.05.005
65. Mariani, L., Parisi, S.G., Cola, G., Lafortezza, R., Colangelo, G., Sanesi, G.: Climatological analysis of the mitigating effect of vegetation on the urban heat island of Milan, Italy. Sci. Total Environ. **569–570**, 762–773 (2016). https://doi.org/10.1016/j.scitotenv.2016.06.111
66. Yao, X., Chen, Y., Zhang, Q., Mou, Z., Yao, X., Ou, C.: Assessment of the urban expansion and its impact on the eco-environment – a case study of Hefei municipal area. Sustainability **14**(17), 10613 (2022). https://doi.org/10.3390/su141710613
67. Sebastiani, A., Marando, F., Manes, F.: Mismatch of regulating ecosystem services for sustainable urban planning: PM10 removal and urban heat island effect mitigation in the municipality of Rome (Italy). Urban Forest. Urban Green. **57**, 126938 (2021). https://doi.org/10.1016/j.ufug.2020.126938
68. Zidar, K., et al.: A framework for multifunctional green infrastructure investment in Camden, NJ. Urban Plan. **2**(3), 56–73 (2017). https://doi.org/10.17645/up.v2i3.1038
69. Meerow, S., Newell, J.P.: Spatial planning for multifunctional green infrastructure: growing resilience in Detroit. Landsc. Urban Plan. **159**, 62–75 (2017). https://doi.org/10.1016/j.landurbplan.2016.10.005

70. Bodnaruk, E.W., Kroll, C.N., Yang, Y., Hirabayashi, S., Nowak, D.J., Endreny, T.A.: Where to plant urban trees? A spatially explicit methodology to explore ecosystem service trade-offs. Landsc. Urban Plan. **157**, 457–467 (2017). https://doi.org/10.1016/j.landurbplan.2016.08.016
71. Di Leo, N., Escobedo, F.J., Dubbeling, M.: The role of urban green infrastructure in mitigating land surface temperature in Bobo-Dioulasso, Burkina Faso. Environ. Dev. Sustain. **18**(2), 373–392 (2015). https://doi.org/10.1007/s10668-015-9653-y
72. Amir Siddique, M., Boqing, F., Dongyun, L.: Modeling the impact and risk assessment of urbanization on urban heat island and thermal comfort level of Beijing City, China (2005–2020). Sustainability **15**(7), 6043 (2023). https://doi.org/10.3390/su15076043
73. Tayebi, S., Mohammadi, H., Shamsipoor, A., Tayebi, S., Alavi, S.A., Hoseinioun, S.: Analysis of land surface temperature trend and climate resilience challenges in Tehran. Int. J. Environ. Sci. Technol. **16**, 8585–8594 (2019). https://doi.org/10.1007/s13762-019-02329-z
74. Osborne, P.E., Alvares-Sanches, T.: Quantifying how landscape composition and configuration affect urban land surface temperatures using machine learning and neutral landscapes. Comput. Environ. Urban Syst. **76**, 80–90 (2019). https://doi.org/10.1016/j.compenvurbsys.2019.04.003
75. Apicella, L., Quarati, A., Martino, M.D.: Analysing the surface urban heat island effect with Copernicus data. In: Kö, A., Francesconi, E., Kotsis, G., Tjoa, A.M., Khalil, I. (eds.) Electronic Government and the Information Systems Perspective. EGOVIS 2021. Lecture Notes in Computer Science, vol. 12926. Springer, Cham (2021). https://doi.org/10.1007/978-3-030-86611-2_5
76. Ronchi, S., Salata, S., Arcidiacono, A.: Which urban design parameters provide climate-proof cities? An application of the Urban Cooling InVEST Model in the city of Milan comparing historical planning morphologies. Sustain. Cities Soc. **63**, 102459 (2020). https://doi.org/10.1016/j.scs.2020.102459
77. Shih, W.-Y., Mabon, L., Green infrastructure as a planning response to urban warming: a case study. In: Ito, K. (ed.) Urban Biodiversity and Ecological Design for Sustainable Cities, pp. 335–352. Springer, Tokyo (2021). https://doi.org/10.1007/978-4-431-56856-8_15
78. Zhao, W., Li, A., Huang, Q., Gao, Y., Li, F., Zhang, L.: An improved method for assessing vegetation cooling service in regulating thermal environment: a case study in Xiamen, China. Ecol. Indic. **98**, 531–542 (2019). https://doi.org/10.1016/j.ecolind.2018.11.033
79. Pace, R., Chiocchini, F., Sarti, M., Endreny, T.A., Calfapietra, C., Ciolfi, M.: Integrating Copernicus land cover data into the i-Tree Cool Air model to evaluate and map urban heat mitigation by tree cover. Eur. J. Remote Sens. **56**(1), 2125833 (2023). https://doi.org/10.1080/22797254.2022.2125833
80. Greene, C.S., Millward, A.A.: Getting closure: the role of urban forest canopy density in moderating summer surface temperatures in a large city. Urban Ecosyst. **20**(1), 141–156 (2019). https://doi.org/10.1007/s11252-016-0586-5
81. Zhang, Y., Murray, A.T., Turner, B.L.: Optimizing green space locations to reduce daytime and nighttime urban heat island effects in Phoenix, Arizona. Landsc. Urban Plan. **165**, 162–171 (2017). https://doi.org/10.1016/j.landurbplan.2017.04.009
82. Sun, R., Chen, L.: Effects of green space dynamics on urban heat islands: mitigation and diversification. Ecosyst. Serv. **23**, 38–46 (2017). https://doi.org/10.1016/j.ecoser.2016.11.011
83. Mills, G.: Luke Howard and the climate of London. Weather **63**(6), 153–157 (2007). https://doi.org/10.1002/wea.195
84. Howard, L.: The Climate of London. IAUC edition (2007)
85. Voogt, J.: How Researchers Measure Urban Heat Islands (2007). https://www.epa.gov/sites/default/files/2014-07/documents/epa_how_to_measure_a_uhi.pdf. Accessed 05 March 2024

86. Deilami, K., Kamruzzaman, M., Liu, Y.: Urban heat island effect: a systematic review of spatio-temporal factors, data, methods, and mitigation measures. Int. J. Appl. Earth Obs. Geoinf. **67**, 30–42 (2018). https://doi.org/10.1016/j.jag.2017.12.009
87. Shahraiyni, H.T., et al.: A comprehensive statistical study on daytime surface urban heat island during summer in urban areas, case study: Cairo and its new towns. Remote Sens. **8**(8), 643 (2016). https://doi.org/10.3390/rs8080643
88. Rasul, A., Balzter, H., Smith, C.: Spatial variation of the daytime surface urban cool island during the dry season in Erbil Iraqi Kurdistan, from Landsat 8. Urban Clim. **14**(Part 2), 176–186 (2015). https://doi.org/10.1016/j.uclim.2015.09.001
89. Rasul, A., Balzter, H., Smith, C.: Diurnal and seasonal variation of surface urban cool and heat islands in the semi-arid city of Erbil, Iraq. Climate **4**(3), 42 (2016). https://doi.org/10.3390/cli4030042
90. Baranka, G., Bozó, L., Ciglič, R., Komac, B.: Urban heat island gold standard and urban heat Island Atlas. In: Musco, F. (ed.) Counteracting Urban Heat Island Effects in a Global Climate Change Scenario, pp. 41–70. Springer International Publishing, Cham (2016)
91. Li, Z.-L., et al.: Satellite-derived land surface temperature: current status and perspectives. Remote Sens. Environ. **131**, 14–37 (2013). https://doi.org/10.1016/j.rse.2012.12.008
92. Irons, J.R., Dwyer, J.L., Barsi, J.A.: The next Landsat satellite: the Landsat data continuity mission. Remote Sens. Environ. **122**, 11–21 (2012). https://doi.org/10.1016/j.rse.2011.08.026
93. Loveland, T.R., Dwyer, J.L.: Landsat: building a strong future. Remote Sens. Environ. **122**, 22–29 (2012). https://doi.org/10.1016/j.rse.2011.09.022
94. Atkinson, P.M.: Downscaling in remote sensing. Int. J. Appl. Earth Obs. Geoinf. **22**, 106–114 (2013). https://doi.org/10.1016/j.jag.2012.04.012
95. Jiménez-Muñoz, J.C., Sobrino, J.A.: Error sources on the land surface temperature retrieved from thermal infrared single channel remote sensing data. Int. J. Remote Sens. **27**(5), 999–1014 (2006). https://doi.org/10.1080/01431160500075907
96. Krehbiel, C., Henebry, G.M.: A comparison of multiple datasets for monitoring thermal time in urban areas over the U.S. upper midwest. Remote Sens. **8**(4), 297 (2016). https://doi.org/10.3390/rs8040297
97. EEA (European Environmental Agency): Adaptation in Europe: How Cities and Towns Respond to Climate Change. EEA Report No. 12/2020. European Environmental Agency, Copenhagen (2020). https://doi.org/10.2800/324620
98. General Assembly of United Nations: Resolution Adopted by the General Assembly on 25 September 2015. Transforming our world: the 2030 Agenda for Sustainable Development (2015). https://sdgs.un.org/sites/default/files/publications/21252030%20Agenda%20for%20Sustainable%20Development%20web.pdf. Accessed 05 March 2024
99. Valencia, S.C., et al.: Adapting the sustainable development goals and the new urban agenda to the city level: initial reflections from a comparative research project. Int. J. Urban Sustain. Dev. **11**(1), 4–23 (2019). https://doi.org/10.1080/19463138.2019.1573172
100. Aguiar, F.C., et al.: Adaptation to climate change at local level in Europe: an overview. Environ. Sci. Policy **86**, 38–63 (2018). https://doi.org/10.1016/j.envsci.2018.04.010
101. Gancheva, M., Lundberg, P., Vroom, I.: Climate Adaptation: Measuring Performance, Defining Targets and Ensuring Sustainability (2022). https://cor.europa.eu/en/engage/studies/Documents/QG0322225ENN.pdf#search=measuring%20performance. Accessed 05 March 2024
102. Observatory on Non-State Climate Action: Global Synthesis Report on Local Climate Action. Climate Chance (2022). https://www.climate-chance.org/en/comprehend/global-synthesis-report-local-climate-action/. Accessed 05 March 2024

103. Directorate-General for Climate Action: Adapting When the Climate Crisis hits Close to Home (2023). https://climate.ec.europa.eu/news-your-voice/news/adapting-when-climate-crisis-hits-close-home-2023-08-01_en. Accessed 05 March 2024
104. Mendizabal, M., et al.: RESIN Publication. Library Structure Online. Adaptation Options Database Model (2015). https://www.researchgate.net/publication/297376100_Library_structure_online_Adaptation_options_database_model. Accessed 05 March 2024
105. Pio, D.: Mitigare l'isola di calore urbana: il ruolo dei campus universitari. Tesi di Laurea Magistrale (2023). https://webthesis.biblio.polito.it/27301/1/tesi.pdf. Accessed 05 March 2024

Transformation of Disused Railways Based on ES Assessment: Sassari Case

Bilge Kobak[1,2](✉)

[1] IUSS Pavia, Pavia, Italy
bilge.kobak@iusspavia.it
[2] DICAAR Università degli Studi di Cagliari, Via Marengo 2 CA, 09123 Cagliari, Italy

Abstract. Ecosystem services (ES) knowledge leads to action and is essential for improving the relationship between humans and environment, as it provides crucial information for decision-makers. This paper showcases abandoned railway landscapes as valuable resources for Green Infrastructure strategies, focusing on their potential to enhance ES at the provincial level using Sassari Province as a case study. Employing an adapted ESTIMAP model, it evaluates cultural ES by mapping recreation potential in Sassari, using areas of interest and slow mobility network as parameters to emphasize the potential of disused railways as ES sources. The results underscore the importance to integrate ES assessment into decision-making process while demonstrating the potential of disused railways to improve nature-based recreation. By advocating for the reuse of existing structures, and particularly focusing on a single typology of brownfields such as railways, it allows for the identification of specific actions and provides municipalities with actionable insights to determine starting points for implementation.

Keywords: Ecosystem Services · Green Infrastructure · Disused Railways

1 Introduction

Landscape is a dynamic process between society and environment that constantly changes as a result of human activity and environmental influences. The territorial landscape has been significantly shaped by urbanization and industrial development, reflecting the dynamism and growth of cities. As cities expanded, the demand for mobility grew, leading to the emergence of transportation landscapes. Logistics networks, including extensive railway lines, began to integrate into vast industrial areas, shaping not only the urban landscape but also regions. While grey infrastructure continued to grow at an accelerated rate, environmental concerns arose due to unsustainable use of resources. A huge portion of industrial landscapes, including transportation landscapes has been abandoned due to various factors such as economic crisis, emergence of new technologies etc. These abandoned areas, which were formerly thriving centres of urban and industrial activity, present problems with regard to urban decay, unsustainable land use, and environmental degradation. This underscores precisely why they can serve as valuable assets and integral components of green infrastructures. Rather than seeking to

O. Gervasi et al. (Eds.): ICCSA 2024 Workshops, LNCS 14818, pp. 370–386, 2024.
https://doi.org/10.1007/978-3-031-65273-8_24

compensate for adverse impacts on ecosystem services elsewhere, a more strategic approach entails the revitalization of these abandoned sites where environmental degradation was initially occurred through their integration to Green Infrastructure network.

Numerous studies have explored brownfields within the framework of green infrastructure (GI) strategies, recognizing their potential to provide ecosystem services (ES). However, there exists a gap in the literature regarding specific measures to harness this potential effectively. Hence, there is a need for research focusing on the distinctive typologies of brownfields, as comparing, for instance, a quarry to a railway is not feasible. Consequently, instead of treating brownfields as a singular typology, it is imperative to categorize them for more nuanced analysis and strategic planning. This paper, by focusing on railways as part of transportation category, provides a framework to address and mitigate this gap.

Typically, disused or abandoned railways are transformed into greenways. Making them a part of the slow mobility network is a common method in their transformation however their ES values are not highlighted in the literature. Although there many examples of railway regeneration and greenway projects, their relation and potential place in GI framework is not much studied. While studies started to adopt the idea of using brownfields for regeneration in the context of GI, research specifically focusing on railways within this topic remains limited. One notable example is the work of Gasparrini and Terracciano [1], who discuss "The disused railway as a greenway for urban regeneration" along the Messina Palermo railway line in their study for the Green and Blue Infrastructure Messina. However, there is currently a lack of in-depth studies exploring why each of these typologies of brownfields could serve as valuable assets for GI networks, particularly in terms of their potential to provide ES. Moreover, there is a need to elucidate the specific interventions that can be implemented on these areas to maximize their contribution to the GI network.

To provide specific actions, there is a need to improve the ES knowledge because in order to support new policies that specifically take into account the effects on ES, the most direct results are anticipated when ES knowledge is purposefully used to generate actions [2]. To do so, this paper proposes an adapted version of ESTIMAP-Recreation model to assess the recreation potential in abandoned railway landscapes through Cultural ES.

Cultural ES which are associated with intangible values and heritage becomes important in post-industrial areas as these areas reflect the past activities of society. Cultural ES is described as "non-material" benefits that society gets from ecosystems [3]. These services include recreational, educational, cultural, spiritual values, knowledge systems, social relationships, sense of belonging, tourism/ecotourism etc. [4]. Hence, out of the 4 groups of ES, Cultural ES are considered and found the most useful since certain characteristics of railways allow for recreation, promote culture and heritage, and connect different communities.

2 Case Study

2.1 Location

Sassari is one of the four provinces of Sardinia region in the south of Italy, as seen in Fig. 1. The province has 491.571 habitants as of 1 January 2019. Its capital is the city of Sassari, and it is characterized by industrial activities, historic centres, and tourism thanks to its coastal characteristics. It is selected for the application of the methodology because it is characterized by industrial activities and hosts several railway lines on the provincial scale, including abandoned lines and partially operating lines.

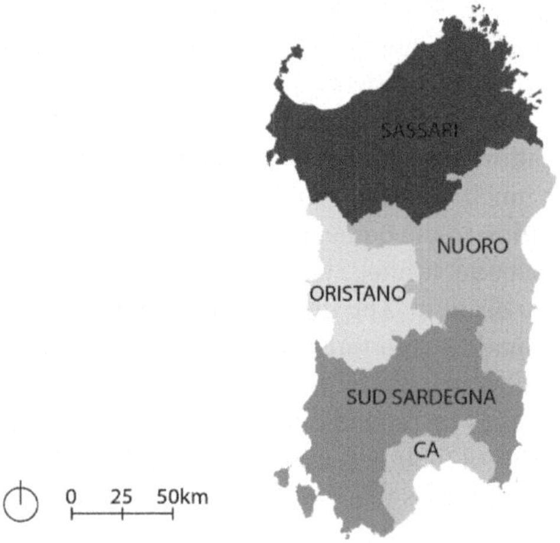

Fig. 1. Four provinces of Sardinia region and Metropolitan City of Cagliari (CA)

To gain a deeper understanding of the existing measures and actions regarding reutilization of disused railways, urban planning documents of Sassari are examined in the next section.

2.2 Planning Documents of Sassari

Strategic Plan 2009 (Provincia di Sassari).[1] The provincial strategic plan mentions industrial areas in abandoned condition such as decommissioned mining sites but neither propose a general objective for "abandoned industrial areas" nor specify "railway". While the plan includes certain objectives regarding environment, cultural areas, green areas, mobility networks etc, it lacks a comprehensive intervention strategy or specific projects aimed at revitalizing abandoned industrial areas.

[1] Provincia di Sassari (2009) Piano Strategico: Provinciale Rapporto Integrale [Provincial Strategic Plan: Full Report] https://ptpc.provincia.sassari.it/Pianificazione/PATTONO-PSP/Rapporto%20integrale.pdf (last accessed 5 March 2024).

Guidelines for the development of the Provincial Strategic Plan (Linee Guida Per L'elaborazione Del Piano Strategico Provinciale).[2] This document discusses general concepts concerning projects tied to strategies tailored to the environmental landscape. It refers to territorial characteristics and high environmental quality and only mentions "abandoned industrial areas" with no specific reference to "railways". It highlights challenges regarding the significant resources required for environmental rehabilitation by providing cost estimations in broad terms, however, it does not specify the typologies.

Strategic Plan 2007 (Comune di Sassari).[3] The Strategic Plan of Sassari (2007) defines 4 themes in order to achieve the goal "Sustainable Sassari (Sassari Sostenible)" which are environmental system (urban parks, green, landscape etc.), new energy scenarios and waste management, heritage public and private housing, urban and urban mobility. The directives include transformation and regeneration plans more in relation with historic centers. Most importantly, it mentions "greenways" in "Direction D2: Environment". The direction of intervention "environment" follows the ecosystem approach that consider the environmental system in the multiplicity of its aspects and possible forms of use, and poses the emphasis on the functions of the ecosystem, on the interconnections between the different uses and on the rules to be respected, according to the principles of sustainability deriving from recognition that the territory contains a finite set of resources, for commensurate activity levels with nature's carrying capacity and need to conserve biodiversity. This directive includes the strategy to support the sustainable use of resources with activities related to ecotourism, nature tourism and networking of areas with a strong naturalistic and cultural value. It also includes the project "La Strada Del Verde" which defines a system of routes dedicated to non-motorized traffic that connect the natural and historical-cultural resources of the area with urban settlements. The project acknowledges the use of disused railway routes as well following the objectives of Italian Greenways Association. The actors involved in this project are listed as follows: Environment and Public Green Sector, Maintenance Sector, ATP (Azienda Trasporti Pubblici Nuoro), FdS (Ferrovie della Sardegna), Province of Sassari, neighbouring municipalities.

While the project definition of "La Strada Del Verde" is very important to reclaim disused railways and for their integration into green infrastructure network considering their ES supply, its focus should expand beyond urban settlements as railway lines cross peri-urban areas as well. Mentioning neighbouring municipalities in actors paves the way for collaborated efforts to achieve the ultimate goal of sustainable cities, however, since 2007, there is still a long way to go. The Directive 2 focusing on environment provides important goals towards areas with environmental and cultural value, however, to show concrete progress, there is a need for identification of intervention areas in order to provide specific actions.

---

[2] Provincia di Sassari (n.d) Guidelines for the development of the Provincial Strategic Plan [Linee Guida Per L'elaborazione Del Piano Strategico Provinciale]. Resource document. https://ptpc.provincia.sassari.it/Pianificazione/PATTONO-PSP/Linee%20guida.pdf (last accessed 5 March 2024).

[3] Comune di Sassari (2007) Piano strategico della Città di Sassari: Progetti—Allegato 1.3 [Strategic plan of the City of Sassari: Projects—Annex 1.3]. Resource document. porto / https://www.sardegnaterritorio.it/documenti/6_83_20100526092101.zip. (last accessed 5 March 2024).

Environmental Report 2012 (Comune di Sassari).[4] Environmental Report of 2012 has several mentions of interventions related to dismissed railways. In article "3.7.5 The unresolved areas within the compact city" it specifically mentions "railways" as part of industrial areas that can be transformed. According to the report, artisanal and disused railways cover a total area of 25 hectares and are mainly located in the railway yards (fascio ferroviario) and in the Via Verona area. The report also mentions a railway regeneration project for Porto Torres a potential ecological corridor in relation with "neglected valley system" of Sassari in article "3.7.2 The valley system and the linear park". It proposes a linear park of 6km along the Railway line of Porto Torres. This is a project that the PUC (Piano Urbanistico Comunale) considers to be a "backbone" with priority for intervention and the report mentions its cornerstones as follows: the connection between the valleys, with their recovery and their introduction into the urban system, ensuring their continuous pedestrian use and the creation of a linear park parallel to the railway for Porto Torres. According to the report, this is an opportunity to regenerate along these settlement situations lacking in terms of urban quality (service provision, lack of equipped public spaces, usable green areas, etc.).

Urban Planning Sector-The Integrated Programme: Framework Document 2019, Comune di Sassari (Settore Pianificazione Urbanistica - Il Programma Integrato: Documento Di Inquadramento 2019).[5] Following the PUC objectives, the Integrated Program Framework Document of March 2019 mentions the project for Porto Torres which was also mentioned in Environmental Report of 2012. The document highlights that it links the peripheral system of the city and with its diverse functions, it has the potential to become a significant and vital attraction for the entire Sassari suburban area. It also mentions archaeological complexes and industrial heritage areas that connects to Porto Torres. This document acknowledges the railway areas being decommissioned, the potential of disused railway lines to connect diverse areas of cultural value and proposes areas of interventions.

Outcomes of Planning Documents. Examining these planning documents reveals that at the provincial level, integrated reports or pacts are signed by various municipalities. However, specific actions at the provincial level are not explicitly mentioned. Notably, documents at the capital city level are more detailed, outlining a few areas of intervention. Since 2007, municipal reports and documents have been aligning in terms of goals related to repurposing dismissed railways, with each one focusing on different aspects of particular projects. They have been developing frameworks and becoming more precise in identifying intervention areas. However, the process is progressing slowly and still lacks specific actions. The proposed methodology in this paper is beneficial in this

---

[4] Comune di Sassari (2012) Piano Urbanistico Comunale di Sassari. Rapporto Ambientale. Resource document. [Municipal Urban Plan of Sassari. Environmental Report]. Resource document. https://puc.comune.sassari.it/pucweb/elab/Allegati/Allegato_M1_VAS_rapporto_ambientale.pdf (last accessed 5 March 2024).

[5] Comune di Sassari (2019) Settore Pianificazione Urbanistica. Il Programma Integrato: Documento di Inquadramento – Oggetto 7.1. [Urban Planning Sector. The Integrated Programme: Framework Document – Object 7.1]. Resource document. https://www.comune.sassari.it/.galleries/doc-documenti_old/ALL.-A-PIRU_DOCUMENTO-DI-INQUADRAMENTO.pdf (last accessed 5 March 2024).

context since it identifies areas with high recreation potential based on environmental and cultural value, examines their connectivity levels within existing slow mobility routes and determines the intervention areas through the superposition of disused railway routes. Disused railways hold immense potential for revitalization, as they offer opportunities to enhance slow mobility networks and connect areas of significant cultural value.

3 Application of Methodology: Dismissed Railways of Sassari

The focus is on three railway lines, shown in Fig. 2; one is the main line from Sassari to Palau (Sassari-Tempio-Palau or Sassari-Luras-Palau) which is 150.2 km long and is a part of Trenino Verde. The other one which is connected to this line as well is the 28 kms stretch from Luras to Monti. These two stretches are connected in Luras and they differ in certain characteristics. The third one is Tirso-Chilivani which is 78.48 km. "The Ferrovie Abbandonate[6]" offers a comprehensive database of abandoned railways in Italy, including these stretches. According to their data, the shorter stretch from Luras to Monti is partially paved. While the part between Luras and Calangianus is transformed into an ordinary asphalted road, the part from Calangianus to a little before Monti is transformed into a path that can be followed by bike or on foot. Along the route, abandoned or partially collapsed service buildings exist which are considered as potential service areas for visitors in this study.

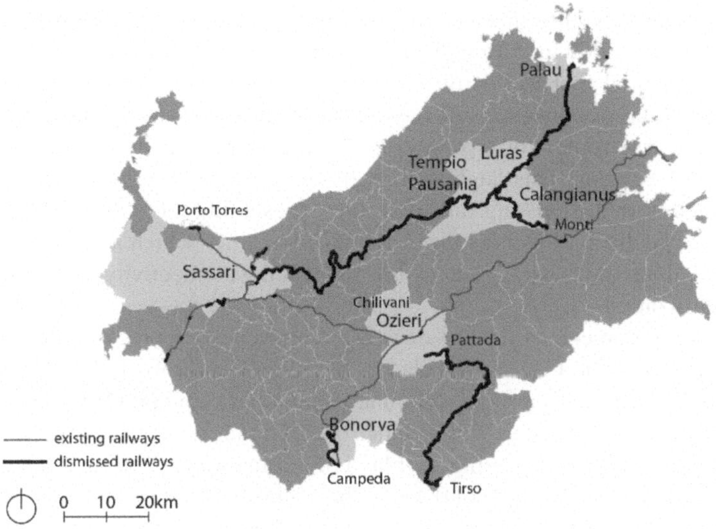

Fig. 2. Dismissed Railway of Sassari

The proposed methodology outlined in the following paragraphs will reveal the precise intervention points and elucidate the "how" aspect of implementation. While

[6] Ferrovie Abbandonate. Linee ferroviarie chiuse al traffico. Retrieved from https://www.ferrovieabbandonate.it/dismesse.php?t=dismessa (last accessed 5 March 2024).

numerous research and municipal documents provide guidance on what needs to be done in general terms, they often lack detailed information on the execution process.

3.1 Methodology

To assess and evaluate the nature-based recreation potential in Sassari province, an adjusted version of ESTIMAP Recreation model is adapted which was initially created by the European Commission's Joint Research Centre to map ecosystem services across Europe [5, 6]. The original model consists of 2 modules. First module is mapping the Recreation Potential (RP) which is referred to as ecosystem-based potential map, the second module includes human inputs which are available infrastructures that is shown with a proximity map considering distance from roads and residential areas. While the original model uses 3 parameters to map RP which are i. degree of naturalness in land cover, ii. Presence of protected areas, iii. Presence of water related elements; these parameters are sometimes tailored to fit different contexts in different research studies such as the works of Cortinovis and Genelletti as well as Isola et al. [7–9]. As shown in Fig. 3 that explains the adaptation of the methodology, the input parameters used to map RP in this study are as follows: i. degree of naturalness in land cover, ii. Presence of protected areas, iii. Presence of areas of interest. For the second module, the human inputs, slow mobility network components such as biking, hiking and trekking routes are considered instead of roads. This strategic choice aligns with the emphasis on promoting nature-based activities, thereby discouraging reliance on automobiles. The second module results show us the availability of slow mobility networks. After obtaining the two modules, through cross-tabulation analysis, areas with significant recreational potential that lack sufficient connectivity to slow mobility routes can be identified. The final step involves overlaying the abandoned railways layer to assess the potential of existing routes for repurposing. All the analysis is done withing the QGIS Desktop 3.20.3 version. This analysis reveals the cultural ecosystem services supply potential inherent in the existing railway routes, presenting them as valuable resources to be utilized effectively. Final results enable us to identify which areas to prioritize for improvement along the existing railway line, enhancing connectivity to facilitate greater access and utilization.

For the first parameter of RP, the degree of naturalness, scores ranging from 0–9 is assigned to Corine Land Cover 2018 classes based on Hemeroby Index values outlined in Isola et al. [9], originally proposed by Paracchini and Capitani [10]. For the purpose of this study, the codes that start with 5 which shows water bodies are given the values of 1 or 0 to differentiate between being an inland or marine water. Marine water codes 521, 522, and 523 are assigned a score of 0 due to their minimal contact with land boundaries, whereas codes 511 and 512 receive a score of 1 as they experience comparatively more disturbance being situated within land boundaries. Land cover categories are typically assigned a single value, except for agricultural and forest areas were depending on the degree of management practice intensity they can get one of the three values presented in Table 1 showing correlations between hemeroby index and CLC classes. Following the approach presented in Isola et al. [9], for the agricultural areas which are represented by codes starting with 2, nutrient inputs and the density of livestock is considered to define which of the three values are given showing low-medium-high intensity. For forest areas,

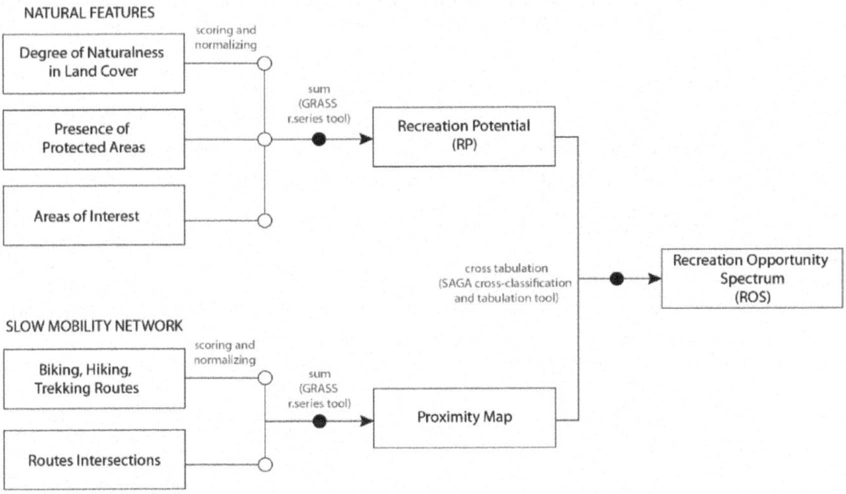

Fig. 3. Adaptation of ESTIMAP-Recreation Model for Sassari. Modified after Zulian et al. (2013)

which are represented by codes starting with 3, the difference between the actual existing vegetation and the potential vegetation presence in the case of no human activity is used as a proxy.

Table 1. Hemeroby Index (HI) values of CLC classes. Adapted after Paracchini and Capitani (2011) and Isola et al. (2022)

CLC codes (artificial)	HI	CLC codes(agricultural)	HI	CLC codes(forest – seminatural)	HI	CLC codes(wetlands)	HI	CLC code(water bodies)	HI
111	9	211	5–6–7	311	2–3–4	411	2	511	1
112	9	212	5–6–7	312	2–3–4	421	2	512	1
121	9	213	5–6–7	313	2–3–4	422	6	521	0
122	9	221	4–5–6	321	2–3–4			522	0
123	9	222	4–5–6	322	2			523	0
124	9	223	4–5–6	323	2				
131	8	231	3–4–5	324	2				
132	8	241	4–5–6	331	2				
133	8	242	4–5–6	332	1				
141	8	243	4–5–6	333	2				
142	8	244	3–4–5	334	6				

The nitrogen levels in kg per province as well as the livestock density (livestock unit/ha) per municipality of Sardegna is obtained from the extensive data presented in

Isola et al. [9] in which they utilize data from the National Census Institute for the information on nitrogen and the data from the National Zootechnical Register for the number of livestock.

For the nitrogen input, Sassari's nitrogen level from 2019 which is 1569000kg divided by the total ha of agricultural areas which make a sum of 448047 ha gives us the number 3.5 which scores as low according to the standards given by Paracchini et al. [10] shown in Table 2. Then for all Sassari province, nitrogen level was considered low. Consequently, in municipalities where livestock density is also low, the intensity of management practices is likewise regarded as low, leading to the assignment of the smallest value among the three options. Vice versa, in areas with higher livestock density, management practices tend to intensify accordingly.

Table 2. Intensity of management degree according to Nitrogen input and Livestock density classes. Sourced from Paracchini et al. (2011).

	Low input	Mid input	High Input
N input (kg/ha)	>0–30	>30–150	>150
Livestock density (LU/ha)	>0–0.5	<0.5–1.2	<1.2

For forest and seminatural areas, represented by the codes starting with number 3, as suggested in the Sardinian context, the distance between the current vegetational situation and the potential vegetation state is proposed to describe the disturbance level. For this purpose, Isola et al. [9] resort to Bacchetta et al. [11] which conducted mapping to determine the potential distribution of vegetation series and geoseries in Sardinia. By comparing the actual forest vegetation with the potential distribution through spatial intersection, patches classified as "311," "312," "313," or "321" were assigned hemeroby index values of 2, 3, or 4, depending on how well the current land cover matched the potential vegetation, with lower values indicating better compatibility. Looking at the results of the forest disturbance levels provided by the work of Isola et al. [9], one could simplify the calculations in the case of Sardegna since the scores generally are the same within each code. Which means the hemeroby index for forest areas are considered as following: 311 – "2", 312 – "4", 313 – "3", 321 – "2".

After assigning the hemeroby index values from 0 to 9 to each CLC codes, the degree of naturalness is obtained by normalizing the values between 0 to 1 in equal intervals, with 1 representing max degree of naturalness.

To obtain the RP, 3 parameters are needed as raster inputs with scale from 0 to 1. After obtaining the first one which is degree of naturalness in land cover, the other two parameters are rasterized. The second and third parameters, indicating the presence of protected areas and areas of interest, are scored as either 0 or 1. A value of 1 indicates presence, while 0 indicates absence.

For the protected areas, national parks, regional parks, natural reserves, Natura 2000 sites and areas defined as Siti di Importanza Comunitaria (SIC), Zone Speciali di Conservazione (ZSC), Zone di Protezione Speciale (ZPS), which refers to "Sites of Community Importance, Special Conservation Zones, Special Protection Areas" are considered. The protected areas are mapped using the following two layers. First layer is «Art. 142 - Parchi e riserve nazionali o regionali (dati indicativi)» downloaded from Sardegna Geoportale[7] and appearing as "ART142_PARCH_RIS_NAZ_REG_INDICPolygon" in GIS environment and the second one is «N2000_spatial_IT_2023_12_15_region_WGS_1984_UTM_Zone_32N» downloaded from the website of Ministero dell'Ambiente e della Sicurezza Energetica (MASE)[8] showing N2000 Sites including SIC, ZSC and ZPS.

For the areas of interest, cultural sites, caves, landscape assets, identity assets, archaeological assets, monumental trees, natural monuments which are mapped using 7 layers downloaded from Sardegna Geoportale: «luoghi_della_cultura», «grotteCaverne», «BP» (Beni paesaggistici), «BI» (Beni Identitari), «ARCHEO» (BeniArcheologici), «alberi_monumentali_agg_18_09_2023», «monumentiNatististituitiLr31_89».

After obtaining the three input rasters, the sum of them represents our RP map. In this map, a pixel value of 1 indicates areas where all three input layers overlap, signifying the highest recreation potential as seen in Fig. 4.

Fig. 4. Recreation Potential Map (RP) of Sassari.

The second module, as explained before, is tailored to fit into the context of sustainability as a strategic decision to promote nature-based activities, aiming to reduce dependence on motorized traffic. Hence, the emphasis shifts from conventional road networks to components of slow mobility networks such as biking, hiking, and trekking

[7] Sardegna Geoportale [Official Website for geospatial data of the Sardinia Region]. https://www.sardegnageoportale.it/webgis2/sardegnamappe/?map=aree_tutelate (last accessed 5 March 2024).

[8] Ministero dell'Ambiente e della Sicurezza Energetica (MASE). https://www.mase.gov.it (last accessed 5 March 2024).

routes. To create the final output, which is proximity map, two components are considered. First are the slow mobility routes and the second is the intersection points of routes since they symbolize areas that has higher connectivity. Hiking and trekking routes are sourced from the "Sardegna Sentieri[9] "database which is available online and the biking routes are sourced from "Sardegna Ciclabile[10]" which are visible online, however, not available for download. Hence the biking routes data was provided through the courtesy of the head of the group responsible for designing the regional cycling plan.

After obtaining all the routes that constitutes the slow mobility network, the next step is to perform the Proximity Analysis. For the purpose of this study, two parameters are of importance, the density of routes in the area and the intersection points of the routes since they are a proxy of connectivity. Hence two proximity analysis are going to be performed in order to the create the final Proximity Map. For the first parameter, the "line density" of QGIS is utilized to calculate the density of routes within the area which then is used as an input for the proximity tool. For the intersection points, the proximity tool of QGIS is used to show the distances and the spatial relationship between features. The sum of the two raster maps, one showing the routes density and the other showing the intersection points of the routes, gives us the final Proximity Map as seen in Fig. 5, with value 1 showing maximum proximity.

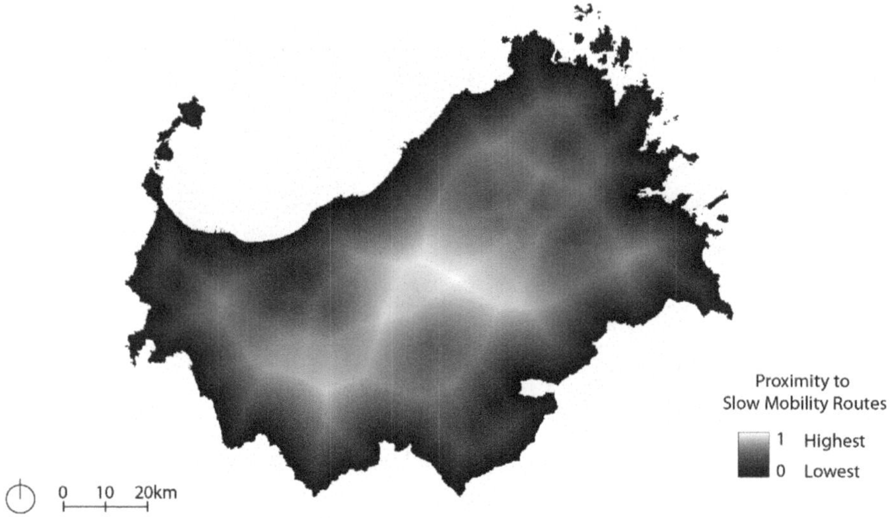

Fig. 5. Proximity Map based on slow mobility network in Sassari.

After obtaining the RP map and the Proximity map, the Recreation Opportunity Spectrum (ROS) map can be created through cross tabulation analysis. To do so, the

---

[9] Sardegna Sentieri [Online catalogue for pedestrian paths]. https://www.sardegnasentieri.it/en/sentieri (last accessed 5 March 2024).

[10] Sardegna Ciclabile [Official website showing all cycling routes in Sardinia] http://www.sardegnaciclabile.it/site/itinerari/ (last accessed 23 February 2024).

SAGA Cross-classification and tabulation tool is used. The final output presents a cross-tabulation of nine values, reflecting the relationship between recreation potential and proximity to slow mobility networks. On one axis, the levels of recreation potential are categorized as high, medium, or low. On the other axis, proximity to slow mobility networks is similarly categorized. As seen in Fig. 6, each cell in the resulting table showing the combination of these two factors represented on the ROS map, providing a comprehensive understanding of which areas have high recreation potential versus less connectivity to slow mobility networks. For instance, the blue colour represents the areas with the highest RP and highest proximity to slow mobility networks while the red colour represents the lowest RP with lower proximity. The green colour represents areas with high RP with lower proximity, suggesting that the accessibility could be improved by utilizing disused railway lines.

The final step of the methodology is the overlay of abandoned railway lines on the ROS Map. For creating the abandoned railways layer, combination of different sources is used. First in Open Street Map (OSM), using the Overpass Turbo tool, the tag "railway:disused" gave the shapefile data which then was compared to other sources such as Ferrovie abbandonate (the project showing all disused railways lines in Italy). After comparison, the data obtained from OSM matching with existing sources is used together with «DGBT_10K_22_V03_01_ELEMENTO-FERROVIARIO_DISUSO» downloaded from Sardegna Geoportale.

The superposition of the disused railways layer on the ROS Map, shown in Fig. 7, shows us the potential intervention areas considering the high recreation potential and less connectivity, and the railway routes' relation with these areas.

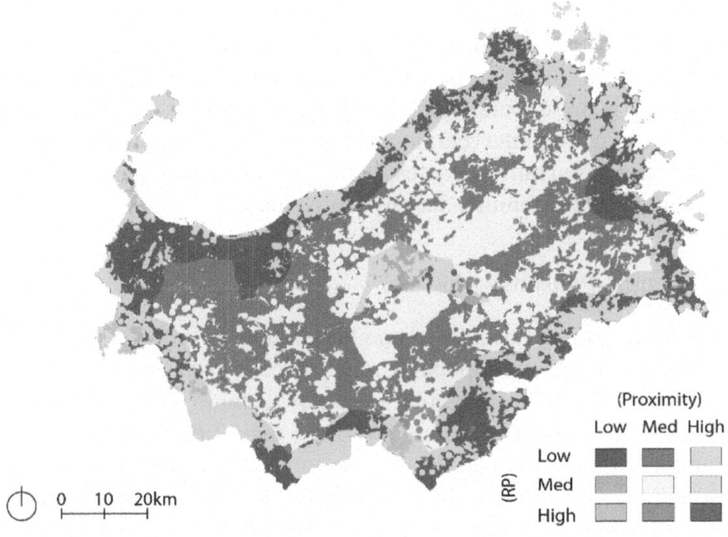

Fig. 6. Recreation Opportunity Spectrum (ROS) map of Sassari.

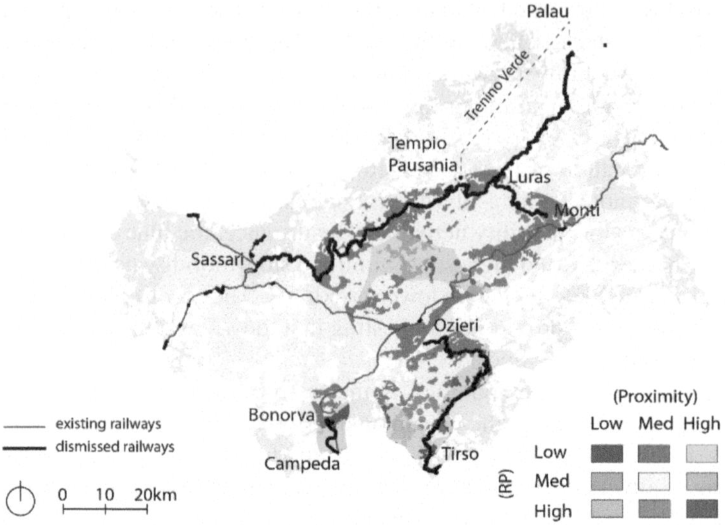

Fig. 7. Cultural ES potential of disused railways in Sassari (Areas of interest are highlighted by reducing the opacity level of the remaining portions of the map)

4 Findings and Recommendations

Figure 7 shows areas with high recreation potential along the railway routes which needs better connectivity. There are bike routes available in 3 stretches which are Palau – Tempio Pausania – Sassari route, Luras Monti and Tirso – Ozieri routes. However, the frequency of use is considered, utilizing Strava Global Heatmap[11], we can see that the routes are not frequently used on foot for hiking, trekking or walking. This might be because the services in these routes might be limited to accommodate hikers and trekkers.

ROS map shows us the intervention areas where there is potential for cultural ES provision along the disused railway routes. Using the information regarding the condition of these areas, following recommendations can be made.

4.1 Palau – Tempio Pausania – Sassari

The 150.2 km long rail line has a large number of areas of interest along its direction. It is a well-connected and most popular route since it connects the west to the east in Sassari province. The route is partially used by Trenino Verde which is a touristic train that operates in summer months. Part of the main line (Sassari to Palau) that is a part of Trenino Verde is between Palau-Tempio which is about 59 kms. An example of a transformation of disused railway site where there are few trains operate can be found in Mexico. The project "Parque Lineal Ferrocarril De Cuernavaca / Linear Park

[11] Strava Global Heatmap. Retrieved https://www.strava.com/maps/global-heatmap?style=dark&terrain=false&sport=All&gColor=hot&gOpacity=100&labels=true&poi=true#8.78/40.9203/8.8602 (last accessed 5 March 2024).

Cuernavaca Railroad" that was awarded in the 10th International Landscape Biennale in Barcelona[12]. This can be a good example because of the strategies they have such as "system of public spaces", "weaving neighbourhoods" and "muti-scale impact".

There is already a biking route following the railway line. Since it connects to the capital of the province, it may provide more connections. In this route, only the part between Tempio Pausania and Luras can be accessed by bike or on foot. Since Luras is also the starting point of the shorter stretch that is mentioned earlier, the route Tempio Pausania – Luras – Monti can be better connected and improved. Although there are areas that connect to hiking and trekking trails, in terms of pedestrian access, there are currently limited connections. Given that this stretch serves as a crucial hub for multiple routes, it's imperative to suggest interventions that enhance user experience. These interventions may include implementing signage elements for better navigation, enhancing security measures, ensuring regular trail maintenance to minimize disruptions, providing an adequate number of stops, and offering essential services at these stops to cater to the needs of users. Since the Trenino Verde offers touristic activities, public areas for people to enjoy areas with views can be improved with adequate number of seatings, shade providers, offering facilities for food and water to cater to the needs of visitors. Finally, the fact that it crosses through a large number of "comune" which are local administrative districts that constitutes provinces, its capacity to weave communities should be acknowledged. As demonstrated by the success of Cuernava Linear Park, it is imperative to ensure active participation from communities in the creation of various public spaces along the route, particularly in areas where the stretch traverses through more urbanized areas.

4.2 Luras – Monti

This is a 23 km long stretch and in Luras there is connection to the previously mentioned route. The Trenino Verde route stops here before it arrives to its final stop, Tempio Pausania. Therefore, since it is already accessible by the touristic train ride, it offers a good historical context considering that the visitors can enjoy the actual train ride before they can have a walk along the route. After the analysis 15 areas of interest are identified that are of max 15 min distance from the route. These are 3 cultural places from "luoghi della cultura" layer, and 12 landscape assets which are a part of the "beni paesaggistici" layer according to the categorization of Sardegna Geoportale. The cultural places consist of 2 museums and 1 tomb which are Tomba di Giganti Di Pascaredda, Museo del Sughero and Museo Diocesano di Arte Sacra Santa Giusta. The landscape assets that exist along the route are 10 rural settlements typical to the north of Sardinia that are called "stazzo", 1 church and 1 "nuraghe" which are a typology of ancient megalithic buildings typical to Sardinia. 12 of these points of interest lcoates in Calangianus such as Stazzo Montada, Stazzo Vaccileddu, Stazzo Bal di Trau, Stazzo Frati Satta, Stazzo Mastruledda, Stazzo Venapiccina, Stazzo Pitreddu, Stazzo Ambrogio Mossa, Stazzo Lu Rustu and Chiesa della Madonna delle Grazie, 1 in Berchidda such as Stazzo Alzola Dei Reste and 1

[12] International Landscape Biennale in Barcelona [Official Website of the Biennale]. Parque Lineal Ferrocarril de Cuernavaca. (2018). Retrieved from https://landscape.coac.net/parque-lineal-ferrocarril-de-cuernavaca (last accessed 5 March 2024).

nuraghe locates in Monti boundaries. More areas of interest exists if the diameter is expanded, for the scope of this study, the diameter of 750 m is considered an adequate distance since it can walked in 15 min.

4.3 Tirso – Ozieri and Bonorva-Campeda

Tirso-Ozieri is a 78.48 km long stretch where there is a cycle route available as a part of the project "Ciclovia Sardegna". Hence, it is already recognized as a slow mobility route by authorities. Although it is a very important step towards reusing disused railway sites for recreational activities there is still room improvement. According to Ferrovie Abbandonate, the railway site does not have the tracks anymore. It is invaded by spontaneous vegetation which increases its value. Between Chilivani and Pattada it has recently been converted into a bridle path and the line offers some works of art along the line which are in good condition. The small houses and buildings of the former stations still exist which can provide necessary services to recreational users such as bike maintenance, assistance, food and water etc. This route is well connected to previously mentioned routes however there is no slow mobility route that connects it to Bornova-Campeda route where another abandoned railway line exists with abandoned buildings. Since the line in Bornova is rapidly deteriorating according to the information received from Ferrovie Abbandonate, the connection to the existing cycle route would enhance the conditions through regenerative strategies. There is an operating line near this site, hence it is accessible by surrounding towns. From Oziero to Monti, which is the start of the route that is previously mentioned, there is a biking route. This route connects the two abandoned railway route (Tirso to Ozieru and Monti to Luras). Similarly, a connection can be provided between Bonorva – Campeda route and Tirso – Ozieri, possibly through a slow mobility route between Bonorva and Ozieri. This would allow for a more connected system in the larger scale.

10 areas of interest are identified within the 750 m distance along the Bonorva-Campeda route. 2 Nuraghe, 1 church, 3 necropolis, 1 archeological site (recinto), 1 cave, 1 museum and 1 not classified. This shows great opportunities for recreation opportunities. After providing a connection to larger routes, several shorter routes can be tailored within the area of the stretch for each typology of user depending on the visitor profile. These can be thematic routes that connects historical sites or sports routes that has level of difficulties depending on the topology. In this kind of historical context, it is important for the visitor to be able to read the historical layers of the site and learn about its story. Hence, guiding elements and highlighted features to make the history visible to the user should be designed accordingly.

5 Conclusion

Railways hold immense potential to be a part of the GI strategies thanks to their capacity to provide multiple ES. This study specifically focuses on cultural ES within the context of Sassari province. The methodology provides a comprehensive analysis by mapping the relationship between areas suitable for nature-based recreation based on their cultural ES provision and the availability of slow mobility networks. The focus on slow mobility networks demonstrates a commitment to encouraging non-motorized activities and promoting a sustainable way of living. Identifying areas based on this method, and then delving into detailed analyses for each route to identify potential interventions, serves as a guide for promoting action and furnishes decision-making authorities with valuable information.

Sassari province was selected in this study due to its characteristics that blends history, culture, industrial heritage and natural landscapes. The high number of abandoned railways in diverse conditions, some of which are partially operating for touristic purposes, some are a part of the slow mobility network, some lacks connectivity etc. provides a range of intervention possibilities and lesson to learn from. Therefore, the significance of this study extends far beyond the confines of Sassari province. It serves as a model for regions worldwide grappling with similar challenges of revitalizing disused railways and brownfields while preserving cultural identity and enhancing environmental sustainability.

However, there are some limitations. The available data is crucial for this method; presence of areas of interest, available slow mobility routes, especially their current condition is not easy to find in each region. In this study, all available data pertaining to the mentioned parameters is utilized. However, to further enhance the quality of this research, it is essential to ensure that the input data is not only up to date but also as comprehensive as possible. This will contribute significantly to the accuracy and completeness of the findings.

Due to the paper's scope, statistics regarding beneficiaries have not been included. However, integrating beneficiary demands into the equation, along with user analysis including user typologies and their respective needs, would yield more precise results and be more effective in encouraging authorities to take action.

The research underscores the significance of typology-specific work in reutilizing brownfields as part of green infrastructure strategies by focusing on disused railways. This highlights the necessity for further investigation into alternative typologies of brownfields and exploration of additional groups of ecosystem services to effectively showcase their potential.

Acknowledgments. This article was produced while attending the PhD program in Sustainable Development and Climate Change at the University School for Advanced Studies IUSS Pavia & University of Cagliari, Cycle XXXIX, with the support of a scholarship financed by the Ministerial Decree no. 118 of March 2nd 2023, based on the NRRP - funded by the European Union - NextGenerationEU - Mission 4 "Education and Research", Component 1 "Enhancement of the offer of educational services: from nurseries to universities" - Investment 4.1 "Extension of the number of research doctorates and innovative doctorates for public administration and cultural heritage".

Disclosure of Interests. The authors have no competing interests to declare that are relevant to the content of this article.

References

1. Gasparrini, C., Terracciano, A.: Messina. Green and Blue Infrastructures for the Re-urbanisation of the City. In: Arcidiacono, A., Ronchi, S. (eds.) Ecosystem Services and Green Infrastructure. Cities and Nature. Springer, Cham (2021). https://doi.org/10.1007/978-3-030-54345-7_14
2. Posner, S., McKenzie, E., Ricketts, T.: Policy impacts of ecosystem services knowledge. In: Proceedings of the National Academy of Sciences **113**(7), 1760–1765 (2016) https://doi.org/10.1073/pnas.1502452113
3. MA: Millennium Ecosystem Assessment. Ecosystems and Human Well-being: Synthesis. Island Press, Washington, DC (2005)
4. O'Brien, L., de Vreese, R., Kern, M., Sievänen, T., Stojanova, B., Atmiş, E.: Cultural ecosystem benefits of urban and peri-urban green infrastructure across different European countries. Urban Forestry & Urban Greening **24**, 236–248 (2017). https://doi.org/10.1016/J.UFUG.2017.03.002
5. Zulian, G., Paracchini, M., Maes, J., Liquete Garcia, M.: ESTIMAP: Ecosystem services mapping at European scale. EUR 26474. Luxembourg (Luxembourg): Publications Office of the European Union (2013). https://doi.org/10.2788/64713
6. Paracchini, M.L., Zulian, G., Kopperoinen, L., et al.: Mapping cultural ecosystem services: a framework to assess the potential for outdoor recreation across the EU. Ecol. Indic. **45**, 371–385 (2014). https://doi.org/10.1016/j.ecolind.2014.04.018
7. Cortinovis, C., Geneletti, D.: Mapping and assessing ecosystem services to support urban planning: A case study on brownfield regeneration in Trento. Italy. One Ecosystem **3**, e25477 (2018). https://doi.org/10.3897/oneeco.3.e25477
8. Geneletti, D., Cortinovis, C.: Identifying Ecosystem Service Hotspots to Support Urban Planning in Trento. Cities and Nature, Part **F335**, 153–165 (2021). https://doi.org/10.1007/978-3-030-54345-7_12
9. Isola, F., Lai, S., Leone, F., Zoppi, C.: Strengthening a Regional Green Infrastructure through Improved Multifunctionality and Connectedness: Policy Suggestions from Sardinia. Italy. Sustainability **14**, 9788 (2022). https://doi.org/10.3390/su14159788
10. Paracchini, M.L., Capitani, C.: Implementation of a EU wide indicator for the rural-agrarian landscape. Water (2011). https://doi.org/10.2788/26827
11. Bacchetta, G., Bagella, S., Biondi, E., Farris, E., Filigheddu, R., Mossa, L.: Vegetazione forestale e serie di vegetazione della Sardegna (con rappresentazione cartografica alla scala 1: 350.000) [Forest vegetation and serial vegetation of Sardinia (with map at the scale 1:350,000], Fitosociologia **46**(1), 3–82 (2009)

Exploring Approaches to Define Cultural Ecosystem Services in Planning Support System

Francesca Perrone[1] and Rachele Vanessa Gatto[2(✉)]

[1] Department of Planning, Design and Technology of Architecture (PDTA), Sapienza University of Rome, Rome, Italy
francesca.perrone@uniroma1.it
[2] School of Engineering, University of Basilicata, Viale dell'Ateneo Lucano 10, 85100 Matera, Potenza, Italy
rachelevanessa.gatto@unibas.it

Abstract. Cultural ecosystem services (CES) play a crucial role in enhancing human well-being by providing intangible benefits such as spiritual enrichment, aesthetic enjoyment, and cultural identity. However, defining and categorizing CES presents a significant challenge due to their subjective nature and multifaceted dimensions. This paper critically examines various approaches adopted in defining and conceptualizing CES within the realm of ecosystem services research. Through a review of existing literature, we explore the diversity of perspectives and methodologies used to identify, assess, and classify cultural services provided by ecosystems. Various approaches exist for spatial identification of CES, ranging from territorial indicator-based methods to philosophical theories. However, literature analysis reveals the difficulty in universally defining evaluation methods applicable to all cases. While macrolevel guidelines aid in recognizing potential CES, contextual analysis of structural elements is necessary. Integrating a system for assessing immaterial values can track both material and non-material benefits of CES within the territory. This suggests a context-based approach for both identification and evaluation of CES, essential for effective planning processes. By synthesizing diverse conceptualizations and methodological approaches, this paper aims to contribute to a deeper comprehension of the complexities inherent in defining and valuing cultural ecosystem services, ultimately fostering more effective conservation and sustainable management strategies.

Keywords: Cultural ecosystem services · sustainable development · territorial analysis

1 Cultural Ecosystem Services in Policy Making: The Debate

The Cultural Ecosystem Services (CES) thinking represents a significant advancement in academic debate, particularly within the field of planning. This framework offers a robust methodological approach that holds the potential to greatly enrich our understanding of various aspects of planning for sustainable exploitation of territorial cultural, landscape

O. Gervasi et al. (Eds.): ICCSA 2024 Workshops, LNCS 14818, pp. 387–395, 2024.
https://doi.org/10.1007/978-3-031-65273-8_25

and intangible assets [1–4]. One of its key strengths lies in its ability to highlight environmental performance, allowing for a more comprehensive evaluation of the impacts of land use and transformation policies. Moreover, CES thinking facilitates the ex-ante assessment impacts resulting from these policies. By incorporating considerations of cultural and natural resources, it provides a nuanced understanding of sustainability and efficiency, which are crucial concepts in contemporary planning practices [2, 5, 6]. This aspect has gathered increasing attention within the planning discipline, as it aligns with the growing emphasis on sustainability and responsible resource management. The necessity to incorporate a recognition of cultural ecosystem services into the policy cycle is crucial.

Firstly, the rigidity inherent in traditional planning methods severely requires the adoption of a holistic approach [7]. Instead of acknowledging the components system in the human-territory environment, conventional planning tends to focus strictly on assessing the suitability of specific areas based on assumptions. This perspective overlooks the complex relationships between human activities and the environment (numbering multiple perspectives and scopes cfr. [8–18]), thereby neglecting the potential contributions of cultural ecosystem services. Secondly, the fragmentation process prevalent in the traditional planning system presents a significant obstacle. With plans and responsibilities dispersed across different levels of hierarchy and sectors, efforts become fragmented, leading to inconsistent outcomes. This fragmentation not only undermines the coherence and effectiveness of planning initiatives but also limits the identification and integration of cultural ecosystem services into policy frameworks. Without a cohesive approach, the full range of cultural benefits derived from ecosystems may remain overlooked or undervalued in decision making processes.

Lastly, the traditional planning paradigm struggles with evaluating its own performance effectively [19–21]. Despite the existence of goals and targets, there is often a lack of robust mechanisms for assessing whether these objectives are being met. This deficiency extends to the evaluation of resource utilization efficiency and the quality of territorial transformations. Without adequate evaluation mechanisms, it becomes challenging to recognize and incorporate the diverse selection of cultural ecosystem services into policy cycles, continuing the limitations of the conventional planning paradigm. In summary, integrating a comprehensive understanding of cultural ecosystem services into the policy cycle is essential to address the shortcomings of traditional planning approaches. By recognizing and valuing the cultural benefits derived from ecosystems, policymakers can foster more inclusive, resilient, and sustainable planning practices that prioritize the interconnected well-being of both human societies and the environment.

The Millennium Ecosystem Assessment (MEA) stands out as one of the most relevant frameworks that laid the groundwork for understanding CES within the broader context of ecosystem services.

The MEA categorizes cultural services into several dimensions, including spiritual and religious values, aesthetic values, and cultural heritage, emphasizing the diverse ways ecosystems contribute to human well-being beyond instrumental benefits [22].

Additionally, the Common International Classification of Ecosystem Services (CICES) offers a comprehensive classification system that includes cultural services as one of its core categories [23]. CICES recognizes cultural services as essential components of human-environment interactions, encompassing elements such as cultural identity, cultural heritage, and recreational experiences.

Moreover, the Cultural Ecosystem Services Framework (CESF) provides an integrated understanding of cultural services by emphasizing the socio-cultural context and perceptions associated with ecosystem benefits. CESF highlights the importance of incorporating stakeholder perspectives and local knowledge systems in assessing and valuing cultural services, recognizing the inherent diversity and complexity of human nature relationships. Overall, these conceptual frameworks offer valuable insights into the multifaceted nature of CES, guiding researchers and practitioners in their efforts to recognize, evaluate, and manage cultural contributions of ecosystems to human wellbeing.

The research focuses on the evaluation of CES through a systematic literature review in order to identify suitable domains for concepts related to landscape, cultural heritage, identity, aesthetic and spiritual values.

The paper is structured into three main sections. First, the method & material illustrates the structure of the scientific literature selected for the analysis. Second, it examines the limitations and challenges of using territorial indicators for spatially identifying Cultural Ecosystem Services (CES) and attempts to identify CES based on territorial aesthetic and immaterial values, drawing from philosophical theories and other relevant literature. Third, it discusses various methods and approaches for spatially identifying CES, highlighting the difficulty in establishing a universally applicable evaluation method and the importance of considering contextual elements. The paper concludes by advocating for a context-based approach to CES identification and evaluation, emphasizing its relevance for effective planning processes.

2 Method and Materials

This paper starts with a synthetic appraisal of selected scientific literature on how CESs are evaluated by assessing selected research papers discussing the methodological directions taken in the CES field.

The method adopted to set up the research work involved a simple and systematic approach. A literature review was conducted using Google Scholar (https://scholar.google.com/), a freely accessible search engine that uses specific keywords to locate texts from the academic literature. The above-mentioned search engine was adopted because it enables the retrieval of articles from a wide range of publishers and from the set of scientific and cultural articles distributed on the Web.

Studies written in languages other than English were not considered at this initial stage of bibliographic evaluation. Studies without open access or not available through the subscription offered by the authors' university library system were not considered. In addition, articles focusing on a very specific area were discarded to avoid complicated comparisons between case studies that differ in type, relevance and historical, cultural and social context.

The procedure adopted for the elaboration of this contribution involved five consequential research phases:

1. phase 1 was aimed at identifying articles focused exclusively on conceptual analysis and an initial evaluation of CES only;
2. phase 2 envisaged a deepening of the analysis conducted on the CES thanks to the identification and selection of specific search terms in the titles, abstracts and keywords of the articles, including "conceptual analysis of cultural ecosystem services", "analysis of cultural ecosystem services linked to the concept of cultural heritage", "analysis of cultural ecosystem services linked to the concept of landscape", "analysis of cultural ecosystem services linked to the concept of identity"; "analysis of aesthetic and spiritual values of ecosystems";
3. phase 3 focused on the deepening and clarification of certain approaches and methods of analysis and evaluation identified within the previously selected articles;
4. phase 4 involved the description of the analysis method adopted for the evaluation of a specific CES linked to the concepts of landscape, cultural heritage, identity, aesthetic and spiritual values
5. phase 5 involved a comparison of the different approaches described above, adopted for the analysis and evaluation of CES.

2.1 Analytical Framework: Selected Methodological Approaches for the Evaluation of CES

The articles chosen for the analysis of CES evaluation approaches were limited to a selection of publications from a time span of 2012 to 2020 [24–29]. In this initial phase of bibliographic evaluation, the search was conducted by adopting English-language terms of reference. The relevance of the publications was checked by checking titles, keywords, abstracts and the full text.

A focus on the above-mentioned period is justifiable since the contribution intends to propose a comparison of recent methodological approaches. In this regard, it should be borne in mind that 2005 was the year of publication of the results of the Millennium Ecosystem Assessment (MEA), an international research project focused on the identification of the conditions, trends in ecosystems, the services they provide and the assessment of the changes they undergo and the consequences for human well-being, with the aim of identifying scenarios for the future through conservation actions and sustainable use of ecosystems. Since then, the concept of CES has gained particular attention, as evidenced by numerous scientific publications.

The search for specific terms in the titles, abstracts and keywords of the articles made it possible to select individual publications on the CES evaluation methods linked to the concepts of cultural heritage, landscape, identity, aesthetic and spiritual values of ecosystems. For example, a number of papers focusing on the entirety of the ES evaluation methods were excluded due to the complexity of the indications and information provided.

One of the advantages of a selective review of documents is that it could give rise to a specific comparison and contrast between some of the methodological directions taken in the ESC field, with potential operational utility.

In the following section two issues are pointed out as preliminary discussion of the ongoing research: how to connect CES to the concept of heritage field and cultural landscape and to the concept of Aesthetic and spiritual values.

3 Framework Results

3.1 CES Linked to the Concept of Heritage Field and Cultural Landscape

The approach adopted for the analysis of CES within the concept of role of cultural ecosystem services in heritage research through a set of indicators included a number of consequential research steps:

1. consideration of relevant indicators for the supply-side assessment of CES, i.e. indicators that measure the potential of an area for cultural practices carried out;
2. identification of relevant CES evaluation indicators for decision-making processes; The evaluation of "services" (e.g. cultural and historical sites) generate "*cultural practices in a given ecosystem/landscape*" [24] subject to generally unbiased and easier-to-standardise measurement methods. The service-oriented perspective emphasises the need to produce indicators that are comparable and clear in their measurement, interpretation and interpretation.

These indicators measure the potential of an area for cultural practices (e.g. the number of tourist accommodations). "*Supply-side CES indicators will always be focused on measuring landscape features as it is from them that CES are generated*" [24]. The main objective is to understand the extent to which Cultural Landscapes can serve as social, cultural and economic resources.

The study identifies a list of the most highly evaluated CES indicators cited in the literature in relation to their effectiveness in decision-making processes. The survey shows that: most indicators relate to the aesthetic value and recreational/touristic potential of CES; there is a trend towards the development of non-monetary indicators; the level of consolidation of CES indicators is relatively low, since the existing literature shows highly heterogeneous ways of evaluating CES due to the great variety of terms used and associated measurement methods.

One contribution of the analysis conducted highlights "*Once a CES-rooted synthetic indicator catalogue is obtained, its usability for heritage management and planning is assessed. […] The evaluation is based on an inclusive combination of two criteria: "**communicability**" and "**relevance for Cultural Landscape context**"* [24]. This means that the ability to transfer the results from the indicators to the decision-making process was verified with regard to the Cultural Landscape context, i.e. an area whose main characteristic is that it has been shaped by a relevant process of interaction between man and nature. Thus, the possibility of using indicators in heritage management and planning of Cultural Landscapes was defined.

The main finding of this research is the empirical demonstration of the potential application of a CES-rooted set of indicators in heritage management and planning of Cultural Landscapes. The main indicators selected are listed in the Table 1 below.

"*There are good examples in the scientific literature of the application of CES indicators in Cultural Landscapes or in related contexts, but there is a great variety of methods and approaches for addressing similar purposes, which leads to a still unclear panorama concerning concepts, definitions and understanding of the processes to be measured*" [24].

Table 1. CES-rooted set of indicators for heritage management and planning of Cultural Landscape

Main CES indicators for heritage management and planning of Cultural Landscapes
Land cover complexity
Land cover naturalness
Distance areas/points of interest - urban settlements
Presence and variety of historically relevant material elements and spaces without protection
Presence of disturbing elements in the viewshed areas
Distance areas/points of interest - arrival points
Presence of water bodies
Presence of mobility infrastructure
Presence of protected natural areas
Presence of facilities
Presence and variety of protected elements and spaces
Landform diversity
Presence of spaces with a current recreation-related use
Presence of vegetation
Accessibility of the areas/points of interest
Presence of viewpoints
Presence of significant elements and species
Presence of view-landmarks in the viewshed areas
Presence of recreational/educational/cultural-related activities
Landscape openness
Land cover rareness
Landscape openness
Land cover rareness
Time Depthness
Presence of scenic backgrounds in the viewshed areas
Presence of walking itineraries
Noise level
Presence of active management system
Viewshed area of viewpoints
Species richness
Presence of networking, relations between activities and/or spaces

(continued)

Table 1. (*continued*)

Main CES indicators for heritage management and planning of Cultural Landscapes
Habitat richness
Presence of landform peaks

3.2 CES Linked to the Concept of Aesthetic and Spiritual Values

The approach aiming to identify Cultural Ecosystem Services (CES) for immaterial aesthetic territorial values represents a significant tool in understanding the connection between the environment and human experiences. This approach focuses on identifying and valorizing the services offered by the ecosystem, which goes beyond material benefits and includes aesthetic, emotional, and spiritual aspects. Through a series of methodologies and tools, such as participatory assessment and discourse analysis, scholars seek to capture people's perceptions and experiences regarding the beauty and aesthetic quality of the landscape. This approach not only recognizes the importance of immaterial values in the context of environmental management and conservation but also contributes to promoting greater awareness of the connection between landscape, culture, and human well-being. As Ryfield et al. claim, aesthetic and spiritual values "*are often socially shared values, frequently intersubjective, and the outcome of historical processes in shared cultures*" [24]. Identifying, through a standardized method, the values of each culture related to CES proves problematic given the perceptual-historical-social nature of each context. Therefore, the identification of CES requires a placed-based approach incorporating tools for local information acquisition. Human science studies analyzed and theories on value aggregation show how devising an aggregation of cultural preferences is ineffective and how a specific analysis, starting from participatory processes to sociological studies in an integrative form, is functional to the evaluation of CES but not to their standardized identification.

4 Discussion and Preliminary Conclusions

The findings presented in this study highlights two critical aspects regarding Cultural Ecosystem Services (CES) and their linkage to both the heritage field and aesthetic/spiritual values. Firstly, the analysis delineates the methodological complexities involved in identifying and evaluating CES within the context of heritage research. Through a systematic approach based on a set of indicators, the study explores the intricate landscape of CES assessment, emphasizing the significance of indicators that measure the potential of an area for cultural practices. However, despite efforts to standardize measurement methods, the study reveals a heterogeneous landscape in the evaluation of CES, indicating challenges in achieving a unified understanding. In conclusion, this research underlines that adopting a comprehensive approach to CES assessment, may reinforce the structure of sustainable policy making in cultural exploitation that encompasses both tangible and intangible aspects of cultural heritage. By recognizing the complex interrelation between aesthetic, spiritual, and material values, policymakers and conservation practitioners can devise holistic strategies for the sustainable management and preservation of Cultural Landscapes. Wider investigation has to extend the specific issues linking planning purposes to CES answering the general question: how cultural ecosystem services may demonstrate the value (or the impact) of planning choices on main territorial assets both in an ex-ante decision making phase and on a scenario building design? Cording to main thematic literature [30, 31].

References

1. National Research Council: Valuing Ecosystem Services: Toward Better Environmental Decision-Making. The National Academies Press, Washington, DC (2005). https://doi.org/10.17226/11139
2. Scorza, F., Pilogallo, A., Las Casas, G.: Investigating tourism attractiveness in inland areas: ecosystem services, open data and smart specializations. In: Calabrò, F., Della Spina, L., Bevilacqua, C. (eds.) New Metropolitan Perspectives. SIST, vol. 100, pp. 30–38. Springer, Cham (2019). https://doi.org/10.1007/978-3-319-92099-3_4
3. Leone, F., Zoppi, C.: Conservation measures and loss of ecosystem services: a study concerning the Sardinian Natura 2000 network. Sustainability (Switzerland) **8** (2016). https://doi.org/10.3390/su8101061
4. Wood, S.L.R., et al.: Distilling the role of ecosystem services in the sustainable development goals. Ecosyst. Serv. **29**, 70–82 (2018). https://doi.org/10.1016/j.ecoser.2017.10.010
5. Casas, G.L., Scorza, F.: Sustainable planning: a methodological toolkit. In: Gervasi, O., et al. (eds.) Computational Science and Its Applications – ICCSA 2016. LNCS, vol. 9786, pp. 627–635. Springer, Cham (2016). https://doi.org/10.1007/978-3-319-42085-1_53
6. Las Casas, G., Scorza, F., Murgante, B.: Razionalità a-priori: una proposta verso una pianificazione antifragile. Scienze Regionali, Ital. J. Reg. Sci. **18**, 329–338 (2019). https://doi.org/10.14650/93656
7. Scorza, F., Saganeiti, L., Pilogallo, A., Murgante, B.: Ghost planning: the inefficiency of energy sector policies in a low population density region1. ARCHIVIO DI STUDI URBANI E REGIONALI, 34–55 (2020). https://doi.org/10.3280/ASUR2020-127-S1003
8. Santopietro, L., Solimene, S., Lucchese, M., Di Carlo, F., Scorza, F.: An economic appraisal of the SE(C)AP public interventions towards the EU 2050 target: the case study of Basilicata region. Cities **149** (2024). https://doi.org/10.1016/j.cities.2024.104957
9. Scorza, F., Fortunato, G.: Active mobility-oriented urban development: a morpho-syntactic scenario for a mid-sized town. Eur. Plann. Stud., 1–25 (2022). https://doi.org/10.1080/09654313.2022.2077094
10. Lai, S., Lombardini, G.: Regional drivers of land take: a comparative analysis in two Italian regions. Land Use Policy **56**, 262–273 (2016)
11. Pogliani, L., Ronchi, S., Arcidiacono, A., di Martino, V., Mazza, F.: Regeneration in an ecological perspective. Urban and territorial equalisation for the provision of ecosystem services in the Metropolitan City of Milan. Land Use Policy **129** (2023). https://doi.org/10.1016/j.landusepol.2023.106606
12. Santopietro, L., Scorza, F., Murgante, B.: Multiple components in GHG stock of transport sector: technical improvements for SECAP baseline emissions inventory assessment. TeMA – J. Land Use Mob. Environ. **15**, 5–24 (2022). https://doi.org/10.6092/19709870/8391
13. Scorza, F., Gatto, R.V.: Identifying territorial values for tourism development: the case study of Calabrian Greek area. Sustainability **15**, 5501 (2023). https://doi.org/10.3390/SU15065501
14. Lagonigro, D., et al.: Downscaling NUA: Matera new urban structure. In: Gervasi, O., et al. (eds.) ICCSA 2023. LNCS, vol. 14110, pp. 14–24. Springer, Cham (2023). https://doi.org/10.1007/978-3-031-37123-3_2

15. Florio, E., et al.: SuperABLE: Matera accessible for all. In: Gervasi, O. (ed.) ICCSA 2023. LNCS, vol. 14110, pp. 152–161. Springer, Cham (2023). https://doi.org/10.1007/978-3-031-37123-3_13
16. Nickayin, S.S., et al.: Soil quality and Peri-urban expansion of cities: a mediterranean experience (Athens, Greece). Sustainability (Switzerland) **13**, 1–12 (2021). https://doi.org/10.3390/su13042042
17. De Pascali, P., Santangelo, S., Perrone, F., Bagaini, A.: Territorial energy decentralisation and ecosystem services in Italy: limits and potential. Sustainability (Switzerland) **12** (2020). https://doi.org/10.3390/su12041424
18. Nickayin, S.S., Jahelka, A., Ye, S., Perrone, F., Salvati, L.: Planning for just cities with nature-based solutions: sustainability and socio-environmental inequalities in San José de Chamanga, Ecuador. Land (Basel) **12** (2023). https://doi.org/10.3390/land12030604
19. Gutiérrez-Romero, R., Haubrich, D., McLean, I.: The limits of performance assessments of public bodies: external constraints in English local government. Environ. Plann. C Gov. Policy **26**, 767–787 (2008). https://doi.org/10.1068/c0661
20. Alexander, E.R., Faludi, A.: Planning and plan implementation: notes on evaluation criteria. Environ. Plann. B. Plann. Des. **16**, 127–140 (1989)
21. Campagna, M., De Montis, A., Isola, F., Lai, S., Pira, C., Zoppi, C.: Planning support tools: policy analysis, implementation and evaluation. In: Proceedings of the Seventh International Conference on Informatics and Urban and Regional Planning INPUT2012: Proceedings of the Seventh International Conference on Informatics and Ur (2012)
22. Arico, S., et al.: Millennium ecosystem assessment. In: Ecosystems and Human Wellbeing: Synthesis. Island Press, Washington, DC (2005). https://doi.org/10.1196/annals.1439.003
23. Grima, N., Jutras-Perreault, M.-C., Gobakken, T., Ørka, H.O., Vacik, H.: Systematic review for a set of indicators supporting the common international classification of ecosystem services. Ecol. Indic. **147** (2023). https://doi.org/10.1016/j.ecolind.2023.109978
24. Ryfield, F., Cabana, D., Brannigan, J., Crowe, T.: Conceptualizing 'sense of place' in cultural ecosystem services: a framework for interdisciplinary research. Ecosyst. Serv. **36** (2019). https://doi.org/10.1016/j.ecoser.2019.100907
25. Stålhammar, S., Pedersen, E.: Recreational cultural ecosystem services: how do people describe the value? Ecosyst. Serv. **26**, 1–9 (2017). https://doi.org/10.1016/j.ecoser.2017.05.010
26. Fish, R., Church, A., Winter, M.: Conceptualising cultural ecosystem services: a novel framework for research and critical engagement. Ecosyst. Serv. **21**, 208–217 (2016). https://doi.org/10.1016/j.ecoser.2016.09.002
27. Cooper, N., Brady, E., Steen, H., Bryce, R.: Aesthetic and spiritual values of ecosystems: recognising the ontological and axiological plurality of cultural ecosystem 'services.' Ecosyst. Serv. **21**, 218–229 (2016). https://doi.org/10.1016/j.ecoser.2016.07.014
28. Tengberg, A., Fredholm, S., Eliasson, I., Knez, I., Saltzman, K., Wetterberg, O.: Cultural ecosystem services provided by landscapes: assessment of heritage values and identity. Ecosyst. Serv. **2**, 14–26 (2012). https://doi.org/10.1016/j.ecoser.2012.07.006
29. López Sánchez, M., Tejedor Cabrera, A., Linares Gómez del Pulgar, M.: The potential role of cultural ecosystem services in heritage research through a set of indicators. Ecol. Indic. **117** (2020). https://doi.org/10.1016/j.ecolind.2020.106670
30. Díaz, S., et al.: Assessing nature's contributions to people: recognizing culture, and diverse sources of knowledge, can improve assessments. Science **1979**(359), 270–272 (2018). https://doi.org/10.1126/science.aap8826
31. Chan, K.M.A., Satterfield, T., Goldstein, J.: Rethinking ecosystem services to better address and navigate cultural values. Ecol. Econ. **74**, 8–18 (2012). https://doi.org/10.1016/j.ecolecon.2011.11.011

Beyond City Average: Spatial Analysis to Support Energy-Saving Interventions in Urban Planning

Gerardo Carpentieri, Carmela Gargiulo, Carmen Guida(✉), and Floriana Zucaro

Department of Civil, Building and Environmental Engineering, University of Naples Federico II, Naples, Italy
carmen.guida@unina.it

Abstract. Actions to handle the threats posed by climate change-related events and energy crisis are becoming increasingly needed. The transition from fossil fuels-based development to a more sustainable one has been prioritised in global agendas. The energy-saving issue is the foundation of the social and economic model of transformation for the near future and the turning point for sustainable, smart and energy resilient urbanisation. Thus, integrating energy-saving solutions with urban planning remains a challenge. This study aims at bridging this gap by investigating the relationship between urban characteristics and residential energy consumption using statistics and spatial analysis. Starting from deepening the results of a first GIS-based analysis applied to Naples (Italy) and the methodology behind them, this contribution aims at exploring how to support policy-makers in decision processes by implementing novel representation techniques. This study reveals how urban context dynamically impacts household energy usage, with a focus on green urban areas, by using Geographically Weighted Regression (GWR) and depicting the non-stationary estimation coefficients derived from the investigation. This knowledge could enable local decision-makers to maximize the efficiency of energy-saving actions in building more resilient and sustainable cities by customizing them for particular metropolitan locations.

Keywords: Energy-Saving · Urban Planning · GIS

1 Introduction and Scientific Background

Regardless of the source, energy is a key factor for urban development and, hence, urban areas substantially influence global energy demand and energy-related emissions. In the light of the increasing impact of climate change-related phenomena and the need to boost a transition towards more sustainable models, energy saving is becoming a major concern for urban scientists, stakeholders and planners.

It is worth noting that cities consume about 75% of global primary energy and emit between 50% and 60% of total greenhouse gases [1]. This trend peaks at approximately 80% when also the indirect emissions generated by urban inhabitants are included [2].

O. Gervasi et al. (Eds.): ICCSA 2024 Workshops, LNCS 14818, pp. 396–411, 2024.
https://doi.org/10.1007/978-3-031-65273-8_26

As far as the European scenario, cities take up only 4% of the EU land area yet are home to 75% of citizens, and this number is expected to rise to approximately 84% by 2050 [3].

Cities, as complex macrosystems, consume a significant amount of energy across various sectors, including buildings, transportation, and industry. Buildings are major energy consumers throughout their life cycle, from material extraction and construction to ongoing maintenance and daily operations like heating, cooling, and lighting [4]. Urban sprawl further exacerbates energy demands by increasing commute distances and hindering the effectiveness of public transport systems, leading to greater reliance on private vehicles [5, 6]. Sprawling cities also contribute to the formation of the urban heat island phenomenon: due to the heat absorption of concrete and asphalt, and lack of green spaces, urbanized areas can get significantly warmer than surroundings. Therefore, urban heat islands generally raise the demand for air conditioning, generating a vicious cycle [7]. These are just a few examples of how cities are challenged by an increasingly complex energy management, due to rapid urbanisation.

Climate change adds urgency, demanding practical solutions to make cities resilient against extreme weather events. Moreover, disruptions in power or fuel supply due to increasingly frequent extreme weather events cascade across infrastructure sectors, impacting businesses, essential services as well as the immediate response of urban system to manage emergencies.

The IPCC's Sixth Assessment Report [8] underscores the link between a city's energy performance and its ability to withstand climate extremes like floods, heat waves, and droughts. The report also warns of escalating risks posed by climate change and states that investments are crucial to accelerate the shift to renewables and away from fossil fuels, mitigating greenhouse gas emissions and bolstering overall energy system resilience. The current energy crisis has further exacerbated the need for cities to curb energy use and finally boost the transition to renewable sources [9, 10].

Despite their vulnerabilities, cities may play a crucial role in turning the current development model, based on an unsustainable fossil fuel dependence into a lower energy footprint and renewable energy use. This shift tackles pollution, greenhouse gas emissions, and potential damage from climate events. This confluence of factors necessitates integrating energy concerns into urban development strategies. However, translating these goals into action remains challenging. Even the latest Conferences of the Parties (COPs), designed to implement the Paris Agreement through practical measures, are struggling to find effective solutions.

Despite these challenges, positive developments have been recorded recently. Recovery plans, developed in response to the Covid-19 pandemic's economic crisis, increasingly emphasise environmental sustainability and energy saving in urban planning tools. In Italy, government incentives like the Ecobonus program (currently in dismission) have promoted energy efficiency and building quality improvements. However, these initiatives lack a focus on collective efforts to enhance the quality of broader urban areas. Given the close connection of the energy challenge to climate resilience, a more comprehensive urban approach should be preferred.

To effectively address these combined challenges, urban planning, building regulations, and other policies need to be integrated to enable a deep and cost-effective decarbonization of all urban activities. Historically, the lack of integration between energy and urban planning has resulted in misalignment between urban development and energy conservation efforts.

To bridge this gap, energy planning must be incorporated into the iterative cycle of urban and territorial transformations. This fosters the development of more comprehensive and effective strategies for reducing energy consumption.

Scientific advancements in recent decades have yielded models to investigate, simulate and, most frequently, to predict energy consumption at the city level. These models fall into two broad categories: top-down and bottom-up [11–13]. Top-down models take a macroscopic view of cities, analyzing historical energy data to understand consumption patterns and their relationship with urban features, considering the effects of technology, socioeconomics, and time. Conversely, bottom-up models examine energy use at the individual building level before extrapolating the data to larger regions. This approach relies heavily on detailed data to analyse individual user consumption. Both approaches are usually designed to predict energy demand in urban areas. Both types of models are mainly used to predict urban energy demand.

This contribution presents the first results of a research specifically targeting the decision-making processes of urban planning by building a solid ground-based knowledge framework, aiming to support local authorities in identifying key areas for intervention to reduce energy consumption. Hence, a GIS-based methodology has been designed in order to delve deeper into the relationships between a city's energy footprint and its urban features.

This research breaks new ground by expanding the investigation from individual buildings to the entire urban environment. This shift in scale necessitates the development of new models capable of interpreting the complex relationships between energy consumption and urban features.

The outcomes of the methodology will help local decision-makers identify areas requiring interventions to reduce energy consumption. Moreover, insights gained into the relationships between energy footprint and urban features will inform the development of more effective urban planning strategies.

The methodology, whose first steps are described in the following paragraph, has been designed with real-world implementation in mind, aiming to integrate energy planning seamlessly with urban and land-use planning processes. The model, built on a Geographic Information System (GIS) platform, will primarily rely on open-source data from resources like ISTAT (Italian Statistical Institute) census tracts. This design choice allows urban technicians and policymakers to easily replicate the model in different areas. In order to test the preliminary steps of the methodology, a case-study approach was adopted and the city of Naples was selected as significant application to validate the methodology results.

This contribution sums the first results of a joint research activity carried out within two research project: NEST[1] - Network for Energy Sustainable Transition and "Definition of a guidelines handbook to implement climate neutrality by improving ecosystem service effectiveness in rural and urban areas"[2].

The contribution is organised as follows: after this introduction, which summed some key elements of the main research question, the developed methodology is presented; in the third paragraph, the results of the application of the methodology to the city of Naples are presented; finally, the last paragraph highlights the main findings as well as areas of improvements for further development of the research.

2 Materials and Method

Recognizing the established connection between urban energy consumption and urban features [14, 15], this contribution presents the early stages of the development of a methodology to identify and weight the most significant urban features. The methodology presented in this section will be the ground of a decision-support tool aimed at identifying strategies and interventions for defining a set of actions and interventions in urban areas to reduce energy consumption. The classification of urban areas based on their features (physical, functional, geomorphological, environmental, climatic and socio-economic) represents the final output of the methodology; such classification will be useful to identify the required interventions of physical transformation, for each kind of urban area, to reduce energy consumption.

The methodology has been designed to be performed in GIS environment, in order to be integrated into urban planning decision processes. In particular, the methodology employs an array of GIS-based statistical and spatial analyses to interpret which urban features significantly affect urban energy consumption and where dependencies among variables are more significant. Hence, the results may explain where these relationships are useful for decision-makers in energy planning.

In this contribution, the first two steps of the methodology will be investigated through an application-based approach: (i) data collection and preparation and (ii) spatial statistical analyses.

The first phase concerns the collection of data and preparation of the dataset. First, urban characteristics and corresponding variables that have the greatest impact on energy consumption have to be identified. Scientific literature suggests taking into account several characteristics of urban environment that can be grouped according to the urban subsystem they relate to:

---

[1] NEST is a project funded under the National Recovery and Resilience Plan (NRRP), Mission 4 Component 2 Investment 1.3 - Call for tender No. 1561 of 11.10.2022 of Ministero dell'Università e della Ricerca (MUR); funded by the Euro-pean Union –NextGenerationEU.

[2] "Definition of a guidelines handbook to implement climate neutrality by improving ecosystem service effectiveness in rural and urban areas" (CUP E53D23018970001) is a project funded by EU in the NextGenerationEU Plan through the Italian "Bando Prin 2022-D.D. 1409 del 14-09-2022".

- Physical characteristics. Information such as the number of buildings, residential and not residential, construction type, and surface area, is important to understand the key features of the urban fabric and built environment;
- Socio-economic condition. The social and economic circumstances of a city may have an impact on energy usage. Therefore, we have chosen to collect data on variables such as the number of residents, age ranges and employment status. Since economic data per each census tracts are not available, proxy variables have been introduced, such as the education level (given as the percentage of people that accomplished different educational goals).

Twenty one indicators, falling in both groups were selected for analyses. More details on indicators are given in Table 1. To these variables must be added the energy consumption data. Hence, the dataset will also include residential energy consumption given as kWh consumed annually by urban residents.

Since the methodology has been designed to support effective energy saving-oriented urban planning practice, data will be collected using opensource dataset. Moreover, data will be associated to census tracts, which are the minimum territorial reference area for this study. The output of this first phase is a set of arrays, for each territorial unit, collecting data about dependent variable y (energy consumption) and independent variable x_i (concerning urban features for each territorial unit i).

The second step concerns statistical and spatial analyses to obtain a model defining the strength (magnitude of closeness) and the kind (positive or negative) of relationships between the dependent variable (residential electric energy consumption) and independent variables (concerning urban physical and socio-economic environment). Geographically weighted regression (GWR) analyses were chosen to investigate spatial heterogeneities and relationships between energy usage and urban environment features. Since GWR was first proposed by Brunsdon et al. [16], it is being increasingly used in spatial analyses of social, environmental and urban data. The main idea beneath the GWR method is that whole map (constant-coefficient) linear regression models (LRM), such as those estimated by ordinary least squares (OLS) may produce unreasonable conclusions about the stationarity of the regression coefficients under investigation, assuming that regression relationships are the same, no matter where you measure them within the study area. Hence, GWR can be defined as follow:

$$y_i = \beta_0(u_i, v_i) + \sum_{k=i}^{m} \beta_k(u_i, v_i)x_{ik} + e_i \tag{1}$$

where (u_i, v_i) are the spatial coordinates of the observations i (the centroids of territorial units), and $\beta_k(u_i, v_i)$ are the coefficients estimated at those locations which describe the kind of relationship (positive or negative) between dependent (y_i) and independent variables (x_{ik}). $\beta_0(u_i, v_i)$ are the intercept terms, while the set of e_i is the random error term, which obeys an independent normal distribution with zero mean and variance σ^2. The calibration process of $\beta_k(u_i, v_i)$ is based on kernel distance-decay functions and their bandwidths. In GWR, bandwidth plays a crucial role as it refers to a distance metric or the number of nearest neighbors used for each local regression equation. It essentially defines the spatial scale of influence for estimating the relationship between variables at a specific location. Automated routines exist to determine an optimal bandwidth for calibrating GWR coefficients, such as the Akaike information criterion (AIC). In general, the result of larger bandwidth is that the GWR model tends toward a whole-map LRM.

The second stage of the methodology is therefore a flow of statistical and spatial analyses.

First, an autocorrelation analysis will be performed. Using the Global Moran's I Statistics and given a set of features and an associated attribute, the autocorrelation analysis calculates Pearson correlation coefficients and p-values to indicate whether the null hypotheses can be rejected. In this case, the null hypothesis states that the values of the features are spatially uncorrelated. The degree of correlation between variables is a crucial factor to consider before conducting the two spatial analyses, to avoid data redundance and multi-collinearity. The output of this step is a first set of indicators, selected from the original sample by excluding those whose Pearson correlation coefficient exceeds 0.700. Such result is the input of the second step, which consists of performing a first GWR. In order to perform a GWR, the opensource software GWR4 developed and programmed by Professor Tomoki Nakaya of the Department of Geography, Ritsumeikan University, Kyoto, Japan [17]. Our methodology implies for the selection of bandwidth the golden section search based on AICc (Akaike Information Criterion corrected for small sample sizes), which is a relatively efficient way to narrow down the search for the optimal bandwidth.

To further refine the model and focus on the most statistically significant relationships, we employed an iterative approach. After running the initial GWR analysis, choropleth maps will be created to visualise the spatial variations in the coefficients $\beta_k(u_i, v_i)$ of each independent variable. Variables with coefficients exhibiting very low significance (below 90% confidence level for more than 80% of the case study area) from the subsequent GWR model. This step helped to improve the overall model robustness by focusing on the most relevant spatial relationships influencing the dependent variable. Variables with coefficients exhibiting consistently low statistical significance (e.g., below 90% confidence level for a substantial portion, such as 80%, of the study area) will be excluded from the subsequent GWR model. This approach enhances the overall model robustness by focusing on the most statistically significant spatial relationships influencing the dependent variable. Thus, for our research questions where understanding all potential relationships is crucial, even those with weaker statistical support, a lower significance coverage threshold might be considered. This approach was preferred at this stage of the research for a more comprehensive exploration of the spatial dynamics, although the resulting model may be less robust and require careful interpretation due to the inclusion of potentially less reliable relationships. In order to support interpretations, maps and statistical charts will be developed, employing data classification schemes informed by the (nonspatial) data distribution, diverging colour schemes, and bivariate choropleth mapping, in order to highlight whether the relationships between dependent and independent variables are significant and, where they are, whether they are positive or negative [18]. These thematic maps utilize color shading to represent the values of a specific variable across different geographical units, such as census tracts in our case. In our study, choropleth maps were created to depict the local β_k values estimated by the GWR model at each location. Darker or more intense colors on the map indicate areas where the model explains a larger portion of the variation in energy consumption (higher significance level), while lighter colors represent areas with a weaker model fit.

It's important to acknowledge that alternative spatial regression techniques exist, such as Spatial Autoregressive Models (SAR). SAR models account for spatial autocorrelation, where the dependent variable at one location is influenced by the values of the dependent variable in neighboring locations. However, unlike GWR, SAR models assume a constant relationship between variables across space. In our research question, we were particularly interested in exploring how these relationships might vary geographically. Therefore, GWR, with its ability to capture local variations, was a more suitable choice for investigating the nuanced spatial dynamics influencing energy consumption in urban environment.

Figure 1 below represents the whole methodology workflow, whose first two steps (Data collection and preparation and Spatial and statistical analyses) have been described in this paragraph.

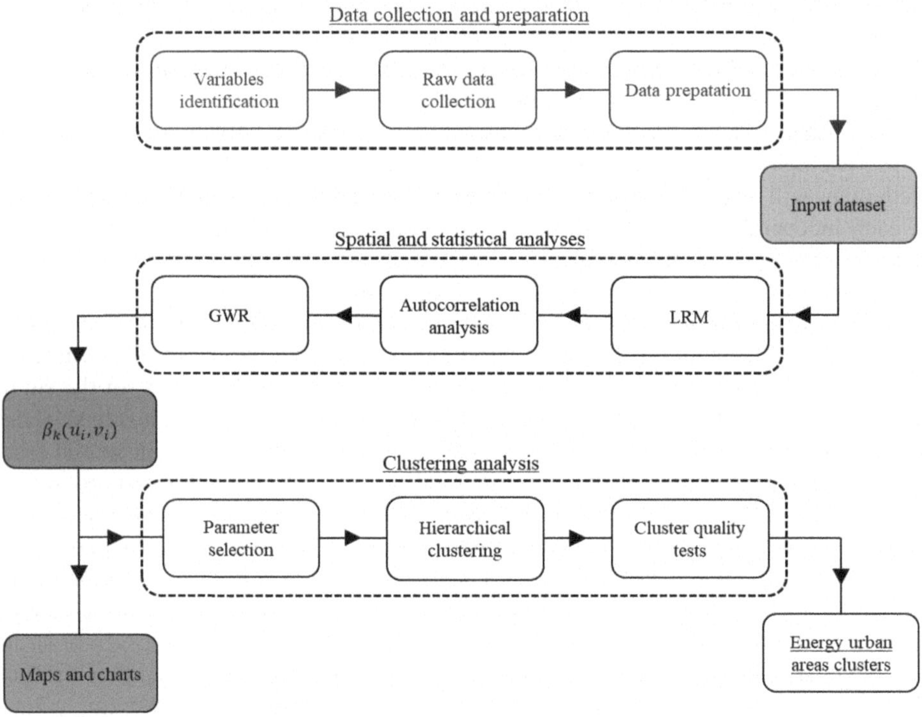

Fig. 1. Methodology workflow.

The following paragraph describes the application of the methodology to the city of Naples and its main results.

3 Application Results

The methodology described in the previous paragraph has been applied to a significant case study in order to be tested. Naples is the third largest Italian city in terms of population, about 909,000 citizens (2022), while it is the most densely populated among major cities. Despite its cultural richness, Naples faces significant challenges regarding energy consumption. The dense urban core, with its older buildings and potentially less energy-efficient infrastructure, is likely to have higher energy demands. Figure 2 below shows how annual electricity consumption per inhabitant varies across the city of Naples. Additionally, economic disparities across different districts might influence energy use patterns. This complex urban environment necessitates a holistic understanding of energy consumption patterns. A one-size-fits-all approach wouldn't be effective. By employing a spatially explicit technique like GWR, we can capture the variations in energy consumption across different districts and develop targeted urban planning and energy management strategies for a more sustainable city.

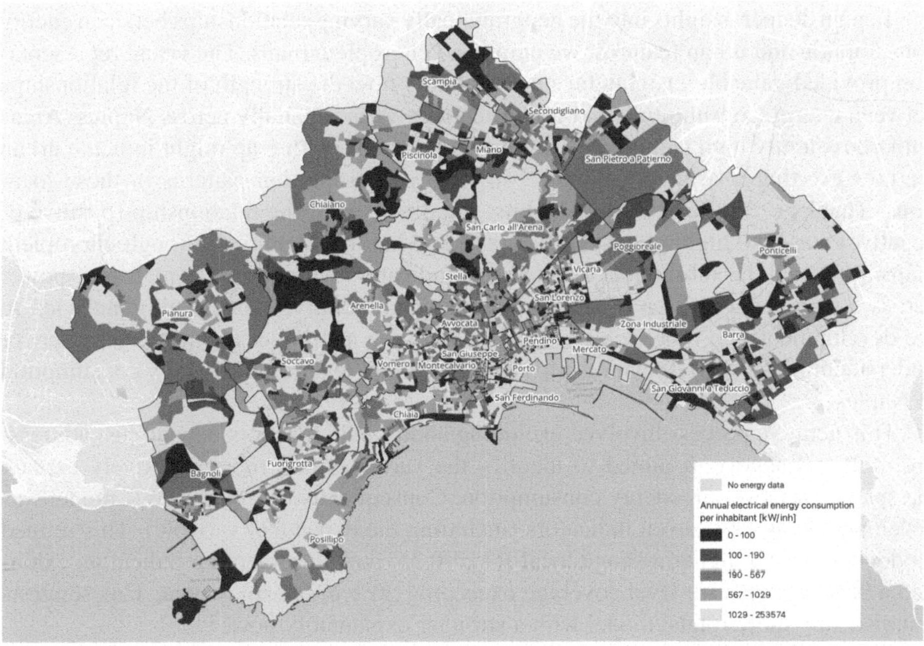

Fig. 2. Annual residential electricity consumption per inhabitant per census tract.

The initial dataset of independent variables was constructed using exclusively open-source data sources to ensure the methodology's replicability across different territorial contexts. Specifically, data was obtained from ISTAT (Italian National Institute of Statistics), Corine Land Cover (CLC), and Città Metropolitana di Napoli (Metropolitan City of Naples). For what concerns energy usage, the annual electrical consumption per census tracts was obtained by Italian revenue Agency (Agenzia delle Entrate). Raw data were then processed in order to be associated to each census tract within the study area.

The second step, concerning spatial statistical analyses, started with a bivariate autocorrelation analysis. We employed Pearson's correlation coefficient to identify and exclude highly correlated independent variables. Variables with a correlation coefficient exceeding 0.70 were removed to mitigate multicollinearity.

Table 1 below presents the selected indicators for the study, along with their descriptions and some initial statistical analyses. The second and third columns, respectively, show the percentage of territorial units where each variable exhibits a significance level exceeding 90%. The third column also highlights indicators with a Pearson correlation coefficient greater than the pre-defined threshold, indicating potential multicollinearity and exclusion from further analysis.

After excluding highly correlated indicators, the initial GWR model based on 16 variables yielded an average local R2 of 0.26. Our primary objective at this stage is to explore the spatial relationships between energy consumption and urban features in greater detail. To achieve this, we examined the significance level of each remaining indicator. We focused on those with a 90% significance level coverage falling below 20% of the entire study area.

To gain deeper insights into the geographically varying relationships between energy consumption and urban features, we employed choropleth maps. The visual representation provided valuable insights for understanding how the strength of the relationships between energy consumption and urban features varies spatially across Naples. Areas with consistently high significance values on the choropleth map might indicate urban features exerting a stronger influence on energy consumption patterns in those locations. Thanks to bivariate colour shading, the direction of the relationship (positive or negative) was also highlighted. These spatial variations, identified through choropleth maps, informed the subsequent exclusion of indicators with limited explanatory power across a significant portion of the study area. This iterative process ultimately led to the development of a more robust GWR model with a stronger explanatory power for understanding the spatial relationships between urban features and energy consumption in Naples.

This iterative process involved excluding additional indicators beyond just addressing multicollinearity. It aimed to identify the variables that most effectively explain the spatial variations in energy consumption. Consequently, a second GWR model was developed using only seven indicators (including the dependent variable). This refined model achieved a higher average local R2 of 0.35, and all remaining indicators exhibited a 90% significance level coverage exceeding 20% of the study area. This suggests a statistically more robust model with a stronger explanatory power.

To visualize the impact of key urban features on energy consumption, we utilized choropleth maps. This contribution focuses specifically on the relationship between the density of urban green areas and energy use.

Table 1 reveals a significant association (greater than 90%) in slightly over 27% of the census tracts analyzed. Interestingly, Image 1 differs from previous choropleth maps exploring energy consumption and other indicators. It shows a wider distribution of census tracts with statistically significant relationships, and these relationships can be either positive or negative.

Table 1. Synthesis of employed independent indicators and some results from statistical analyses.

Indicator	Description	Autocorrelation	GWR 1	GWR 2
		Pearson > 0.70	% 90 significance coverage	
D_P_tot	Density of total population		28,70	64,61
D_rooms	Density of rooms per census tract	x	x	x
I_ut_occ	Ratio between volume of occupied buildings and territorial surface per census tract	x	x	x
P_0_19	Ratio between population aged 0–19 and total population		0,56	x
P_20_64	Ratio between population aged 20–64 and total population		1,51	x
P_over65	Ratio between population aged over 65 and total population		3,53	x
P_grad	Ratio between university graduated population and total population		26,00	20,85
P_high	Ratio between population with high school diploma and total population		29,26	28,51
P_mand	Ratio between population with mandatory school diploma and total population		1,02	x
P_illit	Ratio between illiterate population and total population aged over 6		4,91	x
P_work	Ratio between population in workforce (aged over 15) and total population	x	x	x
P_empl	Ratio between employed population and total population	x	x	x
I_empl	Ratio between employed population and population in workforce	x	x	x
P_unempl	Ratio between unemployed population and total population		11,42	x
I_unempl	Ratio between unemployed population and population in workforce		13,56	x
D_tot_build	Density of building per census tract		70,95	25,09
D_tot_build_res	Density of residential building per census tract		0,00	x

(*continued*)

Table 1. (*continued*)

Indicator	Description	Autocorrelation	GWR 1	GWR 2
		Pearson > 0.70	% 90 significance coverage	
D_notres	Density of not residential building per census tracts		0,00	x
D_mas_buil	Density of masonry buildings	x	x	x
D_con_build	Density of concrete buildings		12,84	24,85
D_sur_green	Density of green areas (ratio between green surfaces and census tract territorial surfaces)		24,23	27,55

The colors on the map (Fig. 3) represent the combination of two information: the red scale, the level of significance of the relationship, while the blue scale how beta coefficient values, estimated by the GWR model, vary their direction, from negative to positive, in the study area. Darker colors likely indicate a positive relationship between green spaces and energy consumption. In these areas, higher green space density might be associated with higher energy use. Lighter colors likely indicate a negative relationship. In these areas, higher green space density might be associated with lower energy consumption.

The map also highlights where energy residential consumption data are missing. Our analysis of green space density and energy consumption in Naples revealed a surprising trend. While initially expecting a negative correlation (green space reducing energy use), approximately 50% of census tracts with a statistically significant relationship (>90% significance level) exhibited a positive association. In simpler terms, these green spaces seem to contribute to higher annual electricity consumption.

This positive relationship is particularly pronounced in peripheral areas, specifically the northeastern and western districts of Naples. These areas tend to have larger, denser green spaces compared to the city center. However, the data also suggests that green spaces are not detrimental to energy savings everywhere, also according to scientific literature on the topic [19–23]. Especially in highly urbanized neighborhoods, both central (Chiaia, Vomero and Posillipo) and peripheral (San Giovanni, Miano and Arenella), green spaces have a significant negative impact on energy consumption. Table 2 provides insightful information for each city neighborhood.

It details the percentage of territorial units where the link between green space distribution and electricity consumption is statistically significant at a 90% confidence level. Additionally, the table shows how the β coefficients vary across these neighborhoods, ranging from negative to positive values.

This result represents the very first attempt to delve deep into the relationships between energy consumption and urban features, and to develop an automated methodology to be integrated into a decision-support tool. While this initial exploration, mainly focused on the extension and distribution of green urban areas, serves as a foundation, it also highlights the potential for further refinement.

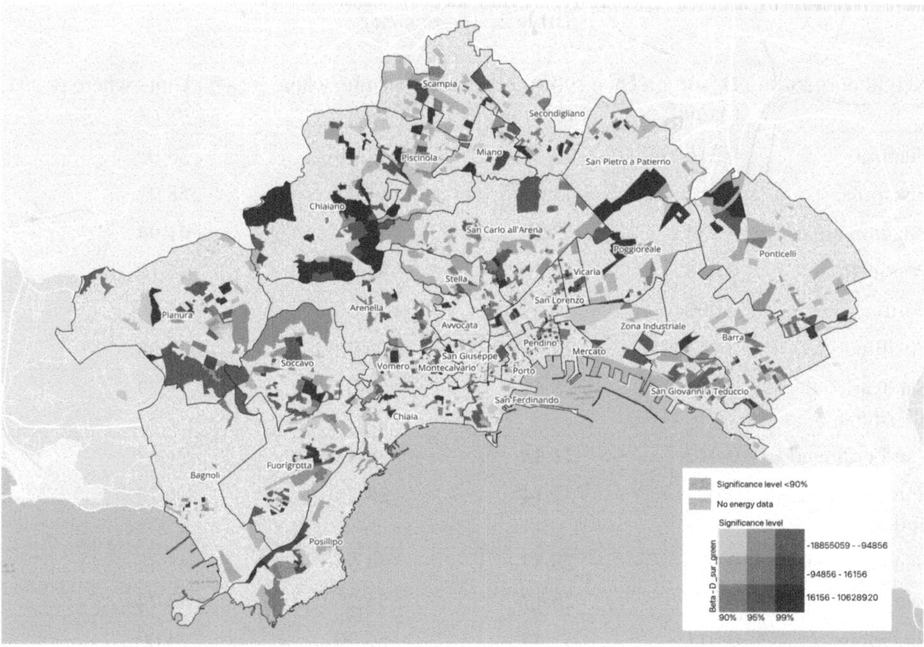

Fig. 3. Choropleth map of main results of GWR model to explain the relationship between density of green urban areas and energy consumption.

Table 2. Main results per neighbourhood.

Neighbourhood	D_sur_green [ha/ha]	90 significance coverage [%]	Units where β > 0 [%]	Units where β < 0 [%]
Arenella	0.225	18.78	32.35	67.65
Avvocata	0.052	11.72	66.67	33.33
Bagnoli	0.146	20.27	40.00	60.00
Barra	0.018	27.09	36.36	63.64
Chiaia	0.097	20.71	31.43	68.57
Chiaiano	0.233	44.55	55.10	44.90
Fuorigrotta	0.141	29.96	46.48	53.52
Mercato	0.000	12.07	71.43	28.57
Miano	0.028	30.77	7.14	92.86
Montecalvario	0.115	17.21	33.33	66.67
Pendino	0.000	39.57	40.66	59.34

(*continued*)

Table 2. (*continued*)

Neighbourhood	D_sur_green [ha/ha]	90 significance coverage [%]	Units where β > 0 [%]	Units where β < 0 [%]
Pianura	0.216	29.59	42.00	58.00
Piscinola	0.062	41.05	41.03	58.97
Poggioreale	0.134	32.29	58.06	41.94
Ponticelli	0.035	23.29	54.90	45.10
Porto	0.000	30.77	25.00	75.00
Posillipo	0.164	29.55	26.92	73.08
San Carlo all'Arena	0.304	28.41	66.67	33.33
San Ferdinando	0.042	21.43	8.33	91.67
San Giovanni a Teduccio	0.023	37.14	28.21	71.79
San Giuseppe	0.000	28.57	54.55	45.45
San Lorenzo	0.003	25.17	55.26	44.74
San Pietro a Patierno	0.011	32.89	56.00	44.00
Scampia	0.064	33.33	41.94	58.06
Secondigliano	0.069	20.99	41.18	58.82
Soccavo	0.075	38.78	61.40	38.60
Stella	0.244	34.38	75.00	25.00
Vicaria	0.000	25.71	88.89	11.11
Vomero	0.079	27.78	46.67	53.33
Zona Industriale	0.000	28.89	23.08	76.92

4 Discussion and Conclusions

This contribution presents the initial stages of a methodology for classifying urban areas based on their energy consumption patterns and key features. We applied the first two steps of the methodology to the city of Naples, Italy.

Physical characteristics, socio-economic conditions, and relevant proxy variables were defined to holistically depict urban environment. In order to allow the replicability of the methodology, only open-source datasets were consulted for the data collection of preparation step. Statistical analyses, including spatial autocorrelation assessment and Geographically Weighted Regression (GWR), were employed to identify significant relationships between energy consumption and various urban features. The initial GWR model provided insights into the spatial variations of these relationships across Naples. Further refinement through the exclusion of highly correlated variables and those with low spatial significance allowed for the development of a more robust model with a stronger explanatory power. Choropleth maps were then generated to visualize these variations, effectively depicting how the strength and direction (positive or negative) of the relationships between energy consumption and urban features differ geographically throughout the city.

For this contribution, urban green areas and how they are related to energy demand in cities have been the focus of the research. The results of the application to Naples show that their effect on electricity consumption vary significantly within the city, both in terms of magnitude and direction. In particular, results showed that a negative effect on the total electricity consumption in cities is exacerbated in the northeastern and western districts of Naples, where larger green spaces are located, compared to the city center. On the other hand, in highly urbanised neighborhoods, both central (Chiaia, Vomero and Posillipo) and peripheral (San Giovanni, Miano and Arenella), green spaces have a significant negative direction on energy consumption (contributing at reducing energy consumption).

However, it is important to acknowledge some limitations of the current research. The first concerns open-source data availability. Future iterations could explore incorporating additional data sources, potentially requiring procedure to update data in real-time. A second concern is about the robustness of the model. While the refined GWR model achieved a higher average local R2, there is still room for improvement. Future steps could explore additional explanatory variables or investigate the influence of spatial factors not yet considered.

This initial exploration paves the way for the next stages of the methodology. These will involve, first, the classification of urban areas in Naples based on their shared features and energy use patterns. Secondly, the methodology will include the development of a decision-support tool that can be used by urban planners and policymakers to identify targeted interventions for reducing energy consumption in different districts of the city. Finally, in order to delve into the influence of green areas on energy consumption, further variables will be investigated to take into account the cooling effect they produce in urban areas.

By understanding the spatial relationships between urban features and energy consumption, this methodology can contribute to the development of more sustainable and energy-efficient urban environments.

Acknowledgments. This research acknowledges the funding received under the National Recovery and Resilience Plan (NRRP), Mission 4 Component 2 Investment 1.3 - Call for tender No. 1561 of 11.10.2022 of Ministero dell'Università e della Ricerca (MUR); funded by the Euro-pean Union –NextGenerationEU. Award Number: Project code PE0000021, Concession Decree No. 1561 of 11.10.2022 adopted by Ministero dell'Università e della Ricerca (MUR), CUPC93C22005230007, according to attachment E of Decree No. 1561/2022, Project title "Network 4 Energy Sustainable Transition – NEST" and also the project "Definition of a guidelines handbook to implement climate neutrality by improving ecosystem service effectiveness in rural and urban areas" CUP E53D23018970001 funded by EU in the NextGenerationEU Plan through the Italian "Bando Prin 2022-D.D. 1409 del 14-09-2022".

Disclosure of Interests. The authors have no competing interests to declare that are relevant to the content of this article.

References

1. United Nations Habitat for a better future. World Cities Report 2022: Envisaging the Future of Cities. https://unhabitat.org/world-cities-report-2022-envisaging-the-future-of-cities. Accessed 20 Dec 2022
2. Su, B., Goh, T., Ang, B.W., Ng, T.S.: Energy consumption and energy efficiency trends in Singapore: the case of a meticulously planned city. Energy Policy **161**, 112732 (2022). https://doi.org/10.1016/j.enpol.2021.112732
3. Steurer, M., Bayr, C.: Measuring urban sprawl using land use data. Land Use Policy **97**, 104799 (2020). https://doi.org/10.1016/j.landusepol.2020.104799
4. Gargiulo, C., Russo, L.: Cities and energy consumption: a critical review. TeMA – J. Land Use Mob. Environ. **10**(3), 259–278 (2017). https://doi.org/10.6092/1970-9870/5182
5. Sharma, S., Saini, G., Kumar, K., Saini, K.: Futuristic approach to energy in smart cities. In: Smart Cities, pp. 293–309. CRC Press (2022). ISBN: 9781003287186
6. Xie, L., Yan, H., Zhang, S., Wei, C.: Does urbanisation increase residential energy use? Evidence from the Chinese residential energy consumption survey 2012. China Econ. Rev. **59**, 101374 (2020). https://doi.org/10.1016/j.chieco.2019.101374
7. Santamouris, M., Cartalis, C., Synnefa, A., Kolokotsa, D.: On the impact of urban heat island and global warming on the power demand and electricity consumption of buildings—a review. Energy Build. **98**, 119–124 (2015). https://doi.org/10.1016/j.enbuild.2014.09.052
8. IPCC: Climate Change 2022: Mitigation of Climate Change. Contribution of Working Group III to the Sixth Assessment Report of the Intergovernmental Panel on Climate Change. Cambridge University Press (2022)
9. Chu, K.: Environmental quality and the role of economic policy uncertainty, economic complexity, renewable energy, and energy intensity: the case of G7 countries. Environ. Sci. Pollut. Res. **29**, 2866–2882 (2022). https://doi.org/10.1007/s11356-021-15666 (2022)
10. United Nations, FCCC – Framework Convention on Climate Change. Adoption of the Paris agreement (2015)
11. Gao, H., et al.: A review of building carbon emission accounting and prediction models. Buildings **13**(7), 1617 (2023). https://doi.org/10.3390/buildings13071617
12. Perera, A.T.D., Hong, T.: Vulnerability and resilience of urban energy ecosystems to extreme climate events: a systematic review and perspectives. Renew. Sustain. Energy Rev. **173**, 113038 (2023). https://doi.org/10.1016/j.rser.2022.113038
13. Yang, J., Deng, Z., Guo, S., Chen, Y.: Development of bottom-up model to estimate dynamic carbon emission for city-scale buildings. Appl. Energy **331**, 120410 (2023). https://doi.org/10.1016/j.apenergy.2022.120410
14. Guida, C.: Energy saving and efficiency in urban environments: integration strategies and best practices. TeMA – J. Land Use Mob. Environ. **15**(3), 517–531 (2022). https://doi.org/10.6093/1970-9870/9549
15. Papa, R., Gargiulo, C., Zucaro, F.: Urban systems and energy consumptions: a critical approach. TeMA – J. Land Use Mob. Environ. (2014). https://doi.org/10.6092/1970-9870/2552 (2014)
16. Brunsdon, C., Fotheringham, A.S., Charlton, M.E.: Geographically weighted regression: a method for exploring spatial nonstationarity. Geograph. Anal. **28**(4), 281–298 (1996). https://doi.org/10.1111/j.1538-4632.1996.tb00936.x
17. Nakaya, T.: Geographically weighted generalised linear modelling. Geocomput. Pract. Primer **217**(20), 201–220 (2015)
18. Besançon, L., Cooper, M., Ynnerman, A., Vernier, F.: An evaluation of visualization methods for population statistics based on choropleth maps. arXiv preprint arXiv:2005.00324. https://doi.org/10.48550/arXiv.2005.00324 (2020)

19. Gargiulo, C., Zucaro, F.: A method proposal to adapt urban open-built and green spaces to climate change. Sustainability **15**(10), 8111 (2023). https://doi.org/10.3390/su15108111
20. Gargiulo, C., Sgambati, S., Zucaro, F.: The analysis of the urban open spaces system for resilient and pleasant historical districts. In: Gervasi, O., et al. (eds.) ICCSA 2023. LNCS, vol. 14106, pp. 564–577. Springer, Cham (2023). https://doi.org/10.1007/978-3-031-37111-0_39
21. Gargiulo, C., Zucaro, F.: Greening networks for smart and resilient cities: from methodology to application in densely built urban contexts. Archivio di Studi Urbani e Regionali, 130–148 (2020). https://doi.org/10.3280/asur2020-127-s1008
22. De Luca, C., Libetta, A., Conticelli, E., Tondelli, S.: Accessibility to and availability of urban green spaces (UGS) to support health and wellbeing during the COVID-19 pandemic-the case of Bologna. Sustainability **13**, 1–13 (2021)
23. Vignoli, F., de Luca, C., Tondelli, S.: A spatial ecosystem services assessment to support decision and policy making: the case of the City of Bologna. Sustainability **13**, pp. 1–19 (2021). Article number: 2787

Greening and Cooling Urban Areas: The Open Space System Contribution for Energy Saving and Climate Change Adaptation

Gerardo Carpentieri, Carmela Gargiulo, Tonia Stiuso, and Floriana Zucaro(✉)

Department of Civil, Building and Environmental Engineering, University of Naples Federico II, P.le Tecchio 80, 80125 Naples, Italy
{gerardo.carpentieri,gargiulo,tonia.stiuso, floriana.zucaro}@unina.it

Abstract. Given the urgency in addressing the climate change acceleration, the international community emphasises the relevance to promote both mitigation adaptation measures to achieving climate neutrality by setting binding targets for member countries in terms of emission reductions. One of the main strategies at the local level is to increase resilience through territorial planning and measures aimed at reducing energy consumption and climate adaptation. In this regard, many studies have shown that the physical characteristics of the urban context are significant factors influencing the achievement of these goals. This paper represents a first step in a larger PRIN-funded research project 'Definition of a handbook of guidelines for implementing climate neutrality by improving the effectiveness of ecosystem services in rural and urban areas', aimed at defining an energy-efficient decision support tool based on urban and open space characteristics. In contrast to traditional studies that tend to address the two issues (energy and climate change) separately and at the scale of individual interventions, this research considers the two types of open spaces as a single network and adopts a holistic approach, privileging the urban and neighbourhood scale. The work has a twofold goal: (i) identifying the most significant urban characteristics and the related descriptive variables mainly affecting thermal comfort and energy consumption, by carrying out an extensive scientific literature review and the authors' previous works and (ii) and classifying the surrounding urban context of open spaces based on the relationships among the variables explaining the urban characteristics identified as relevant to investigate the match between homogeneous characteristics and the reciprocal position of open spaces. This second goal was achieved through a cluster analysis focused on the fabric portions influenced by the cooling effect of existing and potential vegetation.

Keywords: Climate Change Adaptation · Energy Saving · Urban Open Spaces · Green Areas

O. Gervasi et al. (Eds.): ICCSA 2024 Workshops, LNCS 14818, pp. 412–429, 2024.
https://doi.org/10.1007/978-3-031-65273-8_27

1 Introduction

With the acceleration of climate change, the international community emphasizes sustainable measures to address climate-related risks, as exemplified by Sustainable Development Goal 13. The European Green Deal aims for a climate-neutral continent by 2050, covering various sectors such as energy efficiency and biodiversity. The European Climate Law commits to achieving climate neutrality, setting binding targets for member countries in terms of emission reductions and legislative measures to be introduced. One of the strategies being adopted at the local level involves increasing resilience by integrating spatial planning actions and measures to reduce energy consumption and promote adaptation to climate change [1, 26]. Current efforts seem to be more oriented towards promoting the use of renewable sources for clean energy production, reducing waste, and adopting energy-efficient technologies [23, 46], paving the way for a more sustainable and resilient future. Regarding this, there are many studies that have demonstrated that both building and urban physical characteristics significantly influence energy consumption [5, 38, 47]. For instance, urban open spaces, both green and built, play a fundamental role in addressing the challenges of reducing energy consumption. In particular, green areas contribute to reducing the negative effects of rising temperatures by, for example, absorbing excess water, mitigating the effects of urban heat islands, and supporting biodiversity. There is much research that has dealt with the effect of green spaces on the improvement of the urban thermal environment from various perspectives [10, 16, 17, 28, 47]. Furthermore, green spaces can contribute to energy efficiency by providing shade, thus reducing the demand for artificial cooling systems during hot seasons. To this end, some studies aim to develop proposals to optimize the presence of green spaces in consolidated urban contexts [19, 40, 49]. In densely built and layered cities, it becomes strategic to resiliently transform the open space system to strengthen its ecosystemic capacities and improve its adaptive capacity to climate change [16, 18, 41]. This open space reorganisation strategy is also promoted by the European Union, which considers open spaces, both built and unbuilt, as components of an integrated system to mitigate urban warming and enable an adaptation approach "based on the return of nature to cities" [24, 36]. Recognizing the interconnectedness of urban ecosystems, we can create urban environments that not only resist the impacts of climate change but also actively contribute to a more sustainable and energy-conscious future. Section 2 identifies the most significant urban characteristics and the relative descriptive variables that mainly influence thermal comfort and energy consumption, carrying out an extensive review of the scientific literature and the authors' previous works, from which to study methodologies and collect data. Section 3 analyses the surrounding urban context and open spaces of the city of Naples, based on the relationships between the variables that explain the urban characteristics identified as relevant to investigate the correspondence between homogeneous characteristics and the reciprocal position of open spaces. The analysis is conducted through a cluster analysis focused on the fabric sections influenced by the cooling effect of existing and potential vegetation. Section 4 includes the final discussion of the results and conclusions. With the above premises in mind, the main goal of this initial research is to contribute to the reduction of energy consumption from fossil carbon sources and, at the same time, address urban challenges posed by climate change through defining the role (new functions?), dimension, and distribution of open urban spaces. In fact, this work represents a first step of a wider research PRIN-funded "Definition of a guidelines handbook to implement climate neutrality by improving ecosystem service effectiveness in rural and urban area" - aimed at defining a decision support tool that is effective in terms of energy savings, based on urban characteristics and features of open spaces. Unlike traditional studies that tend to address the two issues (energy and climate change) separately and at the scale of individual interventions, this research considers the two types of open spaces as a single network and adopts a holistic approach, favouring the urban and neighbourhood scale. The work was carried out through a (i) literature review to identify the relevant urban characteristics, mainly related to open space system, affecting energy consumption and climate change adaptation and (ii) a cluster analysis to highlight their relationship.

2 Scientific Framework

The scientific community is actively addressing climate change and energy consumption reduction at the urban scale through interdisciplinary and collaborative research. There are an increasing number of studies, which in recent years, are oriented to develop methods and tools to improve the overall organization of urban systems by applying sustainable planning principles and enhancing the presence of greening interventions. The recent research mainly contribute to evidence-based policy recommendations, suggesting regulatory measures and incentives for sustainable practices [6]. In the following paragraphs, the focus will be on a first part of literature review carried out by considering the relationship between urban built environment characteristics and energy saving, in the first part, and climate change adaptation, in the second one.

2.1 Methodology of Literature Review and Research Focus

The literature review process adopted for this study follows a systematic approach aimed at investigating the relation between the urban characteristics, which influences the thermal comfort and energy consumption as described in Fig. 1. The search for relevant literature primarily focuses on the ScienceDirect database, using specific keywords associated with the thermal aspects of urban environments and the impact of urban open spaces, regarding green and built-up areas.

The chosen keywords include "The Cooling Effect In Open Spaces In Urban Areas," "Urban Thermal Comfort," "Climatic Adaptation," "Urban Energy Saving For Climate Change," "Climate Change And Urban Solution," "Green Spaces And Cooling Effect," "Air Quality And Energy Consumption," and "Review Micro-climate Model." The time frame chosen for the literature review ranges from the foundation of the 2030 Agenda for Sustainable Development, adopted by all UN member states in 2015 and the European Green Deal in 2020, to the present [20]. This temporal boundary ensures that the study considers the most recent developments and research in the context of sustainable urban development and climate change mitigation and adaptation. The subject areas covered in the literature review align with the broader theme of sustainable urbanism and include "Sustainable Cities and Society," "Urban Climate," "Urban Greening," and "Urban Planning." Regarding the urban scale, the study encompasses both the city and

Search	Keywords	Time Frame	Subject Areas
ScienceDirect	- The Cooling Effect - In Open Spaces In - Urban Areas - Urban Thermal Comfort - Climatic Adaptation - Urban Energy Saving For Climate Change - Climate Change And Urban Solution - Green Spaces And Cooling Effect - Air Quality And Energy Consumption - Review Micro-climate Model	- 2030 Agenda for Sustainable Development, adopted by all UN member states in 2015 - European Green Deal, founded in 2020 by the European Commission.	- Sustainable Cities and Society - Urban Climate - Urban greening - Landscape and Urban Planning

Urban Scale	Aims	Methodology
- City - Neighbourhood	The aim is to further study the urban microclimate and the evaluation of the thermal effects of urban open spaces, with a focus on green areas.	Methodologies and simulation models to assess the effects of environmental features, urban morphology, green spaces, and other factors on temperature and thermal comfort in urban areas. The methodologies range from using energy balance-based models to numerical simulation approaches like ENVI-met, statistical analyses, and GIS tools.

Fig. 1. Literature Review process.

neighbourhood levels, recognizing that the thermal dynamics of open spaces can vary significantly across different scales [2, 19]. The primary aims of the literature review are to deepen the understanding of urban microclimates and to evaluate the thermal effects of open spaces, with a specific emphasis on green and built areas. In terms of methodology, the review incorporates various approaches, including methodologies and simulation models. These tools are utilized to assess the impact of environmental features, urban morphology, and green spaces on temperature and thermal comfort in urban areas. The methodological spectrum encompasses energy balance-based models, numerical simulation approaches such as ENVI-met, statistical analyses, and Geographic Information System (GIS) tools.

2.2 Definition of the Main Characteristics

Currently, the scientific community is increasingly moving toward a more holistic perspective of the city and its systems [29] with a focus on reducing energy consumption from fossil carbon sources and mitigating the urban impacts of climate change, understanding of the relationships that interact between them and the expected and unexpected shocks they might face, in order to prepare cities to be more flexible and to have the capacity to bounce back in case of acute shocks [13, 31, 37]. This trend reflects a significant change from traditional approaches, which often addressed energy and climate change issues separately. To this end, the most significant urban characteristics to be considered in the analyses and the corresponding descriptive variables were identified, based on the scientific literature and the authors' previous works, which are those that have the greatest influence on thermal comfort and energy consumption [14, 28, 30, 44] (Fig. 2). In particular, according to S. Bouketta [4], plant cover in urban spaces and bodies of water such as basins and fountains act as devices that reduce overheating, create an urban cooling island effect, and provide outdoor thermal comfort conditions very close to ideal for pedestrians [35, 48]. Tree cover, impervious surfaces, and vegetation structure play a crucial role, but other factors such as the building materials, the distribution of open spaces in the different kind of urban fabrics impact cooling, as well as proximity to distances to rivers, lakes, and bodies of water, [3, 9]. Regarding the urban asset, Wang and Gou [45] demonstrated that urban open spaces surrounded by low-rise buildings and low-built density fabrics exhibit the highest thermal comfort, whereas those encompassed by a combination of buildings and green spaces show the lowest level of comfort.

2.3 Research Selected Characteristics and Variables Definition

In this study the energy issue and climate adaptation are considered in relation to open spaces through a systemic approach and in addition, for each indicator, the relationship with energy saving, UHI adaption and flooding adaption was studied, as shown in the Fig. 3. This approach indicates that all elements of the urban system interact with each other. Therefore, in addition to examining the characteristics of open spaces, we also consider the surrounding context. For example, the distribution of buildings can influence cooling effect due to the presence of vegetation, as building height and distances between open spaces and buildings themselves can determine a barrier effect hindering their positive thermal and energy reduction effects [32]. Based on these systemic interactions, we identify four main categories, explicit in the Fig. 4: open space system, urban context, microclimatic and geomorphological, and socio-economic and demographic characteristics.

We focus on open spaces considering their extent, the presence of vegetation, pavement, and other aspects. It is worth noting that these spaces are part of urban context with various characteristics and relationships. Finally, we take into account the relationships between these characteristics and factors such as temperature, heat islands, energy consumption, and solar radiation, included in the category of microclimatic and geomorphological characteristics. For each of these categories, key variables have been identified and assumed to be significant by scientific studies on the subject. For each characteristic there are variables that can be collected taking into account criteria depending on availability in the study area, reliability of sources, suburban scale and spatial data, this implies that for each case study the variables will change or be measured differently.

Indicators Dataset from Literature Review	(Solecki et al., 2015)	(Cruz et al., 2021)	(Pigliautile et al., 2021)	(Das et al., 2022)	(W. Liu et al., 2022)	(Zhang et al., 2022)	(Ma et al., 2022)	(Nimac et al., 2022)	(Shen et al., 2022)	(Bouketta, 2023)	(Best et al., 2023)	(Darbani et al., 2023)	(Zou et al., 2023)	(Diz-Mellado et al., 2023)	(Couret et al., 2023)	(Roest et al., 2023)	(Wang & Gou, 2024)	(Patle & Ghuge, 2024)
Presence of trees, shrubs, grass, greenery.		x			x	x				x	x	x			x	x	x	x
Proportion of green cover.		x		x	x	x				x	x	x		x	x	x	x	x
Grass and permeable surface			x			x		x			x	x			x	x		
Parks, squares, allotments and vegetable gardens.			x	x						x						x	x	
Street orientation and geometry			x				x					x			x	x	x	
Density of courtyards in urban areas.									x					x	x			
Water bodies			x	x				x		x	x					x		x
Materials used in paved surfaces		x	x		x	x	x	x			x	x	x			x	x	x
Manufacturing, industries, business, food and beverage, retail, residential areas, historical area			x	x				x	x			x		x		x		x
Agricultural land, fallow land, open land, and forest coverage.				x				x	x			x						x
Building coverage ratio, floor ratio		x	x	x		x	x	x	x		x	x	x			x	x	x
Infrastructure details such as public transit, transportation modes	x								x						x			x
Shading systems														x				
Building forms						x	x			x		x	x	x	x		x	
Building material		x					x			x		x	x	x			x	
Neighborhood compactness and layout.	x						x		x						x	x		x
Amount of sunlight and shadows			x	x		x		x		x	x	x	x	x	x			
Global solar radiation and Mean Radiant Temperature (MRT)			x	x		x	x			x		x	x			x	x	x
Energy use	x				x				x				x					
Wind direction and speed		x	x			x	x	x		x		x	x	x			x	
Relative humidity		x	x			x	x	x		x	x		x				x	
Precipitations										x						x		
Average temperatures and Temperature condition index (TCI)				x	x	x	x	x		x	x			x			x	
Physiological Equivalent Temperature (PET)						x	x					x		x				
Temperature variations and Land Surface Temperature (LST)		x	x	x		x				x	x							x
Air quality																	x	
Distance to commercial centre, river, and elevation.								x			x				x			x
Municipal boundaries	x			x												x		
Proximity to different urban features	x										x							
Total population	x								x				x		x			
Age	x																	
City users	x																	
Education	x												x					
Lifestyle									x									
Real estate value	x								x									
Workers													x					
Local governament	x																	
Institutions													x					

Fig. 2. Dataset from Literature Review.

Characteristics	Variables	Indicators Dataset from Literature Review	Energy saving	UHI adaptation	Flooding adaptation
1. Open space system	Vegetation and Greenery	Presence of trees, shrubs, grass, greenery	x	x	x
		Proportion of green cover	x	x	x
		Grass and permeable surface	x	x	x
	Surface Area	Parks, squares, allotments and vegetable gardens	x	x	x
		Street orientation and geometry	x	x	
		Density of courtyards in urban areas	x	x	
		Water bodies	x	x	x
	Pavement Type	Materials used in paved surfaces		x	x
2. Urban context	Land Use	Manufacturing, industries, business, food and beverage, retail, residential areas, historical area	x		
		Agricultural land, fallow land, open land, and forest coverage	x	x	x
	Building Density	Building coverage ratio, floor ratio	x	x	x
		Infrastructure details such as public transit, transportation modes	x		
		Shading systems	x	x	
	Building Layout	Building forms	x	x	
		Building material	x	x	
		Neighborhood compactness and layout	x	x	
3. Microclimatic and geomorphology	Solar Radiation	Amount of sunlight and shadows	x	x	
		Global solar radiation and Mean Radiant Temperature (MRT)	x	x	
		Energy use	x	x	
	Wind	Wind direction and speed		x	
	Temperature	Relative humidity		x	
		Precipitations			x
		Average temperatures and Temperature condition index (TCI)		x	
		Physiological Equivalent Temperature (PET)	x	x	
		Temperature variations and Land Surface Temperature (LST)	x	x	
	Geomorphological features	Air quality		x	
		Distance to commercial centre, river, and elevation		x	
		Municipal boundaries			
		Proximity to different urban features		x	
4. Socio-Economic and Demographic Indicators	Demographic data	Total population			
		Age			
		City users			
	Socio-economic	Education			
		Lifestyle			
		Real estate value			
		Workers			
	Governance	Local governament			
		Institutions			

Fig. 3. Indicators relation with energy saving, UHI adaptation and flooding adaptation.

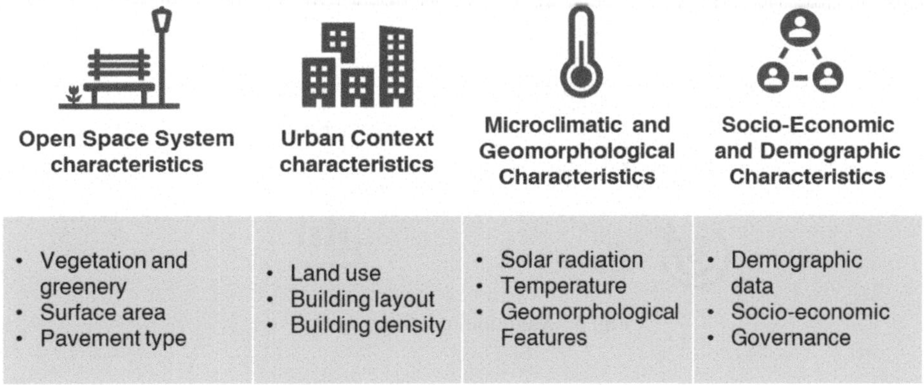

Fig. 4. Definition of urban characteristics.

3 Methodological Process and Data

The aim of this section is to describe the GIS and statistical methodology steps after a literature review, that has enabled the identification of the main dataset addressed in the recent studies, allowing the urban characteristics, the variables, and the indicators to be outlined for input within the analyses. In the next stage, it was necessary to process the data, through calculations and geoprocessing operations, to prepare the dataset for spatial statistical analysis. A cluster analysis was carried out to investigate the match between homogeneous characteristics and the reciprocal position of open spaces, this is focused on the fabric portions influenced by the cooling effect of existing and potential vegetation. This output allows to reach results and conclusions, in order to support decision-makers in defining different greening interventions oriented to energy saving and climate adaptation, according to urban context where open spaces are located. Once the variables of urban characteristics, which emerged from the literature review, have been identified, (a) we proceed to the selection of the indicators describing the data. we need also consider the availability of data for the territorial context under analysis. By first consulting open-source sources, (b) we proceed with the collection, which may involve vector data, raster data or calculation tables. Once the dataset has been collected, it is necessary to make the data continuous and homogeneous, with respect to a single reference unit (census section, buffer area, district, etc.), and normalise it. (c) We will then proceed with correlation analysis, which will identify possible relationships between the indicators and proceed with k-means clustering. (d) This will be concluded by reading the results in the form of both graphs and maps. A summary of the framework methodology is shown in the Fig. 5.

3.1 Case Study Area

The city of Naples (Latitude 40.8517746 and Longitude 14.2681244) is the capital of the Campania Region in Southern Italy with 917.510 inhabitants and covers an area of 117.27 km^2. Naples is one of the oldest cities in Europe, whose contemporary urban fabric preserves the elements of its eventful history. The city has an excessively high

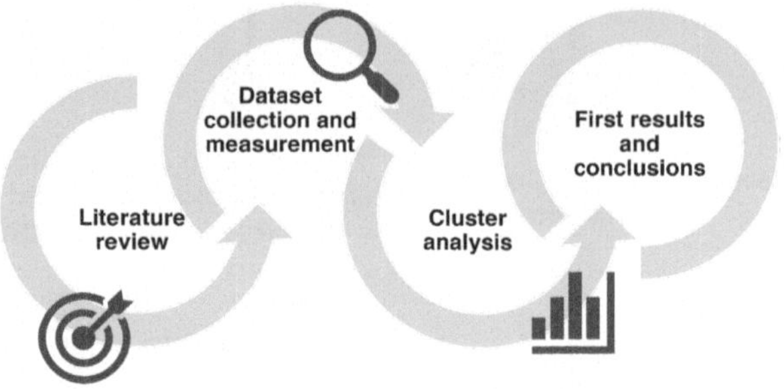

Fig. 5. Methodology framework.

population density, 7,823.91 per km^2, even with reference to the national Fig. (189 inhabitants per km^2). It is marked by a continuous increase in the aging demographic, surpassing national averages [18]. Considering this demographic trend, the impacts on the mobility and well-being of the population, the case study of Naples offers valuable insights. The case study of Naples is also interesting for the heterogeneity of its territory in terms of geomorphologic characteristics, such as its hilly conformation and coastal position, and of urban layouts characterised by densely built-up urban fabrics with different distributions and types of activities. Therefore, as shown in the Fig. 6, the open space system in a densely stratified built city like Naples represents a relevant resource for increasing urban resilience, cooling the built environment, improving stormwater management, and encouraging sustainable mobility.

3.2 Dataset Collection

In this first part of the study, were identified the indicators belonging to the variable's vegetation and greenery, surface area, building density, building layout, temperature, and demographic data, based on the availability of data for the city of Naples. For each characteristic there are variables that can be collected taking into account criteria depending on availability in the study area, reliability of sources, suburban scale and spatial data, this implies that for each case study the variables will change or be measured differently.

They are listened in the Fig. 7 most data utilized for assessing the variables underwent processing within a GIS framework, sourced from open databases such as ISTAT for population data and Urban Atlas for vegetation coverage, greenery, surface area, building density, and layout. Notably, raster image processing was conducted for temperature variables. The indicators specifically concern:

- Open spaces that refer to public spaces with a minimum surface area of 55 m^2 as, below this value, it is no longer a space but an element of street furniture (e.g. a roundabout with vegetation):
  - o Percentage of green cover: considering green urban areas and sports and leisure facilities.
  - o Built spaces: including public squares or courtyards.

Fig. 6. City of Naples vegetation, street network, open space and built areas.

- Building coverage ratio: that considering the building density.
- Building materials: the number of buildings made of reinforced concrete, masonry, wood or steel, based on ISTAT data.
- Microclimatic and geomorphology: LST and overlapping of cooling areas.
- Demographic data: total resident population.
- Vulnerable population: population that would be most affected by the effects of climate change, i.e. children and the old people [7].

3.3 Indicator Calculation

- Land Surface Temperature (LST)

Measurement of temperature that corresponds urban heat island intensity values necessitated the analysis of multispectral and thermal data obtained from Landsat 8 satellite imagery, accessible via the U.S. Geological Survey website. Specifically, a medium-resolution (30 m/pixel) raster image was analyzed to examine the spatial distribution of air temperatures within the urban canopy layer, spanning from the ground surface to 2 m above, particularly focusing on vegetation cover for the date of July 25, 2022, when a cloud-free image of the study area was available. The variables associated with the land surface temperature (LST), i.e. the radiant temperature measured at the interface between the surface of a material (such as a tree canopy, water, soil, ice or snow) and the atmosphere, influenced by solar reflectance, thermal emissivity and thermal storage capacity; the intensity of the urban heat island effect; the urban area benefiting from

Characteristics	Variables	Indicators Dataset	Measurement	Source	Year	Geometry
1. Open space system	Vegetation and Greenery	Proportion of green cover	mq	Urban Atlas	2018	Polygons
	Surface and built area	Parks, squares, courtyard	mq	Urban Atlas	2018	Polygons
2. Urban context	Building Density	Building coverage ratio, floor ratio	mc/mq	Urban Atlas	2018	Polygons
	Building Layout	Building material	n. of building in reinforced concrete	ISTAT	2011	Polygons
			n. of building in brickwork			
			n. of building in steel, wood, etc.			
3. Microclimatic and geomorphology	Temperature	Land Surface Temperature (LST)	Average °C	Landsat 8	2022	Polygons
		Combined temperature effect*	-	-	-	Polygons
4. Socio-Economic and Demographic	Demographic data	Total population	n. of total residents	ISTAT	2011	Polygons
	Volnurable popolation	Children*	n. of residents 0–9 years old	ISTAT	2011	Polygons
		Old people*	n. of residents over 60 years old	ISTAT	2011	Polygons

Fig. 7. Dataset for measuring the open space system variables in Naples's case study.

the cooling effect due to the presence of green spaces of a particular size. Through the process of evapotranspiration, the energy of solar radiation absorbed by leaves is transformed into latent rather than sensible heat flow, thus reducing the temperature of the canopy and surrounding air. Green areas, especially those characterised by the presence of trees, can also lower the air temperature by intercepting solar radiation, thus preventing the underlying surface from absorbing short-wave radiation, a phenomenon known as the shadow effect. Based on these mechanisms, the air temperature in urban green spaces can be 1–3 °C to 5–7 °C lower than in adjacent urbanised areas, and this phenomenon can also extend to the surrounding environment. The proportion of the urban area that experiences the benefit of cooling due to the presence of green spaces is the result of microclimatic simulations integrated within a GIS environment simulations developed in previous works of the authors [16–18] and this variable have allowed us to outline the microclimatic behaviour of three common urban fabric types, characterised by various building densities and sizes of green spaces (from 1,000 square metres to 35,000 square metres), typically found in stratified contexts. From the simulations conducted using the ENVI-met software, it emerges that green areas with a size of around 5,000 square metres are the most effective and efficient, capable of reducing the average temperature by about 1 degree Celsius in the surrounding built-up areas within 100–150 m from them [16, 18]. The cooling area represents the size of the urban area that benefits from the presence of green spaces and is the result of integrated microclimatic simulations in a GIS environment. These simulations made it possible to define the microclimatic behaviour of the three most common urban fabric types, characterised by different building density values and different sizes of green spaces (from 1,000 m^2 to 35,000 m^2), which are likely to be found in stratified contexts. After performing simulations using the ENVI-met software, green areas sized around 5000 m^2, which can lower the average temperature by 1 °C on the surrounding built area between 100–150 m from them, were found to be more effective and efficient.

- Cooling Areas

The cooling areas will be the reference unit over which the variables will be distributed, to study correlation and carry out clustering. Specifically, a buffer area of 110 m, 122 m and 146 m will be created based on the surface area of the green area or square. In fact, in the first case, the relationship between actual green areas and urban characteristics will be assessed, and in the second the possible relationship and benefits of covering built open spaces with green areas will be analysed. The buffers will be defined as follows:

- Buffer 110: area < 5000 mq
- Buffer 122: 5000 < area < 30000 mq
- Buffer 146: area > 30000 mq

- Overlapping Effect

Another indicator considered and proposed in the study concerns the calculation of the overlap of the different cooling airs created around green areas. This variable makes it possible to consider the overall, overlapping effects resulting from the proximity of several areas. It can be the context indicator we propose for consolidated cities where it is common for the distribution of open spaces to be 'concentrated'. At this point was carried out the statistical normalisation, that is the scaling of the distribution of a variable to make comparisons with respect to sets of items and the mean by eliminating the effects of influences. This calculation involved all continuous variables and adjusted to the scale of the cooling areas. Once all variables have been calculated to perform the spatial analysis, it is necessary to merge the normalised indicator fields into a single matrix.

3.4 Cluster Analysis

In the realm of data science, data analysis involves examining, tidying, reshaping, and constructing models with data to unveil insights that guide conclusions and bolster decisions. In this study, cluster analysis was performed, to study the relationship between the spatial variables and to analysing the urban environment near open areas by examining the connections between the factors delineating previously identified urban characteristics, to explore the alignment between similar attributes and the relative positioning of open spaces. Cluster analysis is a statistical method for processing data by grouping the elements of a set, according to their characteristics, into classes that are not assigned a priori. Clustering or cluster analysis is used to classify objects, characterized by the values of a set of variables, into groups. It is therefore an alternative to principal component analysis for describing the structure of a data table. It is used to show relationships among data that are not apparent at first glance, to create homogeneous sets useful for further analysis.

4 First Results and Conclusions

The analysis conducted produced five distinct clusters, whose barycentres are shown in the Fig. 8 and the mapping in Fig. 9, each of which portrays different urban characteristics within the city of Naples. Cluster 1 is characterised by extensive green cover and open spaces, mainly in neighbourhoods object of planned processes, such as Scampia, Bagnoli, Chiaia and Ponticelli. These areas are characterised by mainly masonry buildings, thus of old construction, compared to the other built in more recent periods, with a lower percentage of vulnerable population aged 0 to 9 and over 60. In contrast, Cluster 2 has a high building density, small open spaces and limited green coverage, moreover consistent level of overlapping effect of the cooling areas.as the open spaces are localised in the stratified and consolidated districts of the city, with lower proportion of demographic population and vulnerable population, according to the presence of different urban services and commercial areas. These neighbourhoods, including San Lorenzo, Montecalvario, Pendino and Chiaia, suffer significant urban heat island effects (UHI) due to their consolidated urban fabric. Cluster 3 presents a more balanced distribution between open spaces and green areas, a characteristic common to most neighbourhoods in Naples, with the exception of the central districts. These areas have a lower demographic population, resulting in a lower number of vulnerable individuals. Cluster 4 is characterised by the presence of recently constructed buildings, together with a higher demographic population and a moderate proportion of vulnerable individuals. Mostly squares in established urban fabrics characterized by a medium to high built density value and distributed in different neighborhoods.

Finally, cluster 5 has a substantial proportion of medium-sized green cover and a diverse range of building materials. Despite the dense population, these areas also host a significant percentage of vulnerable individuals and are therefore more exposed to the effects of UHI. Overall, these distinct clusters offer insights into the urban fabric of Naples, providing a holistic understanding of its different neighbourhoods. Such insights are fundamental to the creation of targeted urban interventions aimed at improving the sustainability, resilience and overall quality of life of the city's diverse communities.

N. of elements	31	113	131	82	25	
Cluster	1	2	3	4	5	Average
Land Surface Temperature	0,213	0,381	0,265	0,333	0,256	0,290
Total population	0,366	0,171	0,580	0,628	0,652	0,480
Children	0,073	0,048	0,040	0,165	0,355	0,140
Old people	0,099	0,064	0,062	0,261	0,409	0,179
Brickwork	0,047	0,021	0,080	0,070	0,232	0,090
Reinforced concrete	0,044	0,073	0,048	0,081	0,629	0,175
Steel, wood etc.	0,030	0,005	0,051	0,261	0,246	0,119
Building coverage ratio	0,385	0,631	0,199	0,204	0,207	0,325
Combination effect	0,592	0,689	0,286	0,255	0,405	0,445
Green areas	0,392	0,010	0,075	0,060	0,060	0,119
Squares	0,655	0,961	0,976	0,971	0,986	0,910
Open spaces	0,545	0,027	0,085	0,073	0,066	0,159

Fig. 8. Average Barycentres Pivot Table

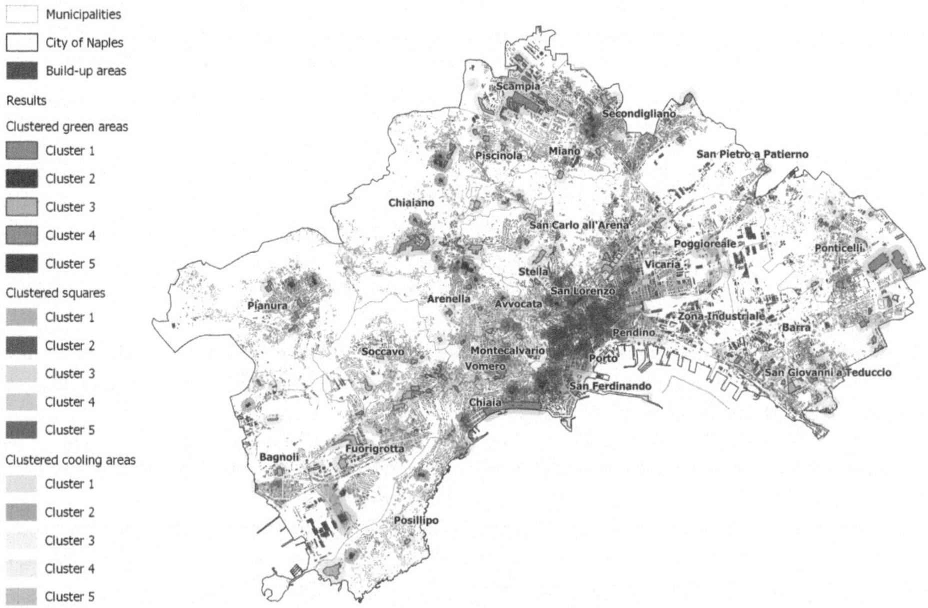

Fig. 9. Cluster Analysis according to the cooling effect areas, green areas, and squares.

These findings are a key step in a broader research effort that focuses on defining guidelines to achieve climate neutrality through increased utilisation of ecosystem services in urban settings. The policy implications of this study are significant, as the paper emphasises the need for integrated approaches to urban planning that simultaneously consider energy reduction and climate adaptation measures. By recognising the role of urban open spaces in increasing resilience, promoting energy savings and improving thermal comfort, policymakers can prioritise investments in green infrastructure and sustainable urban design. By understanding the unique characteristics of different neighbourhoods, policymakers can adapt strategies to improve sustainability and resilience where they are most needed. This approach can help mitigate the urban heat island effect, reduce vulnerability to climate change impacts and improve residents' overall quality of life. However, it is important to recognise the limitations of this research: although cluster analysis offers valuable insights, it may oversimplify the complexity of urban systems. Moreover, the focus on the city of Naples may limit the generalisability of the results to other urban contexts. Future research should aim to replicate these methods in different cities to validate their applicability in different contexts.

Acknowledgments. The research leading to these results has received funding from the National Recovery and Resilience Plan (NRRP), Mission 4 Component 2 Investment 1.3 - Call for tender No. 1561 of 11.10.2022 of Ministero dell'Università e della Ricerca (MUR); funded by the Euro-pean Union –NextGenerationEU. Award Number: Project code PE0000021, Concession Decree No. 1561 of11.10.2022 adopted by Ministero dell'Università e della Ricerca (MUR), CUPC93C22005230007, according to attachment E of Decree No. 1561/2022, Project title "Network 4 Energy Sustainable Transition – NEST" for the section 3.3, while the other ones are also related to the project "Definition of a guidelines handbook to implement climate neutrality by improving ecosystem service effectiveness in rural and urban areas" CUP E53D23018970001 funded by EU in the NextGenerationEU Plan through the Italian "Bando Prin 2022-D.D. 1409 del 14-09-2022".

References

1. Abubakar, I.R., Dano, U.L.: Sustainable urban planning strategies for mitigating climate change in Saudi Arabia. Environ. Dev. Sustain. **22**(6), 5129–5152 (2019). https://doi.org/10.1007/s10668-019-00417-1
2. Bai, X., et al.: Quantifying threshold and scale response of urban air and surface temperature to surrounding landscapes under extreme heat. Build. Environ. **247**, 111029 (2024). https://doi.org/10.1016/j.buildenv.2023.111029
3. Best, L., Schwarz, N., Obergh, D., Teuling, A.J., Van Kanten, R., Willemen, L.: Urban green spaces and variation in cooling in the humid tropics: the case of Paramaribo. Urban For. Urban Greening **89**, 128111 (2023). https://doi.org/10.1016/j.ufug.2023.128111
4. Bouketta, S.: Urban Cool Island as a sustainable passive cooling strategy of urban spaces under summer conditions in Mediterranean climate. Sustain. Cities Soc. **99**, 104956 (2023). https://doi.org/10.1016/j.scs.2023.104956
5. Carpentieri, G., Zucaro, F., Guida, C.: Urban energy consumption in the city of Naples (Italy): a geographically weighted regression approach. In: He, B.J., Jupesta, J., Pignatta, G. (eds.) CCES 2022, pp. 93–104. Springer, Cham (2022). https://doi.org/10.1007/978-3-031-46109-5_8
6. Carra, M., Caselli, B., Rossetti, S., Zazzi, M.: Widespread urban regeneration of existing residential areas in European medium-sized cities—a framework to locate redevelopment interventions. Sustainability **15**, 13162 (2023). https://doi.org/10.3390/su151713162
7. Carra, M., Rossetti, S., Tiboni, M., Vetturi, D.: Urban regeneration effects on walkability scenarios. An application of space-time assessment for the people-and-climate oriented perspective. TeMA – J. Land Use Mob. Environ. Spec. Issue **1**, 101–114 (2022). https://doi.org/10.6093/1970-9870/8644
8. Couret, D., Baldoquín, N.C., De La Paz Pérez, G.A., Guzmán, L.A.R.: Urban variables for adaptation to global warming in a hot-humid climate. Cuban cities as a case study. Urban Clim. **51**, 101633 (2023). https://doi.org/10.1016/j.uclim.2023.101633

9. Cruz, J.A., Blanco, A.C., Garcia, J.J., Santos, J.A., Moscoso, A.D.: Evaluation of the cooling effect of green and blue spaces on urban microclimate through numerical simulation: a case study of Iloilo River Esplanade Philippines. Sustain. Cities Soc. **74** (2021). Article 103184. https://doi.org/10.1016/j.scs.2021.103184
10. Darbani, E.S., Rafieian, M., Parapari, D.M., Guldmann, J.: Urban design strategies for summer and winter outdoor thermal comfort in arid regions: the case of historical, contemporary and modern urban areas in Mashhad Iran. Sustain. Cities Soc. **89**, 104339 (2023). https://doi.org/10.1016/j.scs.2022.104339
11. Das, M., Das, A., Momin, S.: Quantifying the cooling effect of urban green space: a case from urban parks in a tropical mega metropolitan area (India). Sustain. Cities Soc. **87**, 104062 (2022). https://doi.org/10.1016/j.scs.2022.104062
12. Diz-Mellado, E., López-Cabeza, V.P., Roa-Fernández, J., Rivera-Gómez, C., Galán-Marín, C.: Energy-saving and thermal comfort potential of vernacular urban block porosity shading. Sustain. Cities Soc. **89**, 104325 (2023). https://doi.org/10.1016/j.scs.2022.104325
13. Gargiulo, C., Lombardi, C.: Urban retrofit and resilience: the challenge of energy efficiency and vulnerability. TeMA – J. Land Use Mob. Environ. **9**, 137–162 (2016). https://doi.org/10.6092/1970-9870/3922
14. Gargiulo, C., Russo, L.: Cities and energy consumption: a critical review. TeMA – J. Land Use Mob. Environ. **10**(3), 259–278 (2017). https://doi.org/10.6092/1970-9870/5182
15. Gargiulo, C., Tulisi, A., Zucaro, F.: Climate change-oriented urban green network design: a decision support tool. In: Series on Computers and Operations Research, pp. 255–278 (2017). https://doi.org/10.1142/9789813200012_0011
16. Gargiulo, C., et al.: Effect of urban greenspaces on residential buildings' energy consumption: case study in a Mediterranean climate. Smart Plann. Sustain. Mobility Age Change, 109–125 (2018). https://doi.org/10.1007/978-3-319-77682-8_7
17. Gargiulo, C., Zucaro, F.: Greening networks for smart and resilient cities: from methodology to application in densely built urban Contexts. Archivio di Studi Urbani e Regionali, 130–148 (2020). https://doi.org/10.3280/asur2020-127-s1008
18. Gargiulo, C., Zucaro, F.: A method proposal to adapt urban open-built and green spaces to climate change. Sustainability **15**(10), 8111 (2023). https://doi.org/10.3390/su15108111
19. Gargiulo, C., Sgambati, S., Zucaro, F.: The analysis of the urban open spaces system for resilient and pleasant historical districts. In: Gervasi, O., et al. (eds.) ICCSA 2023 Workshops, pp. 564–577 (2023). https://doi.org/10.1007/978-3-031-37111-0_39
20. Gisotti, M.R., Masiani, B.: Promoting a local and just green deal. School open spaces as a strategic opportunity for the city in the ecological transition. TeMA – J. Land Use Mob. Environ. **1**, 97–114 (2024). https://doi.org/10.6093/1970-9870/10298
21. Gluch, R., Quattrochi, D.A., Luvall, J.C.: A multi-scale approach to urban thermal analysis. Remote Sens. Environ. **104**(2), 123–132 (2006). https://doi.org/10.1016/j.rse.2006.01.025
22. Graça, M., Cruz, S.S., Monteiro, A., Neset, T.: Designing urban green spaces for climate adaptation: a critical review of research outputs. Urban Climate **42**, 101126 (2022). https://doi.org/10.1016/j.uclim.2022.101126
23. Hölscher, K., Frantzeskaki, N.: Perspectives on urban transformation research: transformations in, of, and by cities. Urban Transform. **3**(1) (2021). https://doi.org/10.1186/s42854-021-00019-z
24. Lazzarini, L., Mahmoud, I., Pastore, M.C.: Urban planning for biodiversity. TeMA – J. Land Use Mob. Environ. **1**, 45–60 (2024). https://doi.org/10.6093/1970-9870/10197
25. Ma, X., Leung, T.M., Chau, C.K., Yung, E.H.: Analyzing the influence of urban morphological features on pedestrian thermal comfort. Urban Clim. **44**, 101192 (2022). https://doi.org/10.1016/j.uclim.2022.101192

26. Mohtat, N., Khirfan, L.: The climate justice pillars vis-à-vis urban form adaptation to climate change: a review. Urban Clim. **39**, 100951 (2021). https://doi.org/10.1016/j.uclim.2021.100951
27. Natanian, J., Auer, T.: Beyond nearly zero energy urban design: a holistic microclimatic energy and environmental quality evaluation workflow. Sustain. Cities Soc. **56**, 102094 (2020). https://doi.org/10.1016/j.scs.2020.102094
28. Nguyen, C.T., Nguyen, D.T.H., Phan, D.K.: Factors affecting urban electricity consumption: a case study in the Bangkok metropolitan area using an integrated approach of earth observation data and data analysis. Environ. Sci. Pollut. Res. **28**, 12056–12066 (2021). https://doi.org/10.1007/s11356-020-09157-6
29. Nimac, I., Bulić, I.H., Žuvela-Aloise, M.: The contribution of urbanisation and climate conditions to increased urban heat load in Zagreb (Croatia) since the 1960s. Urban Clim. **46**, 101343 (2022). https://doi.org/10.1016/j.uclim.2022.101343
30. Papa, C.G., Zucaro, F.: Urban systems and energy consumptions: a critical approach. TeMA J. Land (2014)
31. Papa, R., Gargiulo, C., Zucaro, F.: Urban systems and energy consumptions: a critical approach. Tema J. L Use Mob. Environ. (2014). https://doi.org/10.6092/1970-9870/2552
32. Papa, R., Gargiulo, C., Zucaro, F.: Towards the definition of the urban saving energy model (urbansem). In: Smart Energy in the Smart City, pp. 151–175 (2016). https://doi.org/10.1007/978-3-319-31157-9_9
33. Liu, W., et al.: Green space cooling effect and contribution to mitigate heat island effect of surrounding communities in Beijing Metropolitan Area. Front. Public Health **10** (2022). https://doi.org/10.3389/fpubh.2022.870403
34. Patle, S., Ghuge, V.V.: Urban fragmentation approach for assessing thermal environment dynamics: a case study of semi-arid city from a comfort perspective. Urban Clim. **53**, 101784 (2024). https://doi.org/10.1016/j.uclim.2023.101784
35. Pigliautile, I., D'Eramo, S., Pisello, A.L.: Intra-urban microclimate mapping for citizens' wellbeing: novel wearable sensing techniques and automatized data-processing. J. Clean. Prod. **279**, 123748 (2021). https://doi.org/10.1016/j.jclepro.2020.123748
36. European Environment Agency (EU body or agency), et al.: Urban Sustainability in Europe, Publications Office of the EU (1970). https://op.europa.eu/en/publication-detail/-/publication/a5e8d85f-f68b-11ec-b976-01aa75ed71a1/language-en. Accessed 25 Mar 2024
37. Resilient City Planning Framework. | download scientific diagram (No date). https://www.researchgate.net/figure/Resilient-city-planning-framework_fig1_266735023. Accessed 25 Mar 2024
38. Quan, S.J., Li, C.: Urban form and building energy use: a systematic review of measures, mechanisms, and methodologies. Renew. Sustain. Energy Rev. **139** (2021). https://doi.org/10.1016/j.rser.2020.110662
39. Roest, A., Weitkamp, G., Van Den Brink, M., Boogaard, F.: Mapping spatial opportunities for urban climate adaptation measures in public and private spaces using a GIS-based decision support model. Sustain. Cities Soc. **96**, 104651 (2023). https://doi.org/10.1016/j.scs.2023.104651
40. Semeraro, T., Scarano, A., Buccolieri, R., Santino, A., Aarrevaara, E.: Planning of urban green spaces: an ecological perspective on human benefits. Land **10**(2), 105 (2021). https://doi.org/10.3390/land10020105
41. Scheiber, S., Zucaro, F.: Urban open and green spaces: is Malta planning and designing them to increase resilience? TeMA – J. Land Use Mob. Environ. **16**(2), 331–352 (2023). https://doi.org/10.6093/1970-9870/9951
42. Shen, Y., Lin, Y., Cheah, W., Lin, J., Li, X., Li, Y.: Influences and pathways of urban form and lifestyle on carbon emission reduction. Urban Clim. **46**, 101325 (2022). https://doi.org/10.1016/j.uclim.2022.101325

43. Solecki, W., et al.: A conceptual framework for an urban areas typology to integrate climate change mitigation and adaptation. Urban Clim. **14**, 116–137 (2015). https://doi.org/10.1016/j.uclim.2015.07.001
44. Wang, M., Wang, W., Wu, L.: Application of a new grey multivariate forecasting model in the forecasting of energy consumption in 7 regions of China. Energy **243**, 123024 (2022). https://doi.org/10.1016/j.energy.2021.123024
45. Wang, M., Gou, Z.: Gaussian Mixture Model based classification for analyzing longitudinal outdoor thermal environment data to evaluate comfort conditions in urban open spaces. Urban Clim. **53**, 101792 (2024). https://doi.org/10.1016/j.uclim.2023.101792
46. Wolfram, M., Frantzeskaki, N., Maschmeyer, S.: Cities, systems and sustainability: status and perspectives of research on urban transformations. Current Opin. Environ. Sustain. **22**, 18–25 (2016). https://doi.org/10.1016/j.cosust.2017.01.014
47. Woo, Y.E., Cho, G.H.: Impact of the surrounding built environment on energy consumption in mixed-use building. Sustainability (Switzerland) **10**(3) (2018). https://doi.org/10.3390/su10030832
48. Yao, J., et al.: Dry deposition effect of urban green spaces on ambient particulate matter pollution in China. Sci. Total. Environ. **900**, 165830 (2023). https://doi.org/10.1016/j.scitotenv.2023.165830
49. Zhang, Q., Zhou, D., Xu, D., Rogora, A.: Correlation between cooling effect of green space and surrounding urban spatial form: evidence from 36 urban green spaces. Build. Environ. **222**, 109375 (2022). https://doi.org/10.1016/j.buildenv.2022.109375
50. Zou, Y., et al.: Comprehensive analysis on the energy resilience performance of urban residential sector in hot-humid area of China under climate change. Sustain. Cities Soc. **88**, 104233 (2023). https://doi.org/10.1016/j.scs.2022.104233

Description of a Pilot Digital Interactive Tool of Territorial Resource' Spatial Projection and Management

Stelios Kouzeleas[1](✉) and Dimitris Goussios[2]

[1] Department of Interior Architecture, International Hellenic University, Serres, Greece
stelios_kouzeleas@yahoo.fr
[2] Department of Planning and Regional Development, University of Thessaly, Volos, Greece

Abstract. A resource as a virtual structure that is directly connected to the space (territorial resource) is synthesized as a multitude of items – ingredients of material or immaterial dimension. The material ingredients have a tangible substance, such a permanent element of nature or a human construction, such as landscape, water, monuments, paths, etc., while the immaterial ingredients deal with any intangible notion such as heritage, sustainability, prosperity, historical event, social relationships, etc. The properties of both material and immaterial ingredients connection of a resource are dramatically important to describe and highlight a territorial resource. This connection should be associated with a spatial projection/visualization allowing actors (decision makers, training local policy, citizens, etc.) who are involved in a process of construction territorial resources to better understand the space and to consult in a more fertile manner in a participatory territorial resource's management process. The research deals with the theoretical description of a digital tool of territorial resources' representation and management that will combine modern simulation technologies and will investigate the search of a functional relationship between virtual environment, territorial resource' elements and local actors' institutional learning cognitive needs, leading in this way, to enhanced sustainable management of territorial resources. Finally, the proposed tool should help the process of identifying and highlighting the territorial resource, the active participation of local actors and the collective local memory in order to project in space the connection of the material/immaterial elements of the territorial resource.

Keywords: Territorial resource's representation · Spatial territorial simulation tool · Territorial resource's management · Consumer's participation · Social learning

1 Introduction

In recent years, in a globalized environment, many regions seeking competitiveness through distinctiveness [1] and benefiting from the shift of consumers towards quality and authenticity, invest in them. All these dynamics lead to the transition from raw

O. Gervasi et al. (Eds.): ICCSA 2024 Workshops, LNCS 14818, pp. 430–442, 2024.
https://doi.org/10.1007/978-3-031-65273-8_28

material to territorial resource (TR) signaling a significant change in the relationship of the produced good or service with the place of origin. The TR, in contrast to the raw material, results from a series of material and immaterial elements anchored in a place [2, 3]. The material (tangible) elements are linked, specifically to agri-food and local alternative tourism, to the soil, plant varieties, animal breeds, processing and packaging materials, etc. The immaterial (intangible) elements refer to heritage [4, 5], practices, knowledge and know-how, in what defines the relationship of local culture with the reproduction and exploitation of the resource.

It becomes obvious that the effort to highlight the uniqueness of the resource and the final product encounters difficulties when is going to project the connection of the material with the immaterial nature of the constituent elements of the resource. For example, if anchoring in space and place is easy to demonstrate (surface, building, cultivation, etc.), the integration of living heritage and local culture is more difficult as it refers to knowledge and know-how that is often informal. International literature supports the view that ultimately the resource does not exist but is constructed by local actors cleverly incorporating heritage through a combination of elements and is therefore virtual [23].

Therefore, both the difficulty of visualizing the connection of the material and immaterial dimension of the resource and the fact that the target group is consumers, argue for the need to support the identification process with special claim guarantee tools.

These tools use new visualization technologies and are part of Participatory Guarantee Systems (PGS) in which local and non-local actors, public services, etc. are activated [6–8]. These systems are usually aimed at uninformed consumers but are also used by local actors to coordinate the management of the resource. In both cases, these systems work in a framework and a learning process for local actors (control and strengthening of guarantee instruments) and consumers (understanding the nature of the elements and the importance of their interconnections within the resource). Finally, active participation in a PGS creates trust, transparency, reciprocity between producers, products and consumers [9–11] while facilitating consumer knowledge [21].

The concept of virtualization is not an abstract notion but it attempts to connect the tangible and intangible dimension of the resource and therefore the liaison of the imprints in the space (territory, monuments, artisanal workshops, landscapes) with the accumulation of an intangible heritage (history, know how, natural resources management practices, etc.). That due to its contact with local economy acquires a new status and a new function in the production process that is proposed by a group of local actors. The introduction of inheritance and anchorage define the territorialization procedure that should be based in a parallel institutional learning process of the actors of these areas aiming at forming new forms of cooperation and 'governance'.

Therefore, new methods and tools should combine the spatial and territorial dimension with the virtual (intangible) nature of the resources and also contribute to the identification and attribution of new properties and uses in these resources. A TR' management system, should, inter-alia:

(1) represent the area that the resources and the actors refer to,
(2) integrate local actors and support them by facilitating observation and identification,

(3) contribute to the interconnection of tangible and intangible ingredients of the resource by elaborating and evaluating any information compatible with the market,
(4) facilitate the distribution of roles within cooperation and coordination forms and draw an action plan for the promotion and utilization of these resources.

The consolidation of a future procedure-methodology for the management or even the construction of TR with the necessary spatial projection of the relations "tangible-intangible" and the correlations of the ingredients of a territorial resource requires, among others, the creation of a digital interactive tool that will be able to simulate spatial entities and enable both their recording and their structure as well as their enrichment by participants.

2 Project Framework

This research work is being proposed on the occasion of 'Terra Thessalia' PGS project, that has been published previously by the same authors [21]. 'Terra Thessalia' cluster was created in 2016 in the framework of Lactimed European program [2012–2015] [34] and aims to strength the production and distribution of innovative dairy products in the Mediterranean and proposes a PGS managed by a territorial administration via an organization of local industries [35]. 'Terra Thessalia' PGS's objectives are, inter alia, to:

(1) enhance more territorialized specifications
(2) highlights the quality and the identity of the final diary product of Feta cheese
(3) integrate Feta cheese to the pastoral resource
(4) brings to the surface the question of the transmission of the qualitative and identity characteristics of the territorial product
(5) create an alternative system of guarantee - certification – traceability
(6) enhance small-scale productions and the local economy
(7) encourage the participation of consumers by sharing knowledge, experiences with producer groups [28, 36], based on the principles of participation, reciprocity, transparency and trust [9–11].

Along with the creation and development all guarantee - certification – traceability processes of 'Terra Thessalia' PGS, a pilot interactive visualization tool of PGS tool was developed. This tool visualizes spatially all the PGS mechanisms from the production process of a diary product, including the whole traceability process, the attachment ('anchorage') of the local product with the unique territory (territorialization) and TR's aspects, until the consumer's participation and the final specific guarantee nature product (guarantee certification of a process, animal feed, animal milk quality, remote monitoring GPS information of animal path and location, etc.) [21].

This work, as it focuses on the management and spatial projection of TRs and their intangibles / tangibles ingredients correlation in order to contribute and facilitate the description of a future platform of TR's construction with consumer's participation, has a complementary research approach to the PGS pilot visualization tool.

The proposed visualization tool description is based on an interactive methodology [5] on TRs construction mechanism and procedure allowing the guarantee of the resource's bonds with space (spatial bonds, social, cultural, economic, etc.). The above methodology refers mainly to the identification, registration and correlation analysis of the tangible-intangible ingredients as a TR construction tool for local stakeholders; one of the main characteristic and further objective of the above methodology should be the detection/development of a mechanism process allowing its adaptation to the corresponding process of spatial projection and visualization tool by correlating the corresponding process / methodology steps.

Thus, the description and the enhancement of a TRs management tool via spatial simulation remains absolutely necessary, irreplaceable and integral part of a holistic TRs management and construction tool, as it should facilitate and contribute drastically to an institutional learning of the territories' actors (decision makers, producers, consumers, etc.).

3 Territorial Resources Aspects

3.1 Territorial Resource's Notion

The territorial approach of the rural area's development focuses on the economic dimension of the construction and emergence of territories and territorial resources. The Territorial Resource is a territorial development model based on a self-organizing movement aiming to:

(1) attach (anchor) a resource (agrifood product, historical event, monument, etc.) to the local territory by introducing a relevant inheritance,
(2) search competitiveness through the distinctiveness and the specific character of the manufactured resources and not through the reduction of the production cost,
(3) favorize a positive external environment created by decentralization,
(4) enhance public motivation policies
(5) create new expectations and institutional learning process of an increasing part of the consumers by forming new cooperation models

Thus, the operational objective that this issue tends to form is based on the territorial dimension that the local resource / product is now assuming due to the bonds with its production area and the incorporation of heritage [4].

3.2 Tangible and Intangible Ingredients

The component's unit of a TR is in essence the element that make up the central 'link' of the whole chain (eg. Agri-food chain). Each 'link' consists of its own corresponding.

units of TRs components. Every TRs 'link' refer to a production chain, structure and process (production unit). Every production unit consists of main components or sub-components associated with intangibles/tangibles ingredients that characterize and enhance it.

For example, in the case of 'Terra Thessalia' PGS the main 'links' of agri-food chain are 'Geo-Production Unit' and 'Artisanship unit' which in their turn consist of main components (such as 'Livestock', 'Grazing Feedstuffs', 'Space Milk', 'Collection Product', etc. Each component, eg. 'Livestock' consist of sub-components such as 'Herd', 'Corral', 'Animal breed' etc. Every sub-component is associated with intangibles/tangibles ingredients. On one hand, the tangibles ingredients have to do with any element that has a tangible substance, is a permanent element of nature, such as landscape, water, flora, microclimate, etc. or even a tangible human construction, such as buildings, paths, bridges, monuments, etc. On the other hand, the intangibles ingredients have to do with any immaterial element or a notion such as heritage, sustainability, prosperity, historical event, social relationships, etc.

The intangibles ingredients can be supported, completed or mainly fall into a central category of intangibles "Principles and Values", such as "Legacy", "Proximity", "Quality", "Solidarity" and "Tradition" etc.

The spatial projection (simulation) and management of TRs components are based mainly on some correlations assuming some main criteria. Any inter-correlation between TRs unit, component assuming a tangible or intangible ingredient can be a key feature of TR spatial projection in the visualization component of the proposed tool and a search item in the management component of the same tool. The proposed TRs visualization and management tool have to be adapted to a given methodology of a TRs construction tool where all conditions, criteria, components and their correlations are identified. So, the main conditions, criteria under which the tool can operate, are:

(1) Both intangibles/tangibles' ingredients should be anchored in space
(2) Incorporation of main components in TRs units
(3) Decomposition of main TRs components in tangibles/intangibles ingredients
(4) Create correlations between TRs components and their ingredients
(5) Create intercorrelations between tangibles/intangibles ingredients
(6) Linking/completion of intangibles ingredients with 'Principles and Values'.

4 Related Systems – Discussions

Territorial Management (TM) is a mode of governance considering actions on social cohesion, economic strategies, environmental protection and their feasibility for the development of an area [12] and is also a mode of public management of these actions [13].

A plethora of territorial management and local development tools aim to encourage the economic and social growth of the territory while considering its specificities and resources, in order to create a development process engaging all local stakeholders [12].

Many 3D spatial representations systems allow the dynamic territorial management simulation with social features [14], while other systems, such as Dynamic Spatial Simulation (DSS) and Participatory System Dynamics modeling (PSD) integrate programming solutions [15] and computer models to simulate territorial changes [16] and to integrate stakeholders into a deliberation process [14].

According to the international literature, rarely, TR's digital spatial simulation tools project TR's ingredients and elements and their interconnection into a given space so that stockholders can perceive the spatial dimension of all ingredients [17] while others project a general territorial space with diverse simulation techniques, such as agent-based modeling, cellular automata, GIS, spatial statistics, real-time remote spatial simulation, etc., allowing researchers and stakeholders to evaluate policy, impact assessment and undertake scenario analysis [18]. Other user-friendly digital tools and visualization methods [19] simulate elements and forms' entries in the space by creating anchors to space and the identity of TR's characteristics [20].

All these spatial simulations tools project the TR's elements into a space or even their interconnection between them based on diverse simulation techniques while they do not focus on the spatial representation and management of TR. This work, highlights the need to describe and design an interactive tool for the spatial simulation and management of TR facilitating in this way the integration, awareness and collaboration of diverse stakeholders into a participative process.

5 Objectives and State of the Art

This research promotes "bottom-up" participation of local actors in a participatory procedure of territorial development, supporting democratic planning and combining at the same time interactive simulation, social research and IT tools with territorial development approaches. This work focuses on digital tool definition that combines simulation of visual spatial information and supports multi-partner procedures and participatory planning of TR. It also promotes a functional connection, correlation and projection of the relationship between tangible and intangible elements functioning as ingredients of a TR in the space.

The final purpose of the research is for the simulation procedure to function not only as an information and visualization tool of the TR's ingredients but also to constitute a tool a) that allows to the consumer-user the interactive enrichment of its needs (information retrieval, online consultation-forum) and their correlation with the relevant spatio-productive sections and the resources contents promoting a social learning and b) visualization that allows constant awareness, education and territorial correlation of development procedures.

The coexistence in an online platform of a multi-environment's complete spatial simulation with correlation and registration of TR's ingredients in a database constitutes an innovative territorial resource's management approach. The spatial projection of the TR's tangible and intangible ingredients as innovative approach can create new uses, properties and development actions to integrate new production systems aiming at the creation of new specific products. The proposed digital tool can contribute in an institutional social learning and participation procedure on TR's management.

Finally, an important contribution field is related to the formation of new cooperation and coordination forms (governance). An operational implementation of the proposed tool could structure and contribute to an optimal development actions governance model.

6 Digital Tool Description of Territorial Resources' Spatial Projection and Management

6.1 Introduction

The definition of a TR' spatial representation and management tool presuppose an identification, a structure and a final spatial projection of the main ingredient's units of the TR's links which are in essence the elements that make up the main links of the

TR' structure. A similar structure of TR' identification can be found in Participatory Guarantee Systems (PGS) visualization where every TR's ingredient is associated with tangible and intangible elements that characterize it [21]. The tangible as well as the intangible elements are correlated ("anchored") in the space, both to each other and to corresponding elements of other ingredients.

Each element as a key feature of identifying and searching a territorial resource, corresponds to a series of multimedia messages (representative images, maps, appropriate descriptions, etc.) which refer to the TR's structure. All these multimedia messages contribute to the simulation of a TR's representation/management tool so that consumers, in their turn contribute to the ongoing feedback and adjustment of TR's structure itself. The correlation of all TR's ingredients usually is based on their relationship in the field, the ability for visualization, the involvement of intangible elements into TR and the ability to attribute new elements' properties [21].

The selection and the visualization of TR's ingredients will contribute to the knowledge and ability of the consumers to recognize the value of the ingredients' correlation and facilitate the consumer integration [22].

6.2 Consumers'/Stakeholders' Participation

The consolidation of a procedure for the TR's management with the necessary spatial projection of the relations "tangible-intangible" and the correlations of the ingredients of a TR requires an active participation of consumers and other stakeholders in all phases of the TR's simulation and management. Researches showed limited awareness and knowledge of consumers on practices connecting TR's ingredients such as production processes and quality attributes [22]. PGS are facilitated by interactive visualized knowledge [23] of tangible and intangible TR's ingredients raise awareness while helping and sharing know-how and good practices [22]. This consumer's integration (1) prepares their future "physical" and/or "virtual" engagement in a bottom-up collaboration for the ongoing negotiation of trust and social control built [22] and (2) supports the importance of social learning and education in establishing trust in certification and guarantee process [24, 25].

The interpretation of the correlation among tangible and intangible features of the TR has a significant effect on consumer perception regarding quality and comprehension of virtual representations by specific digital spatial tools [26]. This increases awareness of the TR's management tool and highlights its benefits while the visualization contributes to the social learning process making more understandable to consumers the complex tangible and intangible relations between local resources, consumers and stakeholders [22].

6.3 Theoretical Approach

The spatial simulation of territorial areas for the purpose of their social and economic development is often found in decision support systems (DSS) while they include, among others, statistical models, forecasting, database storage, etc., as well as the results of digital spatial representation tools [27]. Other systems for forecasting socio-economic development of territorial areas use and apply dynamic spatial simulation models [28] and highly sophisticated "Agent-based modeling" and simulation systems [29].

The description of the proposed pilot digital tool is not a classic case of a decision support system since it aims to activate the participation of local actors in the management of a territorial resource [1, 3]. Thus, it places great importance on the creation of a comprehensive digital spatial simulation and interactive environment, while at the same time the creation of a database would be particularly supportive of descriptions and deliberations around the central core of spatial simulation.

The proposed tool is based on the connection and correlation of the tangible and intangible ingredients of a TR as well as the spatial projection and simulation of all these elements and correlations in space [21].

The correlation and spatial projection of the TR's links with the space creates "anchors" with spatial references which then give new uses to the TR's ingredients and allow corresponding actions to be defined by respective participants with specific roles in order to integrate them into productive systems (e.g. Creative Tourism, Agrifood production, etc.) and to create innovative idiosyncratic local products and services (see Fig. 1).

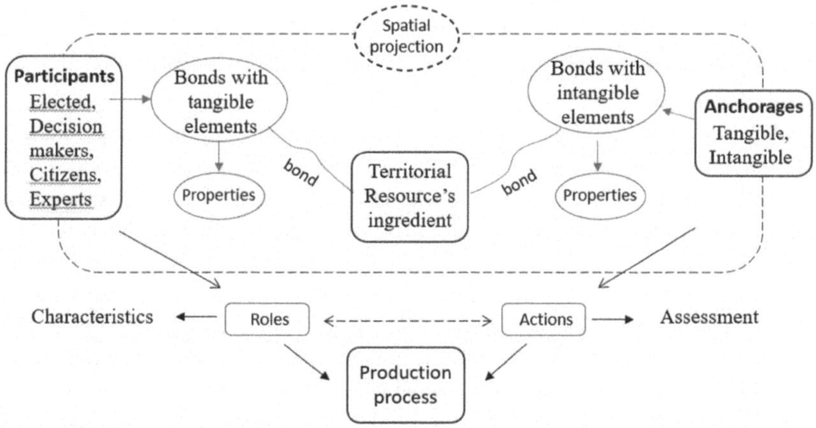

Fig. 1. Schematic representation of spatial projection of Territorial Resource's ingredients management.

6.4 Technical Approach

First, the tool should allow users to be aware of issues of TR's management by presenting either schematically or interactively the respective methodology by operating in an online environment that allows multimedia and interactive representation.

The pilot digital tool should then facilitate and contribute to the participatory deliberation processes of all involved stakeholders. Thus, each participant must be able to consult around a platform, enabling:

(1) a spatial representation of TR's ingredients contributing to the common spatial perception and creating a "common language" of communication, in combination with
(2) the possibility of enriching the TR's ingredients simulation with comments, judgments, descriptions, deliberation results from the participants and of course recording and connecting all of them to an ergonomic database.

To achieve this, the environment of the proposed digital tool, first should allow, among others:

(1) complete and in some cases dynamic simulation of digital background with geographic reference, image and video management (scaling, interaction, deletion, etc),
(2) representation of interactive multimedia applications,
(3) embedding of web spatial multimedia simulation results (GIS background, DTM, 3D models, VR, Virtual tour, Real-time spatial navigation, etc.) as well as the possibility of online hyperlinks of all multimedia elements [30]. At the same time, all simulation data should be able to be structured and linked together in a descriptive and graphical database. It is obvious that the techniques and procedures of 3D modeling and simulation of spatial entities are multiple and will be implemented outside the tool environment using appropriate software and embedded techniques.

The possibility of spatial highlighting and projection of the TR's ingredients links via tangible/intangible elements connection should allow the creation of an environment of institutional learning of the participants [22]. Thus, the proposed tool should use spatial digital simulation techniques and tools to make decisions and manage complex relationships allowing easier understanding of issues and provide the basis for learning conversations [31].

At the same time, the ability to connect each graphical element (spatial view) with a descriptive element, should be recorded in a database. This will allow the retrieval of any type of data (descriptive, graphical, multimedia) based on "smart" queries. Nevertheless, it should be noted that today at the global level the assembly process of graphic and descriptive data involves many problems related to both interactivity and time consumption [32]. Nevertheless, the better use, adaptation and visualization of interdisciplinary spatial data are required as they are directly related to the better presentation of the simulated models as well as the ergonomic operation by the end users [33]. Therefore, the proposed tool tends to put more emphasis on the ability of the environment to integrate digital spatial simulations that will be implemented in suitable external software, rather than on the ability to create an intelligent database that interactively relates graphical and descriptive data.

Finally, the pilot tool should allow the structure of all TR's ingredients in such a way that interconnected and interdependent linked data groups can be created. These data groups should allow:

(1) the creation of new uses of the TR's ingredients and their inclusion in production processes,
(2) a governance model creation of development actions with assessment possibility. To implement this, the proposed online platform should be based on an intelligent and

flexible database (SQL data base) allowing both the data composition into groups as well as the decomposition of data groups into data ingredients by the users themselves (see Fig. 2).

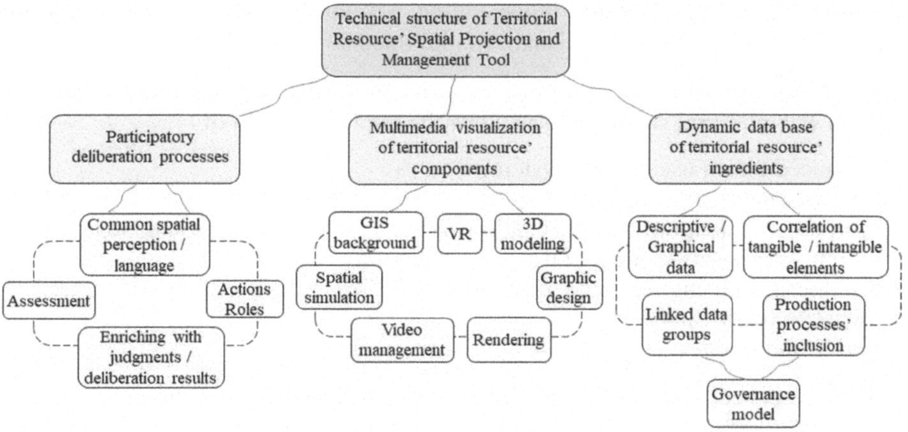

Fig. 2. Technical schematic structure of territorial resource' spatial projection and management Tool.

6.5 Tool's Architecture and Requirements

The proposed tool should be implemented through the creation of a digital online combinatorial platform (e.g. open-source technology) allowing simultaneously 3 main interconnected features' management, such as Multimedia visualization of TRs ingredients, a dynamic online TRs database and finally an online consumer's participatory deliberation forum.

This web environment could be supported by a web server, such as apache 2.4 with mode rewrite and PHP 7.4 with extensions enabled hosts SQL server, JAVA and HTML applications.

Regarding to visualization environment, the web platform (as the above web open-source technology supports), should at least allow the implementation of

(1) 360° virtual tour plugins
(2) 360° video plugins,
(3) multimedia interactive applications in HTML5 and JavaScript,
(4) GPS tracking web technology with corresponding real-time geolocation sharing plugins via web browser allowing instantly view historical tracks and statistics, get notifications and alerts via remotely control devices.

Regarding to the online dynamic TRs database, the above proposed web platform should at least allow the implementation of

(1) Data tables plugins for jQuery supporting any database table size

(2) Form builder creating data apps for back-end and front-end
(3) Query Builder to write SQL queries,
(4) Remote database access to server typers (e.g. MariaDB, MySQL, SQL Server, PostgreSQL, Oracle, MS Access, etc.)
(5) Auto sync CSV, Google Sheets, JSON and XML files, and
(6) Master-detail pages, lookups, inline editing, advanced search features and dynamic hyperlinks connected to the visualized multimedia ingredients.

Regarding to the online consumer's participatory deliberation forum, the above proposed web platform should at least allow the implementation of

(1) Advanced forum user profile system
(2) Live user notification system,
(3) Multi-board and multi-language, allowing separate forum pages and languages,
(4) Forum Access sets (per user / user group / forum)
(5) Multiple search engine (topic, post, tag, etc.)
(6) Social Network Share
(7) Rating and Badges

All interconnected features should interoperate so that every database query's result is linked to the corresponding spatial projected multimedia tangible/intangible ingredient and all its information. The database query should be attempted either through a TRs components structure chart layout or through a unique random search into a classical search field.

7 Conclusions

A description/definition of a digital tool of TR's ingredients representation and management is essential to the territorial diagnosis as it should allow to consolidate a spatial interconnection of TR's ingredients with the territory by the use of modern spatial simulation technology. All data elements will be registered into a descriptive and graphical embedded data base facilitating communication between experts and non-experts' stakeholders.

The theoretical description of the proposed tool facilitates the integration of consumers/users/stakeholder in a common deliberation platform by identifying cooperation and coordination roles and defining action plans for decisions makers.

The impact of such a digital tool should contribute to (1) a better understanding by the local actors of the necessary procedures for the valorization of local resources, (2) the successful completion of actions to increase the value of the goods and services produced and (3) ensure high competitiveness based on the uniqueness of the resources (products).

Finally, the implementation of a TR's representation and management tool which in the future could be transformed as a TR's construction tool should facilitate a more effective application of scientific and research knowledge in sectors and fields that directly concern the dynamic groups of actors and local rural communities by creating convenient conditions for the meeting of traditional knowledge with new scientific knowledge and contributing to enhanced supply of ecosystem services and sustainable management of territorial resources.

References

1. Colletis, G., Pecqueur, B.: Révélation de ressources spécifiques et coordination située. Économie et Institutions, 1er et 2nd semestres, pp. 51–74 (2004)
2. Fonte, M.: Food consumption as social practice: solidarity purchasing groups in Rome Italy. J. Rural. Stud. **32**, 230–239 (2013)
3. Renting, H., Marsden, T., Banks, J.: Understanding alternative food networks: exploring the role of short food supply chains in rural development. Environ Plan A **35**, 393–411 (2003)
4. Goussios, D., Anthopoulou, Th.: Relocalisation de la feta AOP par la coopération des petits territoires laitiers en Thessalie: de la coordination de la filière à la coordination territoriale. Pecqueur, B., and Glon, E., Eds, PUR (Presses Universitaires de Rennes), Rennes (2016)
5. Goussios, D., Faraslis, I.: The driving role of 3D Geovisualization in the reanimation of local collective memory and historical sources for the reconstitution of rural landscapes. Land **12**(2), 364 (2023)
6. Raynolds, L.T.: The globalization of organic agro-food networks. World Dev. **32**(5), 725–743 (2004)
7. IFOAM: Definition of Participatory Guarantee Systems. In: International Federation of Organic Agriculture Movements, IFOAM, pp.1, Bonn Germany (2008)
8. Anthopoulou, T.: Geographical indications and territorial development potentials in Greece. The difficult emergence of collective action. In: Proceedings On the locality and Specificity of Foods. A Territorial Approach to the Development of Rural Areas, pp. 255–279. Papazisis Eds, Athens (2013)
9. Weatherell, C., Tregear, A., Allinson, J.: In search of the concerned consumer: UK public perceptions of food, farming and buying local. J. Rural. Stud. **19**, 233–244 (2003)
10. IFOAM: Participatory guarantee systems. In: International Forum for Organic Agriculture Movements, IFOAM (2005)
11. Paltrineri, R., Spillare, S.: Collaborative consumption and trust building processes in the emerging new food economy. In: Proceedings on Contemporary Collaborative consumption. Trust and Reciprocity revisited, pp. 131–152. Cruz, I., Ganga, R. Wahlen, S. eds., (Springer Fachmedien Wiesbaden GmbH) (2018). https://doi.org/10.1007/978-3-658-21346-6_7
12. Belbachir Wafa, A.: Territorial management: a tool for local development. In: International Conference on Management Economics and Business 2023, IConMEB, vol. 32, pp. 80–84 (2023)
13. Duez, P.: Management territorial des risques et prospective territoriale. Marchés et organisation **2**, 141–169 (2009)
14. Binder, B., Vogl, C.R.: Participatory guarantee systems in Peru: two case studies in Lima and apurímac and the role of capacity building in the food Chain. Sustainability **10**(12), 4644 (2018)
15. Muessner, R., Sousa Pinto, I.: How to halt biodiversity loss? The need for science policy interfaces. Rom. Biodivers. Res. **2**, 11–16 (2005)
16. Scorm. SCORM Explained 201: A deeper dive into SCORM. Homepage, https://scorm.com/scorm-explained/. Accessed 5 Feb 2024
17. Nikolaidou, S., Kouzeleas, S., Goussios, D., Goulas, A.: Participatory Guarantee Systems for small farms and local markets: involving consumers in the guarantee process, In: XXVIII European Society for Rural Sociology Congress 2019, ESRS2019, Trondheim, Norway (2019)
18. Kouzeleas, S.: Real-time remote monitoring urban and architectural 3D digital design and spatial simultaneous simulation. Int. J. Adv. Res. **6**(12), 360–377 (2018)
19. Kouzeleas, S., Mammou, O.: Architectural, urban digital design and spatial simulation tools in digital cities cartography: Contribution in spatial design and perception. Int. Am. J. Contemp. Res. (IAJCR) **2**(8), 237–256 (2012)

20. Rallet, A., Torre, A.: Proximité et localisation. Economie Rurale **280**, 24–41 (2004)
21. Kouzeleas, S., Nikolaidou, S., Goussios, D., Goulas, A.: Pilot interactive visualization tool of a participatory guarantee system: the case of 'Terra Thessalia''s PGS. Int. J. Eng. Innovative Technol. (IJEIT) **9**(9), 1–17 (2020)
22. Nikolaidou, S., Kouzeleas, S., Gousios, D.: A territorial approach to social learning: facilitating consumer knowledge of local food through participation in the guarantee process. Sociologia Ruralis J. **63**(1), 66–88 (2022)
23. Campagne, P., Pecqueur, B.: Le développement territorial. Une réponse émergente a la mondialisation. Charles Léopold Mayer Eds (2014)
24. Onozaka, Y., Nurse, G., McFadden, D.T.: Local food consumers: how motivations and perceptions translate to buying behavior. Mag. Food Farm Resour. Issues **25**(1), 1–6 (2010)
25. Slade, P., Michler, J.D., Josephson, A.: Foreign geographical indications, consumer preferences, and the domestic market for Cheese. Appl. Econ. Perspect. Policy **41**(3), 370–390 (2019)
26. Kouzeleas, S.: Architectural and urban spatial digital simulations. Int. J. Eng. Model. **36**(2), 95–118 (2023)
27. Lychkina, N., Shults, D.: Simulation modeling of regions' social and economic development in decision support systems. In: III International conference 'System identification and control problems' 2004, Moscow (2004)
28. Barkalova, T.G., Kokotkina, T.N., Tsaregorodtsev, E.I.: Challenges and opportunities of using simulation models in forecasting socio-economic development of the region. Int. J. Environ. Sci. Educ. **11**(7), 10661–10671 (2016)
29. Macal, C.M., North, M.J.: Tutorial on agent-based modeling and simulation. In: Winter Simulation Conference 2005, IEEE, Orlando Florida USA (2005)
30. Kouzeleas, S.: Computational design contributions of integrative architectural and urban digital design methodology based on satellite images. Int. J. Res. Archit. Constr. PARC **2**(7), 86–101 (2011)
31. Klerkx, L., van Mierlo, B., Leeuwis, C.: Evolution of systems approaches to agricultural innovation: concepts, analysis and interventions. In: Darnhofer, I., Gibbon, D., Dedieu, B. (eds.) Farming Systems Research into the 21st Century: The New Dynamic. Springer, Dordrecht (2012). https://doi.org/10.1007/978-94-007-4503-2_20
32. Callahan, S.P.: VisTrails: visualization meets data management. In: International Conference on Management of data 2005, ACM SIGMOD '06, pp. 745–747. Chicago USA (2006)
33. Matthiesa, M, Giupponib, C., Ostendorfc, B.: Environmental decision support systems: current issues, methods and tools. Environ Model Softw. **22**(2), 123–127 (2007)
34. Doleisch, H., Gasser, M., Hauser, H.: Interactive feature specification for focus context, visualization of complex simulation data. In: Joint EUROGRAPHICS, IEEE TCVG Symposium (2003)
35. Stave, K.: Participatory system dynamics modeling for sustainable environmental management: observations from four cases. J. Sustain. **2**(9), 2762–2784 (2010)
36. Lychkina, N.: Dynamic simulation of socio-economic systems and its application in information-analytical solutions for strategic management. Bus. Strat. **2**, 44–49 (2013)

Author Index

O. Gervasi et al. (Eds.): ICCSA 2024 Workshops, LNCS 14818, pp. 443–444, 2024.
https://doi.org/10.1007/978-3-031-65273-8

The manufacturer's authorised representative in the EU is Springer Nature Customer Service Centre GmbH, Europaplatz 3, 69115 Heidelberg, Germany. If you have any concerns regarding our products, please contact ProductSafety@springernature.com

Printed and bound by CPI Group (UK) Ltd, Croydon, CR0 4YY
30/09/2026
02225585-0017